The Palgrave Handbook of Gender and Migration

Claudia Mora · Nicola Piper
Editors

The Palgrave Handbook of Gender and Migration

palgrave
macmillan

Editors
Claudia Mora
Center for Technological Society
and Human Future
Universidad Mayor
Santiago, Chile

Nicola Piper
Queen Mary University of London
London, UK

ISBN 978-3-030-63346-2 ISBN 978-3-030-63347-9 (eBook)
https://doi.org/10.1007/978-3-030-63347-9

Cover illustration: uschools/E+ via Getty Images

This Palgrave Macmillan imprint is published by the registered company Springer Nature Switzerland AG
The registered company address is: Gewerbestrasse 11, 6330 Cham, Switzerland

Contents

Editors and Contributors

About the Editors

Claudia Mora is Doctor in Sociology (Ph.D.), Professor at the School of Humanities and a Senior Researcher at the Center Technological Society and Human Future, Universidad Mayor, in Santiago, Chile. Her research focuses on work, labour market dynamics, intersectionality, social inequality, gender, and migration. In her publications, she has addressed the reproduction of gender, social class, and racialization of migrants in labour markets, contributing to a micro-social analysis of boundary formation and inequalities at work. Professor Mora's publications include the edited volumes *Domestic Work: a long road to decent work* (2009, ILO); *Furtive Gender: Interdisciplinary evidence of gender inequality in Chile* (2018); and *Inequality in Chile: the enduring relevance of gender* (2013). She is also the author of "Racialisation of Immigrants at Work: Labour Mobility and Segmentation of Peruvian Migrants in Chile" and "The Gender Deficit: Everyday Practices of Differentiation in the Chilean Labour Market" among a number of articles in Latin American and European journals.

Nicola Piper a political sociologist, is a Professor of International Migration at the University of Sydney (and the Founding Director of the Sydney Asia Pacific Migration Centre). She is currently a British Academy Global Professor Fellow hosted by Queen Mary University of London's School of Law, to conduct research on global governance of labour migration and decent work for migrant workers (January 2019 to December 2022). Nicola is also partner investigator in the UKRI funded Global Challenge Research Hub *on South-South Migration and Inequality* as co-lead of the Gender Work Package (2019–2023). Her publications include the edited volumes *New perspectives on gender and migration: livelihoods, rights, and entitlements* (2008), *South–South Migration: implications for social policy and development* (with Katja Hujo,

2010) and the co-authored book *Critical Perspectives on Global Governance: rights and regulation in governing regimes* (with Jean Grugel, 2007). She is co-founder and co-editor of the Routledge book series *Asian Migration* and one of the chief editors of the journal *Global Social Policy*.

Contributors

Elaine Acosta González Cuban Research Institute, Florida International University, Miami, FL, USA

Lailah Alidu School of Law, Social and Behavioural Sciences, Kingston University, London, UK

Anna Amelina University of Cottbus, Cottbus, Germany

Leah Bassel University of Roehampton, London, UK

Anna Boucher Government and International Relations, University of Sydney, Camperdown, NSW, Australia

Monica Boyd University of Toronto, Toronto, ON, Canada

Heaven Crawley Centre for Trust, Peace and Social Relations (CTPSR), Coventry University, Coventry, England

Petra Dannecker University Vienna, Vienna, Austria

Malathi de Alwis University of Colombo, Colombo, Sri Lanka

Belinda Dodson University of Western Ontario, London, ON, Canada

Jennifer N. Fish Old Dominion University, Norfolk, VA, USA

Lucy Fiske University of Technology Sydney (UTS), Broadway, NSW, Australia

Luisa Feline Freier Universidad del Pacífico, Lima, Peru

Chrisanthi Giotis University of Technology Sydney (UTS), Ultimo, NSW, Australia

Herminia Gonzálvez Universidad Central de Chile, Santiago, Chile

Menara Guizardi National University of San Martín, Buenos Aires, Argentina

Karlijn Haagsman Department of Society Studies, Maastricht University, Maastricht, The Netherlands

K. C. Hari Wilfrid Laurier University, Waterloo, ON, Canada

Jenna Hennebry Wilfrid Laurier University, Waterloo, ON, Canada

Kwanchanok Jaisuekun Research Institute for Languages and Cultures of Asia, Mahidol University, Nakhon Pathom, Thailand

Rose Jaji University of Zimbabwe, Harare, Zimbabwe

Rahel Kunz Université de Lausanne, Lausanne, Switzerland

Kyoko Kusakabe Asian Institute of Technology, Khlong Nuen, Thailand

Johanna Leinonen University of Oulu, Oulu, Finland

Eleonora López Pontificia Universidad Católica de Chile, Santiago, Chile

Helma Lutz Goethe University Frankfurt, Frankfurt, Germany

Rianne Mahon Carleton University, Ottawa, ON, Canada

Julia Maisenbacher Université de Lausanne, Lausanne, Switzerland

Valentina Mazzucato Department of Society Studies, Maastricht University, Maastricht, The Netherlands

Claudia Mora Center for Technological Society and Human Future, Universidad Mayor, Santiago, Chile

Lena Näre Sociology, University of Helsinki, Helsinki, Finland

Camilla Nordberg Åbo Akademi University, Turku, Finland

Julia O'Connell Davidson School of Sociology Politics & International Studies (SPAIS), University of Bristol, Bristol, UK

Antero Olakivi Sociology, University of Helsinki, Helsinki, Finland

Ingrid Palmary University of Johannesburg, Johannesburg, South Africa

Sònia Parella Universitat Autònoma de Barcelona, Barcelona, Spain

Ruth Pearson School of Politics and International Studies, University of Leeds, Leeds, UK

Sandra Pertek School of Social Policy, Sociology and Criminology, University of Birmingham, Birmingham, UK

Jenny Phillimore School of Social Policy, Sociology and Criminology, University of Birmingham, Birmingham, UK

Nicola Piper Queen Mary University of London, London, UK

Parvati Raghuram The Open University, Milton Keynes, UK

Nieves Fernández Rodríguez WZB, Berlin, Germany

Kyoko Shinozaki Department of Sociology, Paris Lodron University of Salzburg, Salzburg, Austria

Isabel Shutes Department of Social Policy, London School of Economics and Political Science, London, UK

Gunjan Sondhi The Open University, Milton Keynes, UK

Thales Speroni Universitat Autònoma de Barcelona, Barcelona, Spain; Pontifical Catholic University of Rio Grande do Sul, Porto Alegre, Brazil

Denise L. Spitzer School of Public Health, University of Alberta, Edmonton, AB, Canada

Carolina Stefoni Universidad Mayor; COES/FONDAP, Santiago, Chile

Magdalena Suerbaum Max-Planck-Institute for the Study of Religious and Ethnic Diversity, Göttingen, Germany

Sirijit Sunanta Research Institute for Languages and Cultures of Asia, Mahidol University, Nakhon Pathom, Thailand

Sirpa Wrede Sociology, University of Helsinki, Helsinki, Finland

List of Figures

List of Tables

CHAPTER 1

An Intersectional and Global Approach to the Study of Gender and Migration

Claudia Mora and Nicola Piper

Abstract This chapter provides an overview of the organizing themes discussed in this state of the art study of gender and migration. Combining multiple and interdisciplinary perspectives, the text runs through two narratives: intersectionality and globality/transnationality. In reflecting upon the complex nature of inequality, contributing authors argue migration is a process shaped by gender, class, race, and nation, and also conditioned by public policies, international regimes, states' and other normative definitions, often rooted in colonialism. Countries connect globally through unequal power relations imbricated in empire building, as evident from studies taking a post-colonial lens in analysing migration law, naturalisation, and citizenship policies. The chapter also brings the significance of 'South–South' migration to the fore beyond western settings, where transnationalism highlights the intricate socio-cultural, political, and economic linkages between the places migrants have left and the places to which migrants move, and the fluidity of movements that may never truly result in 'settlement' or 'integration' as a completed project. Gender and intersectionality characterise those dynamics and define

C. Mora (✉)
Center for Technological Society and Human Future, Universidad Mayor, Santiago, Chile
e-mail: claudia.mora@umayor.cl

N. Piper
Queen Mary University of London, London, UK
e-mail: n.piper@qmul.ac.uk

C. Mora and N. Piper (eds.), *The Palgrave Handbook of Gender and Migration*, https://doi.org/10.1007/978-3-030-63347-9_1

the continuing relations across place and space, as particularly evident in the case of care and caring arrangements.

The intersectional approach to social inequalities has been one of the most important contributions to the social sciences in the last decades. Coined by American legal scholar Kimberlé Crenshaw in 1989, it directed attention to the limits of unidimensional analyses of gender and race as separate categories of social exclusion, which prevented the applicability of anti-discrimination laws to the particular experiences of Black women in the United States. Since then, there have been many elaborations on the fundamental idea that crossing inequalities of gender, race, class, national origin, sexual orientation, among many other categories of social exclusion, configure a particular social space that shapes experiences and life conditions of subordinate groups in many different ways. Later developments in the intersecting inequalities approach have focused on people's lived experiences, rather than on constraining (and reifying) social categories, which are posed as preliminary (and provisional) analytical tools to show the processes of mutual construction and imbrication (Walby, Armstrong, & Strid, 2012).

The intersectional approach to the analysis of social phenomena and inequalities is transversal in the debates laid forward by the thirty-one chapters comprising this Research Handbook on Gender and Migration. Scholars in this collection reflect upon the micro, meso, and macro conditions that enmesh with complex inequalities to create particular challenges to women, men, and families on the one hand with institutions, organisations, and the State on the other, confronting, from multiple ideological starting points, the trials of migration. In this endeavour, they widen our understanding of intersectionality, arguing that crossing social inequalities that affect and finally condition migration not only refer to social categories of gender, class race, and nation, but also to public policies, international regimes, states', and other normative definitions and discourses on gender, sexuality, and colonialism, all of which construe specific groups of people as 'questionable' and hence suspect, in migration processes. The contributions to this Handbook also demonstrate the complexity, even contradictions, in the development of migrant identities and experiences when taking an intersectional lens on social space where multiple patterns of exclusion crisscross with structural, geopolitical, normative, and discursive definitions of the 'Other' and adapt over time or a migrant's life course.

This handbook also grapples with the study of gendered migration from a global perspective, understood as a planetary vision from a world systems or an institutional perspective pitched at the global level (i.e. 'global governance'). The incentive behind highlighting gendered migration as a truly global phenomenon is multifold and includes the desire and need to bring the significance of 'South–South' migration to the fore beyond western settings, in order to avoid theorising migration from the context of South–North flows only. The experience of Europe and classic countries of immigration covers only one segment of the whole 'migratory' picture. The exact nature of specific gender effects or outcomes in each world region or in the context of specific migration corridors may differ, but gender dynamics play a role all around the globe. A global perspective also directs research towards engaging in comparative analysis across a number of countries. While acknowledging the role of the state (and state borders), there are vital actors and processes above and below states that need to be acknowledged—and the gender dynamics implicated in them. Going beyond the confines of the (nation-)state, transnationalism is one key lens by which migration scholars have demonstrated the intricate sociocultural, political, and economic interlinkages between the places migrants have left and the places to which migrants move and the multidirectionality and fluidity of movements that may never truly result in 'settlement' or 'integration' as a completed project. Gender and intersectionality characterise those dynamics and define the continuing (yet changing) relations across place and space, as particularly evident in the case of care and caring arrangements, and also across generations. In the realm of politics and policy, the global level has also become a vital arena where migration in general and gendered migration in particular are being framed, shaped, and responded to in certain ways. Such responses in themselves are based on gendered assumptions, thinking, and behaviour. Globality also has historical roots connecting countries through unequal power relations imbricated in empire building, with lasting effects as evident from studies taking a postcolonial lens in analysing migration law, visa, naturalisation, and citizenship policies.

Theorisations and Perspectives on Gender and Migration

Engaging in different themes and organised in broadly conceived sections, the contributing authors to this volume engage with the multidimensionality of gendered migration to understand how social, political, and economic patterns and contexts interact at various levels to define categories of migrants, their life experiences and chances. In the first section, Theorisations and Perspectives on Gender and Migration, four themes are brought to the fore: the need to incorporate a complex understanding of gendered migration; the questioning of established truths regarding the notion of transnationalism and transnational lives; the quest to recover the fundamental lens of (post)colonialism and coloniality in migration analyses; and the uninterrupted reflexivity necessary in migration studies research.

On the first theme, and using recent data from the United Nations and the International Labour Organisation, Monica Boyd (Chapter 2), spotlights a long-lasting trend of women's migration and the 'feminisation' of migrant stock dating back to the 1960s. The author presents available data that allows for an analysis of micro, meso, and macro level factors that condition women's migration; the type of work available to them; and their contribution to destination economies. However, Boyd appraises data quality issues and critiques the limited topics covered by international organisations' statistics, which still constrain a more nuanced analysis of gendered migration as a changing process, a perspective that can only strengthen social and migration studies. Feminisation of trends, the author argues, represents just the beginning of understanding the international migration of women.

Regarding the contestation of prevailing notions of transnationalism and transnationality, Johanna Leinonen and Lutz & Amelina provide a clear argument for widening the focus on *the nation* and for the rejection of an acritical acceptance of the nation-state as the given frame of analysis (Leinonen, Chapter 3). While recognising that migration policies anchored in nation-states frame the possibilities for transnational connections, Leinonen highlights the need to approach the symbolic dimensions of transnationalism, for example, in the circulation of racialised and gendered notions of certain groups. She also highlights the fields opened to scholars if they were to broaden their attention to a *mobilities paradigm*, prioritising a study of geographical movements, long and short, not tied to nation-states, which would allow for an understanding of the movements significant to migrant themselves. In a similar light, Lutz and Amelina point to Pries' notion of transnational contexts as relational orders (chapter three) as spatialised relations not grounded exclusively in nation-states; to Faist's differentiation of transnational relations according to their durability and formalisation; and to Levitt and Glick Schiller's focus on power struggles in transnational relations. This, to critique 'a sedentary understanding of the nation state and of spatiality inscribed in the classical studies of migration' (Lutz & Amelina, in this volume) and to foster a multi-local frame of reference encompassing geographies grounded on and beyond nation-states.

Another guiding theme explored by Ingrid Palmary (Chapter 5), poses the quest to recover the fundamental lens of (post)colonialism and coloniality in migration analyses, to help understand the enduring connections between current inequalities and their colonial past—which often mark their 'original sin', although they are produced and reproduced in contemporary frameworks. Palmary also argues for an awareness and inclusion in migration scholarship of sexuality issues and identities, particularly women's, as they have been central to colonial projects past and present, and still are a permanent feature in migration dynamics, especially through contemporary forms of migration management.

Concomitant to these general directives and recommendations, Kyoko Shinozaki Chapter (6) places the fundamental issue of power in migration research, advocating for a 'reflexive turn' that would question researchers' positionality in field relations, knowledge production, dissemination and public engagement. While this debate has preoccupied qualitative researchers for decades and has resulted in their awareness of hierarchies in the field, which must be contested and negotiated, Shinozaki broadens the inquiry to quantitative research, providing examples of reflexivity in this methodological approach, illustrating methodological strategies to level the research field in quantitative studies of migration. This move forward may help amend the tendency in feminist migration research to focus largely on qualitative methods.

Intersectionality is embraced by these authors as a path to uncover more complex approaches to gendered migration and migration processes, averting a common overemphasis on ethnicity and on ethnic identification as paramount to migrants' identity (Leinonen, Chapter 3). Gender, class, race, country of origin, sexual identities, etc., are contended as fluid and temporary categories as migrants experience throughout their trajectories different and overlapping social positionings and political/economic contexts that are also not static. Transnational research brings attention to the changing hierarchies of transnational actors and to their contradictory mobility, as well as to the material and symbolic clashes that forge transnational spaces.

Care, Affective, and Emotional Labour

Gendered transnational migration has brought to the centre the discussion of transnational political economies of care, characterising the mobility of migrant women workers to the North, but also South–South (Mora & Piper, 2011). Global inequalities of care overwhelmingly affect women, and care continues to be considered separate from productive, 'real' work. For women migrants, care work is the transnational equivalent of a dead-end job, readily available via a myriad of national policies, international regimes, gendered mores, and labour market preferences for a gendered niche of unqualified and qualified women from the global south. Section two explores the dynamics of care and affective and emotional labour, to problematise relations of inequality in different locations and actors as well as to highlight the struggles of women carers and the crises ensuing for societies and families of origin.

In her very lucid argument, Isabel Shutes (Chapter 7) poses that 'care is a concept that attends to the inequalities of social relations in terms of the gendered division of labour, manifested in the construction of care as "women's work" and as historically undervalued labour' (p. 109). The author points to the inequalities of care regarding other forms of labour, other obligations and other institutions such as costs to family, state, market, and communities in addition to women's own. But perhaps more importantly, she posits a crucial link between migrant workers and wider processes of labour market transformations in ways that connect migration research to what Castles (2010) calls a step-change in social transformation: how labour migration, and particularly gendered migration responds to, shape, and/or challenges labour market structures and organisations in both, societies of origin and destination.

Research on the topic of care has mostly centred on women migrants and their struggles to fulfil gender-defined roles as transnational mothers and carers. Haagsman and Mazzucato (Chapter 8) contribute to a welcomed balance, comparing the effects of separation of migrant mothers and fathers. While confirming the endurance of gender expectations for migrant men as economic and women as emotional providers, they explore what these expectations mean to migrants' ability to parent transnationally. The authors argue that qualitative research has largely shown gender norms as having a definite effect on parents and children's evaluations of separation, but quantitative studies have also shown that there are no significant differences in the well-being of migrant parents or in their relationship with their children when socio-economic and migratory status are controlled for. In addition, in these quantitative approaches, other socio-demographic characteristics make a difference in the experience of separation, such as single motherhood and markers of women's vulnerability, such as divorce and unstable care arrangements.

A focus on men is also pursued by Jaisuekun and Sunanta (Chapter 9), whose chapter based on empirical research in Pattaya, Thailand features the intersection between gendered expectations of Thai women; notions of manhood; faulting/insufficient welfare systems in the Global North; class differences; and forms of neocolonialism in the Global South in their study of marriage and lifestyle migration of German (older) men living in Pattaya (a Thai beach resort and well known for sex tourism). The authors show how lifestyle—marriage migration and sex tourism are closely linked together, sometimes as parts of the same process of transnational mobility, one that it is also deeply gendered. Lifestyle migration is coded by gender, they argue, as older German men, most of whom are in relationships with young Thai women, can live up to their ideals of manhood, have a better quality of life, family, sex, and care through their retirement migration to Thailand.

While studies of gendered migration that focus on men's experiences are a novelty in the field, the same is true regarding the exploration of migrant women juggling paid and unpaid domestic and care labour in destination societies. Pearson and Kusakabe (Chapter 10) take the case study of Burmese garment factory workers in Thailand to engage in the discussion of the strategies migrant women display to fulfil their direct face-to-face gendered care duties with children present in societies of destination and transnational care at home, with parents and other kin. As many migrants experience obstacles to 'mother' in countries of destination, the authors focus on the study of Burmese women workers, who confront hostile environments for care in both, sending and receiving societies and whose care labour, as factory workers, has been largely ignored in the literature, '[which] rarely gives much attention to the fact that that paid work and unpaid care are frequently carried out simultaneously, [and that] … when this occurs across international boundaries the constraints for women are even harsher' (p. 164). The authors glean out strategies of care from their fieldwork data, gathered with lengthy research conducted with Burmese factory workers, to show the relevance of geography (factory location), social protection policies, and border control, in the success (or failure) of women's devised child care solutions.

The role of states' social protection policies in materialising proper care for their population, in particular when women migrate, is another realm of underexplored hardships confronting transnational families. Acosta

(Chapter 11) delves onto the case of Cuba to show the emergence and consequences of a care crisis following Cuban women's migration from the island, which is paired with missing social protection networks and an invisibilisation, by an authoritarian regime, of new and pressing social problems such as an ageing population without guaranteed care. The growing number of women migrating from Cuba in search for better wages has impacted severely the living arrangements and life conditions of the elderly. Acosta argues that emigration and reducing family size should be read as strategies to cope with a generalised care crisis in the country. For women migrants, temporary returns allowed by authorities coupled with remittances to their families from abroad, configure a strategy of care that fills in the care gap for the elderly left in the island.

Gendered Work, Employment, and Skills Mobility

Migrant workers' occupation and status are usually conditioned by gendered expectations of their abilities. Their evaluation is also imbued by notions of race, nation, and social class, among many other social forces, construing migrant workers as social 'Others' which in turn defines their access to employment, working conditions, and mobility. For labour migrants, work trajectories determine the success of their migratory endeavour and the possibilities to fulfil familial duties. Hence, migration scholars have widely research migrants' conditions of work and employment, analysing structural and interactional factors that affect migrants' work trajectories.

The intersectional approach has been useful in this regard, because inequalities in the distribution of material and symbolic resources are founded in the convergence of processes of differentiation that shape migrants' opportunities in the labour market, which in turn determines their life conditions. Intersectionality allows for a better understanding of the historical changes in the labour market affecting women in very different ways, for example, the increasing number of women's paid labour in the Global North which opened a path to the care labour of women from the Global South tailored to often qualified, but racialised, migrant women.

In this section, the focus is on the underexplored experiences of highly qualified migrants, which the intersectional perspective warns as crucial to the analysis of migrant workers: all social categories are to be fully compared not only in their subordinate rungs but also in their 'dominant' positions, as they are relationally tied and explained (McCall, 2005). Following this path, Anna Boucher (Chapter 12) explores skilled migration to pose the issue of skill recognition as a highly gendered process affecting women's chances to obtain work permits, visas, autonomous status, and engagement in the labour market. Boucher argues that the notion of skills is constructed in ways that benefits male versus female occupations, as skill indicators are designed to value men's credentials. The level of work independence and wages, educational qualifications, occupational classifications, employers' defined models, and 'extraordinary talents' are criteria commonly advantageous to highly skilled migrant men.

Belinda Dodson (Chapter 13) explores a related topic, adding that skills definition is not only gendered but that it also bolsters existing gender and geographic inequalities. While skills definition is not usually neutral, migration further complicates the process by incorporating the social positioning of migrants (gender, nation of origin, race, (post)colonial status, etc.), which are haphazardly interpreted in skill assessments processes, assigning value to 'brain skills' over 'embodied skills'. The outcome not only impacts women's migration status as skilled workers, but determines their employment chances and conditions in destination societies, often leading to deskilling and under-employment.

An additional subject, not garnering sufficient attention among scholars of work and migration, is the flow of international students to the global north (defined as migrants by the IOM), status that exerts a significant influence on women's chances of being qualified as skilled migrants. This is the focus of Raghuram and Sondhi's contribution (Chapter 14). The authors point to misconceptions regarding international students' social origin as elite citizens in their countries of origin, given that data show an increasing number of middle-class international students reaching all tiers of global north schools, thanks to national policies seeking to strengthen human capital (for example, via scholarships to advantaged students), as well as to students' own social strategising for mobility. While racism and patriarchy are also part of international students' experiences, the authors find that women's return to their 'home country' is often marred by gendered expectations and assumptions that prompts them, more often than men, to migrate back to their school's country—or to a third country—to find suitable work.

While all three chapters contribute to a more comprehensive view of gendered skilled migration, Denise Spitzer (Chapter 15) broadens the intersectional lens by posing the multiple dimensions affecting well establish truths about migrants' work, to argue that dominant discourses regarding entrepreneurship are highly gendered male and that women entrepreneurs face particular challenges tied to their expected social roles and duties. Lack of social capital, family responsibilities, supportive environments, gendered cultural norms, normative limitations, among others shape women's chances of engagement in entrepreneurial activities, which offer them more independence and autonomy—but also, more risks. The analysis of the constellation of factors determining the possibilities of migrants' entrepreneurship is defined by Spitzer as kaleidoscopic, as there are no a priori guarantees of facilitating factors and/or limitations facing migrant entrepreneurs and their purported contribution to development, but rather, an open field for intersectional exploration.

To close this section, Wrede et al's analysis of health care workers in Finland (Chapter 16), place the emphasis on the compounding effects of spreading economic neoliberalism; discourses of competition and austerity policies that have brought about deregulation; flexibilisation and privatisation; on labour markets and the organisation of work along gendered and racialised lines. The authors provide a critical and global perspective, arguing that spreading neoliberal doctrines preaching the need to 'flexibilise' the allegedly 'bureaucratic' nature of welfare states have contributed to the 'flexibilisation' of care work, which involves heavy reliance on migrant labour. Their discussion offers an intersectional examination of the exploitative 'coping management' that the neoliberal discourse underpins. The rearticulation of care work as work requiring 'flexibility' along gender and racialised talk, legitimises this work as particularly fitting for migrants which in turn has problematic gendered and racialised implications for workplace dynamics.

Gendered Migration as Site of Governance and Development

International migration has been subject to global regulation ever since the establishment of the United Nations. Initially, under one framework and the remit of one organisation, the ILO (which happened to be the first human rights organisation in the history of what became the UN, set up in 1919): soon it was divided into two migration regimes, one for refugees and one for labour migration. Over the following decades, greater and more intensified institutionalisation of migration policy at national, regional, and global level occurred, underpinned by particular discourses on migration linking it to economic development and human rights concerns (Dannecker & Piper, Chapter 19). This has inspired migration scholars to research and analyse the making of migration policy by highlighting specific techniques and tools of regulating migration ('governmentality'). Invariably, 'gender codes' are built into governance and its methods, as demonstrated in the contribution by Hennebry and KC (Chapter 18) who analyse a particular governing tool that has been widely used around the world; bilateral agreements and its gender outcomes. Bilateral instruments that govern the migration of domestic workers, such as in the case of migratory corridors within Asia and between Asia and the Middle East/Gulf region, demonstrate the intersectional nature of such governing tools, channelling lower class, lowly skilled women from less resourced countries to work in households located in higher income countries, rather than facilitating mobility more generally. A similar conclusion is arrived at in the contribution by Kunz and Maisenbacher (Chapter 20) who discuss gender as a governing trope via the example of the financialisation of remittances, a trend which promotes a distinct notion of empowerment that narrows options down rather than nurturing various forms of gendered empowerment.

The process of migrant empowerment could be viewed as a form or outcome of social development. One key frame at the global level under which migration has been discussed is in fact its link to development, albeit most in relation to economic aspects. Scholars of gendered migration have contributed insights into this link from both the classic focus on economics, placing remittances at the centre (see Kunz & Maisenbacher, Chapter 20) but also in terms of the social implications on care giving and care providing, transnationally split families and the stigmatisation of women as primary income earners leaving spouses and children behind (see Dannecker & Piper, Chapter 19).

The global debate on migration and development is part of the emergence of 'global migration governance' which relates to the intensified institutionalisation of migration at the global and regional level which involves various international organisations (IOs), many of whom make up the United Nations. The field of global migration governance is characterised by a multiplication of organisational actors which take more or less of an interest in its gendered effects (see Mahon, Chapter 17). Those IOs with expertise on women affairs are often more marginalised with smaller budgets than the big players (Grugel & Piper, 2007). The emergence of global migration governance is paralleled by expanding civil society engagement and rising migrant rights activism demanding clearer human rights commitments to the treatment of migrants, including greater gender responsiveness. They do so on the basis of the existing body of international human rights instruments but also by advocating additional ones in order to close loopholes or address particular problem areas. When this activism takes place in a globally networked manner, supported by influential actors such as global union confederations, success stories can and do occur, as Jennifer Fish's contribution on the negotiations of the ILO Convention 189 on Decent Work for Domestic Workers shows (Chapter 21). The dynamics between 'top down' and 'bottom up' processes involving state and non-state actors have given rise to contributions to global governance studies which introduce a perspective emanating from various migrant and civil society initiatives, their networks and/or the involvement of a web of facilitators, that is recruiters and brokers (Baas, 2018), to what has otherwise been treated as a state-led arena. What the rising migrant rights activism around the world has demonstrated is that much of the innovative thinking around what migrant rights entail derives from the Global South and also from women migrant's experience (Piper, 2015).

What these scholarships on gendered social development and gendered migration have in common is the emphasis on the equal importance of structure and agency. Celebrating migration indiscriminately as a form of empowerment for women takes on greater nuance by demonstrating the constraints under which such agency takes place, in light of highly selective and restrictive migration policies and governing tools that offer women migrants a narrow range of options (Hennebry & KC, Chapter 18; Dannecker & Piper, Chapter 19). Although collective political activism is the quintessential form of displaying and asserting agency, the struggle for greater freedom of movement is faced with many barriers. Even within the sphere of civil society, there are gender hierarchies and power inequalities replicating the North–South divide that pose barriers to the empowerment of the most marginalised groups of migrants. The world of trade unions has for instance, long closed its doors on female workers, especially those in the informal sector. In this case, the ILO C189 has offered a political opportunity to allow female migrants to enhance their organising efforts and influence, cementing a path to achieving greater levels of agency and empowerment (Chapters 21 by Fish and 16 by Mahon).

Forced Migration, Gender-Based Violence, and Conflict

Gendered migration as a site of governance is perhaps more clearly epitomised by the UN Convention on the Status of Refugees. While the definition of 'refugee' refers to 'someone who is unable or unwilling to return to their country of origin owing to a well-founded fear of being persecuted for reasons of race, religion, nationality, *membership of a particular group*, or political opinion' (Convention Relating to the Status of Refugees 1951, emphasis added), there have been several limiting interpretations of the Convention affecting—and excluding—women and sexual minorities in particular (Crawley, Chapter 22). On the one hand, gender experiences, such as rape and sexual violence, familial (domestic) violence, honour killings, female genital mutilation (FGM), forced marriage, and sexual orientation, have not been generally recognised as persecution, according to the Convention; social and economic rights have been deemed outside its scope and rather, a matter of *economic migration*; and more importantly, there has been an overall privileging of a 'public sphere' of activities, dominated by men, over the 'private sphere' where women's activities and 'sex' take place (Crawley: pp. 359–372).

At the same time, even though men have been the main beneficiaries of international protection as refugees, their social positioning and identities are severely questioned by their refugee status, as Jaji (Chapter 23) and Suerbaum (Chapter 24) pose in this volume. Victimhood is symbolically gendered feminine while men are, or should be, immune to it. Jaji argues that a continuing feminisation of flight precludes the possibility of considering male victims, leading to an incongruent understanding of refugee men, especially from Africa and the Middle East, as both, menacing and not men enough. Jaji delves into conflicting notions of masculinity in Africa as a consequence of colonisation and examines the resulting (shifting) signifier of 'feminised men' in the realms of conflict, war, and flight to Western societies that end up construing refugee masculinities as aberrant. In a similar vein, Suerbaum (Chapter 24) takes on the case study of the Middle East, to show the intersection of class and masculinity in the refugee experiences of Syrian men in Egypt to explain the reasons why men perceive their position as challenged in times of forced migration. Middle-class Syrian men reject the refugee label and struggle to maintain their provider role as their path to counter exclusion and non-recognition and preserving their manhood.

Men, women, and children escaping violence are also vulnerable to violence throughout their journey and in arrival societies. Phillimore et al. (Chapter 25) explore the multiple stages, actors and stressors shaping the experiences of vulnerability of refugees regarding sexual and gender-based violence, that may be inflicted by police and security services, smugglers, other refugees and family members and partners. The authors argue that "experiences [of sexual and gender-based violence] need to be conceptualised as an ongoing and multifaceted experience of trauma, with both immediate and long-term consequences for refugee resettlement" (Phillimore: p. 406). Drawing from an extensive review of the literature, the authors signal to the influence of sexual and gender violence on refugees' mental and physical health and on their ability to integrate in countries of refuge. The vulnerability of people on the move, be they categorised as migrants or refugees, has often been read in the West within a criminal justice scenario of exploitation and abuse. O'Connell Davidson (Chapter 26) argues a thought-provoking understanding of the exploitation of trafficking not only as a form of modern slavery, but as a wider set of (often positive) meanings imbued by the experiences, intentions, and understandings of those undergoing such struggles. O'Connell Davidson poses the notion of 'marronage' as a useful tool to fully explore the experience of trafficking and modern slavery in mainstream discourse.

A common thread in this section is to bring our attention to the triviality of normative categorisations of people on the move, facing common struggles and hurdles irrespective of their legal status and classification. Hence, Fiske & Giotis (Chapter 27) discuss migration at the intersection of gender and disability. Drawing on Critical Disability Studies, the contribution by Fiske and Giotis recognises disability as just one aspect of identity and social location, and introduces the nascent body of studies that have topicalised the complex nexus of refugees with disabilities. Filling the remaining gap by bringing in a gender analysis, the authors argue for a 'generative' rather than 'additive' approach to gender by placing the focus on social location.

Malathi de Alwis (Chapter 28) closes by focusing on large-scale displacement and forced migration of people *within* national borders, as a result of civil war and conflict. While not categorised as refugees, and without the international protection ensuing from that status, she argues for similar dangers and vulnerabilities of collective trauma and suffering. Taking the case study of displacement in Sri Lanka, de Alwis poses, however, that the dispossession of displacement is also paired to new possibilities and change.

Gendering Migrant Rights, Social Welfare, and the Politics of Inclusion

In this final section, the authors bring to the fore enduring shortcomings of political and academic understandings of gender and migration. Recent events have acutely exposed the link between migrant rights (citizenship, human rights), social welfare and the politics of inclusion: the COVID 19 pandemic revealed migrants as both, essential workers and a disposable surplus population, bearing the greatest risks of the health and economic crises facing the world. The pandemic and subsequent movement restrictions have shown the ways in which economic, social, and structural inequalities impact upon some groups of workers and migrants more than others. Gender is one such distinguishing factor. Exposed were also the low rates of pay for women and migrants, many of whom ironically work in 'essential jobs' (e.g. care, cleaning, food). The pandemic has drawn attention to the negative effects of austerity politics: economic nationalism and rising anti-immigration sentiments and existing forms of supranational solidarity crumbling, with dire implications for the politics of inclusion.

Another aspect of the health crisis is the amplification of discriminatory behaviour and practices against migrants as bearers of the disease. This is nothing new: the association of migrants with disease has its roots in eugenics and degeneration narratives that both, sustain and oppose liberal norms of equality and rights. These intrinsic hierarchies erupt into national discourse during periods of economic contraction and uncertainty, when immigration is framed as a threat to the health of the nation. Theories of biological racism are an enduring legacy of European colonialism that persists worldwide with respect to immigrants. The current predicament of migrant workers is a repetition of this pattern. Whether or not marginalisation is rationalised in terms of race, persons lacking the full legal recognition of the state in which they live find themselves identified as a potential risk, and treated as undeserving of protection.

Deservingness in the context of welfare conditionalities that shape the movements and lives of individuals and families who are migrant is a theme that appears in the contribution by

Parella and Speroni (Chapter 29), who explore the looming subject of social protection that too often eludes migrant workers' possibilities in receiving countries. The authors pose the need to think about social protection from a transnational standpoint, which implies a focus on migrants' strategies to respond to social risks such as health, care, education, and social security, by a combination of resources, formal and informal, and from different nation-states and actors. As migrant women shape the provision of social protection and care, they are at the same time, more vulnerable to social risks. Hence, the authors call transnational social protection a 'gendered quest', made possible by the often sacrificial actions and strategies of women. They also draw on the intersectional lens for an analysis of the junctures of formal and informal social protection, gender inequalities and other social hierarchies conditioning the possibilities for women's social protection.

The notion of deservingness is discussed and critiqued by Leah Bassel (Chapter 30) in her analysis of naturalisation and legalisation processes. As part of the experience of 'becoming citizens', migrants negotiate new hierarchies of inclusion and exclusion which, as the author argues, reflect 'broader neoliberal and ethos-based conceptions of citizenship'. At the intersection of gender, sexuality, race, and class, the longstanding power relations at play in the process of becoming a citizen can be identified. As demonstrated by the author, ongoing colonial legacies underpin policy instruments and shape experiences of the citizenship process as well as the kind of citizen a migrant becomes. Challenging the gender binary and manifesting its origins in European 'modernity' shows the arbitrary and particular understandings of genders and sexualities that a migrant, however, needs to meet in order to be deemed deserving of protection and asylum.

Freier and Fernández-Rodríguez (Chapter 31) take their analysis to the effects that visa policies have on the reinforcement of gender roles within the labour market and the household. Their discussion in fact takes the issue of gender effects of visa policies beyond the world of work to include student or study visa and family reunification policies. In combining these various scenarios, the authors relate their analysis to the broader phenomenon of gendered geographies of international migration. One of their key arguments concerns the fact that the effects of visa policies remain understudied in the context of South–South migration where such policies are however highly, if not increasingly, relevant.

Requirements for a visa and intricate visa policies relate to the process of border crossing. In discussing how gender systems are intertwined with the reproduction of borders, Stefoni et al. (Chapter 32) critique the androcentrism that characterises mainstream Border Studies. By pointing to the emerging studies of borders which have begun to incorporate the experience of women, the authors note that the place occupied by women in the construction of national boundaries has been made visible. In addition and importantly, as the authors argue, a gender perspective leads to foregrounding how violence is multiplied and ingrained in the everyday construction of border regions. Critical Border Studies as a subfield has highlighted three asymmetries of power that have become more prominent in recent studies on borders: economic inequalities, national hierarchies, and gender inequalities. Diverse studies on women's experience in border territories make visible the violence that women and non-male gendered persons experience in these spaces. Yet, the act of crossing borders and circumventing rigid control regimes ('migration management') is a sign of defiance and an opportunity for contestation.

The gendered notions of migrant rights, social welfare, and politics of inclusion are vital dimensions of theories dealing with non-discrimination, equality, and equity in relation to structure and agency. These elements are well captured in a combined manner by Nancy Fraser in her discussion of the interwoven and mutually influential nature of the 'three Rs': representation, redistribution, and recognition reflects the relations and interactions between social and institutional structures, the agency of actors and regulatory cultures. The social sciences in general and feminist (including feminist migration and legal) scholars in particular have for long grappled with understanding the relationship between structure and agency and how to acknowledge the importance of both in understanding social action and social change. Gendered migration, in particular its feminised and transgender forms, has also been subject to a debate about its empowering and constraining effects from both structural and individual perspectives.

In exploring how individuals' experiences and identities interact with forms of authority and discipline, including the law, the contributions under this section demonstrate that intersectionality should be situated within a family of analytical frameworks designed to do just that. This larger family of critical methodologies in which intersectionality finds its home includes such well-established schools of critique as black feminism, postcolonialism, and a range of feminist inquiries in both the global North and South concerned with issues of identity and difference, power, and inequality.

References

Baas, M. (2018). *What determines the cost of migration? A perspective from Indian agents facilitating migration to Singapore and the Middle East* (Asia Research Institute Working Paper Series No. 264). Singapore: ARI, National University of Singapore.

Castles, S. (2010). Understanding global migration: A social transformation perspective. *Journal of Ethnic and Migration Studies, 36*(10), 1565–1586.

Crenshaw, K. (1989). *Demarginalizing the intersection of Race and Sex: A black feminist critique of antidiscrimination doctrine, feminist theory and antiracist politics*. University of Chicago Legal Forum: Vol. 1989, Article 8.

Grugel, J. B., & Piper, N. (2007). *Critical perspective on global governance: Rights and regulation in governing regimes*. London: Routledge.

McCall, L. (2005). The complexity of intersectionality. *Signs: Journal of Women in Culture and Society, 30*(3), 1771–1800.

Mora, C., & Piper, N. (2011). Notions of rights and entitlements among Peruvian female workers in Chile. *Diversities, 13*(1), 5–18.

Piper, N. (2015). Democratising migration from the bottom up: The rise of the global migrant rights movement. *Globalizations, 12*(5), 788–802.

Walby, S., Armstrong, J., & Strid, S. (2012). Intersectionality: Multiple inequalities in social theory. *Sociology, 46*(2), 224–240.

Claudia Mora is Doctor in Sociology (Ph.D.), Professor at the School of Humanities and a senior researcher at the Center for Technological Society and Human Future, Universidad Mayor, in Santiago, Chile. Her research focuses on work, labour market dynamics, intersectionality, social inequality, gender, and migration. In her publications, she has addressed the reproduction of gender, social class and racialisation of migrants in labour markets, contributing to a micro-social analysis of boundary formation and inequalities at work. Professor Mora's publications include the edited volumes *Domestic Work: A long road to decent work* (2009, ILO); *Furtive Gender: Interdisciplinary evidence of gender inequality in Chile* (2018); and *Inequality in Chile: The enduring relevance of gender* (2013). She is also the author of "Racialisation of Immigrants at Work: Labour Mobility and Segmentation of Peruvian Migrants in Chile" and "The Gender Deficit: Everyday Practices of Differentiation in the

Chilean Labour Market" among a number of articles in Latin American and European journals.

Nicola Piper a political sociologist, is Professor of International Migration at the University of Sydney (and the Founding Director of the Sydney Asia Pacific Migration Centre). She is currently a British Academy Global Professor Fellow hosted by Queen Mary University of London's School of Law, to conduct research on global governance of labour migration and decent work for migrant workers (January 2019 to December 2022). Nicola is also partner investigator in the UKRI funded Global Challenge Research Hub *on South-South Migration and Inequality* as co-lead of the Gender Work Package (2019–2023). Her publications include the edited volumes *New perspectives on gender and migration: livelihoods, rights, and entitlements* (2008), *South–South Migration: Implications for social policy and development* (with Katja Hujo, 2010) and the co-authored book *Critical Perspectives on Global Governance: Rights and regulation in governing regimes* (with Jean Grugel, 2007). She is co-founder and co-editor of the Routledge book series *Asian Migration* and one of the chief editors of the journal *Global Social Policy*.

Part I

Theorisations and Perspectives on Gender and Migration

CHAPTER 2

Women, Gender, and Migration Trends in a Global World

Monica Boyd

Abstract The international migration of women was a neglected topic prior to the 1980s. However, new data disseminated by the United Nations Population Division on the feminization of migration documents what previously was ignored: namely, females are nearly half of the world's international migrant population and have been for many decades. Analysis of recent data from the United Nations and from the International Labour Organization shows that where female international migrants currently live is associated with their regional origins and specific countries of origin. International migrant women also contribute to the destination economies by their labour force participation, and they are an important source of domestic workers, particularly in the Arab, South East Asia, East Asia, and in Latin America and the Caribbean regions. However, the quality of data is problematic and the topics covered are limited. The concept of gender, rather than binary sex, stimulates the "engendering" of migration and expands understandings of issues faced by migrant women, men, and other groups.

Introduction

Women have always migrated, but female migration has a "herstory" of receiving little attention. In the writing of "history", many accounts of migration for purposes of war, colonising, or nation building efforts either are genderless (and thus gender-blind) or name men as the principal actors. For

M. Boyd (✉)
University of Toronto, Toronto, ON, Canada
e-mail: monica.boyd@utoronto.ca

C. Mora and N. Piper (eds.), *The Palgrave Handbook of Gender and Migration*, https://doi.org/10.1007/978-3-030-63347-9_2

the most part, women and children—if they accompanied men—are viewed as appendages or part of the "baggage" transported by men. To the extent that women's presence was acknowledged, they were viewed as reproducers of the next generation, and as producers of domestic goods to be consumed in the home.

These scripts changed in the second half of the twentieth century. The growth of the female labour force, the renewed attention to feminist perspectives, widespread political and economic interconnections between countries, the international division of labour and its focus on international stratification and neoliberal domination and the rise in supra-organisations such as the United Nations, the International Organization for Migration (IOM), the International Labour Organization (ILO), and others meant that the presence of women in migration flows could no longer be ignored. By the start of the twenty-first century, statistical systems increasingly collected data on migrant women, which confirmed the presence of women in contemporary migration flows. Researchers, civic associations (variously defined as nongovernmental associations or civil society organisations) and policy makers also scrutinised the important relationship between gender, migration and inequality, adding valuable data and insights.

These developments underlie the organisational structure of this chapter. First, the global international migration of women is highlighted using data assembled by leading supra-international organisations, notably the United Nations Population Division and the International Organization for Migration. In this analysis, international migration is defined as a permanent change in residence that involves crossing international boundaries. The phrase "permanent change" potentially has many definitions. Sometimes it includes people who have moved for short periods of time or who have entered a country with a short-term residency visa; sometimes it refers to people who will reside in the destination country and never return to their previous home. Either way, the definition links two countries and depicts people as moving across geographical spaces and crossing borders.

While the global presence of female migration now is firmly established, the feminisation trends represent only the beginning of understanding the international migration of women, where they are and what are their experiences and opportunities. This chapter also demonstrates the regionally specific migration patterns for females and males, and their labour market participation, with emphasis on domestic workers. More knowledge remains to be garnered, however. The remaining sections of the chapter highlight the need to improve data quality, and to extend data collection to other topics. Currently, tension still exists between the use of sex-aggregated data and how gender produces inequalities in the migration decision making, movement, and settlement processes.

The Feminisation of Migration

Despite carefully crafted analyses confirming the international migration of women during past centuries (Donato & Gabaccia, 2015), the dominant image of an international migrant was that of an adult male, ready to labour in a new land. In reviewing the post-world war II era, Casas and Garson (2005) categorise the following stages of sensitivity to the presence of women. First, up to the mid-1970s, there existed an almost complete absence of studies on female migration. Second, beginning in 1973–1974 with the closure of European borders to new workers, the predominance of family migration generated modest attention to the women in these flows. Third, during the 1980s influential works appeared that emphasised women as actors in the migration process but also noted differences between men and women in the migration process and in settlement (Morokvasic, 1983, 1984). Fourth, by the late 1990s and thereafter, the term "feminisation of migration" became a standard term in the migration field.

Three factors underlay the now widespread use of this phrase "feminisation of migration". First, the attempts of the United Nations Statistical office to develop and improve the data collection systems of many new nations after their post-colonial independence also directed attention towards collecting data on migration and on girls and women. The first presentation of data on the feminisation of migration appears in 1995 (Zlotnik, 1995). Later in a 2002 release, the UN Population Division made sex-specific migration data pertaining to destination countries available to the public on its website; releases continued in 2005 and approximately every two years with the most recent release occurring for 2019 (United Nations Department of Economic and Social Affairs, Population Division, 2019a). Second, the migration of women became firmly embedded in the widely recognised processes of globalisation through analytical works on the transformative economic changes occurring in the 1970s and beyond. The 1993 first edition of *The Age of Migration* (Castles & Miller, 1993) listed the feminisation of migration as one of the five features of globalisation. Other writers called attention to the structural adjustment programs in developing nation, the disruptive labour displacement effects and the resultant pressures for female labour migration (Sassen, 1988, 2000). Third, the growing engagement of United Nations agencies with gender equality issues produced four major conferences on women: Mexico City in 1975, Copenhagen in 1980, Nairobi in 1985 and Beijing in 1995. These conferences touched on many gender inequalities issues, emphasising female migration outside of high-income countries. The conferences also stimulated additional investigations within the UN. In 1990 the UN Population Division held a major conference on International Migration Policies and the Status of Female Migrants (United Nations, 1995). In anticipation of the United Nations' 2006 High Level Dialogue on international migration and development, several UN divisions held expert conferences and produced reports that explicitly focused on migrant women around the world (United Nations, 2005, 2006).

The phrase "feminisation of migration" implies that the female share among migrants is increasing; the precise meaning rests on temporal trends either in the numbers of females and/or in the percentage of females as a share of total migrants. Since increasing numbers can simply reflect population growth, the preferred statistic is the percentages of migrants who are female since it is standardised to a base of 100 (for example, 40% indicates that 40 women migrants exist out of 100 migrants, male and female). The statistics generated by the U.N. data collection efforts refer to "stock" data, defined as those who are currently living in a country outside their original place of residence.

Although the UN sex-disaggregated migration data go back only to 1960, they reveal that females have been part of global migration for a very long time (Zlotnik, 1995, 2003). Migration data collected by destination countries from censuses, surveys, reports and administrative records (such as admission visas or population registrars) show that in 1960, females were 47% of all migrants living outside their former countries. The percentages rose to almost half (49.3%) around 2000 and thereafter declined to 48% in 2019 (Fig. 2.1). This downward decline is attributed to increasing education of women in their home countries that improved employment prospects there and to growth in male migration. That said, for almost 6 decades (1960–2019) the overall changes are fairly small, and one can continue to say that nearly half of all migrant stock are women and girls.

The feminisation of migration index serves an important purpose by irrefutably demonstrating that females are an integral part of migrant populations and thus warrant extensive scrutiny in terms of their treatment,

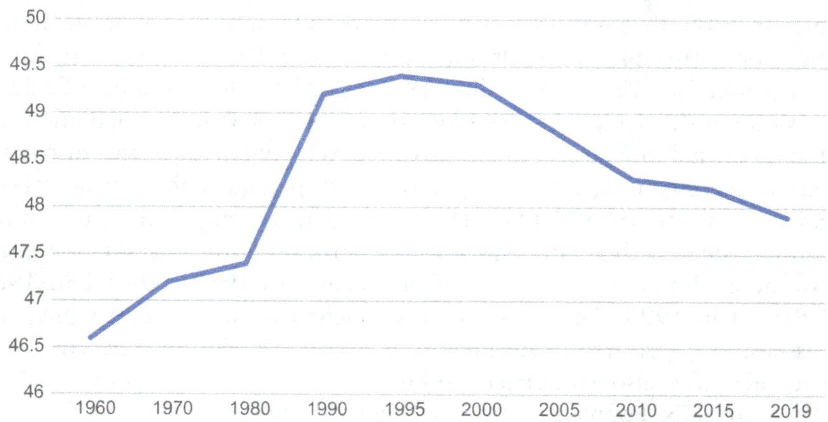

Fig. 2.1 Percent female of international migrants I the world, 1960–2019 (*Source* For 1960–1980, Zlotnik 2003; for 1990–2019, United Nations, Department of Economic and Social Affairs. Population Division [2019b])

experiences, human rights, and opportunities. Indeed, the value of documenting that females are close to half of all migrations appears to be part of the origins of the index (Zlotnik, 1995, 2003). However, analytically, the traditional feminisation of migration index is a blunt instrument for understanding female international migration for several reasons. First, calculated as a percentage of females in the overall migration population, it offers no insight into the causes of high or low values. Second, in the conventional measure of feminisation of migration, women and girls are combined into one sexed group, as are men and boys. This combination over-simplifies distinctions across the life cycle. Women and girls may share some experiences but they also differ in the opportunities for migration, in their migration experiences, how they are treated in migration policies in both sending and receiving countries, and depending on the human rights and social rights contexts of receiving countries, they may differ in rights and entitlements. For example, access to schooling may be far more important for migrant children while the application of labour standards in the workplace may have greater meaning for migrant women. Depending on the admission protocols or how people enter a country (legal or irregular), migrant women may age in place, outliving spouses. Alternatively, they may migrate in old age to reside with, or near, younger family members including sons and daughters. Because women generally live longer than men do, the female population has higher proportions of elderly than do migrant males.

Third, the most recent United Nations Population Division data cover 262 countries plus additional areas such as territories. These countries and regions have different histories, human rights, labour market needs, and treatments accorded women and migrants. Consequently, the feminisation of migration index varies by destination country, by region and by development or wealth status (also see: Asis, 2005; Zlotnik, 2003). In 2019, for example, feminisation was highest for older migrants, particularly in Europe and North America (Table 2.1). Figures were lowest for working age (aged 15–64) international immigrants living in Asia (40%) and in Africa (46%) but in other regions working age women represented half of all migrants.

These age and regional variations can be seen even more easily if the data are expressed as sex ratios, defined by demographers as the number of males per 100 females (Table 2.1, second panel). Males are especially likely to outnumber female migrants in Asia, where 141 migrant males exist for 100 migrant females. The prevalence of males is especially high in Asia for the working age population where there are 154 males for every 100 females. Further investigation shows that this predominance of males in Asia in part reflects the male migration to the Middle Eastern countries. Women migrate to these areas as well, often for employment as domestics or in the health and teaching fields. However, demand for male construction workers also is high.

Table 2.1 Percent female and male/female sex ratios by age and destination region for international migrants, 2019

Geographic destination regions	*Total, all ages*	*Age 0–14*	*Age 15–64*	*Age 65 plus*
Feminization index, % female				
World total	48	48	47	55
Africa	47	51	46	49
Asia	42	46	40	52
Europe	51	48	51	56
Latin America and the Caribbean	50	50	50	52
Northern America	52	50	51	57
Oceania	50	48	50	51
Male/female sex ratio				
World Total	109	107	114	82
Africa	113	97	118	105
Asia	141	116	152	94
Europe	95	107	97	77
Latin America and the Caribbean	100	101	101	91
Northern America	93	101	97	74
Oceania	98	107	98	96

Source Calculated by author from United Nations, Department of Economic and Social Affairs. Population Division (2019b)

Areas of Destination

Once in place, the data banks assembled by the United Nations Population Division can be used to highlight other demographic characteristics of female migration, notably destination areas and connections between origin and destinations. Globally, Asia, Europe, and North America are the three major regions where international migrants reside (Table 2.2, column 1). This is true for both males and females; (regrettably age-specific data are not available by region or country). Compared to males, females are slightly less likely to reside in Asian countries and slightly more likely to live in Europe or in North America. However, for both females and males, the origin area exerts a major influence on their destinations. As highlighted in Table 2.1, over half of females (and males) migrating from Africa remain in the African area. This also is true for those originating in Asia and Oceania, which includes Australia and New Zealand. Over two-thirds of female and male migrants from the European and Latin American and Caribbean regions reside in countries within those regions. Only migrants from North America have greater dispersion, with over one-third of females remaining in North American (one-quarter for males) and approximately one-quarter going to Europe and another quarter migrating to Latin America and the Caribbean.

In actuality, in order to meet the definitional criteria of international migration, people must exit from their countries of birth or countries of citizenship

Table 2.2 Origin and destination corridors for female and male international migrants, all ages, 2019

Sex and Regions	Origin Regions							
	Total	*Africa*	*Asia*	*Europe*	*Latin Am & Caribbean*	*North America*	*Oceania*	*Other Areas*[a]
Female International Migrants								
Numbers, by Origin Region	130,154,101	18,673,531	49,344,735	32,346,605	20,938,981	2,407,950	1,065,936	5,376,163
Percent by Origin Region	100.0	14.3	37.9	24.9	16.1	1.9	0.8	4.1
Destination Areas								
UN development groups	100.0	100.0	100.0	100.0	100.0	100.0	100.0	100.0
More developed regions	60.2	37.1	46.3	84.5	79.6	64.5	91.0	37.3
Less developed regions	39.8	62.9	53.7	15.5	20.4	35.5	9.0	62.7
Geographic regions	100.0	100.0	100.0	100.0	100.0	100.0	100.0	100.0
Africa	9.6	54.0	0.9	1.4	0.1	1.3	1.2	26.2
Asia	26.7	8.3	54.4	11.9	1.0	9.6	4.1	34.8
Europe	32.5	26.7	21.8	68.7	13.9	23.7	18.7	11.5
Latin America and the Caribbean	4.5	0.1	0.4	2.1	19.9	25.6	0.5	2.8
Northern America	23.3	8.8	18.5	11.4	64.6	34.9	26.3	23.3
Oceania	3.5	1.5	4.0	4.4	0.5	4.9	49.2	1.4
Male International Migrants								
Numbers, by Origin Region	141,488,004	21,533,099	62,530,937	28,756,490	19,524,459	2,121,298	1,039,539	5,982,182
Percent by Origin Region	100.0	15.2	44.2	20.3	13.8	1.5	0.7	4.2
Destination Areas								
UN development groups	100.0	100.0	100.0	100.0	100.0	100.0	100.0	100.0
More developed regions	52.1	34.6	35.9	84.5	78.5	56.2	89.4	35.1
Less developed regions	47.9	65.4	64.1	15.5	21.5	43.8	10.6	64.9
Geographic regions	100.0	100.0	100.0	100.0	100.0	100.0	100.0	100.0
Africa	9.9	51.6	1.2	1.6	0.1	1.6	1.0	27.6
Asia	34.5	13.6	63.8	11.4	1.1	13.4	5.1	36.4
Europe	28.3	26.0	18.3	68.3	10.6	24.8	19.1	9.3
Latin America and the Caribbean	4.1	0.1	0.4	2.5	20.9	29.6	0.6	2.7
Northern America	20.0	7.3	13.3	11.2	66.8	25.2	22.8	22.6
Oceania	3.1	1.3	3.0	5.1	0.5	5.4	51.4	1.4

Source Calculated by author from United Nations, Department of Economic and Social Affairs. Population Division (2019b)

[a]Consists of areas outside the 232 countries referenced in this dataset

and enter another country, thereby crossing an international boundary. Origin country characteristics including history, religion, language, governmental involvement in fostering migration, and gender relations all influence where people go. In fact, in many cases, migration becomes path-dependent; earlier migration from one country to another become accepted pathways and "migration corridors" become regularised and accepted. These corridors are highlighted for females from select countries in Table 2.3. Any one of the 232 countries for which 2019 data are available from the United Nations could have been selected as examples, but these eight origin countries are distinctive as sources for female labour, particularly domestic work or health care workers (Indonesia, Philippines, Ecuador, Mexico and Poland), as refugee producing countries (South Sudan and Syrian Arab Republic) or a former immigrant country (Spain) that became a destination for females from Ecuador in the late 1970s and late 1990s and more recently from Morocco. Some of these origin countries have longstanding immigration histories that encourage permanent residence elsewhere alongside more recent short-term migration for work (Philippines, Ecuador, Mexico, Poland).

Table 2.3 Female international migrants from select countries of origin by region of destination and the five top countries of destination, 2019

	Country of origin							
	Indonesia		*Philippines*		*Ecuador*		*Mexico*	
Total World Population	2,040,940		2,881,646		625,963		5,518,534	
Destination Regions								
Total	100.0		100.0		100.0		100.0	
Africa	0.2		0.2		0.0		0.0	
Asia	88.3		35.3		0.0		0.1	
Europe	5.3		12.8		48.1		1.4	
Latin America and the Caribbean	0.1		0.1		8.5		0.7	
Northern America	3.3		44.2		43.1		97.8	
Oceania	2.9		7.4		0.2		0.1	
Major Destination Countries	Saudi Arabia	523,555	USA	1,269,989	USA	261,155	USA	5,351,204
	Malaysia	457,222	Canada	363,567	Spain	223,791	Canada	45,875
	United Arab Emirates	202,404	Saudi Arabia	287,571	Italy	51,924	Spain	31,286
	China, HK SAR	167,627	Australia	169,654	Venezuela	18,574	Germany	9,293
	Bangladesh	107,648	China, HK SAR	141,620	Chile	17,300	France	9,235
	Singapore	91,781	Japan	141,556				
			UAE	137,751				
	South Sudan		Syrian Arab Republic		Poland		Spain	
Total World	1,317,116		3,389,582		2,291,016		778,499	
Destination Regions								
Total	100.0		100.0		100.0		100.0	
Africa	97.9		2.3		0.4		0.7	
Asia	0.6		83.5		1.7		0.6	
Europe	0.0		12.0		85.1		61.6	
Latin America & the Caribbean	0.0		0.3		0.4		21.5	
Northern America	1.2		1.5		11.0		14.3	
Oceania	0.3		0.4		1.4		1.3	
Major Destination Countries	Uganda	565,181	Turkey	1,365,453	Germany	862,161	France	168,952
	Sudan	391,768	Lebanon	602,239	U.K.	481,083	USA	87,451
	Ethiopia	230,532	Jordan	362,253	USA	233,202	Germany	74,808
	Demo. Rep. Congo	49,688	Saudi Arabia	330,329	Italy	90,238	Argentina	57,214
	Kenya	44,363	Germany	213,138	Canada	85,263	UK	56,289

Source Calculated by author from United Nations, Department of Economic and Social Affairs. Population Division (2019b)

These specific examples show how variable female migration is at the country level. Female migration from Indonesia is largely to Asian countries, with Saudi Arabia, Malaysia, the UAE, Hong Kong, and Bangladesh accounting for over half of the 2019 migration. Although not shown in Table 2.3, just under 92,000 additional Indonesian females are in Singapore. Many work as domestics. Females from the Philippines are more diversified in their destinations, with the largest numbers found in the USA and in Canada; here they mostly have rights to permanent residence and their migration reflects both earlier marriage migration (USA) and labour migration. In Canada, many Philippine women entered originally through the Foreign Domestic Worker program (1982–1991) and the Live-in Caregiver Program (1992–2014). Other top destination countries are Saudi Arabia, Australia, Hong Kong. Additionally, as of 2019, Japan has 141,556 females from the Philippines; the United Arab Emirates closely follows with 137,751 females. Female migrants from Ecuador are found mostly in the USA and in Spain; although migration to Spain began in the 1980s, women lead in the outmigration to Spain following the Ecuadorian banking crisis of 1998–1999. The long history of migration from Mexico can be seen in the extreme concentration of female Mexican immigrants in the United States. Many are legal permanent residents, but some are "undocumented" or "irregular" migrants. In contrast, female migrants from war-torn South Sudan and the Syrian Arab Republic are mostly found in nearby countries, the one exception being the location of over 213,000 females from Syria in Germany. Polish migration within Europe increased dramatically with the 1989 dissolution of the USSR and women often worked in care or health occupations. Church groups in Canada and in the USA also sponsored migrants from Poland prior to the 1990s and these early links were reinforced by additional migration in the 1990s and later. Finally, Spain is a country that shifted from sending migrants in the 1950s and 1960, to a receiving country. Nonetheless, females still leave Spain for other countries, mostly in Europe.

Female Migrants at Work

Indicators of the feminisation of migration data and knowing the origin and destination pathways highlight contemporary global trends, but they do not specifically speak to the factors that cause migration or that underlie female–male social and economic inequalities. Finding employment and earning livelihoods are powerful motivators for both women and men. For individuals, other major factors are moving with, or joining family members who already migrated, and moving for marriage. Although researchers and policy makers consider family-based migration and marriage migration as distinct from labour migration, often the labour market outcomes are blurred. For example, quite apart from unpaid work in the home, women (and men) who move to be with spouses or to be married often subsequently enter the labour force for paid employment. Gendered occupational structures, including sex typing of jobs for women and men, can exist in both the sending and receiving countries. When this occurs, migrant women usually are disadvantaged, relative to men and to native-born women.

The International Labour Organization produces a global perspective on the labour market insertion of migrant women and men by aggregating data for individual countries. Wide variability exists in the likelihood of labour force participation for immigrant women and whether or not their participation rates are less than or exceed those of non-migrant women. In Sub-Saharan Africa and in Latin America and the Caribbean, female international migrants have lower rates than their non-migrant counterparts. However, some may be working in the informal economy where their rates are not officially tabulated. Female international migrants also have slightly lower labour force participation rates in Southern Asian countries than do non-migrant women living in the same subregional countries (Table 2.4, columns 1 through 3). Nonetheless, what also is evident is how much both female and male international migrants contribute to the labour force population (Table 2.4, columns 7 and 8). One in five workers in North American and in North, South, and West Europe are international migrants. In the Arab States, nearly 4 out of ten females in the labour force are international migrants. These trends in general are also found in the male migrant population.

What is shown in another earlier ILO analysis of 2013 data is the importance of female migrant labour as a source of domestic workers for select destination regions. As of 2013, domestic work represents the jobs of 11.5 million out of 150.3 million migrant workers (persons aged 15 and older). But, these jobs are disproportionately held by women, who globally represent 73.4% of all migrant domestic workers (Table 2.5, column 3). As is true for labour force participation rates, there are large geographical variations in the feminisation of domestic work. Among domestic migrant workers, over nine out of ten who work in the Southeast and Pacific region or in Eastern Asia or in Latin America and the Caribbean are women (Table 2.5, column 3).

In addition, all work undertaken by migrants in these areas is heavily dominated by domestic employment. In the Arab States, six out of ten female migrant workers are employed as domestics, followed by four out of ten in South East Asia and the Pacific region, which includes Singapore and Malaysia. In Latin America and the Caribbean and in Eastern Asia (which includes Hong Kong), over one-third of all female migrant workers are employed as domestics (Table 2.5, column 1). It is also the case that in some of these regions the numbers of migrant domestics are substantial enough to make this employment sector highly dominated by migrants. In the Arab States, for example, nearly three-quarters (73%) of all female domestics are migrant women. In North, South, and West Europe seven out of ten female domestic workers are migrants (Table 2.5, column 4).

These patterns highlight the global association of being female and a migrant with the provisioning of care work and domestic cleaning, usually within the home. This association rests on powerful stereotypes about the

Table 2.4 Labour force participation rates for women and men by immigrant and non-immigrant status, major world regions, 2015

Area of Destination	*Labour force participation rate*						*Migrant workers as % of all workers*	
	Women			*Men*				
	Migrant	*Non-migrant*	*LF Gap*	*Migrant*	*Non-migrant*	*LF Gap*	*Women*	*Men*
Total	63.5	48.1	15.4	75.5	75.2	0.3	5.0	4.5
Northern Africa	33.0	21.7	11.3	70.9	72.0	−1.1	1.5	1.6
Sub-Saharan Africa	47.3	65.0	−17.7	85.2	73.6	11.6	1.9	3.9
Latin America & Caribbean	47.4	51.6	−4.2	72.2	77.4	−5.2	1.3	1.5
Northern America	64.6	54.4	10.2	73.3	67.4	5.9	20.9	20.4
North. South. West Europe	75.0	48.0	27.0	69.5	63.0	6.5	20.0	16.0
East Europe	70.8	50.9	19.9	66.0	67.9	−1.9	10.1	8.2
Central & West Asia	86.5	41.5	45.0	64.8	74.6	−9.8	16.4	7.6
Arab States	45.7	13.6	32.1	85.9	72.0	13.9	39.9	41.0
Eastern Asia	64.0	60.2	3.8	75.7	75.5	0.2	0.7	0.6
SE Asia & Pacific	66.3	56.3	10.0	71.1	79.9	−8.8	3.6	3.2
Southern Asia	24.1	28.0	−3.9	92.1	79.0	13.1	0.7	1.1

Source ILO Labour Migration Branch and Department of Statistics, 2018. Table 2.13

work considered appropriate for women and men. While it is the case that men also are domestic workers (but to a much lesser extent) they usually are employed as drivers, gardeners or butlers. Domestic work performed by migrant women is of concern because of the abuse and exploitation that can exist. Migrant domestic workers frequently are not covered by labour standards; they may have limited freedom to change employment, or they may work under conditions of debt bondage to recruiters. Entry status to the country of employment frequently is temporary and also can be irregular, which increases vulnerability. Sexual abuse, low pay, and restricted or denied access to the legal system are also problems facing migrant domestic workers in many countries.

Table 2.5 Domestic work force for women and men by immigrant and non-immigrant status, major world regions, 2013

	Migrant domestic workers as % of all Migrant workers		*% Female of all domestic migrant workers*[a]	*Migrant domestic workers as a % of all domestic workers*	
	Female	*Male*		*Female*	*Male*
Total	12.7	3.7	73.4	15.7	22.9
Northern Africa	23.0	3.5	71.4	9.8	5.3
Sub-Saharan Africa	9.4	5.8	53.4	4.9	13.0
Latin America & Caribbean	35.3	2.6	92.0	4.4	2.8
Northern America	3.3	0.3	92.1	71.0	68.3
North, South, West Europe	10.6	1.9	84.2	65.8	28.4
East Europe	0.8	0.3	75.0	23.6	30.3
Central & West Asia	4.5	2.5	69.2	33.4	29.3
Arab States	60.8	10.4	50.6	73.1	95.7
Eastern Asia	33.9	4.5	90.0	7.6	6.6
SE Asia & Pacific	39.2	3.2	90.6	26.9	13.8
Southern Asia	2.8	6.7	22.7	2.5	14.9

Source ILO Labour Migration Branch and Department of Statistics, 2015. Table 2.19
[a]Calculated by author from estimates of the total number of female and male migrants who are domestic workers

Limitations of Global Data on Migrant Women: Measurement Issues and Breadth of Coverage

Since the 1990s, the growing awareness of female international migrants in part is due to compilation and dissemination by international organisations of sex-specific data on migrants. In addition to the United Nations Population Division, compilation and distribution is done by other organisations that have oversight on certain types of migration or migration-related behaviours such as international students (United Nations Educational, Scientific and Cultural Organisation), refugees (United Nations High Commission on Refugees), trafficking and smuggling (United Nations Office on Drugs and Crime), remittances (the World Bank), irregular migration (United Nations Development Programme and the International Organization for Migration), and labour migration (International Labour Organization and the Organisation for Economic Cooperation and Development) (see: Anonymous, 2020; Hennebry, Williams, & Walton-Roberts, 2016; Hennessey, Foresti, & Hagen-Zanker, 2018). However, organisational oversight does not always mean that migrant women and men are treated as distinctive populations in analytical reports or in policy recommendations. Also, population coverage or topic coverage may be incomplete or partial. Two examples suffice here. The

UNODC provides estimates disaggregated by sex, but the estimates are only for reported victims of traffic, rendering invisible those whose victimisation is hidden. Second, the World Bank does not present remittance information by sex. And, even if the World Bank did report remittances by sex, the organisational emphasis is on formal channels of transmission such as bank wires or Western Union transfers (Hennessey et al., 2018). However, analysis of similar surveys in 11 countries confirms that women tend to rely more on in-person case transfer than do men, a feature that would be ignored if only formal banking and financial services are studied (UNWomen, 2020).

In other instances, the resulting data sets can have data quality limitations. First, countries vary in how international migrants are defined and the definition determines who are included (or not) in official statistics. For instance, if irregular migrants, or temporary visa migrants or even women who entered as migrant worker family members are not counted, these groups will be omitted from the summary findings for the specific country in question. Further if one country of interest collects data on refugees or irregular migrants and another does not, this can affect comparative findings. This latter problem plagues both the United Nations and International Organisation data used in Tables 2.1–2.4. For example, UN Population Division estimates for 1990–2019 rest on information supplied by destination countries or from another international agency such as the United Nations High Commission on Refugees. In the 2019 estimates, international migrants were defined as foreign-born in 132 destination countries out a total of the 232 countries (52%); estimates for international migrants included the foreign-born and refugees or those in refugee-like situations for another 49 countries (21%), as reported by the Office of the United Nations High Commissioner for Refugees (UNHCR) or, where appropriate, the United Nations Relief and Works Agency for Palestine Refugees in the Near East (UNRWA). In 26 countries migrants were defied as foreign citizens (11%) and in an additional 16 countries the estimates were derived from data for foreign citizens and refugees or those in refugee-like situations (calculations for 2019 made by the author using United Nations, Department of Economic and Social Affairs, Population Division, 2019b).

Second, procedures to be followed in assembling global datasets can change over time, creating inconsistencies in measurement. For the 2013 ILO data on migrant domestic workers (Table 2.5) did not include refugees but the 2015 labour force participation data on all migrant workers did include refugees where possible (ILO, 2015, 2018). Depending on the topic of concern, shifting population coverages can affect time trend analyses. Measurement changes also challenge over time comparisons. In the 2015 data on workers (Table 2.5), the ILO used a cross-product methodology to calculate female labour force participation, instead of straightforward labour force participation rates used in the 2013 report on domestic workers (ILO 2018: 25). The rationale was that the 2017 data tended to underestimate the labour force participation of women, implying data quality issues were of concern.

Finally, Hennebry et al. (2016) note that different sources of data (censuses versus population registers) can create confusing differences in female–male estimates. As well, different organisations may use different measures for similar phenomena. For example, "marriage" can be defined in multiple ways: a union that occurs under religious auspices, a union that is legal under a designated legal system, or a non-religious non-legal union that simply lasts long enough to be considered a "marriage". In general, compatibility in basic definitions, or agreements are essential when creating harmonised data sets that are used to generate global and regional insights. This compatibility of basic concepts becomes especially important when trying to measure inequalities between migrant women and other groups with respect to indicators of social and economic well-being.

Limitations of Global Data on Migrant Women: Too Little and not Gendered

The efforts since the 1990s to produce global data on the international migration of female and males, women and men elicit two additional criticisms. First, notwithstanding the difficulties of producing compatible data on migrant women and men from multiple countries, overall sex-specific data at the global levels remains sparse, focusing primarily on numbers, location, and labour force participation (Anonymous, 2020; Gammage, 2017; Hennebry et al., 2016). Issues of data comparability and data quality compound the data-dearth problem. The limited availability of sex-specific data is an issue because of the need to provide protection against gender discrimination in all stages of the migration process (Gammage, 2017). Researchers, policy makers, civil society organisations, and individuals note that gender inequalities permeate the migration process and the migration experiences of women and that it is crucial to documents the sites of inequality, the sources, and the role of discrimination. Given this context, the absence of sex-aggregated data compounds the ability to design and implement policies aimed at improving the experiences and opportunities for female international migrants (Hennessey et al., 2018).

Second, when they exist, sex-disaggregated statistics do not automatically speak to gender inequalities. As noted in a United Nations (2015) Gender Statistics manual, sex-disaggregated statistics are a first step. However, according to the manual, gender statistics should reflect gender issues, be based on concepts and definitions that adequately reflect the diversity of women and men, migrants and non-migrants and capture all aspects of their lives; gender statistics require using data collection methods that take into account, and mitigate, sexual stereotypes and other factor that may create gender bias in the data. It requires the conceptualisation and collection of data that can be used in monitoring progress towards gender equality.

These are general goals that say little about the specific sources and types of inequalities experienced by migrant women in relation to men or by immigrant women in relation to female nationals. Statistical systems around the world still collect data on biological sex rather than on gender, defined as "the socially constructed roles and relationships, personality traits, attitudes and behaviours, values, relative power and influence that society ascribes to people based on their assigned sex" (Anonymous, 2020). Gender is also not heteronormative; it includes gender identity and sexual orientation. This inclusion expands the role of gender in migration studies. As the International Organization for Migration (2020) recently observes on International Woman's Day, "the roles, expectations, relationships and power dynamics associated with being a man, woman, boy or girl, and whether one identifies as lesbian, gay, bisexual, transgender and/or intersex (LGBTI) significantly affect all aspects of the migration process and can also be affected in new ways by migration". However, few large surveys or population censuses include direct questions on sexual orientation or identity; typically, sexual orientation is measured indirectly based on the sex of the respondent's partner (Valfort, 2017). Valfort further notes that when direct questions about sexual orientation and identity are asked in surveys, variations in question wording and the survey method (phone, mail-in, interviewer, or electronic) can influence the willingness to answer and affect estimates.

The Fifth Stage and Beyond

Writing in 2005, Casas and Garson focused on the "fourth" stage in the awareness of women in migration. Feminisation of migration and the development of sex-disaggregated databases developed to end the gender-blind approach in which men appeared as labour migrants and women remained mostly ignored. In actuality, the fifth stage already was underway. This stage of engendering migration built on the growing awareness of women in migration but extended agendas and conceptual frameworks. The United Nations participated in, and on occasion, help develop these shifts through conferences and subsequent reports (United Nations, 1995, 2005, 2006) and through general emphasis on gender equality. National governments, academics, and non-governmental association/civil society organisations were even more active.

The content of this fifth stage of "engendering migration" is constantly evolving, but several main tenets exist. The first is that social and economic equality for all migrants is the desired outcome. This second is that the disadvantages of international migrant women, in relation to men or to other women in destination societies result from gender stratification in both the sending and destination societies. This in turn reflects the embeddedness of gender in macro-level societal institutions (financial, political, legal, educational, economic, and religious). Gender also permeates the structure of relationships, behaviours, and norms at the meso level, consisting of families, community associations, trade unions, civil society organisations, and activist groups. And finally, gender in the form of social roles and expectations influences the behaviours and experiences of individuals. These linkages are the foundation for the two major questions asked about gender and migration: first, how does gender influence decisions to migrate, through what channels (recruiters, friends, or autonomous) and how (legal, irregular, temporary, and so forth); second, how does the experience of international migration change gender relations and gendered institutions.

The engendered approach to migration greatly enlarges the topics studied (see Fleury, 2016) and directs attention to the diversity that exists for migrant women, men, and other groups. It also encourages disciplinary perspectives and interdisciplinary approaches. In turn, the growth in different epistemologies and different methodologies expands evidence on the relationships between gender and international migration, albeit at subnational and local levels rather than at a global scale.

References

Anonymous. (2020, June 11). *Gender and migration*. https://migrationdataportal.org/themes/gender-and-migration.

Asis, M. (2005). International migration and prospects for gender equality. In *United Nations, international migration and the millennium development goals: Selected papers of the UNFPA Expert Group Meeting* (pp. 113–123). New York: United Nations Population Fund. https://www.unfpa.org/sites/default/files/resource-pdf/migration_report_2005.pdf.

Casas, L., & Garson, J.-P. (2005, September 26–27). *The feminisation of international migration*. Paper delivered at OECD and European Commission Seminar, Migrant women and the labour market: Diversity and challenges. Brussels. https://www.google.com/url?sa=t&rct=j&q=&esrc=s&source=web&cd=&ved=2ahUKEwiYws272rvqAhVMK80KHQh6Ax4QFjABegQIAxAB&url=http%3A%2F%2Fec.europa.eu%2Fsocial%2FBlobServlet%3FdocId%3D3304%26langId%3Den&usg=AOvVaw2yE_YGxUKgqc4WYOrBaEKS.

Castles, S., & Miller, M. J. (1993). *The age of migration: International population movements in the modern world*. New York: The Guilford Press.

Donato, K., & Gabaccia, D. (2015). *Gender and international migration: From the slavery era to the global age*. New York: Russell Sage Foundation.

Fleury, A. (2016). *Understanding women and migration: A literature review*. Washington, DC: World Bank, Global Knowledge Partnership on Migration and Development (KNOMAD). http://www.knomad.org/publication/understanding-women-and-migration-literature-review-annex-annotatedbibliography.

Gammage, S. (2017). Migrant women. In *Handbook for improving the production and use of migration data for development* (pp. 153–161). Washington, DC: Global Migration Group, Global Knowledge Partnership for Migration and Development (KNOMAD). https://www.knomad.org/publication/handbook-improving-production-and-use-migration-data-development-0.

Hennebry, J., Williams, K., & Walton-Roberts, M. (2016). *Women working worldwide: A situation analysis of women migrant workers*. https://www.unwomen.org/en/digital-library/publications/2017/2/women-working-worldwide.

Hennessey, G., Foresti, M., & Hagen-Zanker, J. (2018). Gender dimensions of migration. *Data Bulletin, 11,* 1–4. https://publications.iom.int/system/files/pdf/data_bulletin_11.pdf.

International Labour Organization, Labour Migration Branch and Department of Statistics. (2015). *ILO global estimates on international migrant workers: Results and methodology, special focus on migrant domestic workers.* https://www.ilo.org/wcmsp5/groups/public/—dgreports/—dcomm/documents/publication/wcms_436343.pdf.

International Labour Organization, Labour Migration Branch and Department of Statistics. (2018). *ILO global estimates on international migrant workers: Results and methodology* (2nd ed.). https://eapmigrationpanel.org/sites/default/files/wcms_652001.pdf.

International Organization for Migration. (2020). *Gender and migration.* https://www.iom.int/gender-and-migration.

Morokvasic, M. (1983). Women in migration: Beyond the reductionist outlook. In A. Phizacklea (Ed.), *One way ticket: Migration and female labour* (pp. 13–32). London and Boston: Routledge and Kegan Paul.

Morokvasic, M. (1984). Birds of passage are also women. *International Migration Review, 18*(4), 896–907.

Sassen, S. (1988). *The mobility of labor and capital: A study in international investment and labor flow.* New York: Cambridge University Press.

Sassen, S. (2000). Women's burden: Counter-geographies of globalisation and the feminisation of survival. *Journal of International Affairs, 53*(2), 503–524.

United Nations. (1995). *International migration policies and the status of female migrants, 1990.* ST/ESA/SER/R/126. New York: United Nations, Department for Economic and Social Information and Policy Analysis, Population Division. https://files.eric.ed.gov/fulltext/ED412144.pdf.

United Nations. (2005). *2004 world survey on the role of women in development: Women and international migration.* New York: UNWomen. https://www.unwomen.org/-/media/headquarters/attachments/sections/library/publications/2005/world-survey-2004-women-and-international-migration-en.pdf?la=en&vs=5138.

United Nations. (2006, May 2–3). *Female migrants: Bridging the gaps throughout the life cycle.* Selected Papers of the UNFPA-IOM Expert Group Meeting. New York. https://www.unfpa.org/sites/default/files/resource-pdf/bridging_gap.pdf.

United Nations. (2015). Integrating a gender perspective into statistics. In *Gender statistics manual.* https://unstats.un.org/unsd/genderstatmanual/Refugees-and-internally-displaced-persons.ashx.

United Nations, Department of Economic and Social Affairs, Population Division. (2019a). *International migration 2019: Report.* ST/ESA/SER.A/438. New York: United Nations. https://www.un.org/en/development/desa/population/migration/publications/migrationreport/docs/InternationalMigration2019_Report.pdf.

United Nations, Department of Economic and Social Affairs, Population Division. (2019b). *International migrant stock 2019: UN dataset.* POP/DB/MIG/Stock/Rev.2019. https://www.un.org/en/development/desa/population/migration/data/estimates2/estimates19.asp.

UNWomen. (2020). *Migrant women & remittances: Exploring the data from selected countries.* https://www.unwomen.org/media/headquarters/attachments/library/publications/2020/policy-brief-migrant-women-and-remittances-exploring-the-data-from-selected-countries-en.pdf?la=en&vs=2913.

Valfort, M.-A. (2017). *LGBTI in OECD Countries: A review* (Working Paper 198). OECD, Directorate for Employment, Labour and Social Affairs, Employment, Labour and Social Affairs Committee. https://www.oecd.org/els/soc/LGBTI-in-OECD-Countries-A-Review-Valfort-2017.pdf.

Zlotnik, H. (1995). The south-to-north migration of women. *The International Migration Review, 29*(1), 229–254.

Zlotnik, H. (2003). *The global dimensions of female migration*. Washington, DC: Migration Policy Institute, Migration Information Source. https://www.migrationpolicy.org/article/global-dimensions-female-migration.

Monica Boyd is full professor and Canada Research Chair in Immigration, Inequality and Public Policy, 2001–2015, University of Toronto. Trained as a demographer and sociologist, Dr. Boyd has written numerous articles, books and monographs on the changing family, gender inequality, international migration (with foci on policy, on immigrant integration and on immigrant women) and ethnic/racial stratification.

Her present research focuses on the 1.5 and 2nd generations, irregular migration policy and consequences, labour market integration of refugees, immigrant re-accreditation difficulties, and the employment of migrant women in Canada's care economy. In addition to her numerous contributions in advisory and consultant capacities, Dr. Boyd has served twice as an elected board member of the Population Association of America, as elected presidents of the Canadian Sociological Association and the Canadian Population Society, as Vice President (representing the Academy of Social Sciences) of the Royal Society of Canada, and as Chair of the International Migration section of the American Sociological Association.

CHAPTER 3

Gendering Transnationalism: Migration and Mobility in *Longue Durée*

Johanna Leinonen

Abstract In scholarship on international migration, the pioneering work of anthropologists in the 1990s ignited a surge of multidisciplinary studies on transnational migration. Subsequently, scholars have examined various types of transnational connections (economic, political, sociocultural, familial and subjective); distinct spheres of transnationalism (micro, meso and macro); and different actors involved in transnationalism (e.g. individuals, families, communities, organisations, companies, nation-states). This chapter explores "the transnational" both as a historiography and as history and provides an overview of research on international migration, transnationalism and gender from the perspective of a historian. It also briefly discusses contributions of women's and feminist history on research on gender, migration and transnationalism since the 1970s, and presents a few examples of historical transnational processes that continue to influence international migration patterns and international relations to this day. It concludes by proposing ideas for future research.

J. Leinonen (✉)
University of Oulu, Oulu, Finland
e-mail: johanna.leinonen@oulu.fi

C. Mora and N. Piper (eds.), *The Palgrave Handbook of Gender and Migration*, https://doi.org/10.1007/978-3-030-63347-9_3

Introduction

In this chapter, I provide an overview of research on international migration, transnationalism and gender from the perspective of a historian. Hence, I explore "the transnational" both as a *historiography* and as *history* (cf. Waldinger & Green, 2016).[1] First, I outline trends in historical research on international migration, focusing on studies that revolve around familial, economic, social and political connections that span borders. Many of these studies were conducted before the popularity of the concept of transnationalism took off in the 1990s and the bulk of them were gender-blind. Second, I briefly discuss contributions of women's and feminist history on research on gender, migration and transnationalism since the 1970s. Finally, I explore a few examples of historical transnational processes—for instance, the circulation of ideas regarding masculinity and femininity dating back to the colonial era—that continue to influence international migration patterns and international relations to this day.

Early Works on Cross-Border Connections

In the 1990s, the field of research on international migration was profoundly influenced by the surge of multidisciplinary studies on transnational migration examining linkages between migrants' countries of origin and destination (e.g. Basch, Glick Schiller, & Szanton Blanc, 1994; Faist, 2000; Vertovec, 1999). The often-cited pioneers in research on transnational migration are anthropologists Glick Schiller, Basch and Blanc-Szanton, who defined transnationalism as the "emergence of a social process in which migrants establish social fields that cross geographic, cultural and political borders" (1992: ix). Proponents of transnationalism criticised the traditional view of migration as a one-time, unidirectional movement from one country to another and highlighted migrants' circular movements and continuing connections to their country of origin. In addition, an important contribution was the introduction of the concept of a transnational social field that people on the move formed together with their families and other social networks left behind (Levitt & Glick Schiller, 2004).

What was typical of the early writings on transnationalism was that they viewed it as a unique feature of the period of globalisation and global capitalism. Studies on transnationalism were based on the premise that the modern means of transportation and digital communication made transnational connections possible. With this argument, scholars made a separation between contemporary international migration and earlier global mobility. "Current transnationalism", Basch et al. wrote in 1994 (p. 24), "marks a new type of migrant experience, reflecting an increased and more pervasive

[1] *Historiography* refers to the body of historical scholarship on a certain topic. *History* is a narrative created by historians; it critically analyses and interprets past events, based on a variety of sources, such as historical records of the studied era or oral histories.

global penetration of capital". Migration historians quickly began to criticise the argument that there was something qualitatively and quantitatively different in modern-day transnationalism, as compared to trajectories and activities of migrants of previous decades and centuries. The critique of migration historians was twofold, both historical and historiographical. First, they argued that transnational connections and consciousness of migrants were not a new phenomenon. As Ngai stated in 2005 (p. 59), "the transnational did not drop from the sky or simply appear as part of the recent interest in 'globalization'". Even before modern communication technologies and modes of transportation, migrants "maintained extensive and intensive, transnational ties and operated in what social scientists now call a transnational social field" (Foner, 1997: 369). Second, migration historians brought to the fore works of early historians who conducted research that would now be considered as transnational in scope. Tyrrell (1999) pointed out that nation-state-centred approaches—that the study of transnationalism criticised—were actually a rather new "invention", peaking in the mid-twentieth century. Gabaccia (1999) noted that nations became fixed with making migrants into members of nation-states only after World War I. As a consequence, in the mid-century, assimilation became the principal focus of research on migration, especially in North America.

In reviews of scholarship on migrant transnationalism, writers often cite Randolph Bourne's 1916 essay *Trans-National America* as the first one to use the concept of transnational in a modern sense, as a critique of the Anglo-Saxon understanding of U.S. culture. Bourne argued that the United States "is coming to be, not a nationality, but a trans-nationality, a weaving back and forth, with the other lands, of many threads of all sizes and colors" (quoted in Thelen, 1999: 967–968). In addition to Bourne, in the early twentieth-century United States, there were also other "experimental" transnational studies, although they did not use the same concept as Bourne. Examples of early sociological studies include W. I. Thomas and Florian Znaniecki's examination of linkages between Polish peasants in Europe and the United States and Louis Wirth's study of the ghetto in Frankfurt and Chicago (Gjerde, 1999; Gutiérrez & Hondagneu-Sotelo, 2008; Thelen, 1999). In Europe too, there were research traditions in which the centrality of the nation-state was challenged. For instance, in France, Marc Bloch, Fernand Braudel and the *Annales* school "bypassed the nation-state and focused on local, subnational communities and on transnational regions such as the Mediterranean world" (Thelen, 1999: 970).

Moreover, there were several historians in the early twentieth-century United States, many of them of migrant origin themselves, who conducted studies that would now be seen as transnational, as they looked at the economic and cultural transformations taking place both in migrants' homelands and in the receiving countries. Gjerde (1999) named these migration historians as the "ethnic Turnerians" (after the influential historian Frederick Jackson Turner). Later in the twentieth century, also a few European scholars examined migration within an "Atlantic, capitalist and world economy" (Gabaccia, 1999: 1118). Scholars such as Theodore Blegen and Marcus Lee Hansen in the early twentieth-century United States, and Frank Thistlethwaite in the United Kingdom after World War II, were among "the first historians to see migration as a part of a world process" (Gjerde, 1999: 41–42). Gerstle (1997: 534) credits Thistlethwaite for introducing a "true break with prevailing interpretations" (i.e. the assimilation paradigm). In 1960, Thistlethwaite presented a new paradigm for the historical study of migration. His approach decentralised the United States as the most important migration destination and instead, highlighted a network of global migration in which the United States played only one part. He encouraged scholars to shift attention from "settler migration" to multiple movements of migrants. Furthermore, he advocated studying migration from its source to its culmination, so that the focus would be on migrants themselves.

At the time of Thistlethwaite's seminal essay in the 1960s, a powerful research tradition was established in the United States, that of New Social History which generated an explosion of interest in "new ethnicity". U.S. scholars increasingly turned their attention to explaining the persistence of ethnicity among Americans of European descent. In this context, histories of migrant women started to elicit interest among scholars, too. However, many of the questions that Thistlethwaite had asked remained unanswered by social historians. As migration historians now concentrated on the particularities of time and space, it became more difficult to discuss larger social processes, such as transnational connections. Thus, Thistlethwaite's object—a new, more holistic approach for migration study that would decentralise the United States as the most important destination of migrants—was not yet realised.

Thistlethwaite was not, of course, alone in proposing an Atlantic perspective. At the same time as New Social History gained ground among historians of European migration to the United States, other fields of study, notably Black studies, Chicano/a studies and Asian American studies were "diasporical from their inception" (Kelley, 1999: 1045). Historians in these fields essentially produced transnational histories before the term was in broader use. According to Thelen (1999: 970), "many people who could not act as full citizens – blacks, women, some immigrants – developed narratives that were not grounded in nation but in other experiences, such as those of race and gender". In other words, groups who were excluded from national narratives framed their histories in a way that made sense to them: by highlighting the contexts of diasporas, borderlands, or global/hemispheric consciousness. For instance, black writers and activists in North America and beyond, marginalised among (white) professional historians, imagined an international black community, an "African diaspora" or the "Black Atlantic" (Gilroy, 1993). Because blacks' belonging to the nation was routinely challenged or blocked, allies were sought internationally. Moreover, black historians examined slavery in a global context and followed forced (and voluntary) migrations of black people through a "hemispheric" lens, providing a truly transnational perspective (Kelley, 1999).

In terms of gender, reflecting the broader trends in historical research on women and gender, the first studies in these fields focused on sex ratios and labour patterns more than on gender ideology. However, since the 1990s, in Black studies, scholars have examined how gender and race intersected to influence the politics of slavery and slaves' resistance, the everyday life of slaves, reproduction and sexuality, black femininity and masculinity and so on (for a review, see Wood, 2010). It is worthwhile to note that the origin of the intersectional approach lies in black feminist studies (e.g. Crenshaw, 1989). In Chicano/a studies, scholars started to consider women's situations and perspectives already in the 1980s (González, 2003). Anzaldúa's *Borderlands/La Frontera: The New Mestiza* (1987) was a seminal study of women in the context of the U.S.–Mexico border. She emphasised the politics of displacement and the conflictual character of living in-between cultures (neither "here" nor "there"). Anzaldúa conceptualised borderlands as a site of shifting identities and a transitional space where a new *mestiza* consciousness and hybridity could be formed. Multi-sited identities are, of course, a common theme in research on migrant transnationalism. Hence, Anzaldúa's borderlands can also be argued to represent a study on migrants' "transnational consciousness" before the term was introduced to the study of international migration.

In sum, at the time when the "transnational turn" emerged in social sciences in the 1990s, many racialised minorities were already examined with a transnational lens. At the same time, among U.S. migration historians, the earlier scholarship that incorporated global perspectives on the study of migration had disappeared from the consciousness of younger scholars. Indeed, new generations of historians were more likely to refer to anthropologists' "new" concept of transnationalism than to the legacies of "ethnic Turnerians" or black historians (Gjerde, 1999). By the end of the 1990s, historians such as Gjerde and Gabaccia started to resuscitate the earlier studies on migration that were de facto transnational and ask questions that Thistlethwaite had already raised in the 1960s.

From Women to Gender: A Transnational Context

The (re) introduction of the concept of transnationalism to the study of international migration coincided with an increase of scholarship that understood migration as a gendered process. For most of the twentieth century, the epitomised migrant in migration research had been a working-class man, considered as the head of the household and the main breadwinner (Morokvasic, 1984). Until the 1970s, migration scholarship was characterised by "almost complete absence of studies on female migration" (Oso & Ribas-Mateos, 2013: 11–12). As feminist scholarship in social sciences and humanities expanded in the 1970s, focusing on gender roles in different societal realms, migration scholars also started to inquire about women's migratory experiences. In the United States, this change occurred in the context of the rising interest in social and ethnic history. In Europe, female migrants' rising visibility was related to the restriction of movement of labour migrants following the financial crisis of the 1970s and the subsequent increase in family-based migration. As a consequence, many European countries started to view migration differently—it was no longer seen as temporary movement of guest workers but as permanent settlement of migrant families. However, at this point, migrant women were still primarily considered as tied movers, as "the reunited wife" (Catarino & Morokvasic, 2013).

Hence, in the 1970s, most migration scholarship did not provide a critical gender analysis; instead, they "added" women's experiences to the broader histories of migration. Scholars primarily examined women's reproductive role in the private sphere. In the 1980s, scholars started to examine women in more active roles, for instance their labour patterns (Morokvasic, 1984). Among U.S. historians, scholars such as Gabaccia (1984) focused on women's ties between the Old and New World, anticipating the future interest in transnational connections. At the same time, it was still rare to consider gender as "a way of structuring power in all human relationships, including those among people on the move" (Donato, Gabaccia, Holdaway, Manalansan, & Pessar, 2006: 6). Indeed, Gabaccia argued in 1991 (focusing on U.S. scholarship) that migrant women were "nowhere at home"—the increasing scholarship on migrant women had little impact on either migration history or women's history.

In the past two or three decades, the scholarship on gender and migration has expanded enormously to analyse the migration process itself as a gendered phenomenon (e.g. Donato & Gabaccia, 2015). Even though the bulk of the scholarship focuses on the reproductive role of migrant women (e.g. as family members, caregivers, domestic workers, or sex workers) in the context of globalisation, gender and migration are now examined across different scales from the familial and local to the global (Donato et al., 2006; Hondagneu-Sotelo, 2013; Oso & Ribas-Mateos, 2013). Scholars also examine gender in the lives of both men and women and ask whether migration alters gender relations in both sending and receiving countries. Intersectional feminism and queer studies have influenced migration scholarship considerably, as scholars explore the intersection of gender with other axes of difference and power (e.g. Cantú, 2009; Luibhéid, 2002; Manalansan, 2003).

Early proponents of transnationalism included feminist anthropologists who examined Caribbean, Haitian and Mexican migration to the United States. Nonetheless, in the early 2000s, several feminist migration scholars criticised research on transnationalism for gender blindness. Thus, the scholarly fields of gender and migration and transnationalism did not necessarily intersect as frequently in the 1990s as one might expect. Indeed, Mahler and Pessar lamented in 2001 (p. 441) that "gender has rarely been a principal focus of studies on transnational spaces and processes, including transnational migration". Pratt and Yeoh argued in 2003 that the scholarship of transnational migration was "implicitly gendered" as masculine—men were the actors of the transnational era, traversing the transnational space as workers, entrepreneurs, political actors, or adventurers. Feminist scholars criticised, for instance, Arjun Appadurai's "post-national diasporic subjects" for being, in reality, masculine subjects, capable of transcending territorial boundaries in a way that was not possible for women, often understood to be tied by familial links (Kondo, 1997).

To be sure, there were scholars who produced early works on gender, migration and transnationalism, but typically the study context was the family and household (as in earlier studies on gender and migration; e.g. Grasmuck & Pessar, 1991; Hondagneu-Sotelo, 1994; Ong, 1999; Parreñas, 2001). In addition, some scholars examined gender in transnational social movements (e.g. Keck & Sikkink, 1998). An important contribution to the study of gender, migration and transnationalism was the expansive scholarship on Italian migration, conducted in the network dubbed as "Italians Everywhere", headed by Gabaccia and Ottanelli. This international group of researchers placed Italian migration in a global and transnational perspective and studied Italian men's and women's experiences in Europe, Australia and the Americas. The network produced numerous studies on, for example, Italian workers and labour radicalism, the impact of migration on gender, class and ethnic identities, on migration and nation-building as well as marriage, sexuality, emotions and transnational family economies (e.g. Baldassar & Gabaccia, 2011; Cancian, 2010; Gabaccia & Iacovetta, 2002; Guglielmo, 2010; Zanoni, 2018). In the bulk of these studies, gender analysis has been at a central stage.

An additional early critique on transnationalism was the observation that the emergence of transnational movements, networks and social fields did not mean that the power of the nation-state was diminishing. Indeed, several researchers showed how differential inclusions and exclusions of migration policy set the parameters within which transnational connections were created and maintained (Briggs, McCormick, & Way, 2008). This critical scholarship soon also included the study of the gendered nature of transnational processes. Since the early 2000s, research on gender, migration and transnationalism has expanded considerably with gender having been used as both a theoretical and analytical tool (e.g. Mahler & Pessar, 2001). While gender analysis still often focuses on women's reproductive roles either in transnational families or in the labour market, gendered transnational processes have been examined in numerous sites and scales (Erel & Lutz, 2012). In addition, gender is understood to refer to both men and women and scholars increasingly study the formation of masculinities and femininities in a transnational context (McIlwaine, 2010; Ye, 2013).

Transnational Discourses and Colonial Legacies

The differential treatment of international migrants depending on their intersectional position (their gender, race, sexuality, class, age, etc.) highlights the enduring power of the nation-state to control its borders. There is important historical scholarship that reveals the centrality of gender, race and sexuality in policing the nation's borders in the context of nation-building since the nineteenth century (e.g. Lee, 2003; Luibhéid, 2002; Ngai, 2004). At the same time, international movements of migrants and states' desire to control these movements also point to the enduring global economic, political, social and cultural hierarchies that derive from the era of colonialism, imperialism and pseudo-scientific racism. In the context of migration to the United States, a majority of migrants originate from areas that have been targets of U.S. economic, military or territorial interventions (Gabaccia, 2012). As Hondagneu-Sotelo (2013: 233) notes, "Gender is deeply implicated in imperialist, military and colonial conquests, which are widely recognized as the roots of global international migration flows". In other words, disparities between nations, originating in colonial times, affect the global economy because of the continuing economic exploitation and military-political dependency. These processes, in turn, influence the globalisation of (labour) mobility in gendered ways. Labour mobility patterns are not "random", but follow the logics of (neo)colonialism, capitalism and trade (Briggs et al., 2008; Ehrenreich & Hochschild, 2002; Oishi, 2005). The Global South increasingly produces expendable labour for the Global North and women are actively participating in these migration circuits. The demand for migrant women's labour in the Global North is due to "the transfer of social reproduction fomented by the demand in developed countries for women willing to take on work classified on the bottom rungs of the labour market in terms of its social value (domestic service, personal care services and sex work)" (Oso & Ribas-Mateos, 2013: 9). At the same time, it is important to note that the forces of globalisation are increasing the need for cheap female labour not only in the highly industrialised nations of the Global North but also in the newly industrialised economies and even in some developing countries (Constable, 2014; Oishi, 2005). Of course, the colonial past does not only affect labour mobility; for example, U.S. imperialism, military interventions and the following economic divides influenced also refugee movements (Espiritu, 2003). For instance, research has shown that transnational and transracial adoption to the United States has its roots in "hot" and "cold" wars of the twentieth century, U.S. refugee policy of the Cold War, U.S. military occupation and the following neo-colonial relationships (Briggs, 2012; Choy, 2013; Kim, 2010; for transnational marriages borne out of this context, see Yuh, 2002).

Scholars in Asian American studies have been particularly active in bringing forth the impact of U.S. imperialism on gendered and racialised migration patterns (see also Briggs, 2002; Whalen, 2001). Choy (2003), for instance, discusses the impact of colonial legacies on migrations of Filipino nurses to the United States in the post-World War II period. The racialised and gendered hierarchies originating from the imperial relationship between the United States and the Philippines continued to impact migrants' experiences in the new country. Hence, she shows how the transnational legacies of U.S. imperialism shaped migration patterns and migrants' position in society. In a similar vein, Espiritu (2003) reveals the historical entanglements of migration and imperialism in her study of Filipino Americans, highlighting how "Filipino American racial formation is determined not only by the social, economic and political forces in the United States but also by U.S. (neo) colonialism in the Philippines and capital investment in Asia" (p. 1).

More broadly speaking, scholars in Asian American studies and historians of Asian migrations have been central in analysing the transnational forces shaping racialised and gendered imaginings and relationships between nations. Choy and Espiritu's study points to the ways in which "understanding *transnational dimensions* of racial formation is vitally important" in migration scholarship (Lee, 2005: 237, original italics). Lee (2005), for example, has examined the "global dynamics of Orientalism", described as the "ways in which Asian migrants were racialized as dangerous and unassimilable foreigners around the world" (p. 237). Lee describes the transnational circulation of racial ideologies, focusing particularly on the racialised and gendered notions regarding Chinese in the United States, Canada and Mexico (see also Camacho, 2012; Delgado, 2012; Hu-DeHart, 2005). Importantly, transnational racialised stereotypes imposed on the Chinese (and some other Asian migrant groups) were strikingly similar throughout the Western hemisphere: anti-Chinese rhetoric exhibited social tensions about class status, race relations, gender and sexual norms (Lesser, 1999; Leong, 2006).

There is also an important line of scholarship that shows how intimate relationships are formed in the context of global hierarchies based on gender, sexuality and race. The burgeoning scholarship on transnational marriages has brought forward how they seem to follow "cartographies of desire", where relationships formed in a transnational context "both reflect and are propelled by fantasies and imaginings about gender, sexuality, tradition and modernity" (Constable, 2005: 7; see also Leinonen, 2017). For instance, Kelsky (2001) illustrates how racialised, gendered and sexualised images attached to Asian female bodies are rooted in histories of Orientalism as well as in transnational movement of these images. At the same time, Kelsky, in her study on Japanese women's "occidental longings", underlines the women's active role in finding ways to fulfil their desire for the "modern West" through, for instance, romance with Western men. Japanese women, too, attached racialised and gendered meanings to male bodies, meanings that are "deeply imbricated in histories of modernity, colonialism and white hegemony in the West and globally" (p. 154). Hence, these gender-specific patterns of transnational relationships suggest that their formation is linked to larger historical, geopolitical, economic and cultural contexts.

Ultimately, many of the scholars cited above point to the ways in which gender and sexuality have been in key roles in negotiations of relations between nations from the colonial times. In fact, relations often deemed as domestic or private are not just “the passive fallout” of unequal transnational relations; they are a “constitutive part of them” (Briggs et al., 2008: 641). Postcolonial feminist scholars have produced critical analyses of colonialism that gendered empire and revealed how ideas regarding sexuality, gender, race and class profoundly shaped the colonial discourse and justified the colonial conquests (McClintock, 1995; Stoler, 2002). These studies highlighted how “sexual practice, discourse and representation lay at the cornerstone of the ‘colonial order of things’” (Cowan, Guidotti-Hernández, & Ruiz, 2015: 2). Moreover, as the examples above illustrate, scholars have revealed the lasting power of these gendered and racialised hierarchical relations during and after decolonisation. For example, scholars such as Klein (2003) and Uno (2003) argue that orientalist and gendered images of Asia allowed the United States to justify its imperial ambitions by placing Asian nations in a subservient position. Today, when observing migration debates and restrictions in Europe or the United States, they often seem to echo “a colonial past where discourses of sexuality, racial thinking and nationalist rhetoric intersected in the construction of the national subject and its others” (Raissiguier, 2003: 94).

Conclusion

As the examples provided in this essay exemplify, scholarship on gender, migration and transnational processes is wide and expanding continuously. Transnational historians (also beyond migration studies) have rejected “the tyranny of the national” and striven to write history “above, below, within, or outside individual nations” (Gabaccia, 1999: 1123). Considering the central role of historians in nation-building projects of the twentieth century, this has been an important corrective. In addition, migration scholars from various disciplines have taken heed of the important criticism levelled against methodological nationalism, i.e. scholars’ tendency to accept the nation-state as a given frame of analysis (Wimmer & Glick Schiller, 2003). Because of this criticism, scholarly interest in transnationalism, diaspora and critical border studies has expanded.

At the same time, another important criticism towards migration studies, the problems related to the (over) use of the “ethnic lens”, has perhaps been less influential. This refers to scholars’ tendency to downplay or ignore migrants’ “non-ethnic” forms of identification. As Fox and Jones (2013: 386) argue, “ethnicity is naturalised as a taken-for-granted fixture of the migration and post-immigration landscape”, overshadowing other forms of belonging:

> When we study transnationalism and diaspora, we surreptitiously posit the existence of collectivities whose members we assume to understand themselves ethnically. When we study ethnic niches, ethnic networks and ethnic capital, we

> give ethnicity a certain structural fixity (...) in order to account for the (important) ways in which "it" structures social relations. Ethnicity may indeed be important, but its importance should not go unquestioned, its relevance should not be assumed. (Fox & Jones, 2013: 386)

Of course, the intersectional approach already gives us tools to think beyond ethnicity in migration scholarship. Still, intersectional feminism has influenced only a fraction of migration studies (Hondagneu-Sotelo, 2013). A majority of historical studies on migration still focus on a single ethnic group and transnationalism means studying the group's connections to "here" and "there" (and more often "here", meaning, in the Global North). Comparative perspectives could bring out how gender relations operate across global scales and challenge exceptionalist interpretations of national histories (Zanoni, 2020). They could also reveal intergroup solidarities and mobilisation across groups, for example, between racialised groups, historically and today.

A research field that clearly needs interventions by historians of gender, migration and transnationalism is the study of forced migration (including refugees, asylum-seekers, deportees and internally displaced peoples). In the United States, historians have been surprisingly reluctant to study refugee movements (Zanoni, 2020), but also in Europe and elsewhere, forced migration is primarily studied in social sciences, where the focus is on contemporary issues of refugee protection and resettlement. As Marfleet noted in 2007, "there is a dual problem of disinterest among historians in refugee matters and an aversion among specialists in forced migration vis-à-vis history" (p. 136). Existing historical studies look at nation-states' and non-governmental organisations' acts during "refugee crises", and tend to depict refugees as indistinguishable victims (Bon Tempo, 2014). However, history matters for forced migrants too, and they may also use histories and memory for different purposes. Existing social scientific studies show that gender affects the refugee experience crucially, but we have little historical data on this. It is important to note that the tendency to disregard past forced migrations is not only a scholarly issue, but can lead to a situation where the arrival of refugees is repeatedly treated as unprecedented. This, in turn, leads to ad hoc policy responses, as we witnessed in Europe during and after the so-called "refugee crisis" of 2015.

However, I would also state a word of caution against categorically treating forced and voluntary migration separately, as is typical in migration research. At the level of motives and within family systems over a migration trajectory sometimes lasting for several years, it is often difficult to draw a clear line between forced and free-will movements. Social scientists have pointed out how categories used in migration governance tend to bleed into research, creating separate lines of analysis for different groups of migrants, such as labour, family or humanitarian migrants (Näre, 2016). This is an important criticism, as different categories of migrants are often researched as if the categories were mutually exclusive. In reality, migrants often occupy different roles and statuses in different stages of their life (see O'Connell Davidson, in this volume). These intersecting and profoundly gendered statuses and vulnerabilities need more research, not only in history but also in social sciences.

Finally, migration historians may in the future be inspired by the recent scholarship produced within the "mobilities paradigm" that takes movement—or, more specifically, meanings attached to movements, i.e. mobilities—as the focal point of research. The paradigm incorporates new ways of theorising about how mobilities lie "at the center of constellations of power, the creation of identities and the microgeographies of everyday life" (Cresswell, 2010: 551). In migration scholarship, an implicit priority is typically given to movements that take place across international borders and involve crossing long geographical distances. This analytical focus ignores the fact that, for example, most of the world's displaced people move only short distances. The mobilities paradigm accentuates the need to analyse mobility in different geographical scales—from long-distance international moves to everyday mobilities. Taking inspiration from the mobilities paradigm, historians and social scientists could blur the division between the study of internal and international migration and examine "from the ground up" what kinds of mobilities matter most for migrants themselves. They could ask how gender along with race, class, sexuality and other factors shape mobile lives locally and globally. This approach could help researchers avoid the pitfalls of methodological nationalism and the ethnic lens. By allowing migrants to articulate what mobilities and relations are most meaningful to them, we could, perhaps, liberate them, at least for a moment, from the label of a "migrant" that people often find themselves stuck with.

References

Anzaldúa, G. (1987). *Borderlands/La Frontera: The new mestiza*. San Francisco: Spinsters/Aunt Lute Books.

Baldassar, L., & Gabaccia, D. R. (Eds.). (2011). *Intimacy and Italian migration: Gender and domestic lives in a mobile world*. New York: Fordham University Press.

Basch, L. G., Glick Schiller, N., & Szanton Blanc, C. (1994). *Nations unbound: Transnational projects, postcolonial predicaments and deterritorialized nation-states*. Langhorne: Gordon and Breach.

Bon Tempo, C. (2014). Review of the *Making of the modern refugee. The American Historical Review, 119*(5), 1650–1651.

Briggs, L. (2002). *Reproducing empire: Race, sex, science and U.S. imperialism in Puerto Rico*. Berkeley: University of California Press.

Briggs, L. (2012). *Somebody's children: The politics of transracial and transnational adoption*. Durham: Duke University Press.

Briggs, L., McCormick, G., & Way, J. T. (2008). Transnationalism: A category of analysis. *American Quarterly, 60*(3), 625–648. https://doi.org/10.1353/aq.0.0038.

Camacho, J. M. S. (2012). *Chinese Mexicans: Transpacific migration and the search for a homeland, 1910–1960*. Chapel Hill: University of North Carolina Press.

Cancian, S. (2010). *Families, lovers and their letters: Italian postwar migration to Canada*. Winnipeg: University of Manitoba Press.

Cantú, L. Jr. (2009). *The sexuality of migration: Border crossings and Mexican immigrant men* (N. A. Naples & S. Vidal-Ortiz, Ed.). New York: NYU Press.

Catarino, C., & Morokvasic, M. (2013). Women, gender, transnational migrations and mobility: Focus on research in France. In L. Oso & N. Ribas-Mateos (Eds.), *The international handbook on gender, migration and transnationalism: Global and development perspectives* (pp. 246–267). Cheltenham, UK and Northampton, MA: Edward Elgar.

Choy, C. C. (2003). *Empire of care: Nursing and migration in Filipino American history*. Durham: Duke University Press.

Choy, C. C. (2013). *Global families: A history of Asian international adoption in America*. New York: New York University.

Constable, N. (2005). *Cross-border marriages: Gender and mobility in transnational Asia*. Philadelphia: University of Pennsylvania Press.

Constable, N. (2014). *Born out of place: Migrant mothers and the politics of international labor*. Berkeley and Los Angeles: University of California Press.

Cowan, B. A., Guidotti-Hernández, N. M., & Ruiz, J. (2015). Sexing empire. *Radical History Review, 123*, 1–8.

Crenshaw, K. (1989). Demarginalizing the intersection of race and sex: A black feminist critique of antidiscrimination doctrine, feminist theory and antiracist politics. *University of Chicago Legal Forum, 1989*(1), Article 8, 139–167.

Cresswell, T. (2010). Towards a politics of mobility. *Environment and Planning D: Society and Space, 28*, 17–31.

Delgado, G. P. (2012). *Making the Chinese Mexican: Global migration, localism and exclusion in the U.S.-Mexico borderlands*. Stanford, CA: Stanford University Press.

Donato, K. M., & Gabaccia, D. R. (2015). *Gender and international migration*. New York: Russell Sage Foundation.

Donato, K. M., Gabaccia, D. R., Holdaway, J., Manalansan, M., & Pessar, P. R. (2006). A glass half full? Gender in migration studies. *International Migration Review, 40*(1), 3–24.

Ehrenreich, B., & Hochschild, A. R. (Eds.). (2002). *Global woman: Nannies, maids and sex workers in the new economy*. New York: Henry Holt and Company.

Erel, U., & Lutz, H. (2012). Gender and transnationalism. *European Journal of Women's Studies, 19*(4), 409–412. https://doi.org/10.1177/1350506812461466.

Espiritu, Y. L. (2003). *Home bound: Filipino American lives across cultures, communities and countries*. Berkeley: University of California Press.

Faist, T. (2000). Transnationalization in international migration: Implications for the study of citizenship and culture. *Ethnic and Racial Studies, 23*(2), 189–222. https://doi.org/10.1080/014198700329024.

Foner, N. (1997). What's new about transnationalism? New York immigrants today and at the turn of the century. *Diaspora: A Journal of Transnational Studies, 6*(3), 355–375.

Fox, J. E., & Jones, D. (2013). Migration, everyday life and the ethnicity bias. *Ethnicities, 13*(4), 385–400. https://doi.org/10.1177/1468796813483727.

Gabaccia, D. R. (1984). *From Sicily to Elizabeth street: Housing and social change among Italian immigrants, 1880–1930*. Albany: State University of New York Press.

Gabaccia, D. R. (1991). Immigrant women: Nowhere at home? *Journal of American Ethnic History, 10*(4), 61–87.

Gabaccia, D. R. (1999). Is everywhere nowhere? Nomads, nations and the immigrant paradigm of United States history. *The Journal of American History, 86*(3), 1115–1134. https://doi.org/10.2307/2568608.

Gabaccia, D. R. (2012). *Foreign relations: American immigration in global perspective*. Princeton: Princeton University Press.

Gabaccia, D. R., & Iacovetta, F. (2002). *Women, gender and transnational lives: Italian workers of the world*. Toronto: University of Toronto Press.

Gerstle, G. (1997). Liberty, coercion and the making of Americans. *The Journal of American History, 84*(2), 524–558. https://doi.org/10.2307/2952569.

Gilroy, P. (1993). *The black Atlantic: Modernity and double consciousness*. London: Verso.

Gjerde, J. (1999). New growth on old wines. The state of the field: The social history of immigration to and ethnicity in the United States. *Journal of American Ethnic History, 18*(4), 40–65.

Glick Schiller, N., Basch, L. G., & Blanc-Szanton, C. (1992). Towards a definition of transnationalism: Introductory remarks and research questions. In N. Glick Schiller, L. G. Basch, & C. Blanc-Szanton (Eds.), *Towards a transnational perspective on migration: Race, class, ethnicity and nationalism reconsidered* (pp. ix–xiv). New York, NY: New York Academy of Sciences.

Gonzáles, D. J. (2003). Gender on the borderlands. *Frontiers: A Journal of Women Studies, 24*(2/3), 15–29.

Grasmuck, S., & Pessar, P. R. (1991). *Between two Islands: Dominican international migration*. Berkeley, CA: University of California Press.

Guglielmo, J. (2010). *Living the revolution: Italian women's resistance and radicalism in New York City, 1880–1945*. Chapel Hill: University of North Carolina Press.

Gutiérrez, D. G., & Hondagneu-Sotelo, P. (2008). Introduction: Nation and migration. *American Quarterly, 60*(3), 503–521. https://doi.org/10.1353/aq.0.0025.

Hondagneu-Sotelo, P. (1994). *Gendered transitions: Mexican experiences of immigration*. Berkeley, CA and London: University of California Press.

Hondagneu-Sotelo, P. (2013). New directions in gender and immigration research. In L. Oso & N. Ribas-Mateos (Eds.), *The international handbook on gender, migration and transnationalism: Global and development perspectives* (pp. 233–245). Cheltenham, UK and Northampton, MA: Edward Elgar.

Hu-DeHart, E. (2005). On coolies and shopkeepers: The Chinese as *Huagong* (Laborers) and *Huashang* (merchants) in Latin America/Caribbean. In W. W. Anderson & R. G. Lee (Eds.), *Displacements and diasporas: Asians in the Americas* (pp. 78–111). New Brunswick, NJ: Rutgers University Press.

Keck, M. E., & Sikkink, K. (1998). *Activists beyond borders: Advocacy networks in international politics*. Ithaca: Cornell University Press.

Kelley, R. D. G. (1999). "But a local phase of a world problem": Black history's global vision, 1883–1950. *The Journal of American History, 86*(3), 1045–1077. https://doi.org/10.2307/2568605.

Kelsky, K. (2001). *Women on the verge: Japanese women, Western dreams*. Durham: Duke University Press.

Kim, E. J. (2010). *Adopted territory: Transnational Korean adoptees and the politics of belonging*. Durham: Duke University Press.

Klein, C. (2003). *Cold war orientalism: Asia in the middlebrow imagination, 1945–1961*. Berkeley: University of California Press.

Kondo, D. (1997). *About face: Performing race in fashion and theatre*. New York: Routledge.

Lee, E. (2003). *At America's gates: Chinese immigration during the exclusion era, 1882–1943*. Chapel Hill: University of North Carolina Press.

Lee, E. (2005). Orientalisms in the Americas: A hemispheric approach to Asian American history. *Journal of Asian American Studies, 8*(3), 235–256. https://doi.org/10.1353/jaas.2005.0051.

Leinonen, J. (2017). Hierarchies of desirability: Racialized cartographies in media discourses on relationships between Finns and foreigners (1982–1992). *Scandinavian Studies, 89*(2), 217–239.

Leong, K. J. (2006). "A distinct and antagonistic race": Constructions of Chinese manhood in the exclusionist debates, 1869–78. In D. R. Gabaccia & V. L. Ruiz (Eds.), *American dreaming, global realities: Rethinking U.S. immigration history* (pp. 141–157). Urbana: University of Illinois Press.

Lesser, J. (1999). *Negotiating national identity: Immigrants, minorities and the struggle for ethnicity in Brazil*. Durham: Duke University Press.

Levitt, P., & Glick Schiller, N. (2004). Conceptualizing simultaneity: A transnational social field perspective on society. *The International Migration Review, 38*(3), 1002–1039.

Luibhéid, E. (2002). *Entry denied: Controlling sexuality at the border*. Minneapolis: University of Minneapolis Press.

Mahler, S. J., & Pessar, P. R. (2001). Gendered geographies of power: Analyzing gender across transnational spaces. *Identities, 7*(4), 441–459. https://doi.org/10.1080/1070289X.2001.9962675.

Manalansan, M. F., IV. (2003). *Global divas: Filipino gay men in the diaspora*. Durham: Duke University Press.

Marfleet, P. (2007). Refugees and history: Why we must address the past. *Refugee Survey Quarterly, 26*(3), 136–148.

McClintock, A. (1995). *Imperial leather: Race, gender and sexuality in the colonial contest*. New York: Routledge.

McIlwaine, C. (2010). Migrant machismos: Exploring gender ideologies and practices among Latin American migrants in London from a multi-scalar perspective. *Gender, Place and Culture, 17*(3), 281–300.

Morokvasic, M. (1984). Birds of passage are also women. *The International Migration Review, 18*(4), 886–907. https://doi.org/10.2307/2546066.

Näre, L. (2016). Työperäisen maahanmuuton illuusio ja työn uudet hierarkiat. In E. Heikkilä & T. Martikainen (Eds.), *Maassamuutto ja siirtolaisuus kehityksen moottoreina?* (pp. 42–55). Turku: Siirtolaisuusinstituutti.

Ngai, M. M. (2004). *Impossible subjects: Illegal aliens and the making of modern America*. Princeton: Princeton University Press.

Ngai, M. M. (2005). Transnationalism and the transformation of the "other": Response to the presidential address. *American Quarterly, 57*(1), 59–65. https://doi.org/10.1353/aq.2005.0015.

Oishi, N. (2005). *Women in motion: Globalization, state policies and labor migration in Asia*. Stanford: Stanford University Press.

Ong, A. (1999). *Flexible citizenship: The cultural logics of transnationality*. Durham: Duke University Press.

Oso, L., & Ribas-Mateos, N. (2013). An introduction to a global and development perspective: A focus on gender, migration and transnationalism. In L. Oso & N.

Ribas-Mateos (Eds.), *The international handbook on gender, migration and transnationalism: Global and development perspectives* (pp. 1–41). Cheltenham, UK and Northampton, MA: Edward Elgar.

Parreñas, R. S. (2001). *Servants of globalization: Migration and domestic work.* Stanford: Stanford University Press.

Pratt, G., & Yeoh, B. (2003). Transnational (counter) topographies. *Gender, Place and Culture, 10*(2), 159–166. https://doi.org/10.1080/0966369032000079541.

Raissiguier, C. (2003). Bodily metaphors, material exclusions: the sexual and racial politics of domestic partnership in France. In A. J. Aldama (Ed.), *Violence and the body: Race, gender and the state* (pp. 94–112). Bloomington, IN: Indiana University Press.

Stoler, A. L. (2002). *Carnal knowledge and imperial power: Race and the intimate in colonial rule.* Berkeley: University of California Press.

Thelen, D. (1999). The nation and beyond: Transnational perspectives on United States history. *The Journal of American History, 86*(3), 965–975. https://doi.org/10.2307/2568601.

Thistlethwaite, F. (1960, August 21–28). Migration from Europe overseas in the nineteenth and twentieth centuries. In Comité International des Sciences Historiques, XIe Congrès International des Sciences Historiques, Stockholm, *Rapports, vol 5: Histoire contemporaine* (pp. 32–60). Stockholm: Almqvist and Wiksell.

Tyrrell, I. (1999). Making nations/making states: American historians in the context of empire. *The Journal of American History, 86*(3), 1015–1044. https://doi.org/10.2307/2568604.

Uno, K. (2003). Unlearning orientalism: Locating Asian and Asian American women in family history. In S. Hune & G. Nomura, G. (Eds.), *Asian/Pacific Islander American women: A historical anthology* (pp. 42–57). New York: NYU Press.

Vertovec, S. (1999). Conceiving and researching transnationalism. *Ethnic and Racial Studies, 22*(2), 448–456. https://doi.org/10.1080/014198799329558.

Waldinger, R., & Green, N. L. (2016). Introduction. In N. L. Green & R. Waldinger (Eds.), *A century of transnationalism: Immigrants and their homeland connections* (pp. 1–31). Urbana, Chicago, and Springfield: University of Illinois Press.

Whalen, C. T. (2001). *From Puerto Rico to Philadelphia: Puerto Rican workers and postwar economies.* Philadelphia: Temple University Press.

Wimmer, A., & Glick Schiller, N. (2003). Methodological nationalism, the social sciences, and the study of migration: An essay in historical epistemology. *International Migration Review, 37*(3), 576–610.

Wood, K. E. (2010). Gender and slavery. In M. M. Smith & R. L. Paquette (Eds.), *The Oxford handbook of slavery in the Americas.* Oxford: Oxford University Press. https://doi.org/10.1093/oxfordhb/9780199227990.013.0024.

Ye, J. (2013). Migrant masculinities: Bangladeshi men in Singapore's labour force. *Gender, Place and Culture, 21*(8), 1012–1028. https://doi.org/10.1080/0966369X.2013.817966.

Yuh, J.-Y. (2002). *Beyond the shadow of Camptown: Korean military brides in America.* New York: New York University Press.

Zanoni, E. (2018). *Migrant marketplaces: Food and Italians in North and South America.* Champaign: University of Illinois Press.

Zanoni, E. (2020). Global women: Migrants and refugees, 1850s–2000. In N. Hewitt & A. Valk (Eds.), *A companion to American women's history.* Malden, MA: Wiley Blackwell.

Johanna Leinonen holds a Ph.D. Degree in History from the University of Minnesota (2011). Currently Leinonen is Academy Research Fellow at the University of Oulu in the project *Refugee Journeys: Narratives of Forced Mobilities* (Academy of Finland, 2018–2023). She also leads the Academy of Finland funded project *Postmemory of Family Separation: An Intergenerational Perspective* (2019–2023) and the Nordic collaborative project *Histories of Refugeedom in the Nordic Countries* (The Joint Committee for Nordic Research Councils in the Humanities and Social Sciences, 2020–2022). Her research fields include transnationalism, migration and refugee history, transnational families and international marriages, and critical race and whiteness studies. Her publications include articles in the *International Migration Review*, *Journal of American Ethnic History*, *Journal of Ethnic and Migration Studies*, and *Social Science History*.

CHAPTER 4

Intersectionality and Transnationality as Key Tools for Gender-Sensitive Migration Research

Helma Lutz and Anna Amelina

Abstract In most countries of the global North, the portrayal of migrants in public discourses is strongly influenced by stereotypical representations in which gender relations play a central role. For an appropriate analysis, insights from Gender Studies and from Migration Studies need to be taken into account. Both disciplines produce critical perspectives to deconstruct gender images in the context of migrant Othering. We argue that two paradigms, namely a transnational and an intersectional perspective, are necessary to provide these critical insights. This chapter deals with the question of how intersectional and transnational perspectives can be used for the analysis of gender relations in the context of migration processes. We start this article by providing an overview over (the development of) intersectionality as one of the most broadly discussed theories/methodologies in Gender Studies. We describe variations and disputes around its conceptualisation in the context of social inequality analysis and confront/combine them in the second part with transnational migration research which is currently an important theoretical frame in Migration Studies.

H. Lutz (✉)
Goethe University Frankfurt, Frankfurt, Germany
e-mail: lutz@soz.uni-frankfurt.de

A. Amelina
University of Cottbus, Cottbus Germany
e-mail: Anna.amelina@b-tu.de

C. Mora and N. Piper (eds.), *The Palgrave Handbook of Gender and Migration*, https://doi.org/10.1007/978-3-030-63347-9_4

Introduction

Public debates around migrant masculinities and femininities have received remarkable public attention during the last years. Debates around the role of the headscarf in Western societies (Korteweg & Yurdakul, 2014), sexual assaults in the context of the refugee crisis (Dietze, 2019) and migrant care workers as a solution for care scarcity in the context of demographic change (Lutz, 2011) are only a few among many relevant themes to be mentioned in this regard. Both migration research and gender studies do not only carefully reflect upon the respective discourses that have come to dominate, they also produce multiple critical perspectives to deconstruct the publicly sedimented gender images in the context of migrant Othering. Transnational and intersectional perspectives are the two paradigms that provide a sound basis for these critical perspectives. Therefore, this chapter deals with the question of how intersectional and transnational perspectives can be used for the analysis of gender relations in the context of migration processes. We start by summarising debates on one of the most intensely discussed topics in Gender Studies: intersectionality (i.e. Collins, 1990). We describe variations and disputes around its conceptualisation in the context of social inequality analysis. In the second part we turn to another innovative theoretical frame in migration studies: transnational migration research (Faist, 2000; Glick Schiller, Basch, & Szanton Blanc, 1995). We also show how this body of literature has benefited from inequality-sensitive intersectionality research (Anthias, 2012). The third part proceeds with the application of both intersectional and transnational paradigms in our own research on complexities of cross-border care relations in the context of migration. The chapter is finished by a conclusion that addresses the future developments in the field of gender-sensitive migration research.

Key Developments of Intersectionality Research Over Time and Place

Intersectional perspectives focus on the interconnectedness, interdependencies and mutual co-construction of key categories of social marking and positioning, also called axes of difference/inequality: race, class and gender (Anthias, 2001; Lutz, Herrera Vivar, & Supik, 2011; Walby, 2009). As early as 1977, the Combahee River Collective (1981: 213), a US-based Black socialist lesbian feminist collective called for the abandonment of a single category 'feminism' by arguing that in their daily lives they could not separate race, class and 'sex' oppression because they often experienced them simultaneously. While the London based sociologists Floya Anthias and Nira Yuval Davis (1992) attracted attention to what they called 'racialised boundaries' ('race, nation, colour and class'), US sociologist Patricia Hill-Collins developed the model of a 'matrix of domination' and the concept of 'interlocking systems of oppression' (Collins, 1990), which sparked a debate on the entanglement and simultaneity in hegemonic projects of sexism, racism and classism. The feminist legal scholar and critical race theorist Kimberlé W. Crenshaw (1989) used the metaphor of a street crossing (intersection) for the descriptions of institutional discrimination against Black female labourers working for General Motors and introduced the concept of intersections of gender, race and class in the context of the American legal system.

While this critical-political project initially focused primarily on analysing the classic inequality triad race–class–gender, during the last decades social scientific researchers expanded the debate by adding more categories of social positioning (Hancock & Tyler, 2007; Lutz & Wenning, 2001; Winker & Degele, 2009: 15). In addition to the classical triad, categories like sexuality, age/generation, disability/health, national belonging and space (e.g. North/South) (for a detailed description, see Amelina & Lutz, 2019: 11) were included. Furthermore, debates on an appropriate typology of intersectional studies were of high relevance.

Leslie McCall (2005), e.g. proposed a differentiation between inter-categorical, intra-categorical and anti-categorical approaches. According to McCall, the proponents of an inter-categorical approach (used predominantly in quantitative studies) analyse social categories less in terms of their social constitution than in terms of their interactions. The advocates of the intra-categorical approach are portrayed as users of a qualitative perspective to identify the effects of different overlapping or interwoven categories of inequality. Although they analyse social positionings as historically specific social products, they nevertheless regard them as largely stable. In contrast, the proponents of anti-categorical approaches use a poststructuralist perspective to reveal that intra- and inter-categorical approaches convey essentialising views, whereas in their view social life 'is considered too irreducibly complex – overflowing with multiple and fluid determinations of both subjects and structures' (McCall, 2005: 1773). In a similar vein, Paula Villa (2011: 182) argues for taking into account 'processes of embodiment because' categories can never account 'for the concrete complexity of both norms/structures and practices. No one is ever only a gender. No gender-norm is ever solely gender-related. No class position is ever untouched by gender and race issues'.

From our perspective, the model of three different methodological approaches is somewhat simplistic. While the inter-categorical approach which McCalls favours, takes three social positionings: gender, class and race and explores their mutual imbrication; in our view, this approach tends towards an essentialisation as gender, ethnicity/race and class are considered attributes of groups that are regarded as natural and static. The intra-categorical approach insists on the usage of categories as a guiding tool for research questions, but at the same time tries to do justice to the fluidity and transformation processes of categories; in other words, it does not do away with the three key categories, but (a) amends them with others, like sexuality, nation/belonging, religion, (dis)ability, generation and space (see Amelina & Lutz, 2019: 11) and (b) acknowledges their modifications over time, location and space. Both these approaches are challenged by anti-categorical approaches: ones who insist that every research needs to take into account the difference between structure and action and therefore, the result is always unpredictable. As we will show later, these approaches cannot be neatly separated as McCall suggests, as e.g. in our own work intersectionality is used in several variations.

The concept of intersectionality made a global career, not only within Gender and Migration Studies but also in anti-discrimination and human rights law, in education, in the sociology of inequality, in diversity studies, health and studies on ageing and disabilities, etc. Current debates focus on various points of conflict, e.g. the question of whether intersectionality can be considered a theory or a hermeneutic device (Lutz, 2016), or if race or gender must always function as a master-category, or whether new categories can be added without doing harm to the concept as a whole (Crenshaw, 2011). The authors of this chapter have used and developed an intersectional methodology and consider it indispensable for their research on migration and gender (Amelina, 2017; Amelina & Lutz, 2019; Lutz, 2018c).

Summing up, intersectionality allows one to approach processes of post-migration as embedded in multiple and interconnected projects of domination and subordination and to better understand migrant subjects' potential forms of resistance against these processes.

Studying Transnational Migration and Intersectional Inequalities

Dominant Approaches in Transnational Studies in Migration

At the beginning of the 1990s, transnational approaches to migration brought a similar emancipatory potential for migration studies, as intersectional theory for Gender Studies some years before. Transnational approaches emerged as a critique of the traditional assimilation-focused perspectives on migration. They became increasingly influential in the 1990s (Glick Schiller et al., 1995) and remain widely visible in current studies. While the classical assimilation research has approached migration as a one-way, one-time relocation from a sending country to a receiving country, transnational approaches view migration as an unfinished life-cycle related process. In particular, they pay attention to the temporary, bi- or multidirectional mobilities, as well as to ongoing, cross-border relationships between mobile and immobile individuals. We shortly present three particularly significant conceptualisations of transnational relations.

Ludger Pries' relational approach to transnational spatiality (2008) conceptualises transnational contexts as relational orders. This implies that social relations and practices have not to be bound to territorial (nation state) spaces. Instead, Pries argues that multilocal configurations are spatialised by the social relations of individuals and collectives in personal networks, families, associations and diasporic communities. The main advantage of this approach is that it de-privileges nation-state-oriented perspectives in migration research.

Thomas Faist's concept of transnational social spaces (Faist, 2000) provides a typology of transnational social spaces that emerge on the basis of movers' transnational practices between the sending and receiving states. He particularly differentiates four types of transnational spaces: (1) transnational contact arenas, (2) theme-centred networks, (3) transnational kinship groups and (4) diasporic and transnational communities and organisations. Faist uses the term space metaphorically to provide a typology of various forms of transnational relations characterised by their durability and formalisation degree. According to this typology, short-term and less formalised practices form less formalised transnational contact arenas (such as migrant labour street markets in global cities of London, Berlin or Paris), while more formalised, but also short-term cross-border practices generate the so-called *theme-centred networks* (i.e. networks of transnational experts). With regard to more formalised, but relatively short-term transnational formations, Faist considers 'transnational kinship groups'. According to him, kinship groups and diaspora communities are reproduced on the basis of various forms of reciprocity and solidarity derived from the social and cultural capital that migrants invest in transnational relationships. Once again, differentiating between more or less durable and more or less formalised transnational formation, allows for a detailed and comprehensive analysis of cross-border relations.

In contrast to the above approaches, Peggy Levitt and Nina Glick Schiller's (2004) theory of transnational fields pays more attention to power struggles in the context of (long-term) transnational relations. This concept draws on the notion of multilocality and on the idea that migration research must simultaneously consider mobile and immobile individuals and their involvement in organisations and institutions. They argue that many movers maintain multi-local, long-term social relationships across national borders through communication. By drawing on Pierre Bourdieu's field theory (1985), which suggests that all social relationships are structured by specific rules of game, they analyse the respective powerful principles underlying specific logics of social fields such as economics, politics, science, etc. These logics guide the social practices within the transnational fields organised across nation-state borders; these logics also allocate a certain social position to all actors involved, each with a greater or smaller degree of power. Understood in this way, the transnationalisation of such fields is characterised by power struggles around geographic mobility and around the maintenance of multi-local, network-like relationships.

All in all, approaches presented share a critique of the primacy of sedentariness, which goes back to a sedentary understanding of the nation state and of spatiality inscribed in the classical studies of migration. Various analyses on, e.g. transnational mother- and parenthood and families, care chain and care circulation (Hochschild, 2000; Hondagneu-Sotelo, 2001; Lutz, 2011; Parreñas, 2001) have sought to overcome these premises and to ensure appropriate empirical sensitivity to various multi-local social relationships and multidirectional unfinished dynamics of migration (Amelina & Faist, 2012; Amelina, Nergiz, Faist, & Glick Schiller, 2012). Moreover, transnational studies in migration (at least partly) pay particular attention to the multidimensional quality of inequality relations as it will be shown in the next section.

Studying Intersectional Inequalities in Transnational Settings

What are the key assumptions of the transnational research that is inspired by the intersectional analysis? First, it argues that transnational social actors assess their social position within a multi-local frame of reference—that spans between the sending, the receiving and sometimes a third state/locality (Anthias, 2012, 2020). Second, this research suggests that transnational actors are hierarchically located along the multiple dimensions of social inequality, such as gender, ethnicity/race or class (Goldring, 1998; Parreñas, 2001); and that they consequently may simultaneously be confronted with multiple hierarchical attributions (Barglowski, Bilecen, & Amelina, 2015; Nieswand, 2011).

The phenomenon of contradictory social mobility convincingly illustrates both assumptions. The studies of Parreñas (2001) and Goldring (1998) reveal simultaneous, contradictory social positionings of migrants in the social structures of the sending and receiving countries: transnationally located individuals experience downward mobility in the receiving country because of open, hidden, formalised and institutional discrimination, but they simultaneously achieve a higher social and economic status (upward mobility) in their sending country or, as in the case of many female transnational migrants from Eastern Europe, achieve to maintain their middle-class status and that of their children despite disastrous unemployment and poverty in their country of origin (Lutz, 2018c). However, there are also many transnational migrants whose low social positioning includes both the sending and the receiving country.

Many researchers of transnational social inequalities refer to the concept of intersectionality only implicitly through using multidimensional references to class, gender and ethnicity/race, thereby challenging the classical reduction of social inequality to the class dimension. In contrast, Floya Anthias (2012, 2020) employs an intersectional perspective in an explicit way by proposing the concept of 'translocational frames of reference'. The latter offers an intersectional understanding of contradictory social mobility and refers to the ambivalent interplay of a number of potentially contradictory, interrelated, context-specific and situational (dis)locations of social positionings in relation to gender, ethnicity/race, class, etc., in and through cross-border relationships. Anthias' approach is characterised by its sensitivity to multiple positionings of mobile actors and their responses to the complexities of hegemonic projects.

Care as a Paradigmatic Research Field for Studying Transnational Migration from an Intersectional Perspective

Intersectional approaches to gender and migration have hitherto dealt with areas such as care, transnational families and transnational mother and fatherhood, addressing the emergence and reproduction of gender-specific hierarchies and inequalities in the context of migration movements (Hochschild, 2000; Hondagneu-Sotelo, 2001; Lutz, 2011; Parreñas, 2001).

Care Economy

Over the last 20 years, care giving in private households[1] has developed into the largest employment sector for migrant women worldwide and to a much lesser extent for men. The global demand for caregivers has boosted the market of transnational care migration that accounts for the increasing feminisation of migration. The International Labour Organisation estimates that a total of 67.1 million people worldwide is employed in private households, of which 73.4% are women (ILO, 2015). Although in many countries a majority of workers are internal migrants (e.g. China), the number of international migrant workers is also growing and according to the ILO (2015) comprises 8.5 million women and 3 million men, but given the inadequacy of sources it could also be twice as much because reliable figures are difficult to raise (ILO, 2013: 32).

Caring for oneself and others is a basic precondition of life, and hence for social existence. A useful definition comes from the German sociologist Ute Gerhard: 'Care is work, is not a commodity whose material value could be measured, but rather an activity that requires practical, mental and emotional skills (...) It includes paid and unpaid care activities; however, it remains an open question and it is not at all clear to what extent it is commodifiable at all, since activities of caring practice cannot be reduced to assessment criteria of economic exchange processes'. (Gerhard, 2014: 75, translation H.L.) Thus, care refers to activities involving childcare, nursing and other types of care within private households or in care-related institutions (kindergartens, hospitals, nursing homes).

As Hannah Arendt (1953) argued, the debate over the value of care—or in Marxist terms reproductive work—has been an ongoing theme since economists such as Adam Smith in the eighteenth century and Karl Marx in the nineteenth century introduced the distinction between productive and unproductive work. Both economists agreed on the idea that care work was to be identified as unproductive activity belonging to the sphere of consumption in the private home, which since the beginnings of bourgeois society

[1]Migrant care givers are also found in huge numbers in hospitals, elderly homes or care centres for disabled and convalescents (Kigma, 2005).

was characterised as women's work. This identification of unproductive as low and productive as highly valued work, was also connected to a demarcation of private and public space and the distinction between work performed in the household (female) and work outside of the home (male): this 'value-abjection' (Müller, 2016) of care and household work can be retraced from the colonial and the civil servant society of the eighteenth/nineteenth centuries to the post-Fordist service society of the twenty-first century (Lutz, 2010: 26). In colonial households, servants were considered status symbols: the exertion of power over (black) servants facilitated the visibility of the conquerors as the ruling class (Von Trotha, 1994: 216). Internal (from the country side) and international migrant domestic workers characterised the period of the nineteenth and the beginning of the twentieth century's middle-class households in the Global North, while high-class households in Asia, Africa and Latin America drew mostly on internal workers.

Despite technical improvements in housekeeping and care giving, the beginning of the twenty-first century is characterised by a return of migrant maids and care givers in private households of the Global North (Lutz, 2011), while their presence continues in the well-off families of the Global South. Reasons for the growing presence of female migrant care workers in the Global North are diverse. They include the increasing state withdrawal from institutional provision of care, in particular care for the elderly; the introduction of cash for care policies into private households in classic European welfare states; and the continuing absence of state involvement in institutional care giving in countries where welfare state supply has been traditionally low (USA, many Middle Eastern, Asian, Latin-American and African states).

Rapid demographic ageing, and a growing number of female citizens entering the labour market are also cited as factors that stimulate growth. Some regions and states treat the recruitment of female migrant care workers as a core element of their national labour market policy, designed to enable female citizens to engage in waged work, for example in Singapore (Teo, 2014), or less explicitly in the USA and Canada (Michel & Peng, 2012). The new 'Care Economy' is now considered to be one of the main trends in future global development as one-third of the global workforce is predicted to be employed in this sector (ILO, 2015: 23). The pattern that links these migration flows is: (a) a stark contrast of income levels between sending and receiving countries; (b) a wilful ignorance of the value of care work; (c) a general tendency to a much lower wage level in comparison to the average income of state citizens and (d) as care work is located in the privatised employment sector, in many countries, it is rarely subjected to state control or regulation.

The Levels of Intersectional Analysis

A suitable analysis of transnational migrant care work needs a multilevel framework based on an intersectional prospect (Lutz & Palenga-Möllenbeck, 2011). This includes the investigation of three levels: (a) the macro level where national care regimes are governed and intersected by labour and care market regulations, laws and cultural codes, (b) the meso level where the organisation of care labour is facilitated by the implementation of racially/ethnically segregated care work through formal organisations (placement agencies) and informal social networks and (c) the micro level where the actors transnational practices and social positionings are analysed as entanglements of gender, class, race/ethnicity. In this chapter we can only hint at the practical implementation of these three levels.

At the macro level, it is important to identify three different intersecting regimes: first, gender regimes based upon gendered cultural scripts, where care/household work is identified as predominantly female and craftwork as male work; second, care regimes as element of welfare regimes, materialised in the form of governmental regulations, distributing the responsibility for the good of the national population between the state, the family and the market; and third, migration as part of labour regimes promoting (as in Singapore and Hong Kong) or prohibiting the employment of migrants in private households, either officially or as silent accomplices of undocumented and informal care work in households (see Lutz, 2017a; Lutz & Palenga-Möllenbeck, 2011). Far from being coordinated with one another, these regimes reveal fault lines, when for example, the gender regime undergoes transformations because adult national women are encouraged or obliged to work full-time while the care regime continues to rely on them as primary cares: the solution often comes from the migration regime, which delivers the 'repair work' (Sciortino, 2004: 32) through which the care gaps in private households are plugged.

At the meso level, employers recruit migrant care workers either through informal networks (friends, colleagues, doctors, pastors, etc.) which implies the risk that undocumented, often racialised/ethnicised workers may be detected and expelled, or through globally and binationally acting private placement agencies. Research on home care agencies which are hardly subjected to state control is fairly new (for Germany see Benazha & Lutz, 2019; Rossow & Leiber, 2017; for Switzerland see Steiner, Schwiter, & Anahi, 2019; for Austria see Aulenbacher & Leiblfinger, 2019). In the ongoing three-country project Decent Care Work–Transnational Home Care Arrangements,[2] attention is paid not only to the growth of this market in strong welfare state societies, but also to the ways in which precarious work situations of so-called 24-hours migrant care givers from sending countries in Eastern Europe, are legitimated

[2] http://decent-care-work.net/wp-content/uploads/2018/04/Project-description_Decent-Care-Work.pdf.

by European transnational service labour law, which often comes down to an illegal staff-leasing agreement (Rossow & Leiber, 2017).

At the micro level, the Global Care Chain Concept (Hochschild, 2000; Parreñas, 2001) is the most prominent among theoretical explanations: it emphasises the social cost for migrant workers and their non-migrating family members. Care-providing chains are characterised by both, a care drain, because migrant workers, women in particular, leave gaps in the care provision for their stay-behind children and elderly parents, and a care gain on the side of the employer families. The latter win an extremely flexible and low-cost group of employees, and profit from the 'emotional added value' (Hochschild, 2001) when migrant women project and transmit emotional feelings and practices to the care receiving children and elders in their employers' home.

Many studies have focused on 'children left behind' as those paying the social costs of the migration of their mothers (Gamburd, 2000; Parreñas, 2005). However, it is striking that many studies do not critically reflect upon the asymmetrical responsibility of male and female migrants with regard to care replacement. While the absence of migrant fathers has hardly been problematised, in many sending countries public discourses have been blaming and shaming (temporary) absent mothers (see Lutz, 2017b). In contrast, fathering and fatherhood of stay-behind fathers to date is hugely neglected. Studies confirm that the majority of fathers, most of whom do not give up their employment, leave child care to their female relatives, in particular to grandmothers (Carling, Menjívar, & Schmalzbauer, 2012). Exceptional fathers, who do take over the mothering of their children, often suffer from the double work burden and from the absence of recognition for their fathering practices (Lutz, 2018c). So far, female migration seems to intensify the pressure on gender orders and the gender regime of the sending countries. Gender norms and the gendered division of labour become the key elements in the analysis of transnational mother—and fatherhood—performances and care arrangements for stay-behind family members.

Arguing that the 'chain' metaphor creates the wrong impression that care and emotions are reduced to a unilateral relationship, where care travels from mothers to children and not in the opposite direction, Baldassar and Merla (2014) designed the 'care circulation approach'. They underline the power of connectivity and emotional support as part of the moral economy of kin, which travels both ways, is multidirectional and often a precondition for mothers to endure long-term absence and adverse employment conditions. Care circulation requires the ubiquitous accessibility of new technologies in order to create a virtual co-presence, a situation that is not always granted. Both concepts, global care chains and care circulation, can be fruitfully used in the study of transnational families and transnational parenthood. However, for a sophisticated understanding of care migration, a third perspective is needed: the transnational social inequality approach (Lutz, 2018a). As Amelina (2017: 33) has shown, this approach examines how a relational, multi-local setting, co-produces specific stratification orders. Here, the availability of care as a desirable commodity at both ends of the care-giving–care-receiving nexus, produces emotional and care inequalities as its presence on one side implies physical absence at the other (Lutz, 2018a: 583). In addition, an intersectional analysis of transnational inequality requires an analytical focus on the race–migration nexus. Many migration policies are marked by institutional racism at the macro level, but racism also plays a role at the micro level. In the case of care migrants from Eastern to Western Europe, it can be argued that they are desired workers because of their shared whiteness and Christianity, but simultaneously othered as backward and uncivilised members of what German fascism called the 'Slavic worker race' (Lutz, 2011).

Transnational Families

One more concept, prominently used in the transnational migration research is the 'transnational family approach', which defines transnational families as social units in which family members maintain relationships with one another across national borders in the course of migration movements (e.g. via care, remittances or symbolic ties) (Baldassar & Merla, 2014; Bryceson & Vuorela, 2002). Today, this family form can no longer be understood as a deviation from the ideal of the supposedly naturally settled middle-class nuclear family, which is considered obsolete in view of increased expectations of professional mobility. Moreover, current family research draws on a dynamic and open family concept because of the 'diversity of family forms' (Stacey, 1991), which particularly understands the connection of mutual responsibility as a constitutive premise of families. In the transnationally organised family, whether monogamous or polygamous, the household and the family are no longer identical, especially because they are simultaneously distributed and linked across different countries (Charsley & Liversage, 2013; Mazzucato & Schans, 2011).

In essence, the research on transnational family-making highlights that the (re)production of transnational connections is actively accomplished by migrated as well as remaining family members; and it is based on the social construction of the family unit in interactions and discourses beyond biologic foundations. The everyday (re)production of the family is at the same time bound to ideas about what constitutes a proper family and must accommodate both consensus and dissent with regard to (power) hierarchies among its members. In many cases, the production and reproduction of transnational families are influenced by legal discourses and power relations of national migration regimes in both, the sending and the destination countries of migration (Brennan, 2004; Parreñas, 2005). Legal regulations rarely correspond to transnational practices (Barglowski et al., 2015), which often leads to the stigmatisation of transnational families (Lutz, 2017b; Lutz, 2018b).

Studies on transnational families are also closely related to the research on 'transnational motherhood'. The latter addresses a specific constellation in the context of transnational family relationships, in which mothers who have migrated as (care) workers to a country of arrival, leave their children behind in the sending country. The decision to do so is often motivated by a desire to enable their children both, to advance socially and to maintain the household in their country of origin. Analyses of transnational maternity arrangements (Hondagneu-Sotelo & Avila, 1997) should pay particular attention to four thematic areas: first, they should consider how local care is organised in the sending country and how it impacts on children, because the care gap in the sending country due to the mothers' absence is often compensated for by other (usually female) family members. Second, the analysis of gender relations should be fine-grained in the context of such transnational care arrangements, because migrating fathers and mothers are confronted with different gendered expectations and demands regarding the organisation of care (financial vs. emotional support). Third, research in this field should investigate potential forms of redistribution (and reaffirmation) of gender-specific care work in the households of transnational families (Lutz & Palenga-Möllenbeck, 2011). Despite losing their 'breadwinner role', stay-behind fathers do not assume the responsibilities of their migrated spouses. Instead, grandmothers, aunts, older daughters or friends and neighbours provide care. Fourth, the frequency and quality of contact between mothers and children through digital media such as text messages, Skype, WhatsApp and the like should also be considered and investigated. The virtual co-presence made possible by these new media not only creates closeness and connectedness, but it is also used by mothers to control their children. Some studies have referred to maternal care provided in this way through media communication as 'Skype mothering' (Lutz & Palenga-Möllenbeck, 2011).

It should be noted that the approaches presented here are concerned with revealing the power effects of heteronormative (transnationally organised) family and care projects, which are reflected in the form of (intra-family) hierarchies and asymmetries. As we have demonstrated, in both fields, care and transnational families, an intersectional approach is indispensable, because the analysis of the interconnection of social inequality's key categories needs elaborated instruments as theoretical guidance and heuristic devices.

Conclusion: Challenges and Perspectives of Intersectionality-Sensitive Transnational Migration Research

Regardless of the innovative and wide-ranging inspirations from recent studies on care, migration and gender relations, intersectionality-sensitive migration research will have to face a number of further theoretical and methodological challenges in the future. One of these challenges is the continued development of anti-essentialist perspectives. Possible ways forward would be approaches that follow on from the previously discussed anti-categorical ways of thinking (Walby et al., 2012) and that understand gender, ethnicity/race, class and other axes of difference, as well as their intersections, as socially constructed and historically specific and mutable. The continuation of the tendencies towards de-essentialisation and de-centering plays a major role in this process, because it focuses on previously neglected phenomena, such as the (re)production of non-binary gender identities and struggles over belonging.

Another challenge is the need to make the various forms of self-reflexivity required in studies of intersectionality and migration research compatible with one another (see Shinozaki, in this volume). This means, first of all, the significance of the positional reflexivity, since researchers must become aware of their position in the complex gendered and racialised/ethnicised hierarchies inherent in a research process (Shinozaki, 2012). In addition, theoretical reflexivity is crucial here to ensure that migration scholars position their analyses within the appropriate field of theory (Haraway, 1988). Finally, this also concerns political reflexivity (De Genova, 2016), which necessitates researchers not only to reflect scientifically on migration policy debates, but also to position themselves in relation to these debates. Political reflexivity has an ethical dimension in that it requires a stronger awareness that scientific research results, if adopted by politics, influence the lives of mobile and non-mobile individuals. This also includes the need to reflect on the intersectionality-sensitive concepts and methods that scholars can use to analyse political dynamics and their own role in these political processes.

References

Arendt, H. (1958). *The human condition*. Chicago, USA: University of Chicago Press.

Amelina, A. (2017). *Trannationalizing inequalities: Sociocultural boundaries, assemblages and regimes of intersectionion*. London, England: Routledge.

Amelina, A., & Faist, T. (2012). De-naturalizing the national in research methodologies: Key concepts of transnational studies in migration. *Ethnic and Racial Studies, 35*, 1707–1724. https://doi.org/10.1080/01419870.2012.659273.

Amelina, A., & Lutz, H. (2019). *Gender and migration: Transnational and intersectional prospects*. London, England; New York, USA: Routledge.

Amelina, A., Nergiz, D., Faist, T., & Glick Schiller, N. (2012). *Beyond methodological nationalism: Research methodologies for cross-border studies*. London, England. Routledge.

Anthias, F. (2001). The concept of 'social division' and theorising social stratification: Looking at ethnicity and class. *Sociology, 35*, 835–854.

Anthias, F. (2012). Transnational mobilities, migration research and intersectionality: Towards a trans-locational frame. *Nordic Journal of Migration Research, 2*, 102–110. https://doi.org/10.2478/v10202-011-0032-y.

Anthias, F. (2020). *Translocational belonging: Identities, inequalities, intersectionalities*. London, England: Routledge.

Anthias, F., & Yuval Davis, N. (1992). *Racialized boundaries: Race, nation, gender, colour and class and the anti-racist struggle*. London, England and New York, USA. Routledge.

Aulenbacher, B., & Leiblfinger, M. (2019). The “fictitious commodity” care and the reciprocity of caring: a Polanyian and neo-institutionalist perspective on the brokering of 24-hour care. In R. Atzmüller, B. Aulenbacher, U. Brand, F. Décieux, K. Dörre, K. Fischer, & B. Sauer (Eds.), *Capitalism in transformation, movements and countermovements in the 21st Century* (pp. 245–260). Cheltenham: Edward Elgar.

Baldassar, L., & Merla, L. (2014). *Transnational families, migration and the circulation of care: Understanding mobility and absence in family life*. London, England and New York, USA: Routledge.

Barglowski, K., Bilecen, B., & Amelina, A. (2015). Approaching transnational social protection: Methodological challenges and empirical applications. *Population, Space and Place, 21,* 215–226. https://doi.org/10.1002/psp.1935.

Benazha, A., & Lutz, H. (2019). Intersektionale Perspektiven auf die Pflege: Geschlechterverhältnisse und Migrationsprozesse. In C. Rudolph & K. Schmidt (Eds.), *Interessenvertretung und Care. Voraussetzungen, Akteure und Handlungsebenen* (pp. 146–160). Münster: Verlag Westfälisches Dampfboot.

Bourdieu, P. (1985). *Sozialer Raum und >Klassen<: Zwei Vorlesungen*. Berlin, Germany: Suhrkamp.

Brennan, K. (2004). Advances in health care organization theory. *Contemporary Sociology, 33,* 311–312. https://doi.org/10.1177/009430610403300325.

Bryceson, D., & Vuorela, U. (2002). Transnational families in the twenty-first century. In D. Bryceson & U. Vuorela (Eds.), *The transnational family: New European frontiers and global networks* (pp. 3–30). Oxford: Berg Bloomsbury.

Carling, J., Menjívar, C., & Schmalzbauer, L. (2012). Central themes in the study of transnational parenthood. *Journal of Ethnic and Migration Studies, 38,* 191–217. https://doi.org/10.1080/1369183X.2012.646417.

Charsley, K., & Liversage, A. (2013). Transforming polygamy: Migration, transnationalism and multiple marriages among Muslim minorities. *Global Networks, 13,* 60–78. https://doi.org/10.1111/j.1471-0374.2012.00369.x.

Collins, P. H. (1990). *Black feminist thought: Knowledge, consciousness, and the politics of empowerment*. London, England: HarperCollins Academic.

Combahee River Collective. (1981). A black feminist statement. In C. Moraga & G. Azaldua (Eds.), *Writings by radical women of color* (pp. 210–218). New York: Kitchen Table, Women of Color Press.

Crenshaw, K. W. (1989). Demarginalizing the intersection of race and sex: A black feminist critique of antidiscrimination doctrine, feminist theory and antiracist politics. In University of Chicago Legal Forum, *Feminism in the law: Theory, practice and criticism* (pp. 139–168). University of Chicago Press. https://chicagounbound.uchicago.edu/cgi/viewcontent.cgi?article=1052&context=uclf.

Crenshaw, K. W. (2011). Demarginalising the intersection of race and sex: A black feminist critique of anti-discrimination doctrine, feminist theory and anti-racist politics. In H. Lutz, M. T. Herrera Vivar, & L. Supik (Eds.), *Framing intersectionality: An introduction* (pp. 25–42). Farnham: Ashgate.

De Genova, N. (2016). The European question: Migration, race, and postcoloniality in Europe. *Social Text, 34,* 75–102. https://doi.org/10.1215/01642472-3607588.

Dietze, G. (2019). *Sexueller Exzeptionalismus*. Bielefeld: Transcript.

Faist, T. (2000). *The volume and dynamics of international migration and transnational social spaces*. Oxford, England: Claredon Press.

Gamburd, M. (2000) *The kitchen spoon's handle: Transnationalism and Sri Lanka's migrant housemaids*. Ithaca, USA and London, England: Cornell University Press.

Gerhard, U. (2014). Care als sozialpolitische Herausforderung moderner Gesellschaften – Das Konzept fürsorglicher Praxis in der europäischen Geschlechterforschung. In B. Aulenbacher, B. Riegraf, & H. Theobald (Eds.), *Sorge: Arbeit, Verhältnisse, Regime* (pp. 69–88). Baden-Baden: Nomos.

Glick Schiller, N., Basch, L., & Szanton Blanc, C. (1995). From immigrant to transmigrant: Theorizing transnational migration. *Anthropological Quarterly, 68*, 48–63. https://doi.org/10.2307/3317464.

Goldring, L. (1998). The power of status in transnational social fields. In M. P. Smith & L. E. Guarnizo (Eds.), *Transnationalism from below*. Transaction: New Brunswick, NJ.

Hancock, P., & Tyler, M. (2007). Un/doing gender and the aesthetics of organizational performance. *Gender, Work & Organization, 14*, 512–533. https://doi.org/10.1111/j.1468-0432.2007.00369.x.

Haraway, D. (1988). Situated knowledges: The science question in feminism and the privilege of partial perspective. *Feminist Studies, 14*, 575–599. https://doi.org/10.2307/3178066.

Hochschild, A. (2000). Global care chains and emotional surplus value. In W. Hutton & A. Giddens (Eds.), *On the edge: Living with global capitalism* (pp. 130–146). London: Jonathan Cape.

Hochschild, A. (2001). Globale Betreuungsketten und emotionaler Mehrwert. In W. Hutton & A. Giddens (Eds.), *Die Zukunft des globalen Kapitalismus* (pp. 157–176). Frankfurt am Main: Campus-Verlag.

Hondagneu-Sotelo, P. (2001). Immigrant women and paid domestic work: Research, theory, and activism. In J. R. Blau (Ed.), *The Blackwell companion to sociology* (pp. 423–436). Oxford: Blackwell.

Hondagneu-Sotelo, P., & Avila, E. (1997). "I'm here, but I'm there": The meanings of Latina transnational motherhood. *Gender & Society, 11*, 548–571.

ILO. (2013). *Domestic workers across the world: Global and Regional statistics and the extent of legal protection*. Geneva: International Labour Organization. http://www.ilo.org/wcmsp5/groups/public/—dgreports/—dcomm/—publ/documents/publication/wcms_173363.pdf.

ILO. (2015). *ILO global estimates on migrant workers: Results and methodology: Special focus on migrant domestic workers*. Geneva: International Labour Organization. http://www.ilo.org/wcmsp5/groups/public/@dgreports/@dcomm/documents/publication/wcms_436343.pdf.

Kigma, M. (2005). *Nurses on the move: Migration and the global health care economy*. Ithaca, USA: Cronell University Press.

Korteweg, A. C., & Yurdakul, G. (2014). *The headscarf debates: Conflicts of national belonging*. Palo Alto, USA: Stanford University Press.

Levitts, P., & Glick Schiller, N. (2004). Conceptualizing simultaneity: A transnational social field perspective on society. *International Migration Review, 38*, 1002–1039. https://doi.org/10.1111/j.1747-7379.2004.tb00227.x.

Lutz, H. (2010). Unsichtbar und unproduktiv? Haushaltsarbeit und Care Work – die Rückseite der Arbeitsgesellschaft. *Österreichische Zeitschrift Für Soziologie, 35*, 23–37. https://doi.org/10.1007/s11614-010-0052-1.

Lutz, H. (2011). *The new maids: Transnational women and the care economy*. London, England: Zed Books.

Lutz, H. (2016). Intersectionality's amazing journey: Toleration, adaptation and appropriation. *Rassegna Italiana Di Sociologia, 3,* 421–437. https://doi.org/10.1423/84371.

Lutz, H. (2017a). Care as a fictitious commodity: Reflections on the intersections of migration, gender and care regimes. *Migration Studies, 5,* 356–368. https://doi.org/10.1093/migration/mnx046.

Lutz, H. (2017b). Euro orphans—The stigmatization of migrant motherhood. In Y. Ergas, J. Jenson, & S. Michel (Eds.), *Reassembling motherhood: Procreation and care in a globalized world* (pp. 247–268). New York: Columbia University Press.

Lutz, H. (2018a). Care migration: The connectivity between care chains, care circulation and transnational social inequality. *A Global Sociology of Care and Care Work, Current Sociology Monograph, 66,* 577–589. https://doi.org/10.1177/0011392118765213.

Lutz, H. (2018b). Masculinity, care and stay-behind fathers: A post-socialist perspective. *Critical Sociology, 8,* 31–32. https://doi.org/10.1177/0896920517749707.

Lutz, H. (2018c). *Die Hinterbühne der Care-Arbeit. Transnationale Perspektiven auf Care-Migration im geteilten Europa.* Wiesbaden: Beltz Juventa.

Lutz, H., Herrera Vivar, M. T., & Supik, L. (2011). *Framing intersectionality: Debates on a multi-faceted concept in gender studies.* Farnham, England: Ashgate.

Lutz, H., & Palenga-Möllenbeck, E. (2011). Care, gender and migration: Towards a theory of transnational domestic work migration in Europe. *Journal of Contemporary European Studies, 19,* 349–364. https://doi.org/10.1080/14782804.2011.610605.

Lutz, H., & Wenning, N. (2001). *Unterschiedlich verschieden. Differenz in der Erziehungswissenschaft.* Wiesbaden, Germany: Springer.

Mazzucato, V., & Schans, D. (2011). Transnational families and the well-being of children: Conceptual and methodological challenges. *Journal of Marriage and Family, 73,* 704–712. https://doi.org/10.1111/j.1741-3737.2011.00840.x.

McCall, L. (2005). *Do they know and do they care? Americans' awareness of rising inequality.* Berkeley, USA: University of California.

Michel, S., & Peng, I. (2012). All in family? Migrants, nationhood, and care regimes in Asia and North America. *Journal of European Social Policy, 22,* 406–418. https://doi.org/10.1177/0958928712449774.

Müller, B. (2016). *Wert-Abjektion: zur Abwertung von Care-Arbeit im patriarchalen Kapitalismus – am Beispiel der ambulanten Pflege.* Münster, Germany: Westfälisches Dampfboot.

Nieswand, B. (2011). *Theorising transnational migration: The status paradox of migration.* New York, USA: Routledge.

Parreñas, R. (2005). Long distance intimacy: Class, gender and intergenerational relations between mothers and children in Filipino transnational families. *Global Networks, 5,* 317–336. https://doi.org/10.1111/j.1471-0374.2005.00122.x.

Parreñas, R. S. (2001). *Servants of globalisation: Women, migration, and domestic work.* Stanford, USA: Stanford University Press.

Pries, L. (2008). *Rethinking transnationalism.* London, England and New York, USA: Routledge.

Rossow, V., & Leiber, S. (2017). Zwischen Vermarktlichung und Europäisierung: Die wachsende Bedeutung transnational agierender Vermittlungsagenturen in der häuslichen Pflege in Deutschland. *Sozialer Fortschritt, 66,* 285–302. https://doi.org/10.3790/sfo.66.3-4.285.

Sciortino, G. (2004). Immigration in a mediterranean welfare state: The Italian experience in comparative perspective. *Journal of Comparative Policy Analysis, 6*, 111–129. https://doi.org/10.1080/1387698042000273442.

Shinozaki, K. (2012). Transnational dynamics in researching migrants: Self-reflexivity and boundary-drawing in fieldwork. *Ethnic and Racial Studies, 35*(10), 1810–1827. https://doi.org/10.1080/01419870.2012.659275.

Stacey, J. (1991). *Brave new families: Stories of domestic upheaval in twentieth century America*. New York: NY: Basic Books.

Steiner, J., Schwiter, K., & Anahi, V. (2019). Unsichtbare Care-Arbeit. Transnationale Sorgenketten für Schweizer Senior*innen. *GeoAgenda, 1*, 14–17.

Teo, Y. (2014). Population problems, family policies, and the naturalization of differentiated deservedness. In K. M. Nasir & B. S. Turner (Eds.), *The future of Singapore: Population, society and the nature of the state* (pp. 64–82). London and New York: Routledge.

Villa, P. (2011). Embodiment is always more: Intersectionality, subjection and the body. In H. Lutz, M. T. Herrera Vivar, & L. Supik (Eds.), *Framing Intersectionality: Debates on a multifaceted concept in Gender studies* (pp 171–186). Farnham: Ashgate.

Von Trotha, T. (1994). *Koloniale Herrschaft. Zur soziologischen Theorie der Staatsentstehung am Beispiel des "Schutzgebietes Togo"*. Tübingen: Mohr Siebeck.

Walby, S. (2009). *Globalization and inequalities: Complexity and contested modernities*. London: Sage.

Walby, S., et al. (2012). Intersectionality: Multiple inequalities in social theory. *Sociology, 46*, 224–240. https://doi.org/10.1177/0038038511416164.

Winker, G., & Degele, N. (2009). *Intersektionalität. Zur Analyse sozialer Ungleichheiten*. Bielefeld: Transcript.

Helma Lutz has been trained as a sociologist and an educationalist. As a researcher she worked at the Centre for Race and Ethnic Studies, University of Amsterdam and later at the European Centre on Migration and Ethnic Relations (ERCOMER), University of Utrecht.

Since 2007, she has been a professor of women's and gender studies at the Department of Social Sciences at the Goethe University Frankfurt, Germany. Lutz has been acting director of this university's Cornelia Goethe Centre for Women's and Gender Studies since 2015. Her teaching and her research interests are concerned with gender, (transnational) migration, ethnicity, nationalism, racism and citizenship. She has a long record of research located at the intersection of gender, migration, ethnicity and nationality in European societies. Lutz has widely published on these issues in three languages (Dutch, German, English) and her work has been translated into Spanish, Greek, Polish and French.

Her publications include 7 monographs, 14 (co-) edited books, 10 edited special journal issues and more than a hundred articles in national and international journals and edited collections.

Anna Amelina is a Professor for Intercultural Studies at the University of Cottbus-Senftenberg. Her research areas in the field of sociology are transnational studies, migration and citizenship studies, gender and intersectionality, cross-border social inequalities and European studies. Her recent publications include: Boundaries of European Social Citizenship (with Carmel, E., Runfors, A., Scheibelhofer, E.) Routledge (2020); Gender and Migration: Transnational and Intersectional Prospects (with H. Lutz), Routledge (2019); Transnationalizing Inequalities in Europe: Sociocultural Boundaries, Assemblages and Regimes of Intersection, Routledge (2017). She is currently one of the leaders of a research centre on "Migration, Conflict and Social Change" (MIKOWA) (www.b-tu.de/mikowa).

CHAPTER 5

Gender, Sexuality and Migration: Global Questions and Their Colonial Legacies

Ingrid Palmary

Abstract In this chapter, I review the intersecting literature on gender, sexuality and migration to critically reflect on its present state and the emerging possibilities for future research. I begin this chapter with a brief overview of some of the developments that have taken place in the study of gender, sexuality and migration over the past three decades. I then go on to look at the relevance of notions of coloniality for the study of gender, sexuality and migration. I do this because coloniality has become increasingly influential in Southern scholarship in recent years and it is useful for migration studies because of its focus on global power dynamics and how they have changed, or not, in the postcolonial era. However, I argue that postcolonial studies have also been guilty of ignoring the ways in which the colonial project has been shaped by the regulation of gender and sexuality and that attention to gender and sexuality will assist in understanding how coloniality is a feature of contemporary concerns over migration. I analyse how theories of coloniality can help shape a more critical understanding of trafficking as just one area of migration studies where gender and sexuality have shaped coloniality.

Introduction

In this chapter, I review the intersecting literature on gender, sexuality and migration to critically reflect on its present state and the emerging possibilities for future research. I begin this chapter with a brief overview of some of

I. Palmary (✉)
University of Johannesburg, Johannesburg, South Africa
e-mail: ipalmary@uj.ac.za

C. Mora and N. Piper (eds.), *The Palgrave Handbook of Gender and Migration*, https://doi.org/10.1007/978-3-030-63347-9_5

the developments that have taken place in the study of gender, sexuality and migration over the past three decades. I consider the different theoretical influences that have shaped the way that gender and sexuality have been attended to in migration studies and how these have changed over time. I then go on to look at the relevance of notions of coloniality for the study of gender, sexuality and migration. I do this because coloniality has become increasingly influential in Southern scholarship in recent years and, as I argue, it is useful for migration studies because of its focus on global power dynamics and how they have changed, or not, in the postcolonial era. If migration represents a significant fault line of global conflict and inequality, then the concept of coloniality can be a useful analytic tool for understanding how this conflict and inequality emerges out of the conditions of coloniality.

However, I argue that postcolonial studies have, in themselves been guilty of ignoring the ways in which the colonial project has been shaped by regulation of gender and sexuality and that attention to gender and sexuality will assist in understanding how coloniality, meaning ongoing colonial practices and ideologies, are a feature of contemporary concerns over migration. As an example, to illustrate this argument, I analyse how theories of coloniality can help shape a more critical understanding of trafficking. I use trafficking as an example because it represents how gender sexuality and migration are intertwined. It is also a topic that has garnered both a widespread popular and policy concern stimulating much globally focused activism as well as a focus of research. Using the theoretical resources on the gendered nature of the colonial project, I argue that we can better critique trafficking as a form of coloniality in ways that open new avenues for understanding the relationship between gender, sexuality and migration.

The Study of Gender Sexuality and Migration: A Brief Historical Overview

Since the 1980s, research on gender and migration has increased dramatically (see Manalansan, 2006; Willis & Yeoh, 2000). What is noteworthy is that much of the early attention to women's migration was inspired by the critique that migration studies had been blind to the migration experiences of women and, as a result, had constructed an artificially universalised perspective on migration rooted in the experiences of men (Morokvašic, 1984). For example, in this early research, family decisions were described without attention to the gendered roles of family members (see De Jong, 2000; Janes, 1990; Smits, Mulder, & Hooimeijer, 2003; for examples) reflecting a gender blindness in the broader study of household decision-making (Monk & Hanson, 1982; Semyonov & Gorodzeisky, 2005). If mentioned at all, women were often depicted as 'trailing' spouses or partners, rather than as primary migrants. For example, in his 1993 review and consolidation of migration theories, Massey et al. (1993) only mention women in passing, alongside teenagers, as the group that has typically held temporary and precarious occupations. The research on gender and migration that was inspired by this critique focused on comparing the migration of women to that of men with attention given to how women's migration was impacted by economic change and the demand for feminised labour (Morokvaśic, 1984) as well as sex roles in the home (Lamb & Bougher, 2009; Thadani & Todaro, 1984; Williams, Jobes, & Gilchrist, 1986). Much of the priority during these years was to make women's migration visible and to understand it in relation to men's movement. As Curran, Shafer, Donato, and Garip (2006) note, this line of inquiry continued through to the mid-1990s.

The project of integrating women and their experiences into migration studies can be considered successful to the extent that, although work produced throughout the 1980s and into the 1990s lamented the absence of women and/or gender from migration studies, it would not be possible to say that this is currently the case. Indeed, in their introduction to a special issue on gender and migration Donato, Gabaccia, Holdaway, Manalansan, and Pesar (2006) begins by posing the question 'Another special issue on gender? Haven't there been enough of those?' (3). This perhaps more accurately flags the increasing attention to gender across the social sciences rather than only in the context of research on migration. Nevertheless, by the end of the 1990s it was not possible to say that women were not visible in migration studies. Rather the more important question was how were they made visible—on what basis and via what methods? And related to this, the kind of representations of women that were emerging in the study of migration (Palmary, 2006).

Thus, from the early 1990s, work on women and migration began instead to critique the ways that women had been added to the study of migration without challenging its underlying concepts and methods more thoroughly; often referred to as the 'add women and stir' approach (Boyd & Grieco, 2003; Curran et al., 2006; Nawyn, 2010). Drawing from broader debates across the social sciences, the 1990s saw an increasing attention to gender rather than sex and, thus, to a relational understanding of gender in migration studies. Thus, Donato et al. (2006), in answering the question of whether we need another special issue on gender, shifts the focus from sex to gender, reflecting broader shifts in the social sciences at the time towards understanding gender as the socially constructed consequence of biological sex.

Alongside this turn to a relational notion of gender, Donato et al. (2006) note that there have been two complementary changes in migration studies that have reflected broader shifts in social theory. The first such change has been the increasing use of interdisciplinary approaches. Nevertheless, disciplinary boundaries have been maintained in migration studies at the same time that they were subject to critique (see Curran et al. [2006] and Sinke [2006 in sociology, Silvey [2006] in geography). Although many authors still wrote from an explicitly disciplinary position during this time, this belies the considerable overlap in their sources and argument, and the naming of a disciplinary focus is perhaps more superficial that has been the reality. The second change has been the growth of qualitative methods across multiple disciplines alongside this notion of gender as construction. For example, within household studies, there has been increased attention to power relations within households leading to productive debates on the extent to which migration can free women from patriarchal family conditions as well as exploitative labour relations (see Dannecker, 2005; Tsikata, 2009). The increasing mainstreaming of qualitative research and an accompanying attention to the politics of knowledge production has led to increasing focus on including perspectives from migrants themselves and a focus on lived experiences (Jung, 2014; Michaels, 2010). Finally, I would add that the mainstreaming of notions of intersectionality have allowed for more contextualised and nuanced analysis of the category women and migrant in ways that have furthered the understanding of difference amongst categories. The critiques developed within feminist praxis have been useful precursors to a body of literature that is attuned to the intersections of gender and sexuality with race, class and other social divisions (see Cho, 2013; Crenshaw, 1990).

Whilst these theoretical resources have been useful in broadening the focus of studies on gender and migration, the study of sexuality and migration has perhaps been less well developed. A basic search of the literature indicates a proliferation of titles on migration and sexuality throughout 2018 and 2019 in particular (Espin, 2018; Luibhéid, 2014; see Manalansan [2006] for an early exception). Prior to this, much of the focus on migration and sexuality took a public health approach to understanding how migration facilitates the transmission of sexually transmitted infection, especially in Sub-Saharan Africa where the connections between migration and HIV infection have been a preoccupation (Collinson, Wolff, Tollman, & Kahn, 2006; Vearey, 2012).

A number of themes are beginning to develop from this emerging focus on sexuality. Firstly, there is a growing body of literature that has focused on making visible queer migrant experiences (Carrillo, 2004; Husakouskaya, 2015, 2017; Lewis & Naples, 2014) in much the same way that the early literature on women and migration made visible women's migration experiences. For example, there has been considerable work that focusses on the asylum experiences of migrants fleeing persecution on the basis of sexual orientation or gender identity and a substantial body of academic work has focused on the asylum system and the success of such claims (see Carrillo, 2004). However, in more recent times, a tension within this literature has been the extent to which legal protection for sexual minorities risks undermining the fluidity that queer theory introduced through its critique of identity politics, given that the legal system inevitably requres fixed sexuality for the purposes of assessing migrant status (Giametta, 2017).

Where legal status requires an unambiguous and stable identification with a specific sexual category, this risks undoing some of the work of queer theory (see Palmary, 2006, for more) which has rested on a more fluid notion of sexuality (see Lewis, 2019). In a special issue on Queer/migration, Luibhéid (2008) suggests that queer theory allows us to expand beyond the important topic of queer experiences of migration to also consider the role of migration and its intersections with capitalism and racism in shaping sexual identities, policies and practices. As Luibhéid (2008) notes, scholars of queer migration engage in a double move by studying LGBTIQ identities without presuming them to be stable or essential, thereby transcending space and time. Migration studies has also benefitted from much feminist work that has shown how the heterosexual family is an inherent element in nationalism (see Yuval-Davis [1993] for early examples), allowing for analysis of how sexuality frequently defines which migrants are un/welcome. For example, discourses of migrants as hypersexual, sexuality violent, carrying sexually transmitted diseases, being immoral and bearing too many children abound and create the boundaries of the moral nation of citizens.

In reflecting on the theoretical influences in the study of gender, sexuality and migration, I would argue that the current focus on coloniality is a useful future focus area. I argue that it is useful for its ability to give historical perspective to how and why migration has become a global preoccupation in the postcolonial context and how current global inequalities are rooted in past colonial ones. Beyond this however, I would argue that scholars concerned with gender, sexuality and migration need an accompanying attention to the ways in which the colonial project and its legacies are rooted in the regulation of gender and sexuality, in ways that can open up critical reflections on its current gendered forms.

Global (Dis)Connections and the Implications for the Study of Gender, Sexuality and Migration

If the overview above represents some of the key debates and questions that have shaped the study of gender, sexuality and migration to date, then I want to move on to focus on how coloniality, as a concept that is becoming more and more important in the social sciences, can be mobilised for future work on gender sexuality and migration and the questions and debates made possible by this focus.

The emergence of coloniality as a concept has a number of theoretical precursors. One such precursor would be the long-standing debates regarding the politics of knowledge production and particularly the inclusion of scholarly work from the global South (see Comaroff & Comaroff, 2012; Landau & Palmary, 2016). Drawing from this critique, it is noteworthy that the overview of gender, sexuality and migration described above is almost entirely written as if the history of migration studies is the history of Anglo/US knowledge on migration. This is not to suggest that work in other contexts has not taken place. For example, a significant body of literature on migration in Asia, has been developed (Afsar, 2011; Toyota, Yeoh, & Nguyen, 2007) much of it focused on the gendered nature of labour migration and associated changes in family patterns. However, this has seldom figured in the historical overviews of gender sexuality and migration. This problem goes beyond the absence of Southern authors in the body of knowledge on gender sexuality and migration and is embedded in the more complex debates about the global politics of knowledge production.

Whilst some of the empirical research for these theoretical developments has been done outside of the global North (particularly the work on migration decision-making and the 'migration-development-nexus', see Piper, 2013), theories shaping gender, sexuality and migration derive mostly from the global North. Far from being limited to migration studies, this is a concern across the social sciences where the politics of knowledge production is being increasingly challenged—not just for who creates knowledge but for what kinds of knowledge are created and how these map onto global inequalities.

This is not to suggest that we should, or even can, develop a uniquely Southern approach to studying migration. Ndlovu-Gatsheni (2013) too warns against a reactionary position that suggests a return to an imaginary untainted African past. Rather we should recognise that the global contexts are mutually constitutive orders that exist only in relation to their constructed opposition. As Mbembé (2001) notes in the introduction to his book:

> Africa still constitutes one of the metaphors through which the West represents the origin of its own norms, develops a self-image, and integrates this image into the set of signifiers asserting what it supposes to be its identity. (p. 2)

Thus, for Stoler (2002), the colonies are the shadow against which the idea of Empire is built. In postcolonial contexts, these debates on colonial legacies have emerged at a moment when there has been an increasing public discourse on migration since the mid-1990s, and human movement has come to occupy a central place in political, media and social media discourse. This has had profound political consequences. One contemporary example is the way that a discourse on migration facilitated the decision of the UK to leave the European Union; a decision rooted in the mobilisation of British identity constructed in relation to a notion of foreign threat (see Somai & Biedermann, 2016). This illustrates well Mbembé's (2001) point about how Britain has defined itself in relation to the occidental Other. Similarly, in parts of South Africa and many other African cities, violence against foreigners has become commonplace fuelled by a politically expedient narrative of foreigners taking jobs from South Africans and bringing crime to the country (see Landau, 2012). Examples such as these highlight the importance of an analysis of coloniality, that is, the ongoing colonial relationships that exist after the end of formal colonisation (Ndlovu-Gatsheni, 2013). Ndlovu-Gatsheni (2013) identifies three elements of coloniality when he notes that:

> Coloniality of power articulates continuities of colonial mentalities, psychologies and into the so-called postcolonial era and highlights the social hierarchical relationship of exploitation and domination between westerners and Africans... coloniality of knowledge addresses the epistemological questions of how colonial modernity interfered with African ways of knowing...coloniality of being...helps to analyse the realities of dehumanization and depersonalization of colonized Africans. (p. 8)

Thus, through these three elements—power, knowledge and being—colonial practices remain in contemporary society. However, for an understanding of the connections between gender, sexuality and migration, it is necessary to extend Ndlovu-Gatsheni's understanding of coloniality to incorporate the extensive, but often marginal research that has been done on how integral the regulation of gender and sexuality was in the production of colonisation (see Essop-Sheik, 2014). This can create the possibility for integrating a colonial critique into the study of gender sexuality and migration.

For example, Kaler (1999) notes how the African women's home craft movement in colonial Rhodesia aimed to educate African women in the ideologies, knowledge and moral principles of European domestic life and how this imposition of European morals at the family level was central to the validation of the colonial state. Similarly, Summers (1991) notes how demography and public health information, as well as missionary-led moral training, was central to colonial governance in Uganda. Through such examples we see Ndlovu-Gatsheni's illustration of how colonial practices work through power, knowledge and being. In other words, we see a deep and gendered connection between the imposition of colonial forms of governance which were given moral authority through the education, knowledge production and training of colonised populations. In other words, these authors have shown how gendered norms related to family health, hygiene and morality were mobilised by the state to legitimise colonial governance. Indeed, Summers (2002) shows how the violent disciplining of women in the name of development and instilling a patriarchal Christian moral order were deemed necessary by colonial administrators to protect their political and economic interests in the colony.

This idea can be usefully transferred to migration studies to understand how coloniality continues in contemporary migration management. Firstly, migration necessarily involves global relationships in how it is studied, how the activism takes place and what knowledge is produced. As noted above, production of knowledge on migration is far from apolitical and raises questions of who the experts are and which positions/knowledges are marginalised. But more than this, however, migration scholars should be concerned with the gendered nature of how coloniality is a part of migration practices and the gendered ways in which coloniality is reproduced. Drawing on Ndlovu-Gatsheni's (2013) characterisation we need to better understand the ways that knowledge, being and power are reproduced—and to do so understanding the role that gender plays in coloniality.

In the remainder of this chapter I look at how the notion of coloniality can be used to usefully analyse current preoccupations with trafficking. This is just one example of the way that coloniality manifests itself in the study of gender, sexuality and migration. However, in tracing these connections we can then ask how the research on gender, sexuality and migration requires might reproduce or challenge ongoing coloniality.

Global Activism or Coloniality: Reflections on Trafficking Responses

Trafficking offers a useful example of how a focus on gender, sexuality and migration can reproduce or challenge the coloniality inherent in migration practices. Drawing on the conceptualisation of coloniality above and in particular the gendered forms that it takes we can unpack the implications of this for trafficking responses. Whilst trafficking no doubt goes beyond sex trafficking, the ways in which trafficking has been first and foremost driven by concerns about women who sell sex have been well documents (see O'Connell-Davidson, 2010; and also, in this volume; Piper, 2005). Thus, it offers a useful insight into how the element of Ndlovu-Gatsheni's (2013) notion of coloniality as the regulation of knowledge, power and being continues in the current counter-trafficking agenda.

One of the key contestations regarding trafficking has been over who sets the agenda for a counter-trafficking response (see Piper, 2005). Global alliances and power relations are an inevitable part of migration studies in its policy, research and its advocacy. Trafficking is no different and many such global alliances have been built to combat it. Whilst this has the potential to be positive it is also imbued with power relations that need critical reflection as alliance partners are far from equal. the relationships between the global network of organizations that have shaped the agenda for trafficking responses were made evident in a recent research collaboration which focused on the making of gendered migration policy in, South Africa (Palmary & De Gruchy, 2020). Although some lobbying and policy change had happened with regard to trafficking in South Africa, the point at which international organisations took up the advocacy was a moment in which there was a significant move towards developing legislation against trafficking and a shift in focus and funding towards the issue. International advocacy had a catalytic effect on research, advocacy and policymaking, giving impetus to local change and shaping its direction.

In this regard, Betts and Orchard (2014) argues that in order to understand the global nature of migration policymaking we need to attend to the national interests at play in responding to migration, the power dynamics between states, and the ideas that shape the international politics of migration. In doing so it becomes possible to understand these power inequalities and how they shift in the postcolonial context. This is true not just of the influence that different states have but also different non-governmental organisations (see also Bassel, in this volume).

To take just one example, the United States Department of state has emerged as a central body that shapes the knowledge/power/being of contemporary trafficking. The Trafficking Victims Protection Act (TVPA) requires that the U.S. Department of state monitor other countries progress on human trafficking. In South Africa, the U.S. Department of state placing the country on their Tier 2 watchlist was a key moment when migration came onto the policy agenda. The coloniality of power is evident in the ways that the TVPA established global criteria against which progress on combatting trafficking should be monitored across the world. The criteria are universalised and do not account for contexts or priorities of other countries. This is not to suggest that there should not be a concern for trafficking in all countries but rather that the forms it takes might well be shaped by context, thus requiring different responses. The global economic dependence of many poorer countries on the U.S. for aid shapes the decisions that poorer countries make over trafficking. Under the TVPA those countries that fall onto the Tier 3 list risk losing U.S. aid and even sanctions. In South Africa this risk operated alongside the role of international NGOs as a driver of intervention on human trafficking shaping the nature of the response.

Thus, coloniality of power, in the form of state action, continues to shape the ways in which gender, sexuality and migration are responded to globally. What is of significance here is not that the US can pressure countries to implement responses to human trafficking but rather they can impact on the nature and form of policy change. For example, the US has recently lobbied states to refuse the reproductive rights clauses reproductive rights clauses in UN agreements. Most significantly for this debate the U.S. via its PEPFAR funding will not financially support any work that advocates for the decriminalisation of sex work. In many Southern African countries, the decriminalisation of sex work has been a primary call by sex worker organisations who have argued that any form of criminalisation puts them at risk of abuse by police and clients and hinders their access to health services.

This ongoing coloniality of power is deeply connected to the ongoing moral project of colonisation—what Ndlovu-Gatsheni (2013) refers to as the coloniality of being. Through its interventions on trafficking, a U.S. moral injunction is globalised with significant effects. In the South African case this has resulted in trafficking overshadowing other issues through powerful advocacy led by the IOM and many of its partners (see for example their publication 'eye on trafficking'). Richter (2012) rightly notes that what is needed in response is an African feminism that engages more strategically with the global context, not to create an advocacy that is somehow rooted in national borders, but to leverage the inevitable global alliances for greater impact and to shape the agenda.

Central to the coloniality inherent in trafficking responses is that the global practices that put women at risk of trafficking, such as repressive border control are eclipsed. Rather than acknowledging that it is the restrictions on movement (particularly the movement of the poor or 'unskilled') that create a demand for trafficking, the problem is individualised and the legal and social responses focus on individual traffickers without attention to the ways that global inequality as discussed above creates the phenomenon of trafficking. In many of the awareness materials produced as a response to trafficking, it is represented as a problem brought about by individual traffickers eclipsing the ways in which ongoing coloniality structures global inequality and creates the conditions that foster people's movement. I have argued elsewhere that this decontextualised notion of individual traffickers as the problem allows border control to be seen as a solution to trafficking rather than its cause and presents it as a benign act of management rather than one of ongoing coloniality (Palmary, 2006). The privatisation of suffering at the hands of individual traffickers rather than for example state policies has the consequence of displacing the causes of violence from wealthy states onto the black bodies in the global South. It becomes framed as a consequence of violence men from the global South thus requiring moral education much of which resonates with the domestic education described earlier that was so central to the colonial project.

The occlusion of the structural conditions that produce trafficking creates the conditions for a rescue industry rooted in the global North but drawing on old colonial logic of Africans needing saving from themselves (Spivak, 1988). As is the case with the early missionaries this is a logic that occludes the ways that the violence of the global South is produced through unequal state practices such as migration control. Fassin (2012) notes that in recent decades, the politics of compassion feeds Western sense of morality. Chouliaraki (2006) notes that who acts on whose suffering reflects patterns of economic and political agency across global zones of influence—North and South or East and West something she refers to as a politics of pity. She points out from the work of Cohen that it is always 'audiences in North America or Western Europe [that] react to atrocities in East Timor, Uganda or Guatemala, rather than the other way around' (cited in Chouliaraki, 2006: 22).

This is not a new phenomenon and other authors have noted for example how in the asylum systems of Northern countries those who have experienced 'exotic' violence such as female genital mutilation are more likely to be deemed credible asylum applicants. This is (partly at least) because such forms of violence reproduce comforting notions of the non-violence of the global North (Casey, 2012). In this way, states are positioned in hierarchy rooted in gender and sexuality where the global North is positioned as tolerant and benevolent rescuing women from men and less progressive states. Returning to the issue of trafficking, there is a logic of colonisation at work in who shapes the agenda, which kinds of knowledge/analysis of trafficking is deemed valid and what interventions are thus possible.

Often those who have raised such critiques have been written off as trafficking denialists unconcerned for the welfare of women. Yet, as Davey, Borton, and Foley (2013) argue, imperialism was an important context for the development of humanitarianism and one that is often written out of such historical accounts. Her work shows how it is in the peak of Imperial conquests in the nineteenth century that we see an intersection of religious, state and secular notions of humanitarianism coming together. Colonial expansion made suffering in other parts of the world become more visible and the colonies became laboratories for humanitarian interventions.

As such, Butler (2011) is rightly wary of gender politics that draws uncritically from this approach. She notes that in the Western imagination, sexuality is the new marker of modernity in ways that are not always feminist but rather reproduce racisms. Drawing from these arguments, the study of gender sexuality and migration is fraught with the risk of coloniality whilst nevertheless offering opportunities for critical reflections that challenge it. By attending to how the regulation of gender and sexuality has been a central but undertheorised aspect of the colonial project and by drawing on emerging debates on coloniality the coloniality inherent to global migration responses can be challenged. Recognising that the current preoccupation with gender and sexuality is not necessarily a positive thing in the current political context that shapes migration responses can help develop a critical scholarship on the topic.

Conclusions: Productive Discomforts

In this chapter I have attempted to briefly trace the ways that gender, sexuality and migration, have been studied over the past few decades. I have argued that early research tended to focus on making women visible in migration studies through their inclusion in household decision-making whilst also showing them to be migrants in their own right. This operated as a useful challenge to the assumed universality of the male migrant experience. However, shifts within gender studies from a focus on women to a focus on gender as a relational construct, alongside the mainstreaming of qualitative methods resulted in a useful understanding the fluidity of gender relationships as well as the power within them. This shift occurred alongside a growing questioning of the fixed nature of sexuality and how this was inscribed into legal and bureaucratic processes. Each of these shifts in focus has led to productive new ways of understanding the gendered nature of migration.

In this chapter, I have argued that attention to the emerging scholarship on coloniality can be productive for migration studies as it focuses analysis on the globalisation of power, knowledge and being. In addition, it prevents us from imagining that the current inequalities created by a preoccupation with migration are new or without history. However, I have argued that without meaningful attention to how colonial power is structured by gender and sexuality, this analysis will fail to account for the contemporary gendered nature of migration. Theoretically, notions of coloniality allow us to consider how global relationships structure local practice in ways that reflect the global inequalities that are rooted in colonial relationships and how they might be evolving. At a practical level, this critical reflection allows for the creation of global feminist alliances which, I have argued are essential given the global focus on migration policymaking but which are themselves inevitably imbued with the power inequalities associated with colonial legacies. I have argued that the current forms of knowledge and being which individualise the problem of trafficking as being one of evil individual traffickers runs the risk of creating the conditions by which the west becomes the saviour of the global South in ways that reproduces a colonial logic. Whilst I have used trafficking as one example of how coloniality continues in the study of gender, sexuality and migration there are no doubt others. Nevertheless, as gender and sexuality have become more central topics in migration studies, attention to the risks or reproducing colonial moral norms is essential for global activism that meets the needs of those living outside of the global North.

References

Afsar, R. (2011). Contextualizing gender and migration in South Asia: Critical insights. *Gender, Technology and Development, 15*(3), 389–410.

Betts, A., & Orchard, P. (2014). *Implementation and world politics: How international norms change practice.* Oxford, UK: Oxford University Press.

Boyd, M., & Grieco, E. (2003). Women and migration: Incorporating gender into international migration theory. *Migration Information Source, 1*(35), 28.

Butler, J. (2011). *Bodies that matter: On the discursive limits of sex.* New York: Routledge.

Carrillo, H. (2004). Sexual migration, cross-cultural sexual encounters, and sexual health. *Sexuality Research & Social Policy, 1*(3), 58–70.

Casey, M. (2012). Refugee women as cultural others: Constructing social group and nexus for FGM, sex trafficking, and domestic violence asylum claims in the United States. *Seattle Journal for Social Justice, 10*(2), 981–1049.

Cho, S. (2013). Post-intersectionality: The curious reception of intersectionality in legal scholarship. *Du Bois Review: Social Science Research on Race, 10*(2), 385–404.

Chouliaraki, L. (2006). *The spectatorship of suffering.* Thousand Oaks: Pine Forge Press.

Collinson, M. A., Wolff, B., Tollman, S. M., & Kahn, K. (2006). Trends in internal labour migration from rural Limpopo Province, male risk behaviour, and implications for the spread of HIV/AIDS in rural South Africa. *Journal of Ethnic and Migration Studies, 32*(4), 633–648.

Comaroff, J., & Comaroff, J. L. (2012). Theory from the South: Or, how Euro-America is evolving toward Africa. *Anthropological Forum, 22*(2), 113–131.

Crenshaw, K. (1990). Mapping the margins: Intersectionality, identity politics, and violence against women of color. *Stanford Law Review, 43,* 1241.

Curran, S. R., Shafer, S., Donato, K. M., & Garip, F. (2006). Mapping gender and migration in sociological scholarship: Is it segregation or integration? *International Migration Review, 40*(1), 199–223.

Dannecker, P. (2005). Transnational migration and the transformation of gender relations: The case of Bangladeshi labour migrants. *Current Sociology, 53*(4), 655–674.

Davey, E., Borton, J., & Foley, M. (2013). *A history of the humanitarian system: Western origins and foundations*. London: Overseas Development Institute Humanitarian Policy Group.

De Jong, G. F. (2000). Expectations, gender, and norms in migration decision-making. *Population Studies, 54*(3), 307–319.

Donato, K. M., Gabaccia, D., Holdaway, J., Manalansan, M., IV, & Pessar, P. R. (2006). A glass half full? Gender in migration studies 1. *International Migration Review, 40*(1), 3–26.

Espin, O. (2018). *Latina realities: Essays on healing, migration, and sexuality*. New York: Routledge.

Essop-Sheik, N. (2014). African marriage regulation and the remaking of gendered authority in colonial Natal, 1843–1875. *African Studies Review, 57*(2), 73–92.

Fassin, D. (2012). *Humanitarian reason: A moral history of the present*. Berkeley: University of California Press.

Giametta, C. (2017). *The sexual politics of asylum*. London: Taylor & Francis.

Husakouskaya, N. (2015). Becoming a transgender/intersex internal migrant in Urban Gauteng: Challenges and experiences of transition while seeking access to medical services. *"forsch!"-Studentisches Online-Journal de Universität Oldenburg, 1*, 61–65.

Husakouskaya, N. (2017). Queering mobility in urban Gauteng: Transgender internal migrants and their experiences of "transition" in Johannesburg and Pretoria. *Urban Forum, 28*(1), 91–110.

Janes, C. R. (1990). Migration, changing gender roles and stress: The Samoan case. *Medical Anthropology, 12*(2), 217–248.

Jung, H. (2014). Let their voices be seen: Exploring mental mapping as a feminist visual methodology for the study of migrant women. *International Journal of Urban and Regional Research, 38*(3), 985–1002.

Kaler, A. (1999). Visions of domesticity in the African women's homecraft movement in Rhodesia. *Social Science History, 23*(3), 269–309.

Lamb, M. E., & Bougher, L. D. (2009). How does migration affect mothers' and fathers' roles within their families? Reflections on some recent research. *Sex Roles, 60*(7–8), 611–614.

Landau, L. B., & Palmary, I. (2016). Legacies, linkages and limits: Teaching migration studies in a South African University. *Migration Studies, 4*(2), 276–280.

Landau, L. B. (2012). *Exorcising the demons within: Xenophobia, violence and statecraft in contemporary South Africa*. Johannesburg: Wits University Press.

Lewis, R. A., & Naples, N. A. (2014). Introduction: Queer migration, asylum, and displacement. *Sexualities, 17*(8), 911–918.

Lewis, R. A. (2019). Queer migration in homonationalist times. *GLQ: A Journal of Lesbian and Gay Studies, 25*(4), 649–656.

Luibhéid, E. (2008). Queer/migration: An unruly body of scholarship. *GLQ: A Journal of Lesbian and Gay Studies, 14*(2), 169–190.

Luibhéid, E. (2014). Sexuality and migration. In B. Anderson & M. Kieth (Eds.), *Migration: A compass anthology* (pp. 138–140). Oxford: Oxford University Press.

Manalansan, M. F. (2006). Queer intersections: Sexuality and gender in migration studies. *International Migration Review, 40*(1), 224–249.

Massey, D. S., Arango, J., Hugo, G., Kouaouci, A., Pellegrino, A., & Taylor, J. E. (1993). Theories of international migration: A review and appraisal. *Population and Development Review, 19*(3), 431–466.

Mbembé, A. (2001). *On the postcolony*. California: University of California Press.

Michaels, J. E. (2010). Challenging notions of post-wall German identity: Minority and migrant voices. *German Monitor, 72,* 185–206.

Monk, J., & Hanson, S. (1982). On not excluding half of the human in human geography. *The Professional Geographer, 34*(1), 11–23.

Morokvašic, M. (1984). Birds of passage are also women.... *International Migration Review, 18*(4), 886–907.

Nawyn, S. J. (2010). Gender and migration: Integrating feminist theory into migration studies. *Sociology Compass, 4*(9), 749–765.9-223.

Ndlovu-Gatsheni, S. J. (2013). *Coloniality of power in postcolonial Africa*. Dakar: African Books Collective.

O'Connell-Davidson, J. (2010). New slavery, old binaries: Human trafficking and the borders of 'freedom'. *Global Networks, 10*(2), 244–261.

Palmary, I., & De Gruchy, T. (2020). The globalisation of trafficking and its impact on the South African counter-trafficking legislation. *Critical Social Policy, 40*(1), 50–68.

Palmary, I. (2006). *Gender, sexuality and migration in South Africa: Governing morality*. London: Palgrave Macmillan.

Piper, N. (2005). A problem by a different name? A review of research on trafficking in Southeast Asia and Oceania. *International Migration, 43*(1/2), 203–233.

Piper, N. (2013). *New perspectives on gender and migration: Livelihood, rights and entitlements*. New York: Routledge.

Richter, M. (2012). Sex work as a test case for African feminism. *A Journal on African Women's Experiences, 2*(1), 62–69.

Semyonov, M., & Gorodzeisky, A. (2005). Labor migration, remittances and household income: A comparison between Filipino and Filipina overseas workers 1. *International Migration Review, 39*(1), 45–68.

Silvey, R. (2006). Geographies of gender and migration: Spatializing social difference. *International Migration Review, 40*(1), 64–81.

Sinke, S. M. (2006). Gender and migration: Historical perspectives. *International Migration Review, 40*(1), 82–103.

Smits, J., Mulder, C. H., & Hooimeijer, P. (2003). Changing gender roles, shifting power balance and long-distance migration of couples. *Urban Studies, 40*(3), 603–613.

Somai, M., & Biedermann, Z. (2016). Brexit: Reasons and challenges. *Acta Oeconomica, 66*(s1), 137–156.

Spivak, G. C. (1988). Can the subaltern speak? In C. Nelson & L. Grossberg (Eds.), *Marxism and the interpretation of culture* (pp. 271–313). Urbana: University of Illinois.

Stoler, A. L. (2002). *Carnal knowledge and imperial power: Race and the intimate in colonial rule*. California: University of California Press.

Summers, C. (1991). Intimate colonialism: The imperial production of reproduction in Uganda, 1907–1925. *Signs, 16*(4), 787–807.

Summers, C. (2002). Force and colonial development in Eastern Uganda. In J. M Bahemuka & J. L. Brockington (Eds.), *In East Africa in transition: Communities, cultures, and change* (pp. 181–207). Nairobi, Kenya: Acton Publishers.

Thadani, V. N., & Todaro, M. P. (1984). Female migration: A conceptual framework. In J. Fawcett (Ed.), *Women in the cities of Asia: Migration and urban adaptation* (pp. 36–59). New York: Routledge.

Tsikata, D. (2009). Gender, land and labour relations and livelihoods in sub-Saharan Africa in the era of economic liberalisation: Towards a research agenda. *Feminist Africa, 12*(2), 11–30.

Toyota, M., Yeoh, B. S., & Nguyen, L. (2007). Bringing the 'left behind' back into view in Asia: A framework for understanding the 'migration–left behind nexus'. *Population, Space and Place, 13*(3), 157–161.

Vearey, J. (2012). Learning from HIV: Exploring migration and health in South Africa. *Global Public Health, 7*(1), 58–70.

Williams, A. S., Jobes, P. C., & Gilchrist, C. J. (1986). Gender roles, marital status, and urban-rural migration. *Sex Roles, 15*(11–12), 627–643.

Willis, K., & Yeoh, B. (2000). *Gender and migration.* Cheltenham: Edward Elgar.

Yuval-Davis, N. (1993). Gender and nation. *Ethnic and Racial Studies, 16*(4), 621–632.

Ingrid Palmary joined the University of Johannesburg as a Professor in January 2018. Prior to that, she worked at the African Centre for Migration & Society at Wits University from 2005 to 2017. Ingrid completed her Ph.D. (psychology) at Manchester Metropolitan University, UK. Before entering academia, Ingrid worked at the Centre for the Study of Violence and Reconciliation as a senior researcher. Her research has been in the field of gender, violence and displacement. She has published in numerous international journals and is the co-editor of Gender and Migration: feminist interventions published by Zed Press; Handbook of International Feminisms: Perspectives on psychology, women, culture and rights published by Springer; Healing and Change in the City of Gold: Case studies of coping and support in Johannesburg published by Springer. She is the author of Gender, sexuality and migration in South Africa: Governing morality published by Palgrave.

CHAPTER 6

Reflexivity and Its Enactment Potential in Gender and Migration Research

Kyoko Shinozaki

Abstract Migration and feminist researchers have long been engaging in discussions concerning fieldwork and methodology. One of the common, recurring claims made is the importance of reflexivity about (researchers') positionality. Taking the hitherto discussion of positionality as a point of departure, which has so far mainly focused on data collection, the aim of this contribution is threefold: First, it maps out some of the main contours of debates in feminist studies, ethnic and racial studies, and critical migration studies to ascertain the importance given to the issue of power and inequalities between researcher and researched in field relations and knowledge production. Second, this chapter asks how much of this "reflexive turn" is something specific to qualitative migration research. Finally, the chapter explores the co-production of knowledge, dissemination, and the creation of broader engagement as a feedback system of "public science" to enact reflexivity.

Introduction

Among qualitative researchers in social science research it has become fairly common to reflect upon field relationships and researchers' own positionalities in the phase of data collection. Gender and migration scholars using qualitative methods are no exception here. During fieldwork, we experience a range of personal encounters with research participants. Any sound researcher with empirical fieldwork experience knows what we collect is not only tangible data

K. Shinozaki (✉)
Department of Sociology, Paris Lodron University of Salzburg, Salzburg, Austria
e-mail: kyoko.shinozaki@sbg.ac.at

C. Mora and N. Piper (eds.), *The Palgrave Handbook of Gender and Migration*, https://doi.org/10.1007/978-3-030-63347-9_6

in the sense of "contents" we will later analyse, but also emotional blows and more latent feelings, which add to the depth of research data. Ultimately, many moments of field relationships are human interactions of which a power hierarchy between the researcher and researched is an essential part. While the researcher is usually seen as located at the higher end of the hierarchy, this relationship can be more complex. The boundaries, which mark asymmetries, may be mediated through gender, class, "race," migration experience, religion, age, sexuality, and/or a combination of these categories and are well documented in feminist and critical social science literature. Lively discussions, which have mainly taken place in the qualitative research paradigm, have demonstrated these boundaries are far from being structurally predetermined or fixed. Instead, these can be situational, contested and negotiated in the fieldwork setting, which calls for reflexivity as a central heuristic device. I suggest reflexivity has also a lot of potential to offer beyond the situational fieldwork context and to reframe it. Reflexivity can be enacted in other phases of the research process, including co-production of data, dissemination and wider public engagement.

This chapter is organized in three sections. First, it discusses the main contours of a "reflexive turn" in feminist scholarship, ethnic and racial studies, and critical migration studies in relation to the question of power and inequalities between researcher and researched in field relations and knowledge production which has played a big role in these debates. Despite a wide range of discussion in feminist and critical migration scholarship, it is noticeable that the issue of reflexivity is disproportionately debated by scholars employing qualitative methods. Secondly, I ask how much of this reflexive turn is something specific to qualitative inquiry. Finally, the chapter explores potential and creative ways in which reflexivity can be enacted in multiple phases of research.

Reflexive Turn in the Process of Data Collection

The issue of reflexivity has shaped debates around power relationships and inequalities between researcher and the researched among feminist researchers and critical social scientists. They have led extensive discussions around questions concerning power structures, mainly focusing on fieldwork interactions between researchers and the researched (Gill & Maclean, 2002; Hammersley & Atkinson, 2007; Plummer, 2001). A commonly used logic behind the researcher-researched binary has been the "insider vs. outsider" dichotomy. Merton has defined the insider researcher as someone who shares the knowledge of the group studied her/his cultural, racial, gender or national association on an ascribed, rather than acquired basis (Merton, 1972).

Since the late 1970s, there have been lively debates in feminist scholarship as to how to redress (potentially) exploitative power relations between the researcher and the researched by means of various methods, including ethnography (Stacey, 1988). While Hammersley (1992) has claimed that there is no singular coherent program, a "feminist methodology" as such, what feminist scholarship has overall envisaged has been the creation of a more egalitarian and reciprocal relationship with research participants (Mies, 1991). Another central argument has been feminist standpoint theorists' addition to one of the central sociological tenets paralleling Peter Berger and Thomas Luckmann's (1967) classic *Social Construction of Reality*. Harding highlighted the way our societal (gendered) positionality informs how we understand and interpret realities and as such, knowledge is never "neutrally" produced but instead from a specific social gendered location (Harding, 1992). To varying degrees, these concerns directly speak to the issue of reflexivity. Defining reflexivity as a tool to scrutinize "unthought categories of thought which delimit the thinkable," Pierre Bourdieu makes a point to question our conventional academic practice of knowledge production (Bourdieu 1982, cited in Bourdieu & Wacquant, 1992: 40). Exercising reflexivity would serve to "enhance the credibility and rigour of the research process as well as make transparent the positionality of the research" (DeSouza, 2004: 474). More broadly, Margaret Archer (2012) argues that it has become imperative *for us all* to be reflexive in late modernity, especially since the 1980s. Why does the reflexive imperative categorically matter for us all? That is because:

> the emergence of a new conjuncture between the cultural order (ideationally based) and the structural order (materially based) is shaping new situational contexts in which more and more social subjects find themselves and whose variety they have to confront – in a novel manner. […] Increasingly all have to draw upon their socially dependent but nonetheless personal powers of reflexivity in order to define their course(s) of action in relation to the novelty of their circumstances. (Archer, 2012: 1)

Seen this way, reflexivity helps to unsettle the simplified insider vs. outsider binary revealing the complex dynamics in research relationships, something that feminist and gender and migration scholars have problematized for decades along with other critical social scientists (Hammersley, 1992; Plummer, 2001).

In ethnic and racial studies, too, a conventional mode of knowledge production has recently come under attack. Methodological discussions have mainly evolved around challenges that white researchers face when approaching minority populations or concerns among ethnic and racial minority researchers when studying their own or other minority groups (e.g. Adamson & Donovan, 2002). There is a considerable imbalance in the literature on interracial and interethnic research: race and ethnicity are nearly always observed and analysed with a "white gaze" while whiteness and otherness are rarely problematized and examined in a "reversed gaze" (Hoong Sin, 2007: 490).

On a similar note, migration studies have witnessed a reflexive turn (Nieswand & Drotbohm, 2014) in self-critiquing the widespread methodological ethno-national groupism, which supposedly defines clear boundaries between "insiders" and "outsides." Some reject a priori commonalities constructed along a national/ethnic lens while others critically engage with challenges and potential when migrant researchers themselves study migrants, complicating the category of "insiders" emphasizing the role of their female gendered identities (Carling, Erdal, & Ezzati, 2014; Matejskova, 2014; Nowicka & Cieslik, 2014). Their non a priori, intersectional take on face-to-face encounters in fieldwork proves to be conducive to questioning the centrality of an "ethnic lens" (Glick Schiller, Çaglar, & Guldbrandsen, 2006) and a "methodological nationalism" (Wimmer & Glick Schiller, 2003), which presumes that spatial and cultural borders are congruent. For example, in my previous research I have demonstrated that while there was a clear global power hierarchy between the research participants and myself that originated in national belonging (the Philippines and Japan), other positionalities such as gender, class and motherhood (or a lack thereof) at their intersection worked to reverse such an asymmetrical relationship (Shinozaki, 2012). Boundaries between "us" and "them" are not static and are rather created in a situational manner.

The discussions briefly outlined above, albeit in a selective manner, show that a reflexive turn in gender and migration scholarship has helped to challenge dominant binary categorizations. Nonetheless, these vibrant debates have been taking place primarily in qualitative research.

Reflexive Turn: For Qualitative Research Only?

While reflexivity should be an integral part of both qualitative and quantitative methodological traditions (Finlay, 1998), discussion on reflexivity has been scarce in quantitative research to date. Some observers even commented that "many quantitative researchers [...] avoid explicit forms of reflexivity" (Ryan & Golden, 2006: 1191). Yet, it would be erroneous to claim the issue of reflexivity is irrelevant to quantitative research. Indeed, some quantitative methods scholarship explicitly engages with reflecting on their methods in the sense of enhancing robustness and quality of their data. Here I refer to three examples. The first example is survey research practices in social science more generally. A situational and interpretive element of qualitative social research is utilized by way of cognitive interviews and qualitative pretest interviews. This qualitative approach would allow survey researchers to receive feedback from their pretest research participants, which would in turn ensure researchers to ask a "right" kind of questions and to create accurate and mutually exclusive response categories in a survey (Bethmann, Buschle, & Reiter, 2019).

The second example is considering how collecting a face-to-face questionnaire survey through visits not only generates "hard" data, but can also unexpectedly result in communicative field situations. Ryan and Golden (2006), who were at the time commissioned researchers to conduct a face-to-face survey on health and Irish migrants in the UK, discuss how their common cultural heritage, language (i.e. Irish and Irish accent) and migrant status (i.e. Irish migrants living in the UK) resulted in establishing a rapport and mutual trust between the researchers and their respondents. In the course of their survey, the authors began to keep a record of responses they gathered which did not fit to the "tick-the-box" kind of answer. Out of their field notes and the discussion they had on various issues, the authors developed a new research project (Ryan & Golden, 2006).

On a similar note, a number of social science students shared with me their concerns about, if not a doubt in, often unquestioned objectivity in quantitative research, reflecting upon their student job experience as a face-to-face survey interviewer for different major panel and national surveys. They recounted to me how differently their respondents would interpret the questions; how their respondents began to tell their stories rather than directly answering the questions; and how sometimes the interviewers had to rephrase the respondent's initial answer in a shorter form using the pre-given answer categories, in order for them to tick the box. The most alarming aspect for me was that they felt there were not enough opportunities for an exchange between them and research-commissioning agencies to share and discuss these experiences.

We can classify the two preceding articles (Bethmann et al., 2019; Ryan & Golden, 2006) as examples of having effectively adopted a mixed method—by design or as a by-product—going beyond the paradigm argument (cf. Oakley, 1998). By contrast, the third example shows how audit researchers reflect on their methodological approach to measuring *acts* of discrimination based on race, gender, sexuality and educational credentials in the context of hiring and housing. Being a specific type of field experiment, audit research allows researchers to "examine difficult to detect behaviour [... discriminatory] decision-making in real-world scenarios" (Gaddis, 2018: 3). Paralleling better legal provisions, racial and gender discriminatory behaviours, if not beliefs and attitudes, have become less obvious and more covert over the years. Thus, given an obvious bias, audit researchers argue that measuring discriminatory practices through surveys (i.e. observational data) would not permit to capture the "real" extent of discrimination (Gaddis, 2018). Others discuss their thought process for their case selection, reveal difficulties as they conducted an audit, and discuss "lessons learned" (Carbonaro & Schwarz, 2018) as well as audit research's limitations in great detail (Gaddis, 2018).

However, audit research still raises an uneasy question about informed consent. More specifically, when the objective of audit research is to measure uncontrolled discrimination practices based on racial and sexuality-related stereotypes (e.g. employers for job hires or landlords for their tenant selection), obtaining consent prior to the experiment research would stand in the way of obtaining the very results audit research aims to generate, or it might lead to a distortion in the results (Gaddis, 2018). At the same time, in light of ethics, covert research has been controversially debated (Plummer, 2001). There seems to be different strategies used to tackle this problem, ranging from informing employers about the audit research project prior to the field phase (Gaddis, 2018) to discussing the research results with employers.

Co-Production of Knowledge As a Form of Reflexivity

The main focus in reflexivity debates has been placed on the data collection phase. This is understandable for face-to-face communications that occur in real time and the effects are immediate. Fieldwork experiences compel fieldworkers to think of, and over, our ways of doing research. Depending on the social characteristics of the researcher, different empirical data can be generated, which speaks to the issue of "situated knowledges" (Haraway, 1988: 575; Phoenix, 1994: 66). Fieldworkers experience multiple positioning and identification shifts during the fieldwork (Song & Parker, 1995).

Nonetheless, other phases of research should ideally involve reflexivity, and there are multiple *ways* of being reflexive (Archer, 2012). Firstly, writing academic texts is a site of strong manifestations of asymmetrical power relations between the researcher and the researched, as problematized by Clifford and Marcus (1986) in the mid-1980s already. Overlooking these relations has led to reproducing colonial discourses on and colonial knowledge about an ostensibly homogeneous group of a "third world woman" (Mohanty, Russo, & Torres, 1991). Secondly, such a textual representation begs a question about a "traditional" gaze of a researcher, which denotes a substantial imbalance in the literature that "race" and ethnicity are nearly always *observed* and *analysed* (and we can add "imagined") in a "white gaze" (Hoong Sin, 2007: 490), or in a gaze of the dominant society. This is a "taken-for-granted mode of knowledge production" in the academia (Shinozaki & Osanami Törngren, 2019).[1]

A third potential way to reframe this mode of knowledge production is to foster broader engagement in doing research as an *enactment* of reflexivity. Particularly pertinent here is Participatory Research (PR), which denotes

> [A research process] *with* those people whose life-world and meaningful actions are under study. [… T]his means that the aim of the inquiry and the research questions develop out of the convergence of two perspectives—that of science *and* practice. […] The participatory research process enables co-researchers to step back cognitively from familiar routines, forms of interaction, and power

[1] There are potential ways to redress this asymmetrical mode of knowledge production, such as adaptation of a "reversed gaze" Hoong Sin (2007: 490), the "matching" of the social characteristics of the researcher and the researched Archer (2002) as with the case of a stronger presence of migrant researchers conducting research about migrants (Nowicka and Ryan 2015). Nonetheless, these considerations are not free from criticisms (cf. Shinozaki and Osanami Törngren (2019).

> relationships in order to fundamentally question and rethink established interpretations of situations and strategies. (Bergold & Thomas, 2012, emphasis in original)

Just like much of the reflexivity debates outlined in the previous sections, participatory research's premise challenges the insider vs. outsider binary and shows that the boundary is porous. But participatory research does not end here. More importantly, its primary driver is geared towards collaborations and engagement between scientific inquiry, and practitioners and stakeholders, as in the case of Participatory Action Research (PAR). The aspect of engagement and collaborations underscores researching *with*, instead of researching *about* people and social groups, setting itself apart from conventional approaches to scholarly inquiry (Bergold & Thomas, 2012; Bradbury, 2015).

How can then this reflexive mode of knowledge co-production be put into practice in concrete terms? Let me introduce two case studies here. The first example is a Participatory Action Research project "Migrant mothers caring for the future: Creative interventions into citizenship" in East London.[2] The project took the dominant public and political discourse on migrant mothers as a point of departure. It portrays migrant mothers as guardians of a specific ethnic community, and thus as near boundaries of the society they live in. They are seen as recipients of social services, in need of integration. Umut Erel and her colleagues turned this logic around and examined migrant mothers' role in raising their children (i.e. future citizens) as practising new forms of citizenship. They make a home for their families in the UK while relating them to the country of their parents and their own neighbourhood (Erel, Reynolds, & Kaptani, 2020). The project used participatory theatre methods with a diverse group of twenty migrant mothers in terms of the level of education, the time of their entry in the UK, the number of children, and ethnic and racial backgrounds, including Polish, Congolese, Somali, Turkish and Lithuanian backgrounds. Apart from migrants, academics, artists, and practitioners in the field of health, education, migration and family took part in it (Erel, Reynolds, & Kaptani, 2017).

> [The participatory theatre methods in this project] consisted of a series of eight weekly three hour-long theatre workshops, using techniques of Playback (Fox, 1994) and Forum Theatre, as part of Boal's Theatre of the Oppressed (Boal, 1979). The sessions of "Playback" theatre involved the participants sharing their stories, which are then acted out by professional actors, it is an embodied form of biographical narratives in which racialized migrant women can challenge existing societal ways of constructing them as "Other" (cf. Lutz 2018). The performance taking place in the theatre allows participants to see themselves and

[2]"Migrant mothers caring for the future" project was funded by the Arts and Humanities Research Council, UK, led by Umut Erel (PI, Open University) and Tracey Reynolds (CI, London Southbank University at the time of the project, now University of Greenwich), from 2013 to 2015. For project details and video clips, see http://www.open.ac.uk/socialsciences/migrant-mothers/index.php (last accessed on March 30, 2020).

> their interactions with others in a way that creates more reflexive distance than is possible in everyday life. Playback produces meaning by building a triangular conversation between actors, individual tellers and the group of participants. Narratives of individual tellers often relate to each other, elaborating on a particular theme, perhaps refining particular themes, perhaps showcasing alternative or contradicting experiences (Erel/Reynolds, 2014; Erel et al., 2017). (Erel et al., 2020)

How Was the Forum Theatre Run?

> [T]he framework and set of techniques is based on the principles of collective empowerment and emancipation. Boal termed the participating audience member *spect-actor*, for they are not merely spectators or actors but practicing elements of both roles simultaneously. When participants intervene in the scene to change the course of action, forum theatre becomes not only a critical site of negotiation, but a site of active citizenship, collective mobilisation and empowerment by migrant women experiencing marginalisation and inequalities in UK society. In particular, it became an occasion for developing a collective space for the creativity of self, challenging narrow representations of migrant mothers and in this sense, the workshop constituted acts of cultural citizenship. Through the participatory theatre methods, participants reflected on and questioned the boundaries of who can count as a legitimate citizen, whose contributions are valued and who can claim rights. (Erel et al., 2020)

What is noteworthy is the successive nature of these theatre workshops. One of the poignant scenes was when a migrant mother was seeking an emergency appointment at her General Practitioner (GP) and the GP receptionist turned her away (Erel et al., 2017). Over the course of the theatre workshops, participants built on their previous strategies and came up with different strategies to intervene in this difficult situation more effectively. It is a collective way of producing knowledge based on reflections. Erel and her colleagues actively participated in generating knowledge by posing research questions and participating in the theatre workshops and reflections. “In this way, the group built up a collective repository of strategies for challenging exclusion, the denial of access [to emergency surgery] and respect” (Erel et al., 2017: 307). In addition, at the end of the project conference, Forum Theatre was shown in front of the audience who consisted of academics and other professionals, mainly women and ethnic and racial minority. Inviting stakeholders to an academic workshop/conference itself is fairly common, but what I found remarkable was that they were also asked to take part in forum theatre (the GP surgery scene described above). Interaction between the GP receptionist and the patient could considerably alter—as it did during the conference—depending on who embodied the patient in what ways (i.e. white middle-class professional woman with high command of English), and what kind of legitimizing discourse was mobilized in order to access healthcare (i.e. healthcare as her entitlement) (Erel et al., 2017).

While an interpretation of this scene itself is not in the ambit of this chapter, the main advantages of participatory theatre as a research method can be summarized as follows: (1) experience sharing and community building among participants; (2) experiencing concrete situations of oppression and inequality as well as raising awareness about power relations both on the part of participants and researchers; (3) allowing participants to analyse concrete hierarchical power relations while collaborating in sharing subjugated knowledge; and (4) transforming the hegemonic structure through knowledge dissemination to wider audiences beyond a smaller academic circle (on action research in particular see Bradbury, 2015; Erel et al., 2017; see also Kaptani & Yuval-Davis, 2008). The fourth point about the dissemination of co-produced knowledge has been further expanded in a subsequent project "Participation Arts and Social Action in Research (PASAR)": The researchers make a variety of easily accessible materials available on their project website for wider audiences, ranging from conventional journal articles, to policy briefing, toolkit, training materials, photographs and films.[3]

A participatory theatre method (and other methods like walking) lends itself to the co-production of knowledge in an innovative, artistic, affective and corporal way. It lets marginalized voices be heard and allows racialized and sexualized migrant women to collectively develop alternative action. In dissemination, researchers and participants reach out to a wide range of the public (public engagement). To be sure, it is hardly impossible to neatly delineate the beginning and end of research processes of data co-production, data collection and dissemination. In practice, they feed back into each other, questioning the conventional linear or even circular way of conceptualizing and conducting research. This "feedback system" helps to create a kind of reflexive knowledge concerned with public engagement, which connects academics, research participants, students, relevant stakeholders and interested publics. Key questions are "Knowledge for whom?" and "Knowledge for what?" Reflexive knowledge is concerned with a dialogue about *ends* and about the direction of society (Burawoy, 2005).

Reflexive knowledge is a hallmark of public sociology. What is public sociology anyway? According to Michael Burawoy (2005), public sociology "brings sociology into a conversation with publics, understood as people who are themselves involved in conversation. It entails, therefore, a double conversation" (p. 7). One common type of public sociology is "organic public sociology." These sociologists collaborate with a local and active public, aiming to make the invisible visible, make the private to public. They make the seemingly trivial to key issues of our sociological life and inquiry. Good examples are those working with social movements, associations and migrant associations. What characterizes this genre is that they engage in conversation with

[3]http://fass.open.ac.uk/research/projects/pasar (last accessed on March 30, 2020).

a public and there is a process of mutual education (Burawoy, 2005: 7–8).[4] Although Burawoy's discussion is about sociology as a discipline, the question of the purpose and ends of knowledge production can be extended to other disciplines as well ("public science"). We could argue that one such example is a "career" of intersectionality in law and beyond (see also Lutz and Amelina in this volume). Kimberlé Crenshaw (1989) rendered visible the simultaneous discrimination experience of Black women, which was otherwise made invisible. She argued that by either using the white feminist framing or the antiracist framing, we would fail to recognize the structural discrimination pattern of Black women.

The second example stems from my own virtual classroom amid the global Covid-19 outbreak, just as I am writing this chapter. It is about sociology student engagement. The hyper-mobility of the virus has greatly and quickly reduced human mobility. Consequently, many of us are forced to stay home (and to work from home). On the other hand, others have no choice but staying, or being mobile. Of particular relevance here are frontline workers in the home-based care sector, and to some extent the institutional care sector, dominated by migrant women, as documented by gender and migration scholarship (cf. Hochschild, 2000; Itō, 2020; Lutz, 2010; Parreñas, 2001; Shinozaki, 2015).

In my course on social change and mobilities, we are currently turning unprecedented developments affecting migrant care work into student research projects. Given the border shutdown within Europe and travel restrictions, the Federal Ministry of Social Affairs, Health, Care, and Consumer Protection gave a green light to pay out a one-time bonus of 500 Euro to migrant care workers from federal relief funds if she (or he) extends her (his) stay. The state of Upper Austria will add another 1000 Euro on top of it.[5] In the Austrian context, where around the clock care arrangement is legal and needed for certain users, migrant women, mainly from Central and Eastern Europe (CEE), come for two to four weeks, on a rotation basis. Feminist scholars have long fought for better remuneration and recognition of care work, albeit finding no substantial echo in social policy. Only in light of a potential sudden collapse of the 24-hours care system has the "bonus" payment decision been made so quickly.[6] Together with the Chamber of Commerce, some states like Lower Austria, have also chartered a flight to

[4]In contrast, another type of public sociology sets out and stirs up a public discussion on topics, such the nature of societal values, the gap between societal promises and realities (so-called "traditional public sociology"). Typically, we can find these sociologists' work in the opinion papers of prominent newspapers Burawoy (2005: 7).

[5]https://orf.at//stories/3159947/, last accessed on April 1, 2020.

[6]This sudden (but one-time as it seems now) elevation of care labor is all the more remarkable given that just one year ago, the ÖVP (Austrian People's Party)-FPÖ (Freedom Party of Austria) administration effectively cut back on migrant workers' family allowance, who left their children behind in their home countries. This has adversely affected care workers from CCE Kravagna (2019).

bring over 230 care workers from Romania and Bulgaria (John & Schenndl, 2020). Some of the questions we are asking are: Who have been pushing this initiative forward? What logic was used to legitimize the payment, and are there any parallels to feminist problematization about the personal being political? Why do some states give stronger financial incentives than others? What role did trade unions, feminist groups, and commercial placement agencies play in this? Is there any evidence to tell us how sustainable this ongoing (partial) valorization of care work is? How do migrant care workers themselves feel about the sudden change? To what extent will our findings speak to the concept of "caring society" (Tronto, 2019) and solidarity, or a lack of these? What are engaging ways for us to communicate the results?

Conclusions

The issue of reflexivity has been debated in the past four decades. Feminist and migration scholars have been and continue to be actively participating in the debates, primarily focusing on the face-to-face communicative situation during fieldwork. The aim of this chapter has been to propose that reflexivity can be used as more than a tool to critically reflect on past or ongoing field relations. I suggest that it can be enacted in other phases of research process, too, such as co-production of data, dissemination and fostering wider public engagement, paralleling the idea of public sociology (Burawoy, 2005).

However, there are also challenges. An obvious one is that many researchers have neither artistic training nor skills/techniques to implement artistic/experimental methods. Second, researching *with* might result in conflictual situations if researchers and research participants have different interpretations about the data. Third, the co-production and public engagement requires generating additional time for activities beyond the immediate academic environment, placing a higher demand on our time. Finally, expectations in term of accomplishment, which come from different individuals and groups, may not be easy to satisfy, potentially leading to high pressure.

Arts-based research as discussed in this chapter, as well as other examples like the walking method (O'Neill, Erel, Kaptani, & Reynolds, 2019) film ethnography (e.g. Wickwire, 2018), and musical ethnography (e.g. Präger, forthcoming) provide us with a plenty of useful innovative entry points.

References

Adamson, J., & Donovan, J. L. (2002). Research in black and white. *Qualitative Health Research, 12*(6), 816–825. https://doi.org/10.1177/104973230201200606008.

Archer, L. (2002). 'It's easier that you're a girl and that you're Asian': Interactions of 'race' and gender between researchers and participants. *Feminist Review, 72*(1), 108–132. https://doi.org/10.1057/palgrave.fr.9400043.

Archer, M. S. (2012). *The reflexive imperative in late modernity*. Cambridge: Cambridge University Press.

Berger, P. L., & Luckmann, T. (1967). *The social construction of reality*. New York, NY: Anchor Books.

Bergold, J., & Thomas, S. (2012). Participatory research methods: A methodological approach in motion. *Forum Qualitative Social Research*, *13*(1), 191–222. Retrieved from http://nbn-resolving.de/urn:nbn:de:0114-fqs1201302.

Bethmann, A., Buschle, C., & Reiter, H. (2019). Kognitiv oder qualitativ? Pretest-Interview in der Fragebogenentwicklung. In N. Menold & T. Wolbring (Eds.), *Schriftenreihe der ASI—Arbeitsgemeinschaft Sozialwissenschaftlicher Institute. Qualitätssicherung sozialwissenschaftlicher Erhebungsinstrumente* (Vol. 71, pp. 159–193). Wiesbaden: Springer Fachmedien Wiesbaden. https://doi.org/10.1007/978-3-658-24517-7_6.

Bourdieu, P., & Wacquant, L. (1992). *An invitation to reflexive sociology*. Chicago, IL: University of Chicago Press.

Bradbury, H. (2015). *The SAGE handbook of action research*. London: Sage. https://doi.org/10.4135/9781473921290.

Burawoy, M. (2005). For public sociology. *American Sociological Review, 70*(1), 4–28. https://doi.org/10.1177/000312240507000102.

Carbonaro, W., & Schwarz, J. (2018). Opportunities and challenges in designing and conducting a labor market resume study. In S. M. Gaddis (Ed.), *Audit studies: Behind the scenes with theory, method, and nuance* (pp. 143–158). Cham: Springer.

Carling, J., Erdal, M. B., & Ezzati, R. (2014). Beyond the insider-outsider divide in migration research. *Migration Studies, 2*(1), 36–54. https://doi.org/10.1093/migration/mnt022.

Clifford, J., & Marcus, G. E. (1986). *Writing culture: The poetics and politics of ethnography*. Berkeley, CA: University of California Press.

Crenshaw, K. (1989). Demarginalizing the intersection of race and sex: A Black feminist critique of antidiscrimination doctrine, feminist theory and antiracist politics. *University of Chicago Legal Forum, 1989*, 139–167.

DeSouza, R. (2004). Motherhood, migration and methodology: Giving voice to the Other. *Qualitative Report, 9*(3), 463–482.

Erel, U., Reynolds, T., & Kaptani, E. (2017). Participatory theatre for transformative social research. *Qualitative Research, 17*(3), 302–312. https://doi.org/10.1177/1468794117696029.

Erel, U., Reynolds, T., & Kaptani, E. (2020). Migrant women performing citizenship: An intersectional approach. In K. Huxel, J. Karakayali, E. Palenga-Möllenbeck, M. Schmidbaur, K. Shinozaki, T. Spies, L. Supik, & E. Tuider (Eds.), *Postmigrantisch gelesen: Transnationalität, Gender, Care*. Bielefeld: Transcript.

Finlay, L. (1998). Reflexivity: An essential component for all research? *British Journal of Occupational Therapy, 61*(10), 453–456. https://doi.org/10.1177/030802269806101005.

Gaddis, S. M. (2018). An introduction to audit studies in the social sciences. In S. M. Gaddis (Ed.), *Audit studies: Behind the scenes with theory, method, and nuance* (pp. 3–44). Cham: Springer.

Gill, F., & Maclean, C. (2002). Knowing your place: Gender and reflexivity in two ethnographies. *Sociological Research Online, 7*(2), 118-128. https://doi.org/10.5153/sro.710.

Glick Schiller, N., Çaglar, A., & Guldbrandsen, T. C. (2006). Beyond the ethnic lens: Locality, globality, and born-again incorporation. *American Ethnologist, 33*(4), 612–633. https://doi.org/10.1525/ae.2006.33.4.612.

Hammersley, M. (1992). On feminist methodology. *Sociology, 26*(2), 187–206.

Hammersley, M., & Atkinson, P. (2007). *Ethnography: Principles in practice* (3rd. ed., reprinted.). London: Routledge.

Haraway, D. (1988). Situated knowledges: The science question in feminism and the privilege of partial perspective. *Feminist Studies, 14*(3), 575–599. https://doi.org/10.2307/3178066.

Harding, S. G. (1992). *Whose science? Whose knowledge? Thinking from women's lives*. Ithaca, NY: Cornell University Press.

Hochschild, A. (2000). Global care chains and emotional surplus value. In A. Giddens & W. Hutton (Eds.), *On the edge: Living with global capitalism*. London: Vintage.

Hoong Sin, C. (2007). Ethnic-matching in qualitative research: Reversing the gaze on 'white others' and 'white' as 'other'. *Qualitative Research, 7*(4), 477–499. https://doi.org/10.1177/1468794107082304.

Itō, R. (2020). *Kaji rōdō no kokusai shakaigaku: Dīsento wāku o motomete*. Kyōto: Jinbunshoin.

John, G., & Scherndl, G. (2020, April 5). Kritik am Einfliegen von Pflegerinnen aus Osteuropa. *Der Standard*. Retrieved from https://www.derstandard.at/story/2000116444273/kritik-am-einfliegen-von-pflegerinnen-aus-osteuropa?ref=rec.

Kaptani, E., & Yuval-Davis, N. (2008). Participatory theatre as a research methodology: Identity, performance and social action among refugees. *Sociological Research Online, 13*(5), 1–12. https://doi.org/10.5153/sro.1789.

Kravagna, S. (2019, January 10). Kürzung der Familienbeihilfe: Aktion minus 50 Prozent. *Der Standard*. Retrieved from https://www.derstandard.at/story/2000095776931/kuerzung-der-familienbeihilfe-aktion-minus-50-prozent.

Lutz, H. (2010). *New maids: Transnational women and the care economy*. London: Zed Books.

Matejskova, T. (2014). Deep situationality: Interstitial spaces and limits of identity in ethnographies of politics of immigrant integration. *Migration Studies, 2*(1), 16–35. https://doi.org/10.1093/migration/mnt028.

Merton, R. K. (1972). Insiders and outsiders: A chapter in the sociology of knowledge. *American Journal of Sociology, 78*(1), 9–47.

Mies, M. (1991). Women's research or feminist research? The debate surrounding feminist science and methodology. In M. M. Fonow & J. Cook (Eds.), *Beyond methodology: Feminist scholarship as lived research* (pp. 60–84). Bloomington, IN: Indiana University Press.

Mohanty, C. T., Russo, A., & Torres, L. (1991). *Third world women and the politics of feminism*. Bloomington, IN: Indiana University Press.

Nieswand, B., & Drotbohm, H. (2014). *Kultur, Gesellschaft, Migration: Die reflexive Wende in der Migrationsforschung. Studien zur Migrations—und Integrationspolitik*. Wiesbaden: Springer VS. http://dx.doi.org/10.1007/978-3-658-03626-3.

Nowicka, M., & Cieslik, A. (2014). Beyond methodological nationalism in insider research with migrants. *Migration Studies, 2*(1), 1–15. https://doi.org/10.1093/migration/mnt024.

Nowicka, M., & Ryan, L. (2015). Beyond insiders and outsiders in migration research: Rejecting a priori commonalities. *Forum Qualitative Sozialforschung, 16*(2). https://doi.org/10.17169/fqs-16.2.2342.

O'Neill, M., Erel, U., Kaptani, E., & Reynolds, T. (2019). Borders, risk and belonging: Challenges for arts-based research in understanding the lives of women asylum seekers and migrants 'at the borders of humanity'. *Crossings: Journal of Migration & Culture, 10*(1), 129–147. https://doi.org/10.1386/cjmc.10.1.129_1.

Oakley, A. (1998). Gender, methodology and people's ways of knowing: Some problems with feminist and the paradigm debate in social science. *Sociology, 32*(4), 707–731.

Parreñas, R. S. (2001). *Servants of globalization: Women, migration, and domestic work*. Stanford, CA: Stanford University Press.

Phoenix, A. (1994). Practising feminist research: The intersection of gender and 'race' in the research process. In M. Maynard & J. Purvis (Eds.), *Researching women's lives from a feminist perspective* (pp. 49–71). London: Taylor & Francis.

Plummer, K. (2001). *Documents of life 2: An invitation to a critical humanism*. Thousand Oaks, CA, London: Sage. Retrieved from http://site.ebrary.com/lib/alltitles/docDetail.action?docID=10218018.

Präger, U. (forthcoming). *Musical ethnography: Researching and representing migration, expressive culture, and politics of belonging*.

Ryan, L., & Golden, A. (2006). 'Tick the box please': A reflexive approach to doing quantitative social research. *Sociology, 40*(6), 1191–1200. https://doi.org/10.1177/0038038506072287.

Shinozaki, K. (2012). Transnational dynamics in researching migrants: Self-reflexivity and boundary-drawing in fieldwork. *Ethnic and Racial Studies, 35*(10), 1810–1827. https://doi.org/10.1080/01419870.2012.659275.

Shinozaki, K. (2015). *Migrant citizenship from below: Family, domestic work, and social activism in irregular migration*. New York, NY: Palgrave Macmillan. https://doi.org/10.1057/9781137410429.

Shinozaki, K., & Osanami Törngren, S. (2019). *Researching across differences: Unsettling methodological discussions from a minority's perspectives* (Briefs on Methodological, Ethical and Epistemological Issues No. 9). Migration Research Hub. Retrieved from https://migrationresearch.com/.

Song, M., & Parker, D. (1995). Commonality, difference and the dynamics of disclosure in in-depth interviewing. *Sociology, 29*(2), 241–256. https://doi.org/10.1177/0038038595029002004.

Stacey, J. (1988). Can there be a feminist ethnography? *Women's Studies International Forum, 11*(1), 21–27. https://doi.org/10.1016/0277-5395(88)90004-0.

Tronto, J. (2019, September). *The future of caring democracy*. Congress of the Austrian Sociological Association "Alles im Wandel? Dynamiken und Kontinuitäten Moderner Gesellschaften", Salzburg.

Wickwire, M. (2018). *Together apart*. Sacramento: Manifest Media.

Wimmer, A., & Glick Schiller, N. (2003). Methodological nationalism, the social sciences, and the study of migration: An essay in historical epistemology. *The International Migration Review, 37*(3), 576–610.

Kyoko Shinozaki is Professor of Sociology with a focus on Social Change and Mobilities at the Paris Lodron University of Salzburg, Austria. She has widely published on citizenship, care, migration and transnational mobilities of knowledge and academic personnel. Her work looks at the intersection of migration/mobility and gender from a transnational perspective in the (East) Asian and European context. She has been involved in a number of international projects, including the EU project "Integration of female immigrants in labour market and society. Policy assessment and policy recommendations" (2006–2008) and JSPS project "Migrant domestic workers and the ILO Convention No. 189: Inquiry into a new phase of organizing, rights and transnational solidarity" (2015–2018). She is an elected Speaker of the Section "Feminist Theory and Gender Research", Austrian Sociological Association, and takes lead on UN 2030 SDG 10 "Reducing Inequalities" in the Austrian University Alliance of Sustainability.

Part II

Care, Affective and Emotional Labour

CHAPTER 7

Gender, Migration and the Inequalities of Care

Isabel Shutes

Abstract This chapter examines the contribution of research on migration and care labour to understanding the inequalities of care provision. It delves into how we are to understand those inequalities in relation to macro, meso and micro level relationships, policies and processes. At the same time, it considers the contestation of those inequalities, and in ways that make necessary connections between the claiming of rights in relation to migration, paid work and unpaid care. It is argued that an understanding of care inequalities and their contestation provides the basis for a 'social transformation' perspective on migration (Castles in Journal of Ethnic and Migration Studies 36: 1565–1586, 2010), situating migration in relation to wider social processes. Concurrently, in bringing a feminist approach to social transformation to the fore, the importance of understanding the local specificity of intersecting inequalities is also emphasised, to avoid reducing an understanding of global care inequalities to a simplified North/South binary.

Introduction

Across the world, migrant workers play a central role in the provision of care—as domestic workers, nannies, 'au pairs', home care and residential care workers, live-in carers, nurses, and in other care-related types of work—crossing the public and private boundaries of care. In bringing care into focus,

I. Shutes (✉)
Department of Social Policy, London School of Economics and Political Science, London, UK
e-mail: i.h.shutes@lse.ac.uk

C. Mora and N. Piper (eds.), *The Palgrave Handbook of Gender and Migration*, https://doi.org/10.1007/978-3-030-63347-9_7

analyses of migrant care labour from a gender perspective have contributed, not least, an understanding of migrant women's labour in capitalist economies. Beyond this, however, the expansion of research on care and migration has advanced a critical understanding of the transnational relationships and intersectional inequalities that underpin the paid and unpaid labour of migrants in the provision of care.

This chapter examines the contribution of research on migration and care labour to understanding the inequalities of care provision. Drawing on Fiona Williams' analytical framing of migrant care work (Williams, 2011, 2012, 2018), it considers how inequalities in care labour, in the distribution of care, and the conditions of care are produced. At the same time, it considers the ways in which those inequalities are contested, underlining important connections between the claiming of rights in relation to migration, paid work and care. It is argued that an understanding of those inequalities and their contestation provides the basis for a 'social transformation' perspective on migration (Castles, 2010), situating migration in relation to wider social processes. In bringing a feminist approach to social transformation to the fore, the importance of understanding the local specificity of intersecting inequalities is also emphasised, to avoid reducing an understanding of global care inequalities to a simplified North/South binary.

The chapter proceeds as follows. It begins by considering the concept of care, and of care provision, in relation to welfare states analysis in European and North American contexts, and the insights of scholars who have noted the neglect of global South contexts. Second, it draws on Fiona Williams' analytical framing of migrant care labour in the transnational political economy of care to examine the unequal relations bound up with the provision of care by migrant workers. In doing so, connections are made with the meso level of different institutional contexts shaping the provision of care and migration, and with the micro level of care relationships, practices and processes, involving care workers, people in need of care, employers, families and other actors. Third, the chapter considers how, by way of providing a lens onto those transnational and intersectional inequalities, the analysis of migration and care provides the basis for a 'social transformation' perspective on migration (Castles, 2010), connecting debates on migration to wider social processes. Critical to such a perspective, it is argued, is bridging the divide between calls for 'decent work' and 'decent care'.

Care Provision and Its Inequalities

The concept of care in feminist scholarship is one in which social relations are foregrounded. As a relational concept, theorists of care have underlined the limitations of analyses that do not attend to the interconnections and interdependencies of individuals—that is, the social relationships in which people are embedded (Tronto, 1993). While studies of care often focus on the emotional, practical, physical and financial support provided to children, disabled and older adults, broader understandings of care encompass a wider range of social relationships, as well as self-care, as central to human well-being (Kofman & Raghuram, 2009). In this sense, care has been defined as a set of 'social capacities': 'those available for birthing and raising children, caring for friends and family members, maintaining households and broader communities, and sustaining connections more generally' (Fraser, 2016: 99).

At the same time, care is a concept that attends to the inequalities of social relations in terms of the gendered division of labour, manifested in the construction of care as 'women's work' and as historically undervalued labour. In elaborating a concept of social care in the analysis of welfare states—one that considers the social and political economy within which care is embedded—Daly and Lewis (2000) highlight three dimensions. The first considers care as labour, and in doing so draws attention to the relationship between care and other forms of labour, and the conditions of care labour. A second dimension concerns the normative aspect of care in terms of obligation and responsibility, and often familial relations and responsibilities. And third, care provision comprises costs that extend across public and private boundaries, across the state, market, families and communities (Daly & Lewis, 2000: 285). Care provision is shaped by and also shapes inequalities in terms of who provides care, under what conditions, responsibilities and obligations, and at what and whose cost. An analysis of those inequalities thus addresses the distribution of care across the family, community, state and market, at the micro level between individuals within families and communities; as well as the nature of state policies and provision (Daly & Lewis, 2000: 287).

In bringing care into the analysis of western welfare states, scholars such as Jane Lewis identified the gendered assumptions underpinning the development of welfare systems. The 'male breadwinner' model of households treated men primarily as workers, providing the family wage, and women as wives and mothers (Lewis, 1992). With the restructuring of welfare states, a broad shift towards an 'adult worker' or 'universal breadwinner' model was identified (Fraser, 1994; Lewis & Giullari, 2005), with social policies across a range of high-income countries enabling or obligating labour market participation for all, albeit with significant variation in different national contexts (Orloff, 2006). While the promotion of an adult-worker model has been assumed to be gender neutral – in treating men and women 'the same', as citizen-workers—it fails to address women's unequal position in the labour market and the ongoing unequal division of unpaid care (Lewis & Giullari, 2005). In spite of the increase in women's labour market participation, gender differences in hours of work and earnings signify more often a 'one-and-a-half breadwinner' rather than 'dual-worker' (two full-time worker) households (Lewis & Giullari, 2005). Those patterns of employment are strongly shaped by care: women leave the labour market not simply as a result of 'involuntary' unemployment but in order to care. At the same time, their return to work and participation in work is affected by their involvement in care-related activities (Lewis, 2001). Class-based inequalities also affect polarisation between 'dual-worker' and 'male breadwinner' households, with the dual-worker model being more a reality for higher educated women in two-parent households (Hook, 2015).

While the shift towards an adult worker model has been marked by an increased focus on childcare provisions in Europe, there is considerable diversity at the national and local levels in childcare services and parental leave policies. Ciccia and Bleijenbergh (2014) find that only a minority of countries (Denmark, Iceland, Sweden) promote a 'supported universal breadwinner' model, with the state taking responsibility for childcare through high levels of high-quality, publicly financed services. The 'unsupported male breadwinner' model remains more prevalent, with limited provisions, perpetuating class as well as gender-based inequalities in terms of private childcare arrangements (Ciccia & Bleijenbergh, 2014). Likewise, the limits to publicly financed care provision for older people, in terms of coverage and quality, also raise issues for inequalities in terms of who cares, under what conditions, and the distribution of the costs of care (León, Ranci, & Rostgaard, 2014).

Against this backdrop, feminist scholars have emphasised the limitations of a market-oriented model in which care continues to be subordinated to paid work as a valued activity and as the basis for claiming rights (Lister, 2003). A more gender equitable and inclusive policy approach, it is argued, requires facilitating a right to care, including time to care (e.g. through leave policies and the re-organisation of paid work), as opposed to the assumption that all care can be de-familised through state or market provisions (Fraser, 1994; Knijn & Kremer, 1997; Lewis, 2009).

Beyond a European and North American focus, more limited attention to care in global South contexts has been noted (Benería, 2008; Razavi, 2011). Different policy challenges have been highlighted from a global South perspective. This includes the more extensive nature of the informal economy with respect to paid work, including domestic work, and thus the limits to employment-based policies to facilitate time to care, e.g. through leave policies, in contexts of highly irregular types of work. Likewise, the pressures on time for women managing both income earning and care giving is exacerbated by very limited access to state or market provisions (Razavi, 2011). More generally, the de-familisation of care is more limited where family networks form a more pivotal part of people's livelihood strategies (Razavi, 2011). Thus, emphasis is placed on the need for policies to support local care infrastructure, including basic public provisioning as a pre-condition for care giving, such as water and sanitation to support domestic work within households, public transport, as well as local access to childcare and schools (Benería, 2008; Razavi, 2011). Time to care, and not to care (Lewis, 2002), in poorer and more affluent communities, thus needs to be considered in relation to access to income and local care-supporting public infrastructure—in connecting different types of resources derived across market, state, family and community.

Migration, Care Labour and Its Inequalities

Feminist scholarship on the connections between care and migration crucially added an international and transnational perspective to an understanding of the social relations of care, and one that drew attention to global inequalities between North and South, and the intersection of gender with race/ethnicity, citizenship and immigration status. The development of the concept of global care chains (Hochschild, 2000) pointed to the ways in which the political and social economy in which care is embedded needed to be viewed beyond the boundaries of a (comparative) welfare states framework. The analysis of global care chains, or what Parreñas referred to as the international transfer of caretaking (Parreñas, 2001), focused on the transnational inequalities in the distribution of care labour and the costs of care provision. Those inequalities concerned the migration of women from the global South to countries in the global North, whose care labour at the household level as domestic workers facilitated the increased labour market participation of western women (alongside men). The productive labour of higher-waged women and men in the global North was thus seen to be dependent on the outsourcing of unpaid care via the low-waged care labour of migrant women from the South. Migrant women in turn relied on care for their children and other family members to be carried out by female relatives or lower paid women in communities in countries of origin.

The global care chains concept, while crucial in bringing a transnational perspective to care provision, and care inequalities, was criticised for being grounded empirically in studies of domestic work, and migration between particular countries (the USA and Philippines) (Yeates, 2012). The work of Nicola Yeates importantly expanded the analysis to other types of care workers with respect to the migration of nurses, and to different regional contexts, in developing an understanding of the operation of increasingly global care industries (Yeates, 2009). At the same time, the linearity of the care chains concept, focusing on a one-way transfer of care (from North to South) was critiqued. A focus on transnational family relations drew attention to how migrant workers care for family members 'from afar', and how care relations in families and communities in countries of origin are affected by migration (e.g. Parreñas, 2005). This may involve the negotiation of cross-border arrangements of care for family members in countries of origin as well as migrants' own care needs post retirement (Van Walsum & Alpes, 2014). Other ways of conceptualising the multi-directional dimensions of care relations in terms of 'care circulation' were seen as more relevant, including in the analysis of South-South care migration (Kofman, 2012) and internal rural–urban migration contexts (Liu, 2016).

In mapping out how we are to understand the increased role of migrant labour in the provision of care, Fiona Williams set out a framework that connected different scales of analysis (Williams, 2011, 2012, 2018). This involved linking macro level analysis of the wider transnational political economy of care, in which migrant care labour is embedded, to the intersection of state policies relating to care, employment and migration in shaping the configuration of care relations and practices in different contexts. In doing so, it provided the basis for examining intersecting inequalities in relation to migration and care provision, with attention to different institutional contexts (at transnational, national and local levels).

As regards the macro level, Williams draws on the work of Fraser (2016) in connecting the crisis of care, as it has been described, to other manifestations of the global crises of contemporary capitalism. With regard to the contradictions of capitalism and social reproduction, Fraser argues that care is central to the economy and society, yet capitalist accumulation destabilises the care relations on which it relies (Fraser, 2016: 100). Williams notes that the global care crisis in the North is manifested in the tensions between a dual-worker model of households, increasing needs for care (including through population ageing), and the limits to welfare systems to address care needs. The transfer of care labour, between North and South, between higher and lower income countries, as noted previously, has created transnational inequalities in the distribution of care labour but also the costs of providing care. In addition to domestic workers, wider research has addressed the recruitment of migrant workers in public and privately funded care services, in residential care homes, hospitals, home care services, with increased demand for the expansion of care provision and for low-waged care labour in this context (Spencer, Martin, Bourgeault, & O'Shea, 2012). Thus, the unequal distribution of the costs of providing care has involved not simply the strategies of households but the restructuring of welfare states. Indeed, Williams (2011) argues, the labour of migrant women can be seen as enabling welfare states in the North to contain the costs of care.

With reference to the meso level, Williams explores how these macro dynamics relate to state policies towards the organisation of care, and the intersection of those policies with migration and labour market policies (Williams, 2011). This scale of analysis thus draws attention to the different national institutional contexts that have shaped variation in the types of care work in which migrant workers have been employed, and in the conditions of care work. As regards policies associated with the marketisation of care, the use of cash transfers has, for example, shaped the employment of low-waged migrant workers in the home-based care of older people in countries such as Austria and Italy (Roit, Bihan, & Österle, 2007). While a lack of state regulation has often been pinpointed as contributing to the employment of migrant workers in private households (Simonazzi, 2009), it is not so much a lack of regulation but the nature of the regulatory context that has shaped the employment of migrant workers in different institutional settings (Shutes & Chiatti, 2012; Theobald, 2017). That regulatory context involves the intersection of employment and immigration policies, which shape 'choice' for migrant workers regarding their entry to and exit from care work through the differentiation of rights and entitlements attached to worker and immigration status (Shutes, 2012).

Further research has centred more on the institutional context of the countries of origin of migrant workers, revealing the ways in which national state policies have also been instrumental in shaping the 'export' of women's labour as a global commodity in the international care labour market. With respect to the Philippines, both the state and private recruitment agencies have engaged in the racial branding of Filipinos as the ideal global care labour (Guevarra, 2010). Dynamics of race and gender are central to the ways in which migrant workers in this context are marketed as domestic workers, nurses and care assistants for 'export' to countries in North America, Europe, and elsewhere in Asia. These practices, while forming a strategy for promoting the competitive advantage of Filipino labour and of the Philippines in the global economy, at the same time reinforce divisions between Filipino and other migrant workers in the global care labour market (Guevarra, 2010), divisions which likewise materialise at the micro level in terms of hierarchies among care workers.

The micro level within Williams' framework focuses on care relationships, practices and processes involving migrant care workers, their employers and those for whom they care. In doing so, it draws attention to the ways in which gendered and racialised norms shape those relations, and the construction of hierarchies among migrant workers. As regards home-based care work, research has shown how the preferences of employers for migrant domestic and care workers are strongly shaped by race and ethnicity as well as by gender (Anderson, 2007; Shutes & Walsh, 2012; Williams & Gavanas, 2008). Racialised hierarchies and cultural stereotypes operate based on the categorisation of the skills and characteristics of particular care workers according to race, ethnicity and country of origin. Eastern European domestic workers may be favoured over African workers, Latin American and Filipina domestic workers may be seen as more 'loving' and Eastern European workers as more 'hard-working' by individual employers. Employer preferences for migrant workers in terms of their 'willingness' to accept certain terms and conditions, such as working overtime or particular shifts, may also be grounded in the relative control employers have over workers whose labour mobility is restricted by their immigration status (Ruhs & Anderson, 2010; Shutes, 2012).

While analyses of migrant care labour have often tended to adopt a 'North-South' approach to care inequalities at the macro level, studies of different local contexts have provided more nuanced understandings of inequalities in relation to migration and care provision, including the intersection of gender with generation and sexuality. Research on rural–urban migration in China has considered how intergenerational and gendered inequalities are produced in the distribution of care among family members in contexts in which the family is positioned as sole provider of care for older people and children (in the absence of state provisions in rural areas), and where local income-generating opportunities are limited (Liu, 2016). Different types of care related roles are distributed among family members, women and men, including migration to urban areas as a means of providing financial support in relation to care (e.g. to meet the costs of food and medicines), the provision of practical support, emotional support and personal care. These relations and multi-directional exchanges of care giving and receiving involve intergenerational and gender inequalities not simply with respect to who provides care, but what type of care. At the same time, local variation (between different rural areas), in the extent to which local agriculture or urban migration featured as sources of income in the histories of those areas, shape the extent to which sons or both sons and daughters migrate, and in turn who, among the older generation, manages proximate care.

Lived Experiences of Inequalities of Migration, Paid Work and Care, and Their Contestation

Research on migration and care labour has thus expanded understanding of the inequalities of care provision, inequalities that require a transnational and intersectional lens, while attending to the different institutional contexts at national and local levels that shape those inequalities. In addition to developing an understanding of inequalities relating to migration, paid (care)work and unpaid care, it has also been necessary to consider how these inequalities are negotiated and contested.

While transnational inequalities are evident in the provision of care by migrant workers in the global North, those inequalities have been contested by migrant women. The political mobilisation of women as workers has been constrained both by the privatisation and irregularisation of care work in private households. Migrant domestic workers, working in these settings, have limited access to employment protections by virtue of the status of domestic work in labour regulations and the status of domestic workers in immigration regulations, placing them in relations of dependence on their employer. The political mobilisation of migrant women in claiming recognition and rights as workers has, however, been achieved in ways that have brought about changes in national immigration policies (Anderson, 2010), and connected localised forms of claims-making to the transnational and international levels (Thomson, 2009; see also Donnecker & Piper and Fish in this volume). Care workers recruited through programmes for live-in care for older people have likewise involved the organisation of migrant women in claiming rights not only as workers but as carers, that is with respect to their family lives. This has taken place in relation to the Live-in Caregivers programme in Canada, which requires two years on the programme before care workers have access to permanent residence and family reunification rights (Tungohan, 2013). Protesting the ways in which family separation is imposed through migration policies has thus also been important in connecting rights to work to rights to care.

Inequalities in care provision have also been contested by migrant workers providing care alongside older people in need of care support. State and market based sources of care provision in Europe and North American contexts, including residential and homecare provision for older people, have been marked by underfunding, low wages, high turnover and vacancy rates, and, relatedly, ongoing reliance on unpaid care with unmet needs for care support identified both by those in need of care and those providing unpaid care (Brimblecombe, Pickard, King, & Knapp, 2017). Alliances between trade unions, organisations representing care workers, disability groups and other organisations representing older people, have emerged to contest proposed funding cuts at the local level, as Cranford's research on local level alliances in North American contexts has revealed, with implications for cuts to (migrant) workers' wages and to the funding of services provided by family members (Cranford, 2014). These sort of alliances between care workers and care recipients, Cranford argues, have been central to making home care workers visible (Cranford, 2014).

Making these connections between migrants and non-migrants, between the quality of work and care, in the analysis of care inequalities, takes steps towards bridging the longstanding analytical divide between productive and reproductive labour. In doing so, new research directions have opened up that combine historically grounded analyses of gender relations, as underpinning specific national state policies and their development, with attention to what a care-oriented policy might entail in different communities. With respect to Sri Lanka, national immigration policies that restrict the migration of domestic workers have been framed as protecting women as workers from labour exploitation, while reinforcing gendered norms about women's role in the family (Withers, 2019). The need for a 'decent work' agenda, Withers argues, with respect to policy instruments at the international level, thus needs to be coupled with a 'decent care' agenda, one that focuses on the care infrastructure at the local level of the communities in countries of origin of migrant care workers, and supporting an infrastructure that is able to support care needs and enable work (Withers, 2019).

These connections between the rights of migrants and non-migrants, between care, paid work and migration, have to some extent been evident in international policy frameworks as regards the Sustainable Development Goals (Gammage & Stevanovic, 2019; Piper, 2017). Gammage and Stevanovic (2019) have pointed to the importance of the recognition of migrant women within those goals, but also the potential as a framework for linking the goals relating to decent work, the valuation of unpaid care through the provision of public services, infrastructure and social protection, and their relevance to migrant care workers. At the same time, they argue that the SDG framework, in this respect, has relevance to addressing care needs and workers' rights at the level of national governments across so-called developed and developing country contexts. While these debates are beyond the scope of this chapter, such policy frameworks offer potential for linking analytical understanding of the intersection of care, paid work and migration, and to articulating a transnational approach to welfare that places social capacities for care at its core.

The Inequalities of Migration and Care Labour: A Social Transformation Perspective

While research on migration and care labour has contributed to a transnational understanding of care inequalities, there remains a need to consider how migration studies more generally make these sorts of connections. In reflecting on the theorising of migration, Stephen Castles has called for a 'social transformation' perspective (Castles, 2010). While migration theories have often centred on the distinctiveness of migration processes and migrants, Castles argues there is a need to embed migration research in a more general understanding of contemporary society, to see migration as intertwined with wider social processes: 'social transformation can be defined as a fundamental shift in the way society is organised that goes beyond the continual processes of incremental social change that are always at work. This implies a "step-change" in which all existing social patterns are questioned and many are reconfigured' (Castles, 2010: 1576). Migration and labour market change is considered as an example of how migration processes crucially connect to wider processes of labour market restructuring and segmentation, noting the limits of research that focuses solely on migrant workers' employment without making connections to these processes of social transformation.

Analyses of care and migration, I would argue, have been crucially concerned with connecting migration processes to wider dynamics pertaining to the global social and political economy of care. In stressing the importance of linking migration theories to the analysis of social transformation processes at different levels, in recognising the connections between different socio-spatial levels (Castles, 2010), analyses of migration and care appear to be a good starting point for further locating migration in a wider social and political economy framework. At the same time, feminist theories of social transformation are inextricably bound up with the praxis of feminist engagement (Baksh & Harcourt, 2015), in seeking both to understand *and* transform inequalities. While feminist theory and praxis has at times been critiqued for failing to address the intersections of gender with wider inequalities of race/ethnicity, class, sexuality, ability, among other divisions (Hill Collins, 1990), the importance of understanding the local specificity of intersecting inequalities is also emphasised, to avoid reducing an understanding of global inequalities to a simplified North/South binary (Bose, 2012). From a feminist perspective, a call for linking migration to wider social change must, therefore, seek to consider the transformation of global inequalities, but based on an understanding of 'intersectional variation' across regions of the world (Bose, 2012) which does not homogenise or dichotomise experiences of care inequalities.

Conclusion

Research on migration and care labour has been critical to elucidating the global and transnational contexts in which inequalities relating to the provision of care are located. At the same time, analyses of different national and local contexts have provided more nuanced understandings of how those inequalities are shaped by state policies across migration, work and welfare, and by variations in the configuration of state, market, civil society and family roles and relationships in the provision of care. Lived experiences of migration and care work likewise reveal the ways in which gendered and racialised norms and identities are negotiated and (re)constructed in relations between care workers, employers, family members and care recipients, in terms of the provision of paid and unpaid care. Structural inequalities in relation to migration and care have, however, also been challenged at the local level through forms of contestation by care workers and alliances with care recipients. The practice of care on the ground has thus given rise to claims-making by migrants and non-migrants that cut across demands for 'decent work' and for 'decent care', including rights to work and to family life.

Theoretical and empirical work on the inequalities of care and migration contributes to a social transformation perspective within migration studies, in linking migration to the wider transnational political economy of care. At the same time, it also contributes to the ongoing need for care to be brought to the centre of analyses of migration and migrant labour. Research on migration continues, often, to decouple labour from care, migrant workers from the care relations in which they are embedded, and in doing so to reproduce rather than interrogate the division between productive and reproductive labour. This has consequences not simply with respect to failing to grapple with the gendered dynamics of labour migration. With respect to policy implications, arguments for tackling inequalities relating to migration by improving the rights of migrant workers, improving working conditions, and improving employment and income generating opportunities for communities, need to consider the care infrastructure and relations on which those rights, conditions and opportunities depend, at the local level and across national borders.

References

Anderson, B. (2007). A very private business: Exploring the demand for migrant domestic workers. *European Journal of Women's Studies, 14*(3), 247–264.

Anderson, B. (2010). Mobilizing migrants, making citizens: Migrant domestic workers as political agents. *Ethnic and Racial Studies, 33*(1), 60–74.

Baksh, R., & Harcourt, W. (2015). Introduction: Rethinking knowledge, power and social change. In R. Baksh & W. Harcourt (Eds.), *The Oxford handbook of transnational feminist movements* (pp. 1–47). Oxford: Oxford University Press.

Benería, L. (2008). The Crisis of care, international migration, and public policy. *Feminist Economics, 14*(3), 1–21.

Bose, C. E. (2012). Intersectionality and global gender inequality. *Gender & Society, 26*(1), 67–72.

Brimblecombe, N., Pickard, L., King, D., & Knapp, M. (2017). Perceptions of unmet needs for community social care services in England: A comparison of working carers and the people they care for. *Health & Social Care in the Community, 25*(2), 435–446.

Castles, S. (2010). Understanding global migration: A Social transformation perspective. *Journal of Ethnic and Migration Studies, 36*(10), 1565–1586.

Ciccia, R., & Bleijenbergh, I. (2014). After the male breadwinner model? Childcare services and the division of labor in European Countries. *Social Politics: International Studies in Gender, State & Society, 21*(1), 50–79.

Cranford, C. (2014). Toward flexibility with security for migrant care workers: A comparative analysis of personal home care in Toronto and Los Angeles. In B. Anderson & I. Shutes (Eds.), *Migration and care labour: Theory, policy and politics* (pp. 173–191). Basingstoke: Palgrave Macmillan.

Daly, M., & Lewis, J. (2000). The concept of social care and the analysis of contemporary welfare states. *The British Journal of Sociology, 51*(2), 281–298.

Fraser, N. (1994). After the family wage: Gender equity and the welfare state. *Political Theory, 22*(4), 591–618.

Fraser, N. (2016). Contradictions of capital and care. *New Left Review, 100,* 99–117.

Gammage, S., & Stevanovic, N. (2019). Gender, migration and care deficits: What role for the sustainable development goals? *Journal of Ethnic and Migration Studies, 45*(14), 2600–2620.

Guevarra, A. R. (2010). *Marketing dreams, manufacturing heroes: The transnational labor brokering of Filipino workers*. New Brunswick, NJ: Rutgers University Press.

Hill Collins, P. (1990). *Black feminist thought: Knowledge, consciousness, and the politics of empowerment*. London: Unwin Hyman.

Hochschild, A. (2000). Global care chains and emotional surplus value. In W. Hutton & A. Giddens (Eds.), *On the edge: Living with global capitalism* (pp. 130–146). London: Jonathan Cape.

Hook, J. L. (2015). Incorporating 'class' into work–family arrangements: Insights from and for Three Worlds. *Journal of European Social Policy, 25*(1), 14–31.

Knijn, T., & Kremer, M. (1997). Gender and the caring dimension of welfare states: Toward inclusive citizenship. *Social Politics: International Studies in Gender, State & Society, 4*(3), 328–361.

Kofman, E. (2012). Rethinking care through social reproduction: Articulating circuits of migration. *Social Politics: International Studies in Gender, State & Society, 19*(1), 142–162.

Kofman, E., & Raghuram, P. (2009). *The implications of migration for gender and care regimes in the south*. Geneva: UNRISD.

León, M., Ranci, C., & Rostgaard, T. (2014). Pressures towards and within Universalism: Conceptualising change in care policy and discourse. In M. León (Ed.), *The transformation of care in European Societies* (pp. 11–33). Basingstoke: Palgrave Macmillan.

Lewis, J. (1992). Gender and the development of welfare regimes. *Journal of European Social Policy, 2*(3), 159–173.

Lewis, J. (2001). The decline of the male breadwinner model: Implications for work and care. *Social Politics: International Studies in Gender, State & Society, 8*(2), 152–169.

Lewis, J. (2002). Gender and welfare state change. *European Societies, 4*(4), 331–357.

Lewis, J. (2009). *Work-family balance, gender and policy*. Cheltenham: Edward Elgar.

Lewis, J., & Giullari, S. (2005). The adult worker model family, gender equality and care: The search for new policy principles and the possibilities and problems of a capabilities approach. *Economy and Society, 34*(1), 76–104.

Lister, R. (2003). *Citizenship: Feminist perspectives* (2nd ed.). Basingstoke: Palgrave Macmillan.

Liu, J. (2016). Ageing in rural China: Migration and care circulation. *The Journal of Chinese Sociology, 3*(1), 9.

Orloff, A. S. (2006). From maternalism to 'employment for all': State policies to promote Women's Employment across the affluent democracies. In J. D. Levy (Ed.), *The State After statism: New state activities in the era of globalization and liberalization* (pp. 230–268). Cambridge MA: Harvard University Press.

Parreñas, R. S. (2001). *Servants of globalization: Women, migration and domestic work*. Stanford, CA: Stanford University Press.

Parreñas, R. S. (2005). *Children of global migration: Transnational families and gendered woes*. Stanford, CA: Stanford University Press.

Piper, N. (2017). Migration and the SDGs. *Global Social Policy, 17*(2), 231–238.

Razavi, S. (2011). Rethinking care in a development context: An introduction. *Development and Change, 42*(4), 873–903.

Reit, B. D., Bihan, B. L., & Österle, A. (2007). Long-term care policies in Italy, Austria and France: Variations in cash-for-care schemes. *Social Policy & Administration, 41*(6), 653–671.

Ruhs, M., & Anderson, B. (2010). *Who needs migrant workers? Labour shortages, immigration, and public policy*. Oxford: Oxford University Press.

Shutes, I. (2012). The Employment of migrant workers in long-term care: Dynamics of choice and control. *Journal of Social Policy, 41*(1), 43–59.

Shutes, I., & Chiatti, C. (2012). Migrant labour and the marketisation of care for older people: The employment of migrant care workers by families and service providers. *Journal of European Social Policy, 22*(4), 392–405.

Shutes, I., & Walsh, K. (2012). Negotiating user preferences, discrimination, and demand for migrant labour in long-term care. *Social Politics: International Studies in Gender, State & Society, 19*(1), 78–104.

Simonazzi, A. (2009). Care regimes and national employment models. *Cambridge Journal of Economics, 33*(2), 211–232.

Spencer, S., Martin, S., Bourgeault, I. L., & O'Shea, E. (2012). *The role of migrant care workers in ageing societies: Report on research findings in the United Kingdom, Ireland*. IOM: Canada and the United States.

Theobald, H. (2017). Care workers with migrant backgrounds in formal care services in Germany: A multi-level intersectional analysis. *International Journal of Care and Caring, 1*(2), 209–226.

Thomson, M. (2009). Workers not maids—organising household workers in Mexico. *Gender & Development, 17*(2), 281–293.

Tronto, J. C. (1993). *Moral Boundaries: A Political Argument for an Ethic of Care*. London: Routledge.

Tungohan, E. (2013). Reconceptualizing Motherhood, Reconceptualizing resistance. *International Feminist Journal of Politics, 15*(1), 39–57.

Van Walsum, S., & Alpes, M. J. (2014). Transnational households: Migrants and care, at home and abroad. In B. Anderson & I. Shutes (Eds.), *Migration and care labour: Theory, policy and politics* (pp. 151–169). Basingstoke: Palgrave Macmillan.

Williams, F. (2011). Towards a transnational analysis of the political economy of care. In R. Mahon & F. Robinson (Eds.), *Feminist ethics and social policy: Towards a new global political economy of care*. Vancouver, BC: UBC Press.

Williams, F. (2012). Converging variations in migrant care work in Europe. *Journal of European Social Policy, 22*(4), 363–376.

Williams, F. (2018). Care: Intersections of scales, inequalities and crises. *Current Sociology, 66*(4), 547–561.

Williams, F., & Gavanas, A. (2008). The Intersection of Childcare regimes and migration regimes: A three-country study. In H. Lutz (Ed.), *Migration and domestic work: A european perspective on a global theme* (pp. 13–28). Farnham: Ashgate.

Withers, M. (2019). Decent care for migrant households: Policy alternatives to Sri Lanka's family background report. *Social Politics: International Studies in Gender, State & Society, 26*(3), 325–347.

Yeates, N. (2009). *Globalizing care economies and migrant workers: Explorations in global care chains*. Basingstoke: Palgrave Macmillan.

Yeates, N. (2012). Global care chains: A state-of-the-art review and future directions in care transnationalization research. *Global Networks, 12*(2), 135–154.

Isabel Shutes is an Associate Professor in the Department of Social Policy at the London School of Economics and Political Science. Her research examines the interactions of migration and social policies; inequalities relating to citizenship and migration; and the implications for the social relations of paid and unpaid work/care, for social provision and social rights. Related publications include Shutes, I. and Walker, S. (2018) "Gender and Free Movement: EU Migrant Women's Access to Residence and Social Rights in the UK", *Journal of Ethnic and Migration Studies, 44*: 1, 137–153; Shutes, I. and Chiatti, C. (2012) "Migrant Labour and the Marketisation of Care for Older People: the Employment of Migrant Care Workers by Families and Service Providers", *Journal of European Social Policy, 22*: 4, 392–405; and *Migration and Care Labour: Theory, Policy and Politics*, co-edited with Bridget Anderson (2014), Basingstoke, Palgrave Macmillan.

CHAPTER 8

Gendered Transnational Parenting

Karlijn Haagsman and Valentina Mazzucato

Abstract Transnational families in which parents and children reside in different nation states have existed for centuries, yet scholarship has paid attention to this phenomenon only in the past 20 years. The feminization of migration, resulting in more mothers migrating independently and leaving their children in the care of someone else, has contributed to the recent attention. Studies have found that mothers and children are affected negatively in various dimensions of their well-being. Gender norms are often seen as part of the explanation. As mothers are often the primary care providers, their co-presence is considered a prerequisite to healthy child development. If mothers cannot fulfil this parenting norm, this results in stress and guilt for these mothers and emotional loss for children. Men's roles as breadwinners are seen as more compatible with their migration. Yet, there are only a limited number of studies that have looked at fathers and comparisons between migrant mothers and fathers are even scarcer. This chapter will discuss and contrast effects of transnational separation of migrant mothers and fathers. While gender norms are taken into account, this chapter will also look at contextual and structural factors that can explain possible differences.

K. Haagsman (✉) · V. Mazzucato
Department of Society Studies, Maastricht University, Maastricht, The Netherlands
e-mail: r.haagsman@maastrichtuniversity.nl

V. Mazzucato
e-mail: v.mazzucato@maastrichtuniversity.nl

C. Mora and N. Piper (eds.), *The Palgrave Handbook of Gender and Migration*, https://doi.org/10.1007/978-3-030-63347-9_8

Introduction

Global migration has had a significant impact on family life. Migration often results in the separation of family members across borders, at least for some period of time, resulting in what are called 'transnational families'. Historically, such families have always existed, but only in the 1990s did they become a substantive subject of study. Feminist scholars noted that increasingly it was mothers who were migrating, leaving their children behind in the care of someone else. They pointed to the feminization of migration where, due to global inequality and the global division of labour, women were migrating independently (Hochschild, 2000; Hondagneu-Sotelo & Avila, 1997; Parreñas, 2000, 2001a; Schmalzbauer, 2004). As women in the Global North were progressively entering the workforce, women from the Global South were needed to take over their domestic tasks. Feminist scholars argued that capitalist production systems had created a 'care-drain' in the Global South by transferring carers to the Global North.

In the Global North, parenting has often been bound to the concept of biological parenthood, especially for mothers. Biological motherhood is considered especially important for the upbringing of children, and physical proximity is seen as a prerequisite for healthy child development (Erel, 2002). Thus, once transnational families emerged as a focus of research, early qualitative studies focused on the hardship and emotional difficulties mothers faced because they did not conform to the prescribed parenting norm of co-present mothering (Bernhard, Landolt, & Goldring, 2009; Boccagni, 2012; Dreby, 2006; Madianou & Miller, 2012; Parreñas, 2000). Studies also found that children were affected by being separated from their mothers (Parreñas, 2001a, 2005b; Suarez-Orozco, Todorova, & Louie, 2002). However, despite the difficulties imposed by distance, research has shown that mothers remain actively involved in the upbringing of their children and try to parent transnationally through the use of communication technologies, the sending of remittances and gifts, and, if possible, regular visits (Dreby & Adkins, 2010; Madianou & Miller, 2011; Parreñas, 2005b; Zentgraf & Chinchilla, 2012). These acts of transnational parenting have been found to be essential in mitigating the negative effects that transnational parent–child separation has.

Three main criticisms have been made with regard to these early studies of transnational families. First, as feminist scholarship was so focused on the injustices to which migrant women were subject, it overlooked their agency and empowerment (Bernhard et al., 2009; Madianou & Miller, 2012). Second, scholars neglected the fact that extended families have played a key role in the upbringing of children across the world for centuries, i.e. co-present biological parenting is not a global norm and not necessarily detrimental for child development (Mazzucato & Schans, 2011). Third, while migrant men have long been the focus in migration studies and the experiences of migrant women have been silenced, transnational family studies suffered from the opposite: they mostly focused on migrant mothers and their offspring, ignoring the experiences of migrant fathers and their families (Dreby, 2006; Pribilsky, 2004; Waters, 2009). Only since the mid-2000s and early 2010s have studies on transnational fathering emerged, although still in limited numbers. The criticisms above show that normative discourses on gender and parenthood infiltrate academic scholarship and affect what we study and how (Erel, 2002).

This chapter will review the studies on transnational mothering and fathering. It will discuss the differences and similarities in how fathers and mothers parent across borders and the effects transnational parenting has on children in the country of origin and on parents themselves. While gender norms help to explain these differences, this chapter will also look at contextual and structural factors. To contextualize differences between the parenting practices of men and women, we will first briefly explore the main differences in the migration trajectories of men and women. Second, we will dive into transnational parenting and the different expectations that apply to migrant mothers and fathers in transnational caregiving arrangements. Third, we will look at the effects transnational parent–child separation has for both parent and child and whether these effects differ between mother-away and father-away families (see also Shutes, in this volume).

Migration Patterns of Mothers and Fathers

Before turning to the effects of transnational parenting, it is important to recognize the different characteristics of male and female migration because these affect the way fathers and mothers are able to parent from afar. This section will therefore touch upon some of the main distinctions scholars have found and show that both gender norms and structural conditions lead to differences in migration patterns and experiences for migrant men and women.

Women account for almost half of worldwide international migration over the last five decades, if not longer (Donato & Gabaccia, 2016). Yet, migrant women went unnoticed for a long time or were considered 'trailing wives', following their husbands' migration with little agency. Feminist scholars began to change this image by recognizing that even when women follow their husbands abroad, they have agency and are economically active. Moreover, it was not always men who migrated. In some regions (such as Armenia, Georgia, Azerbaijan, Tajikistan and Nepal), female migration clearly dominates, while in others (such as China, Equatorial Guinea, Libya and Gabon) male migration does (United Nations Population Division, 2019). Hence, migration streams are highly gendered and depend on specific contexts of emigration and immigration, such as gender norms and needs for certain types of labour (Bastia & Haagsman, 2020; Bastia & Piper, 2019; Piper, 2003).

Although only a limited number of studies compare male and female migration in any detail, some note important differences between male and female migration, which can affect how men and women operate in the host country but also how they engage with family in their country of origin

(Caarls, Haagsman, Kraus, & Mazzucato, 2018). Studies have found differences between men and women in migration motivations (Curran & Rivero-Fuentes, 2003; Kanaiaupuni, 1999; Massey, Fischer, & Capoferro, 2006), the timing of migration (Kraus, 2019), experiences of migration (Hondagneu-Sotelo, 1999; Wong, 2006), support networks (Curran & Rivero-Fuentes, 2003; Toma & Vause, 2014), economic mobility resulting from migration (Boyle, Kulu, Cooke, Gayle, & Mulder, 2008) and the sending of remittances (Abrego, 2009; Bollard, McKenzie, & Morten, 2009; Lindley, 2009; Petrozziello, 2011). Many of these differences have to do with (conflicting) gender norms in host and origin country (Gallo, 2006; Jolly & Reeves, 2005).

Structural conditions also affect the experiences and opportunities migrant men and women have. Female migrants often have fewer work opportunities in the host country and face worse conditions of employment and greater financial deprivation, irrespective of social class (Boyd & Grieco, 2003; Feliciano, 2008; Malmusi, Borrell, & Benach, 2010). Moreover, migration policies are selective and not neutral to gender (Kofman, 1999; Kraler, 2010; Kraler & Kofman, 2009; van Walsum, 2009). They often reinforce structural inequalities between men and women. For example, one of the reasons it is harder for women to bring their families over is that they earn less than their male counterparts and have greater difficulty meeting the income requirements in host countries (Kraler, 2010; Kulu-Glasgow & Leerkes, 2011; Poeze & Mazzucato, 2016; Strik, De Hart, & Nissen, 2013). The kinds of domestic and care labour women often undertake also restrict their options for family reunification (Poeze & Mazzucato, 2016).

Transnational Parenting Practices and Norms

Parenting roles and migration expectations, opportunities and effects are highly gendered (Avila, 2008; Carling, Menjívar, & Schmalzbauer, 2012; Mazzucato & Cebotari, 2016; Schmalzbauer, 2015). Expectations regarding transnational care are often different for migrant mothers and fathers. In many societies around the world, women are responsible for reproductive care work and domestic labour and are seen as the providers of emotional care. Conversely, the role assigned to men is often that of economic provider and breadwinner. Scholars note that these gender norms greatly affect expectations of transnational parenting (e.g. Carling et al., 2012; Dreby, 2006; Parreñas, 2005b, 2008; Poeze, 2019). For example, mothers' emigration is often highly criticized for undermining traditional gender roles in which mothers are expected to care for children, while male migration is accepted and, in many cases, seen as a sign of masculinity (Boccagni, 2012; Schmalzbauer, 2015). While both migrant fathers and mothers are expected to send remittances, women are also expected to continue providing emotional care to families back home. These expectations affect their ability to parent from abroad. When fathers are economically successful and fulfil what is expected of them, father–child separation is seen as uncomplicated and migration can actually boost the social status of fathers (Avila, 2008; Palenga-Möllenbeck, 2013; Poeze, 2019; Schmalzbauer, 2015). Migrant mothers, in contrast, struggle to provide the emotional care expected of them, which places an extra burden on them compared to fathers, and means they often deal with feelings of guilt and shame (Avila, 2008; Contreras & Griffith, 2012; Dreby, 2010; Parreñas, 2001b).

Some migrant mothers try to make up for their physical absence by engaging in intensive mothering. This can involve the sending of gifts alongside economic remittances (Bernhard et al., 2009; Hondagneu-Sotelo & Avila, 1997; Parreñas, 2001a, 2005b), but it can also take the form of maintaining constant contact, sometimes to the annoyance of their children (Banfi & Boccagni, 2011; Dreby, 2007; Fresnoza-Flot, 2009; Hondagneu-Sotelo & Avila, 1997; Madianou, 2012). Despite these efforts at transnational parenting, people back home may not always see mothers as actively caregiving and judge them for not providing care (Boccagni, 2012).

Some studies suggest that fathers are less involved in transnational caregiving (Avila, 2008; Parreñas, 2005a). Parreñas (2008), for example, finds that transnational fathers did less emotional caregiving and had less frequent contact with their children in the Philippines than transnational mothers. She argues that unlike mothers, who were challenging gender norms, migrant fathers corresponded more to the traditional gender and parenting roles of male authority and main breadwinner. Transnational communication further complicates this as the authoritarian role fathers want to maintain is heightened through telephone communication (Parreñas, 2005b; Pribilsky, 2004). Dreby (2006) also found that men sometimes grew distant from their children because they were not able to conform to the norm of being economic provider. When some fathers did not manage to send remittances, they felt ashamed and consequently diminished contact with their children.

Given that most gender ideologies place women in reproductive and care work, most children are left in the care of women when one of the parents migrates abroad. If not the biological mother, these carers are often grandmothers, aunts, sisters or other female kin (Carling et al., 2012; Dankyi, 2014; Haagsman, 2015). Parreñas (2005b) found that even when fathers remain in the Philippines, they let other female relatives take care of the children. What this means is that when fathers migrate, there is often no significant change in caregiving arrangements as the caregiver is often the mother of the child. However, when mothers migrate, this normally entails a change in caregiver (Mazzucato et al., 2015). In addition, mothers who stay home often help actively in cultivating bonds between the migrant father and children, while this is less the case in alternative care arrangements (Avila, 2008).

In sum, transnational parenting is highly gendered, and while migration changes the practicalities of parenting, expectations and norms are not necessarily altered (Avila, 2008; Dreby, 2006; Parreñas, 2005a). In the next section, we will see that this is important when looking at the effects of transnational separation.

Effects of Transnational Separation

It is often argued, through appeal to attachment theory, that a child has a special attachment to their primary caregiver, often the mother, and that this bond is crucial for healthy development of the child. The bond is not only important in the child's early development, but also affects their social adaptation, ability to build social bonds, educational performance and behaviour later in life (Bernardes, 1999; Erel, 2002; Tizard, 1991). Because the mother–child bond is seen as of utmost importance for a child's development, other caregivers cannot fully substitute for a biological mother. Moreover, it is argued that a child's development is only shaped by people who are proximate, and family members who are geographically far away are therefore considered to be less developmentally relevant for the child (Mazzucato, 2015; Mazzucato et al., 2015; Suarez-Orozco et al., 2002). Migration of parents, especially mothers, and their transnational parenting practices are therefore seen as especially detrimental.

Most of the transnational family studies have shown that the physical separation resulting from parental migration can indeed be challenging and affect the well-being of parent and child, especially in mother-away families. It is difficult to interact across space, and as a result, the parent–child bond suffers (Horton, 2009; Parreñas, 2005b). This can lead to feelings of loss and cause stress, lowering emotional well-being and health for the parent (see e.g. Aranda, 2003; Bernhard et al., 2009; Dreby, 2006; Fresnoza-Flot, 2009; Hondagneu-Sotelo & Avila, 1997; Horton, 2009; Moran-Taylor, 2008; Parreñas, 2001a, 2005b; Schmalzbauer, 2004) and for the child (see e.g. Dreby, 2010; Parreñas, 2005a; Schmalzbauer, 2008). Feelings of estrangement can be exacerbated when the child does not recognize the migrant as their parent (Banfi & Boccagni, 2011). Studies have also found that transnational family life can lead to poor physical health (Asis, 2006), behavioural problems and involvement in violence and gangs (Dreby, 2007; Moran-Taylor, 2008; Schmalzbauer, 2004; Smith, 2006), and poor educational performance (Battistella & Conaco, 1998; Dreby, 2007).

The studies outlined in the paragraph above were all on mother-away families, and most argue that mothers and children evaluate the separation through prevailing gender norms and expectations. Poeze and Mazzucato (2014), for example, showed that Ghanaian children expected their migrant fathers to provide for their material needs and their migrant mothers for their emotional needs. Some scholars argue that children often see mothers as the primary caregivers and fathers as more peripheral (Dreby, 2010; Parreñas, 2005a, 2005b). As a result, some studies suggest that children miss their mothers to a greater extent than their fathers. An absent mother can lead to feelings of resentment in children, especially when they feel they are not taken care of properly (Menjivar & Abrego, 2009; Parreñas, 2005b). Children can reproach their mothers for their absence, increasing the guilt that mothers already feel (Bernhard et al., 2009; Fresnoza-Flot, 2009; Parreñas, 2005a).

While research has shown that children can also feel abandoned by fathers (e.g. Pribilsky, 2004, 2012), some studies find that children reproach mothers more as mothers are seen by their children as the primary caregivers and children are more accepting of fathers migrating in an effort to fulfil their role as breadwinners (Cebotari, Mazzucato, & Siegel, 2016; Menjivar & Abrego, 2009; Parreñas, 2005a). Additional stress for mothers comes from the judgements others make, which sometimes involve accusations that migrant mothers are abandoning their children and not acting as a mother should (Abrego, 2009; Bernhard et al., 2009; Dreby, 2006). In short, migrant mothers and their children appear to suffer more from separation than migrant fathers.

Not only do gender norms affect how parents and children experience separation, these norms have infiltrated academia too. In general, when scholars study men in the realm of migration, the focus is on their work lives. Studies of men in the family sphere are few (Choi, 2019). Similarly, most transnational family research on the effects of migration has focused on women because scholars expect them to be most affected. In addition, when scholars discuss migrant fathers it is often in a negative manner—to suggest, for example, that fathers show little concern for their children's well-being or do not actively provide emotional care or even economic care (Poeze, 2019; Pribilsky, 2004; Waters, 2009). Yet, often these assessments are not based on actual research with migrant fathers. As Waters argues, men are seldom studied themselves but "are examined only to the extent that they impact on women's lives" (2009: 66).

Most of the studies that have been done on migrant fathers and parenting, however, show that fathers are affected by separation and try to actively engage in transnational parenting, rejecting the stereotypes of transnational fathers as indifferent, authoritarian and uninvolved (e.g. Chereni, 2015; Kilkey, Plomien, & Perrons, 2014; Nobles, 2011; Poeze, 2019; Pribilsky, 2004, 2012; Winters, 2014). Their emotional suffering might manifest itself differently. For example, sometimes feelings of loneliness and depression resulting from prolonged separation lead to heavy drinking (Pribilsky, 2012; Schmalzbauer, 2005; Worby & Organista, 2007) or to fathers cutting ties with their families because they are ashamed that they cannot ensure their upkeep (Dreby, 2006).

Studies comparing the effects of transnational child rearing in mother-away and father-away families are limited and mostly quantitative. These studies have shown that there are indeed some gendered effects, but that these are also context dependent. Most studies that compare mother-away and father-way families find that children are worse off when mothers migrate as opposed to fathers (Gao et al., 2010; Jia & Tian, 2010; Jordan & Graham, 2012; Wen & Lin, 2012), thereby supporting the qualitative studies described above. However, other studies do not find such clear differences (Graham & Jordan, 2011; He et al., 2012; Mazzucato et al., 2015; Vanore, Mazzucato, & Siegel, 2015). Of course, overall results also depend on the outcomes that are measured and how they are measured. Most studies measure children's cutcomes through adult assessments of their well-being, assessments which may well be biased. Cebotari et al. (2016) instead based their study on young people's reported well-being and reveal a more nuanced picture.

There are very few studies that have compared the effects of migration on mothers and fathers. Some qualitative studies highlight the differences in gendered expectations on the part of mothers and fathers and find that fathers report emotional suffering less often than do mothers (Avila, 2008; Dreby, 2010; Parreñas, 2005a; Poeze, 2018). Avila (2008), for example, having interviewed both Latina/Latino transnational mothers and fathers in the US, finds that because their migration was less accepted, mothers faced more challenges. Moreover, fathers were given more freedom and more opportunities to cope with the challenges of separation than mothers. Poeze (2018) also finds differences in expectations of Ghanaian parents in The Netherlands, with fathers assessing their roles in transnational caregiving more in financial terms. Yet, she argues that the gender ideologies prescribe but do not dictate how parenting is practised and experienced. She found that, like mothers, fathers provided more than economic care and also tried to stay actively involved in their children's lives through, for example, regular virtual and physical contact. Like the mothers, some fathers also emotionally suffered because of the separation from their families.

Conclusions and Future Research

What we have shown throughout this chapter is that transnational parenting is highly gendered. Parental migration often perpetuates or extends intra-household and gender inequalities, but it is also highly shaped by them. As a result, migrant fathers and mothers face different opportunities and expectations while abroad, which affects how both parents and children experience separation. For example, the fact that mothers are expected to be the prime caregivers often makes their migration more problematic than that of fathers and leads children to reproach them more. And although both fathers and mothers are expected to provide material care, mothers are also expected to provide emotional care, putting an extra burden on them. Hence, parenting norms affect how parents and children experience transnational separation and help explain why differences are found between migrant father- and mother-away families.

As we have highlighted in this chapter, however, most of the research has focused on women and often assessments regarding fathers are not based on actual research with them. Comparisons between men and women are even scarcer. The research that is available on fathers shows that they are certainly affected by separation from their families, as are their children. But we have to be cautious with our interpretations and comparisons, as gender norms also affect how mothers and fathers talk about being separated from their children with interviewers. Finally, as Horton rightfully states, "privileging the exploration of transnational separations through the single analytical lens of gender … reduces the embodied distress of mothers and children to mere 'gender false consciousness'" (Horton, 2009: 21). The same, we would argue, applies to fathers. To get a full understanding of transnational separation, we also have to take structural differences into account in our comparisons.

As only a small number of studies have explicitly compared the transnational family life of mothers and fathers, and most of these have been small-scale, it is hard to come to solid conclusions about the importance of structural factors such as socio-economic status and economic opportunities. Larger survey research does, however, give us some clues. Although studies are scarce, there are some indications that the differences between transnational parenting of migrant mothers and fathers go beyond gender norms and are the result of structural factors. For example, a large survey of Angolans, Nigerians and Ghanaians in Portugal and the Netherlands found no differences in the reported well-being and health of male and female migrant parents or the quality of the relationships they reported to have with their children when the researchers controlled for socio-demographic, socio-economical and migrant characteristics (Dito, Mazzucato, & Schans, 2016; Haagsman, Mazzucato, & Dito, 2015; Mazzucato, Dito, Grassi, & Vivet, 2016). Economic instability and undocumented status explained a large part of the poorer well-being reported by transnational parents compared with migrant parents who live with their children abroad (Dito et al., 2016; Haagsman et al., 2015). As migrant mothers often work in the informal economy, experience more precarious labour conditions and have lower socio-economic status than migrant fathers, this can take an emotional toll, especially when trying to provide for their families (Abrego, 2009; Avila, 2008; Schmalzbauer, 2004).

A recent study by Caarls et al. (2018) showed there were important differences between African transnational mothers and fathers in Europe. They found that migrant mothers in transnational families are more often single, more often have children from different partners and start having children earlier than migrant fathers. Moreover, while the children of mothers were often in the care of a female relative in their country of origin, the children of the fathers were often in the care of the biological mother. The authors argue that the precarious conditions of mother-away families make them more fragile. In addition, Mazzucato et al. (2015) found that children of migrant mothers changed caregivers more often than migrant fathers and that children with unstable care arrangements experienced worse emotional well-being. This complements a study by Avila (2008) which found that transnational mothers were more often single and therefore more often had to rely on alternative care arrangements. Others have found that children seem more affected by parental migration when the parents are divorced (Cebotari et al., 2016; Dreby, 2006; Haagsman & Mazzucato, 2014; Nobles, 2011). Reasons for this could be that the caregiver speaks ill of the migrant parent because they resent them, because single parents have less time for their children, or because they are less able to send remittances, especially when they have started new families (Dreby, 2010; Nobles, 2011).

In sum, although research is limited, structural differences between mothers' and fathers' migration partially explain the different ways transnational separation affects both parents and children. Mothers abroad and their families back home often find themselves in precarious conditions, which affects how mothers can parent from afar and ultimately the well-being of their children. Hence, to really understand how transnational families operate and the effects transnational separation has, it is key for future research to pay attention to structural factors in addition to gender and family norms.

References

Abrego, L. (2009). Economic well-being in Salvadoran transnational families: How gender affects remittance practices. *Journal of Marriage and Family, 71*(4), 1070–1085. https://doi.org/10.1111/j.1741-3737.2009.00653.x.

Aranda, E. M. (2003). Global care work and gendered constraints—The case of Puerto Rican transmigrants. *Gender and Society, 17*(4), 609–626. https://doi.org/10.1177/0891243203253573.

Asis, M. M. B. (2006). Living with migration. *Asian Population Studies, 2*(1), 45–67. https://doi.org/10.1080/17441730600700556.

Avila, E. (2008). *Transnational motherhood and fatherhood: Gendered challenges and coping*. Ph.D. dissertation, University of Southern California.

Banfi, L., & Boccagni, P. (2011). Transnational family life and female migration in Italy: One or multiple patterns? In A. Kraler, E. Kofman, M. Kohli, & C. Schmoll (Eds.), *Gender, generations and the family in international migration* (pp. 287–311). Amsterdam: Amsterdam University Press.

Bastia, T. & Haagsman, K. (2020). Chapter 9: Gender, migration and development. In T. Bastia & R. Skeldon (Eds.), *Routlegde handbook of migration and development* (pp. 103–113): Routlegde.

Bastia, T., & Piper, N. (2019). Women migrants in the global economy: A global overview (and regional perspectives). *Gender and Development, 27*(1), 15–30. https://doi.org/10.1080/13552074.2019.1570734.

Battistella, G., & Conaco, M. C. G. (1998). The impact of labour migration on the children left behind: A study of elementary school children in the Philippines. *Sojourn: Journal of Social Issues in Southeast Asia, 13*(2), 220–241.

Bernardes, J. (1999). We must not define "the family"! *Marriage and Family Review, 28*(3–4), 21–41. https://doi.org/10.1300/J002v28n03_03.

Bernhard, J. K., Landolt, P., & Goldring, L. (2009). Transnationalizing families: Canadian immigration policy and the spatial fragmentation of care-giving among Latin American newcomers. *International Migration, 47*(2), 3–31. https://doi.org/10.1111/j.1468-2435.2008.00479.x.

Boccagni, P. (2012). Practising motherhood at a distance: Retention and loss in Ecuadorian transnational families. *Journal of Ethnic and Migration Studies, 38*(2), 261–277. https://doi.org/10.1080/1369183x.2012.646421.

Bollard, A., McKenzie, D., & Morten, M. (2009). *The remitting patterns of African migrants in the OECD* (World Bank Policy Research Working Paper No. 5260). Retrieved from https://ssrn.com/abstract=1585039.

Boyd, M., & Grieco, E. (2003). Women and migration: Incorporating gender into international migration theory. *Migration Information Source, 1.*

Retrieved from http://www.migrationpolicy.org/article/women-and-migration-incorporating-gender-international-migration-theory.

Boyle, P. J., Kulu, H., Cooke, T., Gayle, V., & Mulder, C. (2008). Moving and union dissolution. *Demography, 45*(1), 209–222. Retrieved from http://www.ncbi.nlm.nih.gov/pmc/articles/PMC2831384/.

Caarls, K., Haagsman, K., Kraus, E. K., & Mazzucato, V. (2018). African transnational families: Cross-country and gendered comparisons. *Population, Space and Place, 24*(7), e2162. https://doi.org/10.1002/psp.2162.

Carling, J., Menjívar, C., & Schmalzbauer, L. (2012). Central themes in the study of transnational parenthood. *Journal of Ethnic and Migration Studies, 38*(2), 191–217. https://doi.org/10.1080/1369183x.2012.646417.

Cebotari, V., Mazzucato, V., & Siegel, M. (2016). Gendered perceptions of migration among Ghanaian children in transnational care. *Child Indicators Research*, 1–23. https://doi.org/10.1007/s12187-016-9407-x.

Chereni, A. (2015). Fathering and gender transformation in Zimbabwean transnational families. *2015, 16*(2). https://doi.org/10.17169/fqs-16.2.2248.

Choi, S. Y. P. (2019). Migration, masculinity, and family. *Journal of Ethnic and Migration Studies, 45*(1), 78–94. https://doi.org/10.1080/1369183X.2018.1427562.

Contreras, R., & Griffith, D. (2012). Managing migration, managing motherhood: The moral economy of gendered migration. *International Migration, 50*(4), 51–66. https://doi.org/10.1111/j.1468-2435.2012.00771.x.

Curran, S. R., & Rivero-Fuentes, E. (2003). Engendering migrant networks: The case of Mexican migration. *Demography, 40*(2), 289–307. https://doi.org/10.2307/3180802.

Dankyi, E. (2014). *Transnational child raising arrangements: An ethnographic study of transnational caregivers in Ghana*. Ph.D. dissertation, University of Ghana, Legon.

Dito, B. B., Mazzucato, V., & Schans, D. (2016) The effects of transnational parenting on the subjective health and well-being of Ghanaian migrants in the Netherlands. *Population, Space and Place*. https://doi.org/10.1002/psp.2006.

Donato, K. M., & Gabaccia, D. R. (2016). *Gender and international migration: From the slavery era to the global age*. New York: Russell Sage Foundation.

Dreby, J. (2006). Honor and virtue—Mexican parenting in the transnational context. *Gender and Society, 20*(1), 32–59. https://doi.org/10.1177/0891243205282660.

Dreby, J. (2007). Children and power in Mexican transnational families. *Journal of Marriage and Family, 69*(4), 1050–1064. Retrieved from http://dx.doi.org/10.1111/j.1741-3737.2007.00430.x.

Dreby, J. (2010). *Divided by borders: Mexican migrants and their children*. Berkeley: University of California Press.

Dreby, J., & Adkins, T. (2010). Inequalities in transnational families. *Sociology Compass, 4*(8), 673–689. https://doi.org/10.1111/j.1751-9020.2010.00297.x.

Erel, U. (2002). Reconceptualizing motherhood: Experiences of migrant women from Turkey living in Germany. In D. Bryceson & U. Vuorela (Eds.), *The transnational family: New European frontiers and global networks*. Oxford and New York: Berg.

Feliciano, C. (2008). Gendered selectivity: U.S. Mexican immigrants and Mexican nonmigrants, 1960–2000. *Latin American Research Review, 43*(1), 139–160. Retrieved from https://muse.jhu.edu/article/232373.

Fresnoza-Flot, A. (2009). Migration status and transnational mothering: The case of Filipino migrants in France. *Global Networks—A Journal of Transnational Affairs, 9*(2), 252–270. https://doi.org/10.1111/j.1471-0374.2009.00253.x.

Gallo, E. (2006). Italy is not a good place for men: Narratives of places, marriage and masculinity among Malayali migrants. *Global Networks, 6*(4), 357–372. https://doi.org/10.1111/j.1471-0374.2006.00149.x.

Gao, Y., Ping Li, L., Kim, J. H., Congdon, N., Lau, J., & Griffiths, S. (2010). The impact of parental migration on health status and health behaviours among left behind adolescent school children in China. *BMC Public Health, 10*(56). https://doi.org/10.1186/1471-2458-10-56.

Graham, E., & Jordan, L. P. (2011). Migrant parents and the psychological well-being of left-behind children in Southeast Asia. *Journal of Marriage and Family, 73*(4), 763–787. https://doi.org/10.1111/j.1741-3737.2011.00844.x.

Haagsman, K. (2015). *Parenting across borders: Effects of transnational parenting on the lives of Angolan and Nigerian migrant parents in the Netherlands.* Ph.D. dissertation, Maastricht University, Maastricht.

Haagsman, K., & Mazzucato, V. (2014). The quality of parent-child relationships in transnational families: Angolan and Nigerian migrant parents in the Netherlands. *Journal of Ethnic and Migration Studies, 40*(11), 1677–1696. https://doi.org/10.1080/1369183X.2013.871491.

Haagsman, K., Mazzucato, V., & Dito, B. B. (2015). Transnational families and the subjective well-being of migrant parents: Angolan and Nigerian parents in the Netherlands. *Ethnic and Racial Studies, 38*(15), 2652–2671. https://doi.org/10.1080/01419870.2015.1037783.

He, B., Fan, J., Liu, N., Li, H., Wang, Y., Williams, J., & Wong, K. (2012). Depression risk of 'left-behind children' in rural China. *Psychiatry Research, 200*(2), 306–312. https://doi.org/10.1016/j.psychres.2012.04.001.

Hochschild, A. (2000). Global care chains and emotional surplus value. In W. Hutton & A. Giddens (Eds.), *On the edge: Living with global capitalism.* London: Jonathan Cape.

Hondagneu-Sotelo, P. (1999). Gender and immigration: A retrospective and introduction. In P. Hondagneu-Sotelo (Ed.), *Gender and U.S. immigration: Contemporary trends* (pp. 3–42). Berkeley: University of California Press.

Hondagneu-Sotelo, P., & Avila, E. (1997). "I'm here, but i'm there": The meanings of Latina transnational motherhood. *Gender and Society, 11*(5), 548–571.

Horton, S. (2009). A mother's heart is weighed down with stones: A phenomenological approach to the experience of transnational motherhood. *Culture, Medicine and Psychiatry, 33*(1), 21–40. https://doi.org/10.1007/s11013-008-9117-z.

Jia, Z., & Tian, W. (2010). Loneliness of left-behind children: A cross-sectional survey in a sample of rural China. *Child: Care, Health and Development, 36*(6), 812–817. https://doi.org/10.1111/j.1365-2214.2010.01110.x.

Jolly, S., & Reeves, H. (2005). *Gender and migration, an overview report.* Retrieved from Brighton https://www.bridge.ids.ac.uk/reports/CEP-Mig-OR.pdf.

Jordan, L. P., & Graham, E. (2012). Resilience and well-being among children of migrant parents in South-East Asia. *Child Development, 83*(5), 1672–1688. https://doi.org/10.1111/j.1467-8624.2012.01810.x.

Kanaiaupuni, S. M. (1999). *Reframing the migration question: An empirical analysis of men, women and gender in Mexico.* Madison, WI: University of Wisconsin-Madison, Center for Demography and Ecology.

Kilkey, M., Plomien, A., & Perrons, D. (2014). Migrant men's fathering narratives, practices and projects in national and transnational spaces: Recent Polish male migrants to London. *International Migration, 52*(1), 178–191. https://doi.org/10.1111/imig.12046.

Kofman, E. (1999). Female 'birds of passage' a decade later: Gender and immigration in the European Union. *The International Migration Review, 33*(2), 269–299. https://doi.org/10.2307/2547698.

Kraler, A. (2010). Civic stratification, gender and family migration policies in Europe. *International Centre for Migration Policy Development (ICMPD).* https://emnbelgium.be/publication/civic-stratification-gender-and-family-migration-policies-europe.

Kraler, A., & Kofman, E. (2009). *Family migration in Europe: Policies vs. reality* (Policy Brief No. 16). Amsterdam: Amsterdam University Press.

Kraus, E. K. (2019). Family formation trajectories across borders: A sequence analysis approach to Senegalese migrants in Europe. *Advances in Life Course Research, 42*, 100290. https://doi.org/10.1016/j.alcr.2019.100290.

Kulu-Glasgow, I., & Leerkes, A. (2011). Playing hard(er) to get: The state, international couples, and the income requirement. *European Journal of Migration and Law, 13*(1), 95. https://doi.org/10.1163/157181611X554267.

Lindley, A. (2009). The early-morning phonecall: Remittances from a refugee diaspora perspective. *Journal of Ethnic and Migration Studies, 35*(8), 1315–1334. https://doi.org/10.1080/13691830903123112.

Madianou, M. (2012). Migration and the accentuated ambivalence of motherhood: The role of ICTs in Filipino transnational families. *Global Networks, 12*(3), 277–295. https://doi.org/10.1111/j.1471-0374.2012.00352.x.

Madianou, M., & Miller, D. (2011). Mobile phone parenting: Reconfiguring relationships between Filipina migrant mothers and their left-behind children. *New Media and Society, 13*(3), 457–470. https://doi.org/10.1177/1461444810393903.

Madianou, M., & Miller, D. (2012). *Migration and new media: Transnational families and polymedia*. Abingdon and New York: Routledge.

Malmusi, D., Borrell, C., & Benach, J. (2010). Migration-related health inequalities: Showing the complex interactions between gender, social class and place of origin. *Social Science and Medicine, 71*(9), 1610–1619. https://doi.org/10.1016/j.socscimed.2010.07.043.

Massey, D. S., Fischer, M. J., & Capoferro, C. (2006). International migration and gender in Latin America: A comparative analysis. *International Migration, 44*(5), 63–91. https://doi.org/10.1111/j.1468-2435.2006.00387.x.

Mazzucato, V. (2015). Transnational families and the well-being of children and caregivers who stay in origin countries. *Social Science and Medicine, 132*, 208–214. https://doi.org/10.1016/j.socscimed.2014.11.030.

Mazzucato, V., & Cebotari, V. (2016). Psychological well-being of Ghanaian children in transnational families. *Population, Space and Place*.. https://doi.org/10.1002/psp.2004.

Mazzucato, V., Cebotari, V., Veale, A., White, A., Grassi, M., & Vivet, J. (2015). International parental migration and the psychological well-being of children in Ghana, Nigeria, and Angola. *Social Science and Medicine, 132*, 215–224. https://doi.org/10.1016/j.socscimed.2014.10.058.

Mazzucato, V., Dito, B. B., Grassi, M., & Vivet, J. (2016). Transnational parenting and the well-being of Angolan migrant parents in Europe. *Global Networks*. https://doi.org/10.1111/glob.12132.

Mazzucato, V., & Schans, D. (2011). Transnational families and the well-being of children: Conceptual and methodological challenges. *Journal of Marriage and Family, 73*(4), 704–712. https://doi.org/10.1111/j.1741-3737.2011.00840.x.

Menjivar, C., & Abrego, L. (2009). Parents and children across borders: Legal instability and intergenerational relations in Guatamalan and Salvadoran families. In N. Foner (Ed.), *Across generations: Immigrant families in America*. New York: New York University Press.

Moran-Taylor, M. J. (2008). When mothers and fathers migrate north: Caretakers, children, and child rearing in Guatemala. *Latin American Perspectives, 35*(4), 79–95. https://doi.org/10.1177/0094582x08318980.

Nobles, J. (2011). Parenting from abroad: Migration, nonresident father involvement, and children's education in Mexico. *Journal of Marriage and Family, 73*(4), 729–746. https://doi.org/10.1111/j.1741-3737.2011.00842.x.

Palenga-Möllenbeck, E. (2013). Care chains in Eastern and Central Europe: Male and female domestic work at the intersections of gender, class, and ethnicity. *Journal of Immigrant and Refugee Studies, 11*(4), 364–383. https://doi.org/10.1080/15562948.2013.822955.

Parreñas, R. S. (2000). Migrant Filipina domestic workers and the international division of reproductive labor. *Gender and Society, 14*(4), 560–580.

Parreñas, R. S. (2001a). Mothering from a distance: Emotions, gender, and intergenerational relations in Filipino transnational families. *Feminist Studies, 27*(2), 361–390. https://doi.org/10.2307/3178765.

Parreñas, R. S. (2001b). *Servants of globalization and domestic work*. Stanford: Stanford University Press.

Parreñas, R. S. (2005a). *Children of global migration: Transnational families and gendered woes*. Stanford, CA: Stanford University Press.

Parreñas, R. S. (2005b). Long distance intimacy: Class, gender and intergenerational relations between mothers and children in Filipino transnational families. *Global Networks—A Journal of Transnational Affairs, 5*(4), 317–336. https://doi.org/10.1111/j.1471-0374.2005.00122.x.

Parreñas, R. S. (2008). Transnational fathering: Gendered conflicts, distant disciplining and emotional gaps. *Journal of Ethnic and Migration Studies, 34*(7), 1057–1072. https://doi.org/10.1080/13691830802230356.

Petrozziello, A. J. (2011). Feminised financial flows: How gender affects remittances in Honduran–US transnational families. *Gender and Development, 19*(1), 53–67. https://doi.org/10.1080/13552074.2011.554022.

Piper, N. (2008). *New perspectives on gender and migration: Livelihood, rights and entitlements*. London: Routledge.

Poeze, M. (2018). *Migration, family separation and caregiving across borders: A multi-sited ethnography of transnational child-raising arrangements between Ghana and the Netherlands.* Ph.D. dissertation, Maastricht University Maastricht.

Poeze, M. (2019). Beyond breadwinning: Ghanaian transnational fathering in the Netherlands. *Journal of Ethnic and Migration Studies, 45*(16), 3065–3084. https://doi.org/10.1080/1369183X.2018.1547019.

Poeze, M., & Mazzucato, V. (2014). Ghanaian children in transnational families: Understanding the experiences of left-behind children through local parenting norms. In L. Baldassar & L. Merla (Eds.), *Transnational families, migration and the circulation of care* (pp. 149–169). New York: Routledge.

Poeze, M., & Mazzucato, V. (2016). Transnational mothers and the law: Ghanaian women's pathways to family reunion and consequences for family life. In M. Kilkey & E. Palenga-Möllenbeck (Eds.), *Family life in an age of migration and mobility: Global perspectives through the Life course* (pp. 187–211). London: Palgrave Macmillan UK.

Pribilsky, J. (2004). 'Aprendemos a convivir': Conjugal relations, co-parenting, and family life among Ecuadorian transnational migrants in New York City and the Ecuadorian Andes. *Global Networks—A Journal of Transnational Affairs, 4*(3), 313–334. https://doi.org/10.1111/j.1471-0374.2004.00096.x.

Pribilsky, J. (2012). Consumption dilemmas: Tracking masculinity, money and transnational fatherhood between the Ecuadorian Andes and New York City. *Journal of Ethnic and Migration Studies, 38*(2), 323–343. https://doi.org/10.1080/1369183x.2012.646429.

Schmalzbauer, L. (2004). Searching for wages and mothering from afar: The case of Honduran transnational families. *Journal of Marriage and Family, 66*(5), 1317–1331.

Schmalzbauer, L. (2005). *Striving and surviving: A daily life analysis of Honduran transnational families*. New York: Roudlegde.

Schmalzbauer, L. (2008). Family divided: The class formation of Honduran transnational families. *Global Networks, 8*(3), 329–346. https://doi.org/10.1111/j.1471-0374.2008.00198.x.

Schmalzbauer, L. (2015). Temporary and transnational: Gender and emotion in the lives of Mexican guest worker fathers. *Ethnic and Racial Studies, 38*(2), 211–226. https://doi.org/10.1080/01419870.2013.857033.

Smith, R. C. (2006). *Mexican New York: Transnational lives of new immigrants*. Berkeley, CA: University of California Press.

Strik, T., De Hart, B., & Nissen, E. (2013). *Family reunification: A barrier or facilitator of integration? A comparative study*. HW Oisterwijk: Wolf Legal.

Suarez-Orozco, C., Todorova, I. L. G., & Louie, J. (2002). Making up for lost time: The experience of separation and reunification among immigrant families. *Family Process, 41*(4), 625–643. https://doi.org/10.1111/j.1545-5300.2002.00625.x.

Tizard, B. (1991). Employed mothers and the care of young children. In A. Phoenix, A. Woollet, & E. Lloyd (Eds.), *Motherhood: Meaning, practices and ideologies* (pp. 178–194). London: Sage.

Toma, S., & Vause, S. (2014). Gender differences in the role of migrant networks: Comparing Congolese and Senegalese migration flows. *International Migration Review, 48*(4), 972–997. https://doi.org/10.1111/imre.12150.

United Nations Population Division. (2019). *International migration stock 2019*. Retrieved from www.unpopulation.org.

Vanore, M., Mazzucato, V., & Siegel, M. (2015). 'Left behind' but not left alone: Parental migration and the psychosocial health of children in Moldova. *Social Science and Medicine, 132*, 252–260. https://doi.org/10.1016/j.socscimed.2014.08.040.

van Walsum, S. (2009). Against all odds: How single and divorced migrant mothers were eventually able to claim their right to respect for family life. *European Journal of Migration and Law, 11*, 295–312.

Waters, J. L. (2009). Becoming a father, missing a wife: Chinese transnational families and the male experience of lone parenting in Canada. *Population, Space and Place, 16*(1), 63–74. https://doi.org/10.1002/psp.578.

Wen, M., & Lin, D. (2012). Child development in rural China: Children left behind by their migrant parents and children of nonmigrant families. *Child Development, 83*(1), 120–136. https://doi.org/10.1111/j.1467-8624.2011.01698.x.

Winters, N. (2014). Responsibility, mobility, and power: Translocal carework negotiations of Nicaraguan families. *International Migration Review, 48*(2), 415–441. https://doi.org/10.1111/imre.12062.

Wong, M. (2006). The gendered politics of remittances in Ghanaian transnational families. *Economic Geography, 82*(4), 355–381. https://doi.org/10.1111/j.1944-8287.2006.tb00321.x.

Worby, P. A., & Organista, K. C. (2007). Alcohol use and problem drinking among Male Mexican and Central American im/migrant laborers: A review of the literature. *Hispanic Journal of Behavioral Sciences, 29*(4), 413–455. https://doi.org/10.1177/0739986307308765.

Zentgraf, K. M., & Chinchilla, N. S. (2012). Transnational family separation: A framework for analysis. *Journal of Ethnic and Migration Studies, 38*(2), 345–366. https://doi.org/10.1080/1369183x.2011.646431.

Karlijn Haagsman is Assistant Professor of Globalisation and Development at Maastricht University, The Netherlands. Her research looks at transnationalism between the Global North and South, with a particular focus on transnational and migrant families, youth mobility and well-being. In her Ph.D. she studied migrant parents geographically separated from their children and investigated how separation affected their overall well-being (parent-child relationship, subjective well-being and job performance) and the effect of structural factors herein.

Valentina Mazzucato is Professor of Globalisation and Development at Maastricht University, The Netherlands. She has led five international and interdisciplinary projects studying the effects of transnational migration between Africa and Europe focusing on networks, families and youth and on the wellbeing of those who migrate and those who remain in the country of origin. She currently leads an ERC Consolidator Grant-funded project on Mobility Trajectories of Young Lives (MO-TRAYL) focusing on the wellbeing and educational pathways of youth who move between Africa and Europe.

CHAPTER 9

German Migrants in Pattaya, Thailand: Gendered Mobilities and the Blurring Boundaries Between Sex Tourism, Marriage Migration, and Lifestyle Migration

Kwanchanok Jaisuekun and Sirijit Sunanta

Abstract Lifestyle migration refers to the relocation of relatively privileged individuals in a pursuit of a better, more fulfilling life. Initially conceptualized to study migration within Europe and the Americas, the concept has later been adopted to study migration from the Global North to the Global South including destinations in Asia and Africa. Gender has been understudied in the literature of lifestyle migration although what constitutes a good life is highly gendered. Studying the migration of predominantly older German men to Pattaya, Thailand, we found that the intimate relationship with Thai women—ranging from casual encounters with bargirls, to boyfriend–girlfriend relationship and to marital partnership, is integral to the migration decision. Embedded in a history of gendered mobilities between Thailand and Germany, the settlement of German men in Pattaya reveals the blurry boundaries between sex tourism, marriage migration, and lifestyle migration. Bringing in the feminist concept of global care chain shifts the emphasis of lifestyle migration from individual migrants to global inequality that underpins care mobility and the global intimate.

K. Jaisuekun · S. Sunanta (✉)
Research Institute for Languages and Cultures of Asia, Mahidol University, Nakhon Pathom, Thailand
e-mail: Sirijit.sun@mahidol.edu

C. Mora and N. Piper (eds.), *The Palgrave Handbook of Gender and Migration*, https://doi.org/10.1007/978-3-030-63347-9_9

Introduction

Migration of citizens from wealthier countries to less affluent ones has recently been conceptualized as "lifestyle migration" characterized by migrants' motivation to move in search of a better way of life. This migration trend is marked by the notion of privilege and agency of individual migrants who choose to move in pursuit of a more fulfilling way of life rather than driven by economic necessity (O'Reilly & Benson, 2009: 2; Torkington, 2014: 102). The concept encompasses retirement migration, leisure migration, counter urbanization, second home ownership, amenity-seeking and seasonal migration (O'Reilly & Benson, 2009: 6). As lifestyle migration research becomes more mature, scholars have pointed out that some migrants decide to move to places where living costs are lower in order to maintain a decent lifestyle not possible at home with their dwindling pensions (Green, 2014; Hayes, 2014; Husa, Vielhaber, Jöstl, Veress, & Wieser, 2014). Initially conceptualized by scholars who studied migration of the British to sunny coastal areas and rural places within Europe (Benson & O'Reilly, 2009), lifestyle migration is now used to analyse migration flows to and from more geographical regions including from the US to Central America (Bastos, 2014; Croucher, 2014; Hayes, 2014, 2015; Zunino, Dattwyler, & Zebryte, 2014) and from Europe and America to Africa (Escher & Petermann, 2014; Therrien, 2014; Van Laar, Cottyn, Donaldson, Zoomer, & Ferreira, 2014) and Southeast Asia (Benson & O'Reilly, 2018; Botterill, 2017; Green, 2014, 2015). Studies of lifestyle migration have not paid much attention to the role of gender although gender exerts a profound and pervasive influence on lifestyle and frequently acts as a signifier for "lifestyle" and vice versa (Croucher, 2014: 21).

Pattaya, a Thai beach resort and world-famous sex tourism destination, has recently become a second home for a growing number of Westerners. German citizens are among the largest groups of Western residents in Thailand (Howard, 2009; Koch-Schulte, 2008) and German residents of Pattaya are relatively visible in this city. An estimate of 7000–8000 German nationals live in the city as a part of the German-speaking community that includes Austrian and Swiss nationals (Kubelka, 2013). Studies of Western (European, Australian, and North American) migrants in Thailand are relatively few but the gendered nature of this migration flow has been noted from the beginning. An online survey by Howard (2008, 2009) reveals that the majority of Western residents in Thailand are male, half of whom are in relationships with Thai women. Koch-Schulte (2008) found that Western migrants in Udon Thani Province in Northeast Thailand (*Isan*) are predominantly older men in marital relationships with much younger local women. In a study of Western men who are in intimate relationships with Thai women in Ubon Ratchathani, also in the Isan Region, Maher and Lafferty (2014) explore the inflated sense of white masculinity and the shift of power to Thai female partners overtime. Thompson, Kitiarsa, and Smutkupt (2016) propose the concept of transient masculinity to analyse the shifting subjectivity of Western male migrants in

Thailand from sex tourist to son-in-law. All the studies point out that the migration of Westerners to Thailand is a predominantly male phenomenon and is closely linked to cross-cultural relationships and marriage migration. Although German migrants represent a large proportion of Western migrants in Thailand, they have been underrepresented in earlier studies due partly to language barriers in research (Howard, 2008). Most studies on Western men who are in relationship with Thai women were carried out in Isan provinces, hometown of the majority of the Thai female partners. In this study, German migrants' decision to settle down in Pattaya, a cosmopolitan tourist destination, further reveals the blurry boundaries between marriage migration and lifestyle migration. The following sections will outline the intersection of gender with lifestyle migration and the linkages between sex tourism, marriage migration, and lifestyle/retirement migration.

Theoretical Framework: Gender and Lifestyle Migration

In this study, we argue that German migration to Pattaya is part of the "lifestyle migration", a relatively new concept in migration studies referring to the "spatial mobility of relatively affluent individuals of all ages, moving either part-time or full time to places that are meaningful because, for various reasons, they offer the potential of a better quality of life" (O'Reilly & Benson, 2009: 2). Work-life balance, warm climate, lower cost of living, and a sense of community are often cited as constituting a better way of life while the decision to move to a chosen destination is often made at vital points in life such as serious illness, a birth or death in the family, a marriage or divorce, and job loss or other employment changes (Benson & O'Reilly, 2009; Hoey, 2009). According to the literature on lifestyle migration, the search for a good life was considered a comparative project between life before and after migration; life before migration is presented negatively while the advantages of life in the destination are often overstated (Benson & O'Reilly, 2009). Practices previously studied as retirement migration, leisure migration, counter urbanization, second home ownership, amenity-seeking and seasonal migration, have been incorporated into a broader concept and study of lifestyle migration (Benson & O'Reilly 2009: 609).

Studies of lifestyle migration tend to neglect the role of gender, kinship, and interpersonal relationship in migration decisions and experiences (Croucher, 2014; Green, 2015). What is not normally recognized is that lifestyle migration is a gendered process and concepts of a good life, home, belonging, community, and ethnicity are highly gendered. Gender, in fact, exerts a profound and pervasive influence on lifestyle and frequently acts as a signifier for "lifestyle" and vice versa (Croucher, 2014: 21). Gender is not only about differences between men and women, but a constitutive element of migration (Hondagneu-Sotelo, 2000). It is an important factor that distinguishes the difference between male and female migrants' lifestyles and experience in a receiving country as well as migration motivations (Croucher, 2014). Pessar and Mahler (2003: 818) introduced gendered geographies of power: "a framework for analysing people's gendered social agency – corporal and cognitive – given their own initiative as well as their positioning within multiple hierarchies of power operative within and across many terrains". In other words, this concept helps in understanding how gender operates or is negotiated across transnational space and different scales and how it intersects with other axes of power. The concept highlights whether mobility across national borders reinforces prevailing gender norms, or, on the other hand, whether transnational migration provides women and men opportunities to challenge hegemonic gender ideologies (Croucher, 2014; Pessar & Mahler, 2003).

Gender informs migration processes and experiences of male and female transnational subjects (Parrenas, 2009). Croucher (2014) posited that the desire to escape from and negotiate with gender norms and expectations in their home country motivate many female lifestyle migrants who seek to enjoy themselves travelling instead of keeping the household. Pruitt and LaFont (1995) explored the negotiation of gender norms by Western female tourists who engaged in emotional and intimate relationships with local men in Jamaica. Away from their usual social circles, the women behaved the way they would never do at home, namely pursuing relationships with men during their travels.

Although existing scholarly works on gender and lifestyle migration focus mainly on female migrants, male migrants' experiences are also gendered. While female lifestyle migrants move to escape from traditional gendered norms, male lifestyle migrants seek to realize traditional masculine roles in their new home (Wilde-Menozzi, 2003). Western men in Thailand attained a sense of empowerment and the status of "real white man" through their role as provider and master of the house, something they had wished for but could not realize in their home country. Although not conventionally attractive by Western standards, Western men in Thailand found that they have an advantage in finding romance and receive full attention from local Thais (Maher & Lafferty, 2014).

Most studies of lifestyle migration portray elderly couples or young families moving together (Casado-Diaz, 2012), with a proportional male to female ratio of migrants. Western migration to Thailand stands out as predominantly male migration. The vast majority of Western residents in Thailand are elderly males from Europe, Australia, and the US who arrived alone without wives or children (Howard, 2008, 2009; Husa et al., 2014; Maher & Lafferty, 2014). The possibility of forming romantic relationships with Thai women is one of the key attractions to Thailand (Howard, 2008, 2009; Husa et al., 2014; Maher & Lafferty, 2014), and many Westerners view Thailand as a more suitable destination for men rather than women (Husa et al., 2014). The history of international sex tourism in Thailand has evidently shaped the current trend of Western migration to the Kingdom.

Between Lifestyle Migration, Sex Tourism, Marriage Migration, and International Retirement Migration

Focusing on individuals' search for a better way of life through leisure and amenity consumption, the lifestyle migration concept tends to overlook the importance of interpersonal relationships in migration decision (Green, 2015). In this chapter, we look at lifestyle migration through the feminist lens of the global intimate—the notion that global interconnectedness is forged by personal relationship and physical and emotional closeness, not just by the abstraction of economic and cultural globalization (Mountz & Hyndman, 2006). The settlement of older German men in Pattaya links lifestyle migration to other categorizations of transnational mobilities—sex tourism, marriage migration, and retirement migration. In the following we will outline different emphases of these mobility categorizations and how they intersect in transnational relocation of German men to Pattaya.

Since the 2000s, the rising number of interracial marriages between Thai women and Western men has attracted popular and academic attention. In 2004, a survey by the National Economic and Social Development Board revealed that about 15,000 Thai women in the Northeast Region (*Isan*) were married to or in partnership with foreign men, mostly European and North Americans. Germans topped the list of foreign husbands married to Thai women (Seepai & Senerat, 2010). In Europe, the feminization of Thai migration has been noted but it is most prominent in Germany where 85% of 57,000 Thai migrants are women and about 60% of these women are married to German husbands (Ruenkaew, 2014). Thailand has become the second most important country of origin of foreign wives of German men from economically weaker countries after Poland (Glowsky, 2007).

Without a shared border and cultural proximity between Thailand and Germany, the high number of Thai wives in Germany is perplexing. The explanation lies in another form of transnational mobilities between the two countries—sex tourism. During the boom of Thai sex tourism in the 1970s and 1980s, Thai women from rural backgrounds, particularly the most economically marginalized region of *Isan*, migrated for work in sex and other service businesses in tourist destinations. Once a Rest and Recreation destination for American military men during the Vietnam War, Pattaya has been a centre of Thai sex tourism since the 1970s. German men ranked first among "sex tourists" from Western countries (Chantavanich, Nittayananta, Rattanaolan-Mix, Ruenkaew, & Khemkrut, 1999) and the area later known as "Little Germany" took roots in Pattaya since around 1975 with a high concentration of German residents and the availability of shops, restaurants, bookstores, and other facilities catering to German-speaking customers (Boonsiri, 2007). The encounter between Thai sex workers and German sex tourists in Thailand was at least partly responsible for the increasing number of Thai woman migrants in Germany. Since 1975, an increasing number of Thai women travelled to Germany for sex work or as wives of German men (Mix & Piper, 2003).

Feminist scholars have observed the connection between sex work and marriage and the blurry boundaries between sex for economic gain and for emotional fulfilment (see also Palmary, in this volume). Mix and Piper (2003) note the back and forth transition between sex work and marriage among Thai migrant women in Germany as sex workers marry their clients while some wives are involved in sex work during or after marriages. In a study of a bride sending village in *Isan*, Sunanta and Angeles (2013) find that some women from the village intentionally enter sex work in tourism destinations in order to meet a prospective foreign husband, a move that they hope will improve their life chances. In Thai sex tourism, Cohen (1996) describes relationships between Thai bargirls and *farang* (white, Western) men as "open-ended prostitution". He observes that these relationships are ambiguous as they often extend beyond short encounters for economic exchange and become medium and long-term relationship as bargirls have regular clients, accompany the men for trips around Thailand, keep in touch after the men return home and in some cases the relationship leads to a marriage (Cohen, 1982).

Cohen attributes Thai bargirl and *farang* men liaisons to something unique and different from prostitution in the West, which he sees as a downright economic exchange. However, feminist scholars have since the 1980s problematized the dichotomy between economic exchange and emotional closeness in sex and other service work involving emotional and physical intimacy. The conceptualization of emotional labour (Hochschild, 2012), bodily labour (Kang, 2010), bodywork (McDowell, 2009), and intimate labour (Boris & Parrenas, 2010) focuses on emotional and physical demand in work performed by domestic workers, caregivers, interactive service workers including sex workers, whose work involves emotional connection and/or physical closeness to another human being. Traditionally carried out at home without pay, commoditized bodywork, and intimate labour are stigmatized and poorly paid. In the global care chain concept, scholars emphasize the tendency for women from low-income countries to take up paid care work in wealthier places, leaving care needs in their own family to be fulfilled by other women in their home country who have even fewer choices. The global care chain concept has predominantly been used to analyse migrant domestic workers especially Filipina domestic workers in Europe and North America (e.g. Ehrenreich & Hochschild, 2004; Parrenas, 2005).

Like the global care chain scholars, Toyota and Xiang (2012) pay attention to the transnational dimension of care in their study of the transnational retirement industry and old age mobility from Japan to Southeast Asia. They also delineate the ways in which national contexts in the sending and receiving states shape elderly care transnationalization. By doing so, they add a macro-scale analysis to the literature on international retirement migration and lifestyle migration, shifting away from the focus on migrants' behaviours and migration motivations. The focus on "old age" in international retirement migration, however, has downplayed other important parts of migrants' life, such as sexuality, intimate, and marital relationships. Certain aspects of sex tourism, marriage migration, and retirement migration are present in the settlement of German migrants in Pattaya and our findings reveal that this transnational flow cannot be reduced to any one category.

German Migrants in Pattaya

The data for this article were drawn from six months of ethnographic study in Pattaya. The first author of this paper interviewed German citizens who met the following criteria: male or female of German nationality, aged 18 and above, had been living in Pattaya for at least three months and was neither an employee of an international corporation nor a diplomat. Twelve in-depth interviews were conducted in German. The interviews lasted about 45 min to 90 min and were tape-recorded. To protect research respondents' confidentiality, all names in this study are pseudonyms. Out of 12 interview respondents, 11 were male. The respondents' ages ranged from 52 to 84 years old. The pool of interview respondents reflects the overrepresentation of men of retirement age among German residents of Pattaya. The participants' length of stay in Thailand ranged from three months to 20 years. Most of the participants described themselves as being middle-class and almost all lived on a pension that ranged from 1 000 Euros to 4,0000 Euros a month. Their previous employment included business owner, government officer, engineer, skilled worker, and others. All male participants had established relationships with Thai women in varying forms: married to a Thai woman, living with a Thai girlfriend, and regularly visiting bargirls. The marital status of the male participants prior to migration to Thailand was: eight of them were divorced, two widowed, and the other was single.

Like British Migrants in Portugal and American migrants in Mexico, German migrants in Pattaya enjoy the warm climate, the beach, and lower cost of living. Sebastian, 84, said he is healthier living in Pattaya, "The weather here makes me healthy. See, it's sunny. Compared to Germany or Sweden where it is cold for seven to eight months a year, I feel better here". Retired and living on a German pension that extends further in Thailand, older German migrants lead a relaxing life in contrast to the life of hard work in Germany. They enjoy leisure activities such as surfing the internet, gardening, walking along the beach, working out, reading, shopping, drinking beer, or simply hanging out. Martin, 76, noted, "I have a better life in Thailand, I am happier than when I was in Germany because I have more freedom which is good for my life".

What is distinctive of German migrants in Pattaya from lifestyle migration in other beach resorts is that all the German men interviewed for this study are in some forms of relationships with Thai women ranging from casual relationships with bargirls to long-term and marital relationships. Narratives of German migrants in Pattaya reveal the blurry boundaries between lifestyle migration, sex tourism, and marriage migration. Some of German migrants in Pattaya today did start the journey as sex tourists. Robert (71), for example, bought a package tour to Thailand for the first time in 1980 and kept visiting

Pattaya every two to three years to visit bargirls. After a few visits, he learned to travel to Pattaya on a self-managed tour and to live in Pattaya at a reasonable price. Nightlife with bargirls in Pattaya was in contrast to the lonely life he had in Germany after the divorce from his German wife. For many years, Robert worked hard in Germany to save for the next trip to Pattaya. He was determined to move to Pattaya permanently once he had saved enough money. He finally moved to Pattaya in 1998 and married a Thai woman seven years later. Enjoying sexual freedom and escaping loneliness in Germany constitute a better life in Thailand for Robert, who eventually gave up transient relationships with bargirls for a long-term relationship and marriage.

For those who refer to themselves as "sex tourists", Pattaya is a place they choose to live. Frank (66) accepted that he is a sex tourist. He first visited Pattaya in 1999 and continued to spend his holidays in the beach city once or twice a year for several years. He could not deny that Thailand is a very beautiful country but the main attractions that brought him back were the women and the bars. In 2011, he bought a condo in Pattaya and started to live in two countries. He now allocates his time between Pattaya and Germany. Every time he is in Pattaya, he visits the bars almost every night.

An advanced age is not an obstacle for German men in forming romantic relationships with younger Thai women. The economic and cultural power associated with Whiteness gives White men a dating advantage in Thailand. As Ruenkaew (1998) pointed out, when German men are in Thailand their physical appearance is not so important; they are treated differently and nicely especially by bargirls. Maher and Lafferty (2014: 435) similarly found that Western men who were marginalized in their home country due to "class, age, body type, or physical appearance" received attention and social respect in Thailand; the men asserted this privilege to access erotic and romantic relationships with Thai women and converting these opportunities into a resource for a better way of living. German men in our study enjoy the freedom to do what is not normally acceptable in German society: drinking all day, visiting bargirls, and dating a young Thai woman. Kevin (70) whose girlfriend is in her twenties stated that he would never be able to find another partner if he still lived in Germany "The advantage of living here is that I am 70 years old now but I can still have a girlfriend. You can't do that in Germany".

Another group of German male residents in Pattaya is men whose Thai wives returned to Thailand after many years of living together in Germany. Phillip, 58, met his current Thai wife while she was visiting her cousin in Germany in 1999–2000. He flew back and forth between Germany and Thailand as the relationship blossomed. In 2003, they finally married and settled down in Germany. Philip knew his wife's life in Germany was not easy. She deeply missed her home country and her family. He would have loved to spend his twilight years in either Germany or the US if not for his Thai wife. After a year in Thailand, he realized that he does not belong to the country. He tries to adjust to life in Thailand for his wife who patiently lived in Germany for him for more than ten years. Although he wants to go back to Germany, he will let his wife decide their future. "Will we move back or will we stay here? I don't know. It depends on my wife. I begged her to live in Germany for 14 years. Now it is her turn". In cases like Philip's, the move to Pattaya is a result of German-Thai couples' negotiation over a place of residence in their advanced age, a kind of retirement migration. German husbands were typically in their late fifties or early sixties when the couples decided to leave Germany. Released from responsibilities in their work life, moving to Thailand is the start of a good life project for some retired German-Thai couples.

Choosing to live in Pattaya instead of in an *Isan* village near their Thai partners' extended family is a strategy German men use to negotiate their position between being an *Isan* son-in-law and a lifestyle migrant with access to expat life. While Thai wives' families in rural Northeast Thailand view *farang* son-in-law as a part of the family and expect them to uphold certain commitments and obligations (Angeles & Sunanta, 2009), German men do not necessarily see things the same way. German migrants in this study, though not all, complain about repeated financial demands from their wives' family members. Living in Pattaya away from the in-laws helps mitigate the tension. After a money feud with his Thai in-laws, Phillip decided to rent a house about 15-minute drive from Pattaya, although his wife is now living in her hometown in *Isan*. Andreas and his girlfriend agreed to live in Pattaya in order to avoid such problems. They had lived together in Germany for eight years before moving to Thailand in 2008. They plan to have a peaceful retirement in Pattaya and travel to visit her family in *Isan* from time to time. German husbands we interviewed are attracted to Pattaya because of the availability of tourist facilities, hospitals, and the German-speaking community (see also Jaisuekun & Sunanta (2016)). A tourist destination and cosmopolitan city, the setting in Pattaya allows German men to navigate their day-to-day life conveniently and more independently compared to German husbands in the village who have to rely on their wives for day-to-day arrangements without a social life of their own.

Conclusion

Gender and sexuality are integral to German migrants in Thailand. The majority of German migrants in Pattaya are male over the age of 55 and in intimate/marital relationships with Thai women. The opportunities to meet and settle down with Thai women, most of the time much younger than themselves, fulfil older male migrants' quest for desired masculinity and sexual relations not available to them in their home country. Like Maher and Lafferty (2014), we found that economic and cultural power associated with being White Westerners in Thailand facilitates the attainment of a good life for German migrants in Pattaya: the fulfilment of sex and romantic life as well as the possibility to form and maintain family life. Our findings are consistent with Croucher's (2014) argument that gender plays a crucial role in lifestyle migration and that the concept of a good life is highly gendered. While female lifestyle migrants are motivated by the desire to break away from traditional gender roles and expectations (Croucher, 2014), German male migrants move to Pattaya in order to reaffirm the sense of traditional masculinity. German men's relationships with Thai women range from casual relationships with bargirls to long term and marital partnerships. Moving to Thailand allows German men to enjoy a good lifestyle with less money and to support their Thai partners and children like a real man.

Against the sex tourist stereotype, the move to Pattaya is sometimes a result of marital negotiation between long-term cross-cultural marriage partners. A number of German men intentionally choose to settle down in Pattaya, instead of relocating to their Thai partners' communities in the Northeast region, to enjoy independent expat life. The relocation of older German men to Pattaya, Thailand, is embedded in a history of gendered mobility between Thailand and Germany. International sex tourism plays an important role in the high proportion of interracial marriages between Thai women and Western men leading to the feminization of Thai migration in Europe (Sunanta, 2020; Statham, Scuzzarello, Sunanta, & Trupp, 2020). While the Thai migration to Germany is dominated by female migration, German migrants in Thailand are predominantly male. Some Thai wives who moved to Germany in their working age decide to return to Thailand for retirement. For Thai-German couples in our study, Germany is a place for work while Thailand is the place they retire to. The "both-way" mobilities between Thailand and Germany can be explained by the feminist concept of the global care chain in which care is outsourced from wealthier countries to lower-income countries (See also Sunanta (2020) and Statham et al. (2020)). The migration of German men to Pattaya reveals the fluidity of transnational mobility categories as sex tourism, marriage migration, and lifestyle/retirement migration are closely intertwined.

References

Angeles, L., & Sunanta, S. (2009). Demanding daughter duty: Gender, community, village transformation and transnational marriages in Northeast Thailand. *Critical Asian Studies, 41*(4), 549–574.

Bastos, S. (2014). Territorial dispossession and indigenous rearticulation in the Chapala Lakeshore. In Janoschka & Haas (Eds.). *Contested spatialities, lifestyle migration and residential tourism* (pp. 47–59). New York: Routledge.

Benson, M., & O'Reilly, K. (2009). 'Migration and the search for a better Way of life: A critical exploration of lifestyle migration. *The Sociological Review, 57*(4), 608–625.

Benson, M., & O'Reilly, K. (Eds.). (2018). *Lifestyle migration and colonial traces in Malaysia and Panama*. UK: Palgrave Macmillan.

Boonsiri, S. (2007, June). Where do you stay in Pattaya? Manager Magazine. Retrieved from http://info.gotomanager.com/news/printnews.aspx?id=59637.

Boris, E., & Parrenas, R. (Eds.). (2010). *Intimate labors: Cultures, technologies, and the politics of care*. Stanford: Stanford University Press.

Botterill, K. (2017). Discordant lifestyle mobilities in East Asia: Privilege and precarity of British retirement in Thailand. *Population, Space, Place, 23*(5), 1–11. https://doi.org/10.1002/psp.2011.

Casado-Diaz, M. (2012). Exploring the geographies of lifestyle mobility: Current and future fields of enquiry. In J. Wilson (Ed.), *The Routledge Handbook of Tourism Geographies* (pp. 120–125). London: Routledge.

Chantavanich, S., Nittayananta, S., Rattanaolan- Mix, P., Ruenkaew, P., & Khemkrut, A. (1999). *The migration of Thai women to Germany: Causes, living conditions and impact for Thailand and Germany*. Bangkok: Chulalongkorn University.

Cohen, E. (1982). Thai girls and farang men: The edge of ambiguity. *Annals of Tourism Research, 9,* 403–482.

Cohen, E. (1996). *Thai tourism: Hilltribes, islands and open-ended prostitution.* Bangkok: White Lotus Press.

Croucher, S. (2014). The gendered spatialities of lifestyle migration. In M. Janoschka & H. Haas (Eds.), *Contested spatialities, lifestyle migration and residential tourism* (pp. 15–28). New York: Routledge.

Ehrenreich, B., & Hochschild, A. R. (2004). *Global woman: Nannies, maids and sex workers in the New Economy*. New York: Henry Holt.

Escher, A., & Petermann, S. (2014). Marrakesh Medina: Neocolonial paradise of lifestyle migrants? In M. Janoschka & H. Haas (Eds), *Contested spatialities, lifestyle migration and residential tourism* (pp. 29–46). New York: Routledge.

Glowsky, D. (2007). Staatsbürgerschaft als Ressource bei der Heirat ausländischer Frauen. Eine Analyse mit Daten des Sozio-oekonomischen Panel. *Zeitschrift für Soziologie* 36(4), 282–301.

Green, P. (2014). Contested realities and economic circumstances: British later-life migrants in Malaysia. In M. Janoschka & H. Haas (Eds.), *Contested spatialities, lifestyle migration and residential tourism* (pp. 145–157). New York: Routledge.

Green, P. (2015). Mobility, subjectivity and interpersonal relationships: Older, Western migrants and retirees in Malaysia and Indonesia. *Asian Anthropology, 14*(2), 150–165.

Hayes, M. (2014). 'We gained a lot over what we would have had': The Geographic arbitrage of north American lifestyle migrants to Cuenca, Ecuador. *Journal of Ethnic and Migration Studies, 40*(12), 1953–1971. https://doi.org/10.1080/1369183X.2014.880335.

Hayes, M. (2015). 'It is hard being the different one all the time': Gringos and racialized in lifestyle migration to Ecuador. *Ethnic and Racial Studies, 38*(6), 943–958.

Hochschild, A. R. (2012). *The managed heart: Commercialization of human feeling.* Berkeley, Los Angeles and London: University of California Press.

Hoey, B. A. (2009). Pursuing the good life: American narratives of travel and a search for refugee. In M. Benson & K. O'Reilly (Eds.), *Lifestyle migration. expectations, aspirations and experience* (pp. 31–50). Farnham: Ashgate.

Hondagneu-Sotelo, P. (2000). Feminism and migration. *The Annuals of the American Academy of Political and Social Science, 571,* 107–120.

Howard, R. W. (2008). Western retirees in Thailand: Motives, experiences, wellbeing, assimilation and future needs. *Ageing & Society, 28*(2), 145–163.

Howard, R. W. (2009). The migration of Westerners to Thailand: An unusual flow from developed to developing World. *International Migration, 47*(2), 193–225.

Husa, K., Vielhaber, C., Jöstl, J., Veress, K., & Wieser, B. (2014). Searching for paradise? International retirement Migration to Thailand—A case study of Hua hin and Cha-am. In K. Husa, A. Trupp, & H. Wohlschlägl (Eds.), *Southeast Asian mobility transitions: Issues and trends in migration and tourism* (pp. 137–167). Vienna: Copydruck KG.

Jaisuekun, K., & Sunanta, S. (2016). Lifestyle migration in Thailand: A case study of German migrants in Pattaya". *Thammasat Review, 19*(2), 89–103.

Kang, M. (2010). *The managed hand: Race, gender and the body in beauty service work*. Berkeley: University of California Press.

Koch-Schulte, J. (2008). *Planning for international retirement migration and expats: A case study of Udon Thani, Thailand*. Winnipeg: University of Manitoba.

Kubelka, T. (2013). Pattaya: vom Fischerdorf zur Redidenzstadt westlicher Auswanderer? Eine Fallstudie zue Internationalen Altersmigration. (Master thesis). Universität Wien, Vienna, Austria.

Maher, K. H., & Lafferty, M. (2014). White migrant masculinities in Thailand and the paradoxes of Western privilege. *Social and Cultural Geography, 15*(4), 427–448. https://doi.org/10.1080/14649365.2014.893703.

McDowell, L. (2009). *Working bodies: Interactive service employment and workplace identities*. West Sussex: Wiley-Blackwell.

Mix, P., & Piper, N. (2003). Does marriage "liberate" women from sex work?—Thai Women in Germany. In N. Piper & M. Roces (Eds.), *Wife or Workers?: Asian Women and Migration* (pp. 53–71). Lanham, Boulder, New York and Oxford: Rowman & Littlefield.

Mountz, A., & Hyndman, J. (2006). Feminist approaches to the global intimate. *Women's Studies Quarterly, 34*(1/2), 446–463.

O'Reilly, K., & Benson, M. (2009). Lifestyle migration: Escaping to the good life? In M. Benson & K. O'Reilly (Eds.), *Lifestyle migration: Expectations, aspirations and experience* (pp. 1–14). Farnham: Ashgate.

Parrenas, R. S. (2005). *Children of global migration: Transnational families and gendered woes*. Stanford: Stanford University Press.

Parrenas, R. S. (2009). Inserting feminism in transnational migration studies. Retrieved from http://lastradainternational.org/lsidocs/RParrenas_InsertingFeminismInTransnationalMigrationStudies.pdf.

Pessar, P., & Mahler, S. (2003). Transnational migration: Bringing gender in. *International Migration Review, 37*(3), 812–846.

Pruitt, D., & LaFont, S. (1995). For love and money—Romance tourism in Jamaica. *Annals of Tourism Research, 22*(2), 422–440.

Ruenkaew, P. (1998). Female Thai migrants in Japan. Retrieved from www.igs.ocha.ac.jp/igs/IGS_publication/journal/5/journal05047.pdf.

Ruenkaew, P. (Ed.). (2014). *Ying Thai lae Chumchon Thai Wannee nai Germany (Thai Women and Thai Community in Germany Today)*. Bielefeld: Network Thais Overseas.

Seepai, S., & Senerat, S. (2010). *Lifestyle of Isan women who marry Western men: A case study in Pimai (Research report)*. Thailand: Roi Et Rajabhat University.

Statham, P., Scuzzarello, S., Sunanta, S., Trupp, A. (2020). Globalising Thailand through gendered 'both-way' migration pathways with 'the West': cross-border connections between people, states and places. *Journal of Ethnic and Migration Studies, 46*(8), 1513–1542, https://doi.org/10.1080/1369183X.2020.1711567.

Sunanta, S. (2020). Globalising the Thai 'high-touch' industry: exports of care and body work and gendered mobilities from and to Thailand. *Journal of Ethnic and Migration Studies, 46*(8), 1543–1561. https://doi.org/10.1080/1369183X.2020.1711568.

Sunanta, S., & Angeles, A. (2013). From rural life to transnational wife: Agrarian transition, gendered mobilities and intimiate globalization in transnational marriages in Northeast Thailand. *Gender, Place & Culture, 20*(6), 699–717.

Therrien, C. (2014). Quest migrants: French people in Morocco searching for 'elsewhereness'. In M. Janoschka & H. Haas (Eds.), *Contested spatialities, lifestyle migration and residential tourism* (pp. 108–123). New York: Routledge.

Thompson, E., Kitiarsa, P., & Smutkupt, S. (2016). From sex tourist to son-in-law | Emergent Masculinities and transient subjectivities of farang men in Thailand. *Current Anthropology, 57*(1), 53–71.

Torkington, K. (2014). Tourism-related mobilities and discursive landscaping in the Algarve. *Dos Algarves, 24*, 40–63.

Toyota, M., & Xiang, B. (2012). The emerging transnational "retirement industry in Southeast Asia". *International Journal of Sociology and Social Policy, 32*(11/12), 708–719.

Van Laar, S., Cottyn, I., Donaldson, R., Zoomers, A., & Ferreira, S. (2014). 'Living apart together' in Franschhoek, South Africa: The implications of second-home development for equitable and sustainable development. In M. Janoschka & H. Haas (Eds.), *Contested spatialities, lifestyle migration and residential tourism* (pp. 190–203). New York: Routledge.

Wilde-Menozzi, W. (2003). Grafting on Italian roots. *The Literary Review, 47*, 157–165.

Zunino H. M., Dattwyler, R. H., & Zebryte, I. (2014). Utopian lifestyle migrants in Pucon, Chile. Innovating social life and challenging capitalism. In M. Janoschka and H. Haas (Eds.), *Contested spatialities, lifestyle migration and residential tourism* (pp. 96–107). New York: Routledge.

KwanchanokJaisuekun is PhD candidate in the Multicultural Studies Program at the Research Institute for Languages and Cultures of Asia, Mahidol University. Currently, she is conducting her doctoral thesis focusing on the migration of Westerners to Pattaya, Thailand, and its implications on local society. Kwanchanok's research interests include transnational migration, gender and popular culture.

Sirijit Sunanta is assistant professor in the PhD Program in Multicultural Studies, Research Institute for Languages and Cultures of Asia, Mahidol University, Thailand. Her research interests include gender and migration, globalization and food cultures, and the politics of diversity in Thailand. Sirijit's current research projects focus on care transnationalization and intimate and bodily labour in the Thai health and well-being industry.

CHAPTER 10

Burmese Migrant Women Workers in Thailand: Juggling Production and Reproduction

Ruth Pearson and Kyoko Kusakabe

Abstract This chapter focuses on the unpaid care work of Burmese migrant workers in garment factories in Thailand, to bring the analysis back to issues concerning women's unpaid domestic and care labour in poor households in developing countries. It tracks the complex strategies migrant women employ to fulfil their care responsibilities, both to their natal families in their countries of origin and their new families in their country of destination. It details the ways in which the state systems in both countries of origin and destination not only fail to contribute support to migrant women's welfare and caring responsibilities, but also often directly obstruct their ability to care for their families and themselves.

Introduction

This chapter is based on empirical research on Thailand's border industries, and illustrates how migrant workers juggle their reproductive roles with their exhausting jobs in garment and textile factories. Whilst extensive attention has been paid to women migrants involved in the global care chain, this is one of the few studies which considers the care responsibilities of migrant workers in other sectors. Utilising the concept of the care diamond, the chapter will

R. Pearson (✉)
School of Politics and International Studies, University of Leeds, Leeds, UK
e-mail: r.pearson@leeds.ac.uk

K. Kusakabe
Asian Institute of Technology, Khlong Nuen, Thailand

C. Mora and N. Piper (eds.), *The Palgrave Handbook of Gender and Migration*, https://doi.org/10.1007/978-3-030-63347-9_10

consider the options open to these workers and the constraints presented by both hostile sending and receiving states and the transborder nature of family networks.

Based on research carried out between 2007 and 2018 which highlights the situation of Burmese migrant women workers in Thailand's export factories (Pearson & Kusakabe, 2012a), this study extends the transnational care analysis by foregrounding the unpaid reproductive work of Burmese women migrants in Thailand who work in non-care sectors of the economy (Pearson & Kusakabe, 2012b). It details the ways in which they undertake paid work in addition to unpaid domestic and care work rather than instead of it. By focusing on the unpaid care work of transnational migrant factory workers in the Global South, specifically Burmese migrant workers in garment factories in Thailand, our study brings the analysis back to issues concerning women's unpaid domestic and care labour in poor households in developing countries. It tracks the complex strategies migrant women employ to fulfil their care responsibilities, both to their natal families in their countries of origin and their new families in their country of destination. It also details the ways in which the state systems in both countries of origin and destination not only fail to contribute support to migrant women's welfare and caring responsibilities, but also often directly obstruct their ability to care for their families and themselves.

Research Methodologies and Challenges

The analysis on which this study is based draws on empirical research carried out by a team of Burmese and Thai researchers led by Ruth Pearson (University of Leeds) and Kyoko Kusakabe (AIT Bangkok), between 2007 and 2010 and is supplemented by a more recent survey which was completed in 2019 (See Kusakabe & Melo, 2019). The team carried out both qualitative in-depth interviews as well as a large-scale structured survey with women and men migrant workers in garment and leather factories. The study was conducted in Mae Sot, Tak province, Three Pagodas Pass of Kanchanaburi Province, as well as Pra Pradaeng of Samut Prakan province. A total of 165 qualitative interviews were conducted (135 women and 30 men) during 2007–2009. The in-depth interviews covered the life-work history of the migrant workers starting with their life in Myanmar, the reasons underlying their decision to migrate to find employment in Thailand, the details of their journeys to Thailand and their experiences of their work and their family/community life once they arrived in Thailand. The majority of the interviews were carried out in Mae Sot (80 interviews of which 68 were with women workers and 12 with male workers. In Three Pagodas Pass 36 women and 7 men were interviewed with a further 31 women and 11 men interviewed in Pra Pradaeng). As far as possible the interviews were conducted in the specific mother tongue of the migrants who were from a range of ethnic groups within Burma/Myanmar, and only those who had been resident for some time in Thailand were able to respond in the Thai language.

The survey respondents totalled 504 over the three locations (371 women and 133 men). The locational breakdown was as follows Mae Sot 302 (211 women and 91 men), Three Pagodas Pass (91 women and 11 men) and Pra Pradaeng (69 women and 31 men). The purpose of undertaking this extensive research in three locations was to explore the extent to which the geographical position in relation to the border affected the kinds of decisions migrant workers make concerning childcare and other care responsibilities. This is explored at length in Pearson and Kusakabe (2012b).

There were a number of challenges we faced in carrying out this research. A major issue was the irregular migration status of many of the migrant workers in Thailand which made them reluctant to talk to outsiders about their personal histories. Our research was facilitated in all the research locations by the support of a range of trade unions and other civil society organisations who trusted our objectives and approach and were able to provide a bridge to the migrant workers themselves.

Migration and Development: Global Care Chains

Discussions about migration in the context of international development are frequently focused on human capital issues in relation to production in the sending or receiving countries, as well as the impact and potential of migrant remittances for sending economies or communities (Hoang & Yeoh, 2015). Much less attention has been given to the issue of reproduction and care relating to the dual roles of migrant workers, especially women workers (see also Dannecker & Piper in this volume). There has of course been an extensive literature on what is termed the Global Care Chain which highlights the flows of predominantly poor women of colour migrating from the global south to so-called developed economies of the global North in order to carry out care activities traditionally the responsibility of women (Yeates, 2012). This feminist literature has thrown light on a very important and complex issue pertaining to contemporary globalisation. Firstly, that a high proportion of migrants are women, who are migrating in order to find paid work, rather than as family members or carers. (Bastia & Piper 2019; Tittensor & Mansouri, 2017). And secondly, that a key part of the international division of labour concerns the ways in which a range of reproductive tasks which historically have been performed within family and kin groups are now being met by wage labour of different kinds (Benería, Deere, & Kabeer, 2012).

The growth in demand for domestic workers who migrate from poorer countries is related to three factors; firstly, the rapid demographic ageing in Northern countries, which means that the number of people requiring different kinds of support and care has grown exponentially. Secondly the long-term decline in the real wage has meant that most working households in the global South are unable to survive on a single ("breadwinner") family wage so that more and more women have been forced to enter paid work to meet household expenses. In addition, austerity policies have seen even Northern European countries retreat from aspirations of a welfare state which would meet the care needs of all citizens from cradle to grave (Lutz, 2016). So whilst in previous eras demand for household care—of children, of older people and of those with illnesses and disabilities might have been met by a combination of inputs from family (mainly women's unpaid labour), state-funded services and privately contracted women from the poorer strata of the population, these inputs have become more and more inadequate. Hence the demand for cheap labour from low-income countries to fill that gap (Kofman & Raghuram, 2015).

Various scholars have described this phenomenon as the Global Care Chain because of the way it links the migration of poor women from the global south with (often badly paid and exploitative) care work in the north (Fudge, 2016). And much attention has been given to the consequence of this for those women who migrate to carry out care work for the families and households in Northern countries for the care needs of their own families—children, ageing parents, disabled family members, etc.—back home. Since it is the case in the overwhelming majority of situations women migrating to do domestic and care work abroad are not allowed to bring their children or other dependents with them because of visa restrictions, they have to make alternative arrangements for care in their own families, often at quite a high cost to those left behind (Pratt, 2012).

The global care chain literature is important because it provides an implicit critique of the general assumption in migration studies that there is a binary divide between the productive and reproductive spheres of the economy. By highlighting the fact that domestic and care workers are employed to undertake reproductive labour it conceptualises the trends in household reproduction noted earlier, focusing on the commoditisation and marketisation of care activities which have previously been assumed to be carried out solely by the unpaid work of family members within the household or private sphere. However, this literature ignores the fact that there are also countless millions of migrant women who leave their countries of origin to do wage work which is not care work, employed as factory workers, agricultural labourers, low paid labour in the restaurant and entertainment sectors. These women, like those involved in global care chains, also have care responsibilities to their own children and other family members but even when they are transnational migrants the literature has tended to ignore this aspect of their lives. Moreover, whilst some of these transnational migrants are South–North migrants—many of them working in North America or Europe, most of them are working in neighbouring countries within their own regions. But whatever their circumstances they also embody the ways in which women workers transcend the division between productive and reproductive work as they struggle to juggle their employment and caring responsibilities.

Beyond the Global Care Chain: Care Work on Non-Care Migrant Workers in Thailand

This chapter interrogates one particular group of South–South migrant workers, Burmese women migrants in Thailand, to illustrate and extend the transnational care analysis by highlighting the unpaid reproductive work of those who work in non-care sectors of the economy (Pearson & Kusakabe, 2012b). We detail the ways in which they undertake paid work in addition to unpaid domestic and care work rather than instead of it. By focusing on the unpaid care work of transnational migrant factory workers in the Global South, specifically Burmese migrant workers in garment factories in Thailand, our study brings the analysis back to issues concerning women's unpaid domestic and care labour in poor households in developing countries. It tracks the complex strategies migrant women employ to fulfil their care responsibilities both to their natal families in their countries of origin and their new families in their country of destination. It also details the ways in which the state systems in both countries of origin and destination not only fail to contribute support to migrant women's welfare and caring responsibilities, but often directly obstruct their ability to care for their families and themselves.

Our research on Burmese factory workers on various parts of the Thai–Myanmar border graphically illustrates this situation. These are women who have crossed the land border between their country and Thailand in search of employment which offers a better life in terms of job prospects, wages and possible futures in the garment and textile factories which have been established since the mid-1990s as part of the Thailand's policy of decentralising industrial production. The growth of migrant workers in manufacturing and export factories in Thailand dates from when Thai exports were increasingly challenged by competition from lower-waged economies in the region such as Vietnam and China. The Thai government sought to make its own exports more competitive by moving production out of the higher cost metropolitan areas such as Bangkok in the centre of the country, to border regions where minimum wage rates were lower and there was access to cheap migrant workers from neighbouring countries.[1]

The numbers of Burmese workers employed in Thailand has grown steadily over the last three decades. From a small base in the 1980s, Thailand became the location of an estimated 2 million migrant workers from bordering countries—more than 80 percent from Burma/Myanmar and a much smaller number from Laos and Cambodia (Arnold & Hewison, 2005). By 2019 The Thai Department of Employment reported that the total number of migrant workers from neighbouring countries was 3,057,954 (Kusakabe & Melo, 2019). These official figures included migrants who entered the country

[1] It should be noted that in some parts of the Thailand there are Cambodian and Laos migrants working in factories but our research focused solely on Burmese factory workers

under a range of migration programmes. The most formal is under the stipulations of Memorandum of Understanding (MOU) between the Myanmar and Thai governments which was concluded with the neighbouring countries of Cambodian Laos and Myanmar. Many migrant workers arrive outside these provisions and have to register under the Nationality Verification Scheme which is a cheaper process. In addition, there is a Border Pass Scheme which allows people living in border areas to cross into Thailand for temporary work for up to 20 days within a designated area along the border. Some 43% of workers registered by these processes were women. Inevitably precise numbers are very hard to ascertain since in addition to an ever-changing and complicated series of measures introduced to document and control migrant workers crossing into Thailand, those that are formally registered have historically been outnumbered by undocumented workers who seeking work without having undertaken the formal (and often expensive) bureaucratic procedures which are required for formal registration. In 2018 some sources indicate that there were at least 5 million migrant workers in Thailand—a much higher number than the 3 million reported through official sources. Although the authorities continuously crackdown on undocumented migrants, with reports that more than 70,000 undocumented migrants were deported back to Myanmar in 2018. Of this total, it is estimated that up to 40% were women employed in the domestic service, agriculture and entertainment sectors as well as the garment and other manufacturing sectors concentrated on the border between Thailand and Myanmar.

The main research on which this chapter is based was carried out in various stages between 2007 and 2018. The research concentrated on the largely female labour force in the garment factories, the majority of which are in Tak province on the Northern East border around the town of Mae Sot, although the research also extended to the border town of Three Pagodas Pass of Kanchanaburi Province, as well as Pra Pradaeng of in the central Samut Prakan province as detailed above. This area around Mae Sot was developed as the main centre for garment production in Thailand since the 1990s with significant growth since the establishment of a formal Special Economic Zone in 2014. Initially the Thai government gave tax incentives to Thai and other producers to locate in this area where the national wage rate is lower than that prevailing in central provinces of Thailand but also this location has the best access to the thousands of mainly undocumented workers who cross the River Irrawaddy between Thailand and Myanmar. The development of the Special Economic Zone was promoted as a measure to "relocate agriculture and labour intensive industries from the centre… and to utilize the cheap labour and resources of the neighbouring countries and finally to stop the inflow of the illegal migrants and the relevant problems" (Tsuneishi, 2005 cited in Kusakabe et al., 2019). The official aspiration was to "provide economic opportunities not only for Thais but also our neighbours, which could help prevent illegal migrants from crossing into inner provinces of Thailand, giving more work opportunities to Thai nationals" (MFA, 2014 cited in Kusakabe & Melo 2019).

It was clear from the early 1990s that a major objective of the border industrial zones was to facilitate access to access cheap transborder labour and nowhere was this more successful than in Tak province in Northern Thailand. The Special Economic Zones (SEZs) which were established in the past decade have accelerated that process. At the same time as Thailand was expanding its production capacity in these production zones, Myanmar (the name which has been universally adopted following the partial democratisation of the country after 2010) has also created SEZs with the aim of competing in international garment ad other export markets on the basis of its own very cheap labour. This is a challenge for the Thai government which at the same time seeks ongoing access to migrant labour for its export factories, but also to prevent Burmese migrants from proceeding further into the central provinces of Thailand, and competing with local Thai workers. Whilst the labour force in the export factories in Tak province are almost exclusively comprised of Burmese migrants, there are still some approximately one million Thai workers employed in garment and other factories in the higher wage and more prosperous areas of central Thailand around Bangkok and other provinces (Clean Clothes Campaign, 2015).

The ways in which the Thai government has sought to manage and document migrant workers in garment factories reveals the tensions inherent in this policy. On the one hand the objective is to ensure a flow of cheap labour but at the same time they are concerned that migrant workers do not acquire the status of permanent residents in Thailand with all the entitlements and protections that might apply to Thai citizens. By decreeing clear limitations on various aspects of migrant workers' lives in the country—work permits of limited duration, political bans on forming trade unions or other associate making it is difficult—if not impossible to legally bring dependent children and other relatives to live with them in their place of employment.

Migrant Workers and Reproduction of Labour in Historical Context

In this way the Thai government has sought to maintain the low cast of Burmese workers by ensuring that it has very limited liability for any potential associated costs of accessing this labour. This is not an unprecedented or unusual situation. There are many examples of migrant labour flows and schemes that have sought to isolate the worker from the social relations in which they have grown and which are necessary for the reproduction of their own labour power—in terms of biological, daily and generational reproduction and survival. Historically, it has been male workers who have been targeted in efforts to provide—and control—migrant workers for specific sectors in different places—for example, the labour force in South Africa's gold mines which was drawn from neighbouring countries such as Lesotho as well as rural areas, confined to dormitories near the mines with no rights to bring wives or children or other family members. In Britain in the 1970s, the textile workforce in the North of England was increasingly staffed by South Asian migrant workers, with extensive restrictions on the possibilities of family settlement. The so-called "Windrush" generation of Caribbean workers who were encouraged to come to the UK in the 1950s as part of the post-war reconstruction—both women, men—frequently had to leave children behind until they were established and could afford to bring them to the UK—and the current issues that have arisen recently as people not born in the UK have discovered that they have no legal right to be here, and some have been subject to arbitrary deportation. There are numerous other examples but the point is that historically, migrant workers have been conceptualised solely in terms of their daily labour power that they are able to sell to their employers without consideration for the conditions under which this working capacity has been produced within families and communities in their countries of origin.

However, the situation of Burmese migrant factory workers in Thailand is particularly instructive in terms of the extraction of the workers' labour power in the context of trying to limit any recognition of the context of the workers' lives. As explained above, the overwhelming majority of the output from the border factories is garments—either knitted or RMC items which are assembled (sewn together) there. Unsurprisingly, the vast majority of the workers employed in these factories are women though it is interesting to note that in the knitted garments factories there is a higher proportion of male workers, possibly associated with the apparent complexity and technical sophistication of the machinery employed. As in many other situations the "ideal" workers in these garment factories are sought from those who are the most docile and dexterous, offering cheap labour in the form of lowest unit costs (Elson & Pearson, 1981). In the making of garments for large-scale low-price cost markets this has generally been found in the form of young women who are newcomers to the industrial labour market—described by one early German consultant appraising the maquiladora industry in Mexico as "industrial virgins with nothing to learn or unlearn" (König, 1975). For many years the standard trope about the cheap labour provided by young women in export garment production focused on the fact that they were young single and docile, signifying that they were able to deliver long hours of work, often in shift times that are usually thought of as "unsocial hours" since they are not compromised by family responsibilities for husbands or children but available whenever production required them. The notion of docility partly reflects the fact that most of the export production in the 1970s and 1980s took place in the so-called Asian tigers, and that women from those regions were widely regarded through a Western orientalist lens as obedient and submissive. But it also reflects what is implicit in the industrial virgin quote above—that young women factory workers—that young female recruits to industrial work are unlikely to resist management methods to increase productivity either individually or by collective action through trade unions or other workers' organisations.

But over time, as export production established itself in a much wider group of countries, it is clear that this standard vision of the ideal women worker needed modification. Firstly, women workers frequently stayed for many years in industrial employment, frequently managing to combine being a worker with being a wife and/or mother. In some countries—for example in the Caribbean where social practice frequently means that women have children at a young age and frequently marry formally at the end of their childbearing years, workers at a Barbados electronics factory in the 1970s were predominantly married and had children. In more recent decades, where production had been located in countries that have achieved substantial industrial development and economic growth—such as Malaysia and Thailand (not-withstanding the Asian Financial Crisis of the late 1990s) the labour situation for export factories has changed. Whereas in earlier decades young women had been willing to submit to the industrial discipline required for such employment, their daughters or granddaughters who had benefited from the improvement in living conditions and education for women sought what was considered superior employment—in retail and service occupations or own account enterprises. Increasing international competition and a so-called "race to the bottom" in search of locations that offer even cheaper options for production has led to the replacement of local workers with migrant workers who can be paid lower wages, and whose discipline can be ensured through migration regulation and other procedures (Pearson, 1998). Parallels can also be drawn with the situation of internal migrants to China's special economic zones in the South-West of the country, whose compliance with intensive management techniques to maximise productivity has also been accompanied by stringent regulations which have prevented migrant workers accessing local services such as housing and schooling for their children under a *huku* (household registration) system which decrees that workers are only entitled to non-wage benefits and services in their communes of birth[2] *(note though there have been recent changes to this ref*) (Ngai, 2005). These migrants who are employed in electronics and garments are also prevented (either de jure or de facto) from forming organisations to bargain collectively for improvements in their conditions and their rights.

[2]The restricted access to a range of welfare services for migrant (or "temporary" workers in Chinese cities) has been the subject of ongoing workers protest in Chine over recent decades. The Chinese government recently announced that such restrictions would be abolished China Briefing (2019).

Burmese Migrant Factory Workers and the Reproduction of Labour

The situation of Burmese migrant workers in garment factories in Tak province is part of this trend where workers from lower-income neighbouring counties (or lower-income parts of the same country as in the case of China) are "imported" to provide the supply of cheap labour deemed necessary for maintaining a market share in global export markets. The fact that garment production has demonstrated a preference for nimble-fingered women workers is in itself a gender issue, since the structural reasons that make women cheap and disposable are a reflection of unequal gender relations. But it is women's ongoing responsibility for the reproduction of that labour power which is most significant. Whilst working in the Thai factories women supply their labour at a cost which is far lower than those paid to Thai workers in factories in other parts of the country—and, according to a more recent survey (Kusukabe & Melo, 2019) is also lower than that paid to the male migrant workers who work alongside them. But the same survey revealed that some 40% of the workers were parents of young children, and that the workers had to deploy a range of strategies in order to secure a means of looking after these children as well as continuing to work in the factories. These strategies included leaving children to be cared for with relatives (usually mother or a sister) in Myanmar, bringing a dependent from Myanmar (usually illegal as most work permit do not permit the worker to be accompanied by a dependent), hiring a local carer or trying to juggle childcare with a partner or friend or, in some cases bringing babies and infants to work which is only possible in the smaller unrecognised, unregulated and house factories which are located within the town of Mae Sot.

The majority of women workers in this survey declared that they sent remittances to the value of up to 40% of their wages to their families at home (Kusakabe & Melo, 2019). Women were found to remit a higher amount and for a longer period than male workers. The data suggests that male workers tended to cease financial support for their families back home after a few years, often when they formed new families in Thailand. However, for women workers, in many cases the objective of the remittances was not just to meet family obligations but also to cover the cost of childcare being carried out by members of their families back home. In previous years the transnational childcare juggling was facilitated by much less controlled entry from Myanmar across the border with Thailand, but in recent years control and surveillance have been stepped up, not least because an increasing number of migrant workers are only allowed a temporary 30-day entry which has to be renewed—a practice which is costly to the worker and also makes it very difficult to escape control. Since many women have worked in the border factories for a number of years—the longest period recorded in the 2018 survey was 12 years, they also have to make provision for the care and education of older children beyond their infancy. A number of migrant organised schools have been established to care for migrant children who are resident in Tak province, for which parents must pay, since they rarely have access to Thai government schools in spite of a nominal "Education for All" policy-introduced in 1990 (Pearson & Kusakabe, 2012b). Save the Children reported that in 2015, migrant children were still excluded from Thai Schools (Save the Children, 2015). Parents are also responsible for the schooling costs of children resident with relatives inside Myanmar.

All the issues raised within the Global Care chain literature with regard to the problems of left behind children, the fact that women migrant workers are assumed to retain responsibility for children and family members at home regardless of their occupation and earnings in their destination country are relevant to the situation of Burmese migrant workers in Thailand. Migrant factory workers generally receive lower wages and work in inferior working conditions, they are rarely permitted to organise or pursue collective bargaining, and they are disciplined not just by the nature of the work itself, but by the structure of migration regulation, costs and penalties exacted by both their employers and the state in the country of employment. This is a common feature of precarious and temporary South–South migration as illustrated by the study of migration in South Asia by Piper, Rosewarne and Withers (2017). Underlying this situation is the stark fact that migrant workers are employed as cheap labour in a situation where neither the employer or the state of the destination country have any responsibility for the reproduction of that labour power—either the reproduction of the next generation of workers or the resources, time and care that have been deployed to (re)produce the workers they find most profitable for their operations.

Other writers have deployed the concept of the care diamond to depict the possible sources of resources and support available for women workers who have care responsibilities for infants and children. According to Razavi (2007, 2011), such resources could be available from four different sources or points on the diamond—the local or national state, the community, the market or the family or the parent her/himself. In the case of Burmese women workers on the Thai border, the care diamond is impoverished in different ways. Neither the local or national state—Thailand—has an interest in investing resources in the care services which are necessary for the women employed in the border garment factories. Indeed, as explained above, the Thai government has gone to considerable lengths to treat migrant workers as temporary residents in the country, and put numerous obstacles in their way if they wish to bring their children into the country from Burma/Myanmar, or if they wish to bring other relatives who might take care of their children.

These hostile policies also include measures to exclude migrant children from attending mainstream Thai schools, or to obtain educational qualifications. In fact, many migrant children in Thailand do not have citizenship in either the country of destination or the country of origin, regardless of whether or not they have been born in Thailand. Nor does the sending state take on these responsibilities whilst the children of migrant workers are in Thailand. In fact, for women workers in Thailand the most support available has come from the community, which in this context refers to the diaspora migrant community organisations rather than Thai civil society or third sector organisations. Since the 1990s there has been a continuous growth of Burmese childcare centres, schools, Buddhist temples and trade unions in Tak province, which have not only provided health and education for migrant workers and their families but also in many cases offered these services to Burmese families still resident in Myanmar who sought access to the comparatively better facilities available to them on the Thai side of the border. In contrast, in spite of various initiatives which have been aimed at allowing migrant workers access to health services—such as the 30 Bhat scheme—most migrant workers and their families have experienced exclusion from such services—because of language problems, because of incompatibility of opening hours with shift working, because of direct discrimination, and because of a fear of being denounced as illegal immigrants and deported. Where they have used local services, this has largely been restricted to childbirth and child immunisation, but no other child support has been available (Kusukabe & Pearson, 2013).

For migrant workers who are trying to juggle childcare and factory employment, a possible source of care might be the market, as it is in other countries. But our research indicated that hiring a Thai childcarer was rarely feasible for these women for a variety of reasons. Firstly, the market rates necessary to pay Thai women to look after their children are way beyond their economic possibilities since their wages are well below the earnings of Thai workers. Moreover, in towns such as Mae Sot there is a very small Thai population most of whom are not willing to accept low paid work such as childcare, and given the anti-immigrant attitudes of much of the local population, they would not be prepared to carry out such services for Burmese women workers. There has been experience of migrant factory workers hiring local private childcare works in central provinces such as Bangkok but research respondents revealed distrust about the level of care offered.

Detailed Analysis of Childcare Strategies of Burmese Migrant Factory Workers

The family, including the worker herself is therefore the default option utilised for the care of migrant children. Our research on childcare strategies of migrant factory workers in three locations in Thailand indicated that a range of strategies were deployed in order to allow workers to manage their jobs and their family responsibilities. The most common strategy, utilised by over 67% of our survey sample, was to send children back to Burma to be cared for by family members. Only 10% of those surveyed were able to organise for family members to come to Thailand to look after their children whilst they were at work, though that option has become increasingly difficult as immigration rules have become tighter. 12% of the sample resorted to paid childcarers within Thailand and the remaining 10% reported an unstable situation whereby they managed to combine working with their spouses or siblings—also migrant workers—by juggling shifts. In addition, the research revealed that though many women tried to manage post-partum infancy by giving up their jobs until the baby could be weaned, this was a very short-term solution and, in most cases, ended when they were able to arrange for their children to be cared for by someone else. As mentioned earlier, women generally remit a large part of their wages to their families in Burma much of which is to cover child care support costs, as well as maintaining their position in the family and their right to call on family support when required.

For migrant workers in the border province of Tak, the option of accessing family assistance by sending children back home is possible, since until recently, there has been less effective in controlling cross-border irregular migration. This study also explored the childcare strategies of workers in the central location of Phrapradaeng, in Bangkok province, where state surveillance is much stricter and prevents migrant workers from accessing any support from either the Thai or Burmese state, as well as making it difficult to utilise assistance from family members back in Burma. In this situation some women reported that they utilised what were known as "baby carriers" who smuggled several small children back to relatives' homes inside Burma—a hazardous journey for adults let alone infants who had to be drugged during the journey. And in the remote border area of Three Pagodas pass in Sangklaburi province migrant workers were able to cross daily from their homes on the Burmese side of the border using informal crossing points, often leaving their children in a number of boarding nurseries and schools which sprung up to meet this demand. But in all locations the research indicated that women were constantly having to shift between different strategies to achieve a childcare solution which would not jeopardise their factory employment (Kusakabe & Pearson, 2013).

At the present time, whilst more factory employment opportunities for garment workers are opening up in the export factories inside Myanmar there is still a steady stream of workers seeking employment in the more developed neighbouring country of Thailand. But even within Myanmar the majority of garment workers are internal migrants who leave more rural and distant home provinces to work in the Free Trade Zones near Yangon and other major cities. And they face similar problems when they have to combine paid work with childcare responsibilities.

Who Cares? Migrant Factory Workers and Reproductive Work

The care responsibilities of migrant workers employed in non-care sectors have hitherto been largely ignored by the Global Care Chain literature although they face similar issues when they have to leave their homes and their countries to seek work elsewhere. Whilst discussion of globalisation has focused directly on the internationalisation of production supply chains, and there has been significant research on the female intensity of the workforce in key export sectors including the garments sector, little attention has been given to the gendered issue of combining women's role in production with their ongoing responsibilities for reproduction. A different contemporary literature which has been concerned with the burden of unpaid work carried out by women in developing and more industrialised economy rarely gives much attention to the fact that paid work and unpaid care are frequently carried out simultaneously, and when this occurs across international boundaries the constraints for women are even harsher. This chapter attempts to redress that balance. The Coronavirus pandemic of 2020 highlighted the ways in which the economy in all parts of the global economy is dependent on and intertwined with reproductive and care work in order to function effectively. The recognition that much of this reproductive work is carried out by migrant workers who also have reproductive responsibilities for their own families, should we hope be considered in future economic and social policy.

References

Arnold, D., & Hewison, K. (2005). Exploitation in global supply chains: Burmese workers in Mae Sot. *Journal of Contemporary Asia, 35*(3), 319–340. https://doi.org/10.1080/00472330580000191.

Bastia & Piper. (2019). Women migrants in the global economy: A global overview (and regional perspectives). *Gender and Development, 27*(1), 15–30.

Benería, L., Deere, C., & Kabeer, N. (2012). Gender and international migration: Globalization. *Development and Governance. Feminist Economics, 18*(2), 1–33.

Clean Clothes Campaign. (2015). *Facts on Thailand's garment industry.* In: https://cleanclothes.org/resources/publications/factsheets/thailand-factsheet-2-2015.pdf.

Elson, D., & Pearson, R. (1981). Nimble fingers make cheap workers: An analysis of women's employment in third world export manufacturing. *Feminist Review, 7*(1), 87–107. https://doi.org/10.2307/1394761.

Fudge, J. (2016). Global care chains: Transnational migrant care workers. In M. Romero, V. Preston, & W. Giles (Eds.), *When care work goes global* (pp. 227–231). New York, NY: Routledge.

Hoang, L., & Yeoh, B. (Eds.). (2015). *Transnational labour migration, remittances and the changing family in Asia.* London: Palgrave MacMillan. https://doi.org/10.1057/9781137506863.

Kofman, E., & Raghuram, P. (2015). *Gendered migrations and global social reproduction: Migration Diaspora and Citizenship.* Basingstoke: Palgrave.

Konig, W. (1975). "Towards an evaluation of international subcontracting activities in developing countries: Interim report upon completion of field work on maquiladoras" in México; borrador para comentarios"," Sede de la CEPAL en Santiago (Estudios e Investigaciones) 32967, Naciones Unidas Comisión Económica para América Latina y el Caribe (CEPAL).

Kusakabe, K., & Pearson, P. (2013). Cross-border childcare strategies of Burmese migrant workers in Thailand. *Gender, Place & Culture, 20*(8), 960–978.

Kusakabe, K., & Melo. (2019). *Jobs in SEZs: Migrant garment factory workers in the Mekong sub-region.* URL: https://www.researchgate.net/publication/336313338_Jobs_in_SEZs_Migrant_garment_factory_workers_in_the_Mekong_Region.

Lutz, H. (Ed.). (2016). *Migration and domestic work: A European perspective on a global theme.* Routledge: London and New York, NY.

Ngai, P. (2005). *Made in China: Women factory workers in a global workplace.* Durham, NC and London: Duke University Press.

Pearson, R. (1998). Nimble fingers' revisited: Reflections on women and Third World industrialisation in the late twentieth century. In C. Jackson & R. Pearson (Eds.), *Feminist visions of development: Gender, analysis and policy* (pp. 171–188). Routledge: London and New York, NY.

Pearson, R. & Kusakabe, K. (2012a). *Thailand's hidden workforce: Burmese migrant women factory workers.* London and New York, NY: Zed Books.

Pearson, R., & Kusakabe, K. (2012b). Who cares? *Gender, Reproduction, and Care Chains of Burmese Migrant Workers in Thailand, Feminist Economics, 18*(2), 149–175. https://doi.org/10.1080/13545701.2012.691206.

Piper, N., Rosewarne, S. & Withers, M. (2017). Migrant precarity in Asia: 'Networks of labour activism' for a rights-based Governance of migration. *Development and Change,* 48 (5), 1089–1110.

Pratt, G. (2012). *Families apart: Migrant mothers and the conflicts of labor and love.* London: University of Minnesota Press.

Razavi, S. (2007, June 3). *The political and social economy of care in a development context: Conceptual issues, research questions and policy options* (Gender and Development Programme Paper). Geneva: UNRISD http://(httpAuxPages)/2DBE6A93350A7783C12573240036D5A0/%24file/Razavi-paper.pdf.

Razavi, S. (2011). Rethinking care on a development context. *Development and Change, 42*(4), 873–903.

Save the Children. (2015). URL: https://thailand.savethechildren.net/news/pathways-better-future-education-migrant-children-thailand.

Tittensor, D., & Mansouri, F. (2017). The feminisation of migration? A critical overview. In D. Tittensor & F. Mansouri (Eds.), *The politics of women and migration in the Global South* (pp. 11–25). London: Palgrave Macmillan.

Tsuneishi, T. (2005). *The regional development policy of Thailand and its economic cooperation with neighbouring countries* (IDE Discussion Paper, 32). Institute of Developing Economies, Japan External Trade Organisation.

Yeates, N. (2012). Global care chains: A state-of-the-art review and future directions in care transnationalization research. *Global Networks, 12*(2), 135–154.

Zoey, Z. Y. (2019). China is relaxing Hukou restrictions in small and medium-sized cities. *China Briefing* https://www.china-briefing.com/news/china-relaxing-hukou-restrictions-small-medium-sized-cities/.

Ruth Pearson is an Emeritus Professor of International Development at the University of Leeds. She is a leading researcher on women's employment in the world economy and has published widely on women workers and migration in Latin America, Asia and Europe. She is a feminist economist and activist with the UK Women's Budget Group.

Kyoko Kusakabe is Professor at Gender and Development Studies at the Asian Institute of Technology, Thailand. Her research focuses on gender and development, labor migration, women in the informal economy. She was a development worker before joining academia and has been conducting research in close collaboration with local NGOs and groups as well as with international organizations.

CHAPTER 11

Migration and Elderly Care When Women Leave, Who Cares for Older Adults? A Case Study of Cuba

Elaine Acosta González

Abstract Various studies have shown that since the late twentieth century, Cuban migration follows global patterns of feminisation. As more Cuban women migrate during their reproductive years, fertility rates have decreased in the island, resulting in the shrinking of the youth population relative to older adults. Both the lack and unequal distribution of care services for older adults and the shrinking welfare services suggest an impending crisis; and the failure of social policy to anticipate these problems and plan accordingly further aggravates it. The result is an uneven landscape, particularly within a general context of social stratification, with some sectors and geographical areas facing greater hardships than others. In this chapter we delve onto these issues to analyse the relationship between international migration patterns and both, population ageing and the crisis of elderly care in Cuba.

Introduction

Cuban population dynamics have shown an annual growth rate that has declined by 90% over the course of a decade, from 3 per thousand persons in 2000 to 0.3 per thousand persons in 2009. As of 2006, the total population began to decrease (−0.4%), and this trend has since continued (Aja &

My thanks and gratitude to Iveris Martínez and Jorge Duany for their comments, proofreading, and style editing corrections.

E. Acosta González (✉)
Cuban Research Institute, Florida International University, Miami, FL, USA

C. Mora and N. Piper (eds.), *The Palgrave Handbook of Gender and Migration*, https://doi.org/10.1007/978-3-030-63347-9_11

Hernández, 2019). As a result of this dynamic, Cuba leads in the proportion of ageing population across the Latin American and Caribbean region, according to an ECLAC-CELADE 2009 report. The National Aging Survey (ONEI, 2019a) and the most recent National Survey of Migration (ONEI, 2018a) implemented by Cuba, both recognise that ageing is the major socio-demographic challenge facing Cuban society today. This challenge is characterised not only by population decline, but also by fertility rates below replacement level, low infant mortality, high life expectancy, and external migration.

Emigration rates from Cuba have experienced growth in recent years, showing a continuously negative migration balance, accompanied by a diversification of migrant destinations, although there is still a high concentration of travel to the United States. Data from the last migration survey in Cuba (ONEI, 2018a) confirm the occurrence of a new phenomenon observed also by previous studies: the growing tendency towards circularity and temporality of migration as a result of the immigration reform of 2013 in the country. As a result of this reform, exit permits and invitation letters were eliminated, reducing costs for people as well as facilitating and speeding up migration procedures. The term to remain outside the country was extended up to two years without being considered "emigrated". In addition, the Cuban government has incorporated a temporary exit permit for children, hence facilitating the family reunification process.

The Cuban migration process follows global feminisation patterns evidenced by life projects in which Cuban women increasingly decide to migrate alone as a survival strategy in the context of an economic and social crisis (Marrero, 2011). However, the decision of women at reproductive age to emigrate inevitably affects population dynamics, causing a decline in fertility rates which eventually reduce the growth of the youngest segments of the population (Nuñez, 2007).

Hence, to understand the global dimension in which Cuban families' care needs are currently addressed is to recognise the care crisis in Cuba as a case study of social dynamics in relation to the crisis of the socialist welfare model. From the beginning of the Cuban Revolution (1959) until 1989, very notable progress has been made in establishing social protection mechanisms (Mesa-Lago, 2012). The Cuban state was committed to women's empowerment and enacted a series of laws and policies to promote women's equal and active participation in all spheres of life (Stavropoulou, Torres, Samuels, Solís, & Fernández, 2020). Through an extensive full coverage network of public institutions or welfare programs, the state was the single protagonist, stopping other (private or non-governmental) alternatives to provide social programs or services (Espina, 2008).

Due to the high cost of social programs, the economic sustainability of the social policy developed during 1959–1989 became questionable in the wake of the economic crisis which erupted in the early 1990s as a consequence of the fall of communism in Europe. This situation was aggravated by demographic ageing. The economic crisis highlighted the weaknesses and challenges of sustainability of a welfare model that had equity and universal coverage as its guiding principles. It became evident that Cuban women had borne the brunt of the crisis and were the most impacted by it, among other reasons, for continuing to support, under unequal conditions, social reproduction.

Targeting the most vulnerable groups, reducing social assistance, and familiarising social policies became essential components of government strategies since the early 2000s. From 2006 to 2015, social assistance to the elderly and disabled persons had experienced profound cuts, i.e. assistance to the elderly decreased by 63%, to mothers with disabled children by 50%, and by 68% to those who need home care (Acosta, Picasso, & Perrotta, 2018). This diminishing of social assistance meant the elderly increasingly relied on their relatives for care (see also Parella & Speroni, in this volume).

Even though Cuban society has made remarkable progress towards women's economic empowerment and gender equality since 1959, notably thanks to the provision of universal social services, maternity protection, and childcare facilities, it is increasingly acknowledged that these measures have been unable to bring about full equality. Care labour continues to pose a challenge to deeply entrenched gendered social norms about women's and men's roles in the family and in society, far from the expected diminishing role of women as caregivers in the domestic space. In addition, official data shows persistent gender inequalities regarding labour participation. In 2018, 49.5% of women were economically active compared to 76.9% of men (ONEI, 2019b). This rate is lower than the regional average for economically active women of 50.3% (ILO, 2018; Stavropoulou et al., 2020).

This chapter addresses the crisis of care in Cuba through the analysis of the growing feminisation of migration and the care of older Cuban adults. The aim is to contribute to a better understanding of the effects on the politics and practice of care in Cuba by highlighting the role of intersecting forces, such as political authoritarianism; economic struggles and crises; migratory policies; and patriarchy. I will argue that despite the thorough social transformation intended by the Cuban socialist model, the *stalled revolution* thesis suggested by Hochschild (1989) in developing countries also applies to Cuba regardless of their declared revolutionary aims regarding gender. The gains from the gender equality movement including gendered work and care, the choice of occupations, equal wage and career trajectories, and the division of household labour are still a promise in Cuba. The gender revolution remains incomplete and "stalled" without redefining hegemonic masculinity.

In this chapter, I start with a brief theoretical discussion on the relationship between gender, migration, and care, using the case study of Cuba as an anchor for current discussions about these topics in Latin America. Secondly, I will summarise the main features of Cuban international migrations affecting the ageing population. Thirdly, I will analyse the deepening problem of the caregiving deficit and its connection to the lack of care facing the ageing of the population in Cuba. Fourth, I propose some hypotheses regarding the strategies of care used by Cuban families as a response to the care crisis which could serve as themes for future research.

Understanding the Linkages Between Gender, Migration, Ageing, and Care

The debates on the feminisation of migration as a phenomenon that addresses not only the quantitative increase in the participation of women in current human mobility, but also their changing positions and roles in these processes, have acquired great significance. This is so because recent scholarship has revealed the key role that contemporary female migrations have played in the reorganisation of care work and, as a result, in the constitution of global networks of care work.

The quantitative growth of international female migrations, besides the qualitative changes in women's migratory models, invites a consideration of the gender variable in the analysis of migration. Global statistics from the United Nations population division show that in 2015 women constituted 48% of international migrants. Other data show that women currently exceed men in the stock of migrants in Europe and North America. The number of migrant women exceeds the number of migrant men in 57 countries of destination and emigration from 80 countries.[1]

Although for most of the twentieth century, women were secondary actors in the debate on international migration, gender analysis has allowed us to identify the presence of a growing plurality of female migratory projects (Vicente & Setién, 2005). However, their effects on sending societies have not been sufficiently explored. When women migrate, families in the society of origin are affected by deeper rearrangements compared to when men leave. Depending on the historical phase and the society in which they occur, such rearrangements will be more or less intense in regards to gender regimes. Pribilsky (2004, 2007) refers to these tensions, resulting from the inevitable rearrangement of responsibilities for family members, as well as the transgression of gender roles that they imply.

The literature on women in migration began to grow rapidly in various disciplines in the 1980s and 1990s (Lube & Gonzálvez, 2019) and gender has been recognised as an important element in the decision to migrate (Hagen-Zanker & Mallet, 2020). Thinking about who, when, how, and where to emigrate is conditioned by gender, depending on the roles and responsibilities associated with the people who migrate (Acosta, 2015). Age and social class are also factors influencing the decision-making process. But gender not

[1] More statistical information on gender and international migration flows is available at http://www.un.org/en/development/desa/population/migration/data/estimates2/estimates15.shtml.

only affects the people who migrate but also the families remaining in the sending countries.

The increasing flows of feminised Latin American migration have triggered extensive research at the beginning of the twenty-first century conducted by scholars based in the United States and Europe. Hondagneu-Sotelo (1994) and Hondagneu-Sotelo and Avila (1997) focused on migration of Latin American women to the United States. In Spain, Parella (2003), Solé and Parella (2003), Oso (2008), Pérez (2009), Martínez (2010), Díaz and Pérez (2011), and Pérez and López (2011), have covered various aspects of the feminisation of today's migratory flows, their characteristics, causes, and effects, among a considerable amount of research conducted ever since.

The role these women, especially Latin American and Asian, have played in compensating for the care deficit in Europe has been documented as crucial, either in the private sphere as well as care work in public institutions (e.g. elderly care homes and hospitals) (see also Wrede et al., in this volume). A study coordinated by Oso and Parella (2012) notes that studies on female migration and domestic service in recent years have shifted to a more specific focus on the concept of care. Vega's (2009) work departs from the analysis of the crisis and reorganisation of care as the starting point to consider the working conditions of women caregivers and the different types of care provided as well as the increasing presence of foreigners in the sector. Martínez's (2010, 2011) focus on the increasing demand for immigrant women caregivers in Spain in recent years, demonstrates that it is precisely immigrant women who accept the most precarious jobs in the domestic sector, thus becoming the first link in the chain of a secondary labour market segment.

At the same time, the social organisation of care in Latin America has taken a challenging transnational dimension, since female migration was largely triggered by a growing demand for domestic care workers in countries of the north as well as in the Global South (Cano, Sofia & Martínez, 2009; Mora, 2008; Stefoni, 2009; Stefoni & Fernández, 2011). Studies on the social organisation of care carried out in the Latin American region (Acosta, 2013, 2015; Arriagada, 2010; Arriagada & Moreno, 2011; Arriagada & Todaro, 2012; Gonzálvez & Acosta, 2015; Herrera, 2012) enabled researchers to scope and clarify the role of international migration in addressing the care crisis, and the role of women within it, thanks to the analytical, political, and ideological weight given to the concept of "care" (Lube & Gonzálvez, 2019). Thus, from the middle of 2000s, the feminist question of who provides care took central space and catalysed interest from researchers in different fields in the region, unveiling new practices and conflicts in which maternity, paternity, and married life were some of the key subjects considered.

It is older adults who often experience a deficit of family care after the migration of women, while at the same time, many older women continue providing care to their grandchildren to facilitate the migratory projects of their daughters. Parreñas (2001) has shown how the development of migratory flows led by women and the consequent formation of transnational households headed by women are responses to the global care crisis. Considering this panorama, Fraser (2016) has concluded that the care crisis is currently one of the main topics of public debate. Ageing and migrations, mainly featuring women, are undoubtedly connected to this crisis. Cuban migration scholars point out to the need to pay more attention to the impact of migration in Cuban society, i.e. take a country of origin perspective, and specifically to its role in demographic dynamics (Casaña, 2003 in Hernández & Foladori, 2012).

In this scenario, it is important to note that on average, Cuban women dedicate 14 more hours per week than men to domestic and care work, and that they do it in an environment characterised by low incomes, difficulties in public transportation and precarious housing (Padrón, 2019). In this larger context, Cuba is going through a crisis of care, but this pressing issue remains invisible because it is immersed in other, deemed more urgent, economic and political matters (Acosta, et al., 2018). Fraser (2016) noted that the care crisis can, however, not be seen in isolation because it is one of the components of a "general crisis", which includes economic, ecological, and political aspects, which intersect with and exacerbate each other: "*the result is a huge crisis, not only of care but also of social reproduction in its broadest sense*" (Fraser, 2016: 112).

In most of the world, caregiving is performed by women's "global care chains"—a term coined by feminist sociologist Arlie Hochschild (2001)—across countries, giving materiality to a way of working in which multiple tasks intermingle at the same time requiring constant management of times and spaces and versatility of knowledge (Gonzálvez & Acosta, 2015).

In the next section, I present the main characteristics of current Cuban international migration and its relationship to the crisis of care of ageing population in the island.

What Is Distinctive About Cuban International Migration?

Currently, the most distinctive population movement in Cuba is towards international migration, characterised by a great diversity of destinations despite maintaining a high concentration in the United States. According to Aja (2006), at the beginning of this century, Cuban migrants were reported in 148 countries, although 98% were concentrated in 20 nations. Besides the multiplicity of destinations, the social composition of Cuban migrants (by sex, level of education, age, and other demographic variables), has gone from being a phenomenon fundamentally associated with sociopolitical variables, to a more complex and multidimensional character (Duany, 1999), not only based on internal factors of Cuban society, but also on a set of dynamics and ways of being in today's world, such as the expansion of life horizons and abandoning political and ideological ties linked to a nation-state (CRI, 2011). Aja (2007) points to the multiplicity of causes explaining current Cuban migration by linking internal factors of society, such as political and economic transformations, social class contradictions and migration chains set up through the development of social networks.

Two topics of Cuban *emigration* have begun to attract the attention of social researchers in recent years. Both are closely connected with the ageing population and the care crisis. The first relates to the impact of high emigration of young people on the ageing population in Cuba; the second concerns the increasing emigration of women compared to men (Hernández & Foladori, 2012) and the growing deficit of caregivers created by this movement. The most recent migration survey (ONEI, 2018a) shows the growing participation of Cuban women in recent migration flows, with a high presence in temporary migration schemes, as well as a high percentage of young individuals among the migrants.

According to Aja (2007), the young adult population is over-represented in Cuban emigration statistics. The typical age range covered by recent Cuban migrants is between 20 and 40 years old. Regarding gender composition, the Center for International Migration Studies (CEMI, now defunct) of the University of Havana points out that an important weakness in analysing Cuban international migrations from a gender perspective is the difficulty of finding statistics regarding sex composition, in addition to the scarcity of publications on this topic (Marrero, 2011). The analysis of several studies carried out by Hernández and Foladori (2012) confirms that the Cuban migration process follows global feminisation patterns since the end of the twentieth century, taking into account that the participation of Cuban women has risen to more than 50% since 1995, maintaining an upward trend in the following years.

Aja (2007) asses the migratory flows in Cuba in the period 2006–2010 pointing out the feminisation of Cuban emigration as a trend, present since 1995 in all categories of emigration from Cuba, except the "abandonment of missions", where men predominate. Previously, migratory balances by sex, between 1960 and 2003, had shown that the number of women had been higher than men in absolute numbers for the entire period. This trend continued in the following years, as confirmed by the last migration survey applied in 2016 (ONEI, 2018b). As a way out of Cuba, women use different alternatives ranging from scholarships, work contracts, marriage with foreigners, family reunification, or individual visits without return. The data confirms that in quantitative terms, there is a feminisation of migration in Cuba since 1960 (Núñez, 2010). However, as in other aspects of Cuban migration, gender relations have not been incorporated systematically in the country's social research agenda (Núñez, 2007).

Regarding men, their greatest participation in migration occurred during the 1994 "rafter exodus", as well as during the "Mariel exodus" in 1980. Women also participated, but only reached 30%, which corroborates the hypothesis of Albizu-Campos and Alfonso (2000) on the relationship of gender migration with the emigration type and possibilities (cited by

Hernández & Foladori, 2012). Years later, the evidence shows that when legal emigration gained space because of the growing flexibility of Cuban migration policy, women increased their participation, reaching more than 55%, with a peak in 2008.

This selectivity of migration is closely related to caregiving, since it is women who are usually left in charge of children and relatives while they can manage safer emigration routes. It has been shown that, in the periods in which definitive departures have been made with a certain order, with certainty to make the trips and with support for family reunification, women emigrated more than men. For example, after 1994, the participation of women rose to 51%, maintaining in the following years the upward trend.

Relating Cuban Women Emigration and the Rise of Elderly Care Deficit

Emigration is impacting the living arrangements of older people in Cuba. About 7% of this group has all of their children living outside of Cuba and 3%, 70,300 people aged 60 and over, have all their children and grandchildren residing outside the country. At the same time, increases in the number of single older adult households are evident, rising from 12.6% in Census, 2012 to 17.4% in the last national ageing survey (ONEI, 2019a).

Older Cuban women's status in their homes makes them doubly vulnerable. The crisis is more acute because, despite the sociocultural transformations fostered by equality policies, Cuba continues to be a family-centred society, and elderly care is typically the woman's responsibility, a characteristic shared with the majority of countries in Latin America. The distribution of the care burden at home is highly unequal. Many of these women provide care for older dependents and children, while continue playing other roles in the labour market and in many cases, themselves present health problems that demand some care. Research shows that Cuban women remain the family primary caregivers and those with paid work responsibilities experience the permanent conflict of trying to reconcile multiple demands (Fleitas, 2014).

In Cuba, budget cuts to social assistance programs have especially affected older adults. The impact of these cuts is worrisome, considering that according to the 2012 Census, 40% of Cuban households include at least one older adult. Increasing care services and improving the infrastructure that facilitates domestic work is one of the biggest challenges facing Cuban society today. Both are unevenly impacting men and women in terms of the *use of time*. Data from the last National Population and Housing Census in 2012 indicate that 1,854,750 Cubans were completely dedicated to household duties. On the other hand, a very high percentage of Cuban women who have some public (state) employment also devote a significant part of their time and efforts to domestic work. Furthermore, the Time Use Survey (ONEI, 2003) found that women dedicate a greater percentage of their time (71%) than men (33%) to unpaid work. More recently, the information provided by the Gender Equality

National Survey (ONEI, 2018b) confirmed that women, on average, spend more time on domestic and care duties, with a difference of a little more than 14 h per week, compared to their male counterpart (9.2 h of the difference correspond to domestic work and 4.9 h to care for people in need or dependents) (ONEI, 2018b).

A corollary to this situation is that Cuban women are over-represented among the poorest of the population (Espina, 2008) particularly older women over 60 (Fleitas in IPS, 2014). The evidence has shown that employment stability has a positive impact on the economic security of older adults. However, family care responsibilities as well as the multiple inequalities that women experience when they participate in the labour market (such as lower-income, lower-level occupations) generate a shorter history of formal work and consequently, smaller pensions in old age. The multiplication of family survival strategies and raising revenue appear as a consequence of this inequality. At the same time, these phenomena have been identified as key features in the understanding of the social re-stratification of Cuba today (Espina, 2008).

The Feminisation of Migration as a Complementary Strategy Facing the Care Crisis

The growing feminisation of migration contributes to deepen the current population crisis in Cuba. Considering this situation, Albizu-Campos and Alfonso (2000) point out that the migratory crisis was circumstantial in specific periods. However, it has become a structural phenomenon since 2006. As an explanation, several authors argue that in Cuba, particularly since the 1990s, international migration has become a *family and personal strategy* to address many social problems (Domínguez, Machado, & González, 2016; Martín, 2006).

Martín (2006, 2007) states that Cuban families often perceive emigration as a survival strategy to face the current economic crisis and the low quality of life. In this context, Cuban families have emerged as key actors in the development of transnational practices and in the formation of transnational communities, although the transnational perspective has not been sufficiently incorporated into the analysis of Cuban international migration. Cuban migrants' current characteristics, composition, dynamics, and transnational networks have become a driving force of information and the circulation of material and cultural goods between origin and destination. This process occurs in a continuous flow resulting from frequent contacts between family members who have migrated and those who have stayed in Cuba (Duany, 2001; Rodríguez & Cumbrado, 2018).

Another research topic is related to the psychosocial impact that emigration produces on the family. The impact can be expressed in different areas, such as family structure, cohesion, and dynamics as well as different care recipient groups, such as children and elderly people. Martín, Aja, Casaña and Martín

(2007) in their research carried out in Havana City, state that for 93% of those interviewed, emigration constitutes *a strategy to face the crisis in current Cuban daily life*, and at the same time, they think it may be the most effective way to deal with daily problems.

The decline in family size and the increase of female-headed households should be considered indirect strategies to face the care crisis. Currently, the size of the Cuban family has decreased, consolidating itself as a small family, with few children (Rodríguez & Albizu-Campos, 2015). In the last sixty years, Cuban households have seen the average number of their members reduced by two people per household as an absolute value and by 40.8% as a relative value (Benítez, 2015). Care for children and the elderly and their multiple needs has become a concern since there are fewer people available to provide care. The social crisis and economic reforms in Cuba since the 1990s, have contributed to the increase of female-headed households, thereby also contributing to weakening the role of men as providers. Currently, about half of Cuban families have female heads of household as they are the ones who contribute most to the home, even if they have a partner (ONEI, 2011).

Addressing the deficit and deterioration of care services for older people on the island (Acosta, et al., 2018), Cuban families seem to combine resources and actions to solve the needs arising from population ageing. The temporary return could be considered one of these complementary strategies, produced from visits by Cubans residing outside the country. Many Cubans temporarily visit their country of origin and return to the place where they have emigrated. It is a peculiarity of the external migration process of Cuba, since according to the immigration law, every person who emigrates definitively, does not have the possibility of returning to live in the country. Return is only possible temporarily through visits to relatives. Although, the same strategy is used by those emigrants who maintain residence in Cuba, because after the migratory reform they can return, without losing citizenship rights, before spending two years abroad.

Martín et al., (2007) point to a sustained growth and interest in travelling to visit and sending remittances to relatives in Cuba. Most interviewees for this study agree that "the family motive is the main reason for being with the family, children and grandchildren" (69%). With the changes incorporated in Cuba's migration law,[2] visitors to the island in 2013 reached a historical record of more than 600,000 persons, considering only Cuban-Americans and

[2]The most important aspects of the Cuban immigration law passed in 2013 in connection with the family topic are a) the elimination of exit permits issued by the government and the letters of invitation; b) allowing more time to staying outside the country up to two years without being considered an emigrant; and c) the granting of the temporary exit permit for children. With this, the family reunification process was reactivated.

Americans passing through the island, and producing again a record in 2016, according to The Havana Consulting Group.[3]

Another migratory category currently in force in Cuban legislation is the Residence Permit Abroad (PRE-Permiso de Residencia en el Exterior). This permit implies a form of temporary migration and returns to the country. People under this migratory category, studied by Martín et al., (2007), are characterised by having a motivational hierarchy where family reasons predominate, more than economic and professional ones.

Remittance sending is one of the most visible connections between Cuban families in their countries of origin and destination, thereby alleviating the costs of social reproduction and care for people in dependency situations, particularly when these are children and parents. With education and health continue being gratuitous in Cuba, the costs of both services have grown, as families must pay many of the costs associated with schooling, hospitalisation, or health care (Acosta, Picasso, & Perrotta, 2018). According to The Havana Consulting Group LLC (2016), remittances to Cuba have had a vertiginous rise since 2008, when they reached 1.251 billion dollars and increased by 1.043 billion dollars for 2011 (Hernández & Foladori, 2012). Approximately 3.5 billion dollars were sent to Cuba in 2016, a figure expected to continue to increase under the new economic change policies. The recipients of these remittances in Cuba are mostly women. Martín et al., (2007) confirm that 83% of respondents send to family members and closest friends some help that could be money, medicines, clothing, or food. In terms of the sending frequency, 67% send money once, twice, or three times a year, while 32% say occasionally; "whenever economically as possible". Most send money with people travelling to Cuba (57%) and through agencies in a lesser proportion (35%).

Unpaid domestic work is part of the delegation strategies of Cuban families, since this work is often moved from the family to another person or social group. With this transfer, Cuban families not only solve their most pressing care needs, but also delegate in other women the care and domestic responsibilities. On the other hand, older women themselves often provide family care, with 7.9% exclusively dedicated to grandchildren care (Romero, 2015). However, in terms of caring for older relatives, men have very little or no direct involvement. In Cuba, the dearth of social services and infrastructure makes domestic work a greater burden than it usually represents in other societies.

The return of paid domestic work could be interpreted as a care and domestic work derivation strategy to the market and other institutions. The economic crisis effects since 1991 have impacted its reappearance as a new legal occupation. As a consequence, home caregivers and domestic workers have begun to be more visible in Cuban homes. It was possible because, in

[3] More information can be found at http://www.thehavanaconsultinggroup.com/es-es/Articles/Article/44.

2010, the government of Raúl Castro made procedures more flexible and expanded the number of allowed self-employed activities (called *trabajo por cuenta propia*). In those days, the authorities granted 211 licences for domestic service in Cuba. By the end of 2013, the figure amounted to 3,149 licences, granted mostly to women, confirming a constant growth in this economic activity (IPS, 2016). Regarding this topic, it is important to remember that domestic service was frowned upon and almost disappeared during the first decades of the Cuban Revolution. As a result, according to Romero (2015) reference to paid domestic work at home in today's Cuban context constitutes a great challenge for social research. Unlike past times, people who now work in these jobs have a wide range of socioeconomic backgrounds. Unfortunately, social research about this topic in Cuba is limited and incipient, in contrast with the higher attention it receives in the rest of Latin America.

Final Remarks and a Note on Future Research Avenues

Various themes related to the close connections between population ageing and the feminisation of international migrations in Cuba in the context of a deep crisis of care have emerged throughout this discussion. The composition and magnitude that have characterised international migration in Cuba in the last few decades, as well as its impact on the family dynamics and life projects of Cuban women, refer to a structural problem more than a temporary one.

Another aspect to be highlighted is the unequal distribution of the care burden in elderly people and the increasing deficit of caregivers, among other reasons, due to migration. The care burden and the elderly situation for Cuban women in their homes make them doubly vulnerable and over-represented within the poorest. As a consequence, migration becomes a survival strategy for the lack of crisis solutions, the low quality of life, the deficit and deterioration of social support by the state concerning care services. The researches have confirmed a lack of support services and the return of paid domestic work as a strategy to transfer to the market of those domestic and care duties that women are not assuming. The costs associated with it will have an unexpected impact on the increase in inequalities in Cuba.

Both, an unequal distribution of care services for older adults and the shrinking welfare services in Cuba suggest an impending crisis. The failure of social policy to anticipate these problems and plan accordingly further aggravates the situation. The result is an uneven landscape, particularly within a general context of social re-stratification, with some sectors and geographical areas facing greater hardships than others. An increasingly feminised international emigration is undoubtedly leading to both a general population ageing and a crisis in elderly care.

This situation demands further research as well as a public debate on the present and future of the social organisation of elderly care and the responsibilities of the welfare state. In order to build a future research agenda in Cuba, one of the first lessons emerging from this research is the need to investigate, disseminate, and increase awareness of the connection between migration, ageing, and care in the island. Gender is another important aspect to incorporate in the research agenda. It is urgent to produce data disaggregated by gender and incorporate the contribution of theoretical discussions about care. This would help in linking the demographic dynamics with the social care challenges from a gender perspective.

Another finding demanding further research is the global/transnational dimension through which Cuban families are currently addressing their care needs in a context of the feminisation of migration. The hypothesis that management of family well-being in Cuba reaches social and material dimensions exceeding the borders of the island needs to be empirically tested, particularly through the incorporation of qualitative methodologies that could document the strategies used by families in the context of the accelerated social changes caused by the aforementioned demographic processes. It would also be useful to deepen the understanding of the motivations as well as the roles of Cuban migrant women.

References

Acosta, E. (2013). Mujeres migrantes cuidadoras en flujos migratorios sur-sur y sur-norte: expectativas, experiencias y valoraciones. *Polis* [En línea], 35. http://polis.revues.org/9247.

Acosta, E. (2015). *Cuidados en crisis y mujeres migrantes hacia España y Chile*. Dan más de lo que reciben, Bilbao, España: Ediciones Universidad de Deusto.

Acosta, E., Picasso, F., & Perrotta, V. (2018). *Cuidados en la vejez en América Latina. Los casos de Chile, Cuba y Uruguay*. Santiago, Chile: Fundación Konrad Adenauer, Programa de Políticas Sociales en Latinoamérica (SOPLA), http://repositorio.mides.gub.uy:8080/xmlui/handle/123456789/1013.

Aja, A. (2006). *Tendencias de la emigración desde Cuba a inicios del siglo XXI*. Organización Internacional de Migraciones (OIM), Mexico, April: Paper presented at the Encuentro sobre Migraciones.

Aja, A. (2007). La migración desde Cuba. *Aldea Mundo: Revista Sobre Fronteras E Integración, 11*(22), 7–16.

Aja, A., & Hernández, W. (2019). Dinámica de la población y sus interrelaciones en Cuba y sus territorios. Recomendaciones para la acción. *Novedades EnPoblación, 15*(29), 56–74.

Albizu-Campos, J. C. & Alfonso, A.C. (2000). *La población cubana en la segunda mitad del siglo XX*. Havana: Centro de Estudios Demográficos (CEDEM).

Arriagada. (2010). *La organización social de los cuidados y vulneración de derechos en Chile*, ONU Mujeres-CEM. http://www.unwomen.org/wp-content/uploads/2011/10/La-organizaci%C3%B3n-social-de-los-cuidados-y-vulneraci%C3%B3n-de-derechos-en-Chile.pdf.

Arriagada, I. y Moreno, M. (2011). "La constitución de las cadenas globales de cuidado y las condiciones laborales de las trabajadoras peruanas en Chile", in Stefoni, C. (edit.) *Mujeres inmigrantes en Chile. ¿Mano de obra o trabajadoras con derechos?*, Santiago de Chile: Ediciones Universidad Alberto Hurtado (149–192).

Arriagada & Todaro. (2012). *Cadenas globales de cuidados: El papel de las migrantes peruanas en la provisión de cuidados en Chile*. Santiago de Chile: ONU-Mujeres.

Benítez, M. (2015). Envejecer en Cuba: Mucho más que un indicador demográfico. *Novedades En Población, 11*(22), 10–19.

Cano, Sofia, & Martínez (2009). "Conocer para legislar y hacer política: los desafíos de Chile ante un nuevo escenario migratorio", *Serie de Población y Desarrollo* 88, Santiago de Chile: CEPAL/CELADE.

CRI. (2011). *La diáspora cubana en el siglo XXI*. Miami: Florida International University.

Díaz, M., & Pérez, A. (2011). La *organización social de los cuidados y vulneración de derechos en España*. ONU Mujeres, http://www.unwomen.org/wp-content/uploads/2011/10/La-Organizaci%C3%B3n-social-de-los-Cuidados-y-Vulneraci%C3%B3n-de-Derechos-en-Espa%C3%B1a.pdf.

Domínguez, E., Machado, L., & González, D. (2016). Nueva inmigración y comunidad cubana en Estados Unidos en los albores del siglo XXI. *Migraciones Internacionales, 8*(4), 105–136.

Duany, J. (1999). Cuban communities in the United States: Migration waves, settlement patterns, and socioeconomic diversity. *Pouvoirs Dans La Caraïbe, 11*, 60–103.

Duany, J. (2001). Redes, remesas y paladares: La diáspora cubana desde una perspectiva transnacional. *Nueva Sociedad, 174*, 40–51.

ECLAC/CELADE. (2009). Proyección de población, Observatorio Demográfico N° 7, Año IV, Santiago de Chile.

Espina, M. (2008). Viejas y nuevas desigualdades en Cuba: Ambivalencias y perspectivas de la reestratificación social. *Nueva Sociedad, 216*, 134–149.

Fleitas, R. (2014). "El discurso invisible del envejecimiento: El dilema de género," *Revista electrónica "Antropológicas"*, Cátedra de Antropología "Luis Montané," University of Havana.

Fraser, N. (2016). Las contradicciones del capital y los cuidados. *New Left Review, 100*, 111–132.

Gonzálvez, H., & Acosta E. (2015). "Cruzar las fronteras desde los cuidados. La migración transnacional: más allá de las dicotomías analíticas." *Las fronteras del transnacionalismo: Límites y desbordes de la experiencia migrante en el centro y norte de Chile*, Menara Guizardi (ed.). Santiago de Chile: Ocho Libros: 126–150.

Hagen-Zanker, J., & Mallet, R. (2020). *Understanding migrant decision-making: Implications for policy*, Working paper, Coventry, UK: MIDEQ: Migration for Development & Equality.

Hernández, J., & Foladori, G. (2012). La dinámica poblacional: Retos del socialismo cubano. *Papeles de Población, 18*(74), 1–29.

Herrera, G. (2012). Repensar el cuidado a través de la migración internacional: mercado laboral, Estado y familias transnacionales en Ecuador. *Cuadernos de Relaciones Laborales, 30*(1), 139–159.

Hochschild, A. (1989). The Second Shift. New York: Avon.Hochschild, A. (2001). "Las cadenas mundiales de afecto y asistencia y la plusvalía emocional", En Hutton, W. y Giddens, A. (Eds.). *En el límite. La vida en el capitalismo global*, Tusquets Editores, Barcelona.

Hochschild, A. (2001). "Las cadenas mundiales de afecto y asistencia y la plusvalía emocional", En W. Hutton y A. Giddens (Eds.). *En el límite. La vida en el capitalismo global*, Barcelona: Tusquets.

Hondagneu-Sotelo, P. (1994). *Gendered transitions*. Mexican experiences of immigration, Berkeley: University of California.

Hondagneu-Sotelo, P. y Avila, E. (1997). "'I'm here, but i'm there': The meanings of Latina transnational motherhood." *Gender & Society, 11*,548–571.

ILO. (2018). *Labour overview 2018: Latin America and the Caribbean*. Peru: ILO Regional Office for Latin America and the Caribbean.

IPS. (2014). "Socióloga mira al envejecimiento con enfoque de género", 29 de mayo, http://www.ipscuba.net/genero/sociologa-mira-al-envejecimiento-con-enfoque-de-genero/.

IPS. (2016). "El trabajo doméstico crece en Cuba, en un escenario atípico", 5 de noviembre, http://www.ipscuba.net/genero/el-trabajo-domestico-crece-en-cuba-en-un-escenario-atipico/.

Lube, M., & Gonzálvez, H. (2019). Women in (Dis)placement: The field of studies on migrations, social remittances, care and gender in Chile. *Revista de Estudios Sociales, 70*, 100–114.

Marrero, G. (2011). "Las migraciones internacionales en Cuba: ¿Una tendencia hacia la feminización?" *El Observatorio*, Havana: Centro de Estudios de Migraciones Internacionales (CEMI). If it is a report, it must be specified the web site where it was taken.

Martín, C. (2006). "Familia y emigración: Memoria histórica y discurso social desde 1959." La Habana: Centro de Estudios de Migraciones internacionales, www.uh.cu/centros/cemi/anuario%200708/.../6Consuelofamilia.pdf.

Martín, C. (2007). "Familias y emigración cubana: Estrategias cotidianas en su contexto psicohistórico." Paper presented at V Reunión del Grupo de Trabajo, Familia e Infancia, CLACSO, October, México. http://www.uh.cu/centros/cemi/cup-content/uploads/2011/11/9-Consuelo-Martin-Familias.pdf.

Martín, C., Aja, A., Casaña, A., & Martín, M. (2007). *La emigración de Cuba desde fines del siglo XX y principios del XXI: lecturas y reflexiones mirando a la Ciudad de La Habana*, Anuario Digital CEMI 2007-2008. Migraciones Internacionales y Emigración Cubana: 137-179.

Martínez, R. (2010). *Bienestar y cuidados: el oficio del cariño. Mujeres inmigrantes y menores nativos*, Madrid: Consejo Superior de Investigaciones Científicas (CSIC).

Martínez, R. (2011). La reorganización de los cuidados familiares en un contexto de migración internacional". *Cuadernos de Relaciones Laborales, 29*(1), 93–123.

Mesa-Lago, C. (2012). *Sistemas de protección social en América Latina y el Caribe: Cuba*. Santiago de Chile: CEPAL, https://www.cepal.org/es/publicaciones/4034-sistemas-proteccion-social-america-latina-caribe-cuba.

Mora, C. (2008). Globalización, Género y Migraciones. *Polis, 7*(20), 285–297.

Nuñez, M. (2007). Género y migraciones externas en Cuba entre 1985 y 2005: Resultados preliminares. *Novedades En Población, 3*(6), 45–70.

Núñez, M. (2010). Cubans abroad: A gendered case study on international migrations. *Cuban Studies, 41*, 105–125.

ONEI. (2003). *Encuesta sobre uso del tiempo*. La Habana: Oficina Nacional de Estadística e Información.

ONEI. (2011). Resultados de la Encuesta Nacional de Envejecimiento poblacional. Primer Informe. Edición 2011, La Habana: Oficina Nacional de Estadísticas.

ONEI. (2018a). *Encuesta Nacional Migración, Migraciones a nivel de estratos de asentamientos: Resultados principales en la Encuesta Nacional de Migraciones*, ENMIG, Havana: ONEI-UNFP-CEPDE, http://www.onei.gob.cu/node/14275.

ONEI. (2018b). *Encuesta Nacional sobre Igualdad de Género- ENIG 2016. Informe de resultados*. Havana: Editorial de la Mujer, http://www.onei.gob.cu/node/14271.

ONEI. (2019a). *Encuesta Nacional de Envejecimiento Poblacional—ENEP 2017*. Cuba: La Habana.

ONEI. (2019b). Anuario estadístico de Cuba 2018. Empleo y Salarios. Edición 2019. Havana: ONEI.

Oso, L. (2008). "Migración, género y hogares transnacionales", En J. García Roca y J. Lacomba (coords.): *La inmigración en la sociedad española: una radiografía multidisciplinar*, Barcelona: Bellaterra (561–586).

Oso, L., & Parella, S. (2012). Inmigración, género y Mercado de trabajo: una panorámica de la investigación sobre la inserción laboral de las mujeres inmigrantes en España. *Cuadernos de Relaciones Laborales, 30*(1), 11–44.

Padrón, C. (2019). "En cifras. La desigualdad de género en Cuba," in *Tremenda Nota*, https://www.tremendanota.com/en-cifras-la-desigualdad-de-genero-en-cuba-graficos/.

Parella, S. (2003). *Mujer Inmigrante y Trabajadora: La Triple Discriminación*, Barcelona: Anthropos.

Parreñas, R. (2001). Transgressing the nation-state: the partial citizenship and "Imagined (global) community" of migrants Filipina domestic workers. *Signs, 26*(4), 11–29.

Pérez, A. (2009). *Miradas globales a la organización social de los cuidados en tiempos de crisis I: ¿Qué está ocurriendo?*, Serie Género, Migración y Desarrollo, Working Paper No. 5, INSTRAW.

Pérez, & López, S. (2011). *Desigualdades a flor de piel: cadenas globales de cuidados. Concreciones en el empleo de hogar y políticas públicas*, Madrid: ONU Mujeres.

Pribilsky, J. (2004). 'Aprendemos a convivir': Conjugal relations, Co-parenting, and family life among Ecuadorian transnational migrants in New York City and the Ecuadorian Andes. *Global Networks, 4*(3), 313–334.

Pribilsky, J. (2007). *La Chulla Vida: Gender, migration, and the family in Andean Ecuador and New York City*. Syracuse, NY: Syracuse University Press.

Rodríguez, G., & Albizu-Campos, J. (2015). La población de Cuba hoy. *Novedades En Población, 11*(22), 1–9.

Rodríguez, M., & Cumbrado, M. (2018). Migración, transnacionalismo y emprendimientos privados en Cuba. Estudio de casos en el Centro Histórico de La Habana. *Novedades En Población, 14*(28), 181–198.

Romero, M. (2015). El cuidado infantil en Cuba: Especificidades de su desarrollo tras la nueva apertura del sector 'cuentapropista'. *Revista Del CESLA, 18*, 255–270.

Solé, C., & Parella, S. (2003). Identidad colectiva y ciudadanía supranacional. *Papeles de Economía Española, 98*, 166–181.

Stefoni, C. (2009). Migración, género y servicio doméstico. Mujeres peruanas en Chile. In C. Mora & M. E. Valenzuela (Eds.), *Trabajo doméstico: un largo camino hacia el trabajo decente* (pp. 191–232). Santiago de Chile: OIT.

Stefoni & Fernández. (2011). "Mujeres inmigrantes en el trabajo doméstico: entre el servilismo y los derechos" in Stefoni, C. (edit.), *Mujeres inmigrantes en Chile. ¿Mano de obra o trabajadoras con derechos*, Santiago de Chile: Ediciones Universidad Alberto Hurtado (43–72).

Stavropoulou, M., Torres, A., Samuels, F., Solís, V., & Fernández, R. (2020). *'The woman in the house, the man in the street' Young women's economic empowerment and social norms in Cuba*, ODI Report, https://www.odi.org/publications/

16681-woman-house-man-street-young-women-s-economic-empowerment-and-social-norms-cuba.

The Havana Consulting Group & TECH. (2016). "Cuba: el mercado de remesas con el crecimiento más dinámico de América Latina", June 23, http://www.thehavanaconsultinggroup.com/es-es/Articles/Article/20.

Vega, C. (2009). *Culturas del cuidado en transición. Espacios, sujetos e imaginarios en una sociedad de migración*, Barcelona: Editorial UOC.

Vicente, T., & Setién, M. L. (2005). "Modelos migratorios femeninos," in J. González y M. L. Setién (eds.), *Diversidad migratoria: Distintos protagonistas, diferentes contextos*, Bilbao: Universidad de Deusto, http://www.deusto-publicaciones.es/deusto/pdfs/migraciones/migraciones01.pdf.

Elaine Acosta González is a Sociologist with a PhD in International and Intercultural Studies from Deusto University, Bilbao. Currently, she is a Visiting Scholar in Cuban Research Institute at Florida International University. She has served as Chair of the M.A. Program in Sociology at the Universidad Alberto Hurtado in Chile (2013-2016). Her research areas include care work, international migration, aging, and welfare policies in Latin America. She is committed to research from a gender and rights-based perspective. Professor Acosta has conducted studies for international agencies (IOM, IADB, UNICEF, UNDP) and served as consultant for several governmental offices in Chile. She is currently co-director of the Interdisciplinary Research Program on Care Work, Family and Welfare and member of the Care, Gender, and Citizenship Studies Network (ESOMI-Spain).

Part III

Gendered Work, Employment and Skills Mobility

CHAPTER 12

Gender Bias in Skills Definition, Labour Market Dynamics and Skills Recognition

Anna Boucher

Abstract The labour market is a key site of gender discrimination. The interaction of immigration and labour regulation amplifies some of the classic forms of employment disadvantage experienced by women, such as horizontal and vertical segregation of the labour market. Policies in these areas can affect women's access to skilled immigration status or other visas, accreditation issues upon arrival, skills recognition and their overall capacity for human mobility. This chapter will provide an overview of the feminist scholarship on labour immigration and gender, including work on the gendered nature of "skill", skills recognition and labour market engagement, access to permanent visas for women compared with men and the way in which gendered paradigms can be reflected in the overall direction of immigration regimes towards a marketised model.

Introduction

Gender discrimination in the labour market is a well-established phenomenon across a variety of labour systems, both in the democratic and non-democratic world. The sources of differential outcomes for men and women in labour force engagement and outcomes vary from structural factors such as obstacles to women's employment, gender differences in training, leading to concentration of women in some sectors and men in others (horizontal segregation) and

A. Boucher (✉)
Government and International Relations, University of Sydney,
Camperdown, NSW, Australia
e-mail: anna.boucher@sydney.edu.au

C. Mora and N. Piper (eds.), *The Palgrave Handbook of Gender and Migration*, https://doi.org/10.1007/978-3-030-63347-9_12

differences in the advancement of men and women into leadership positions (vertical segregation). Intersecting and worsening these effects are women's far higher experiences of sexual harassment and sexual violence in the workplace, an issue that has a long-standing legacy but which has received heightened attention since the rise of the #MeToo movement. Sexual abuse in the workplace carries with it further barriers to female employment and advancement by intimidating women and restricting their willingness to speak out against other forms of gender discrimination (ILO, 2019).

All of these issues that native female workers face are present and potentially heightened for female migrants, given the increased vulnerability that comes with migrant status (Piper, 2008). At the same time as gender discrimination is a central issue in labour markets, labour (economic-based) immigration has also become a more important component of immigration selection. As recent comparative work across thirty major migrant-receiving nations demonstrates that, labour immigration and in particular, temporary labour immigration, now comprises the majority of legal immigration intake globally (Boucher & Gest, 2018). While not all of this labour immigration is skilled, within this broader picture such immigration has also become more important globally.[1] A United Nations World Population Policies Survey indicated that two-thirds of the 172 member states of the UN in 2013 had an explicit interest in developing skilled immigration policies (Czaika & Parsons, 2016). Recent policy changes also reflect this trend with an increase in skilled temporary intake in Australia, proposals for a new points tested model for skilled immigration in the United Kingdom (in light of Brexit) and a revised points test in Korea as a few of many recent examples (SOPEMI, 2019: 48 and 49).

This chapter begins with an overview of the feminist scholarship on labour immigration and gender. It then moves into three issues of particular interest to an analysis of gender and skilled immigration in particular: i) the gendered ways in which "skill" is defined within labour immigration selection regimes; ii) gendered concerns around labour market accreditation and integration and iii) the potential for a gendered hierarchy in how female compared with male immigrants are selected within immigration regimes.

Feminist Scholarship on Labour Immigration and Gender

There is a broad feminist scholarship on skilled immigration and gender. This scholarship is important because it considers how the shift that is occurring globally towards both skilled and temporary labour holds different implications for women than men, and different impacts across and within different ethnic and linguistic groups. Intersecting inequalities can have compounding

[1] Reliable data on the percentage of overall labour flows globally that are skilled is not available as countries differ in their definition of "skill", meaning that no harmonised estimate can accurately be relied upon.

differential effects in this area of regulation, although it is often cast as being "non-discriminatory" or "value-free" and superior to both race-based and even more random lottery driven approaches to immigration selection (Boucher, 2016: 4; Boucher 2019b).

Much of the earlier scholarship in this area identified men's predominance as skilled workers and therefore also as skilled immigrants (see for instance Badkar, Callister, Krishnan, Didham, & Bedford, 2007; Dauvergne 2000; Dobson & Crush 2004; Fincher 1997; Fincher, Foster, Giles, & Preston, 1994; Piper & Yamanka, 2008). There is also emerging evidence that gender interacts with country of origin to influence skilled visa distribution (Kofman, 2014: 122). It is important to note that clear cross-national data on the gender representation of women as skilled immigrants is lacking as many countries do not record these data. For instance, a review of all APEC economies in 2019 found that 13 of 21 had some form of gender disaggregated data but only three out of 31 data differentiated between the principal applicant (the person making the claim who meets the skilled immigration requirements) and the secondary applicant (the trailing spouse) (Boucher, 2019a). Similarly, an overview of 37 OECD skilled immigration visas across twelve countries in 2013, found that only one—Canada—published gender disaggregated data according to primary and secondary status (Boucher, 2016: 38). Yet, such distinctions are crucial to consider female migrant outcomes from a gender perspective. Paucity of gender disaggregated data, particularly that arranged according to visa class, principal status and correlated with ethnicity and occupation, stymies detailed gender analysis in this field by both government officials and academics.

Women's heightened rates as dependent migrants or trailing spouses within skilled immigration routes comprise part of the foundational literature on women and skilled labour (i.e. Boyd, 1997; Zlotnik, 1995). Change to such rates over time is also part of this analysis. For instance, in a study of the effects of gender mainstreaming policies upon gendered outcomes in Canada's skilled immigration policies, Boucher (2010) considers gender trends over a decade. Yet, once this prevailing trend of underrepresentation of women as skilled immigrants is recognised, a more focused question lies in the reasons for this outcome and the potential for variation and even amelioration of gender discrimination through skilled immigration regulatory design. It is here that the scholarship on skill and gender is important.

Gendered Nature of "Skills" Definition

There are different ways of defining "skill" and each carries with it different gendered implications. It is generally clear and accepted that there is no one singular definition of "skill" that exists cross-nationally (Czaika & Parsons 2016). This chapter canvasses several key approaches to skill definition within skilled immigration design and reflects upon their gendered implications: i) human capital; ii) wage driven approaches; iii) educational; iv) occupational;

v) employer-defined and vi) extraordinary talent approaches. This analysis links into existing debates within organisational sociology and industrial relations that are critical of notions of "skill" as objectively determined. These approaches emphasise the way that concepts of skill are socially constructed and the ways in which this social construction has advantaged those in male-dominated trades over female-dominated occupations (i.e. Fincher, 1997; Phillips & Taylor 1980; Steinberg, 1990). A recent contribution of the feminist work in this field is the undervaluation of care work as skilled work, in part related to the strong experiential, non-credentialised nature of such training (Nishikawa, 2011: 115–116).

Human Capital

A central component of many skilled immigration policies in recent years has been their rhetorical focus upon human capital even if the actual theoretical underpinnings of human capital theory as espoused by Gary Becker are not always fully articulated (Boucher, 2016, 5–6). Becker (1964) defined human capital as "the outcome of deliberate investment in educational and work experience on the part of individuals deciding to incur present costs for future benefits". Within immigration selection, this often means focusing on the occupational, educational or linguistic skills of the applicant. Rhetorical substantiation of a human capital approach within skilled immigration selection is evident in government discussion of the Express Entry model in Canada as a form of human capital admission (CIC, 2015) or the skilled stream in Australia in recent years (DHA, 2019). There have also been gendered critiques of human capital in its various forms.

Some theorists note that the process of defining "skill" is in itself an integral step in exercising power "over who can be employed" (Webb, 2015, 268). In particular, concepts of human capital often appear to prioritise individual agents without dependents (Creese, Dyck, & McLaren, 2008). Under skilled migration models that valorise independence, the important role of women's as carers and homemakers who support primary workers, is undervalued and given little weight in visa design.

Wage Driven

Gendered critiques have been levelled at definitions of "skill" that work from the basis of wages. The key argument here is that the gendered wage gap that operates globally, but particularly among higher skilled occupations (Stewart, 2014), renders wages a biased measure of "skill". Various researchers have measured the "motherhood wage penalty" that operates to reduce working mothers' wages by between 5 and 10 per cent per child when compared to working fathers (i.e. Anderson, Binder, & Krause, 2003; Kahn, Garcia-Manglano, & Bianchi, 2014). The wage gap worsens with years out of the workforce (Stewart, 2014: 101). That said demand-side factors such as gender discrimination also explain part of the gender wage gap (Weinberger, 2011: 972). Aside from a motherhood penalty, gender discrimination can play a separate role in influencing wage gaps (Weinberger, 2011: 972). Women may also be less successful at negotiating higher wages given gender discrimination, irrespective of their preference for higher salaries or their fertility choices (Folbre, 2012: 282–283). Wages are therefore far from an objective measure of "skill".

Wage disparities may be particularly high in the occupations in which skilled women predominate, such as health, caring and education (Kofman, 2014). This links to the occupational segregation apparent in occupational selection of migrants, discussed below. As such, the reliance on wages as a proxy for "skill" in immigration selection grids therefore carries with it considerable potential for gender bias. This trend of setting high wage thresholds in order to define skill has increased for the Australian short-term TSS visa and in New Zealand for its skilled migrant category in recent years (SOPEMI, 2019: 48).

Educational

The level of education is sometimes taken as an additional proxy for "skill". International bilateral stock data and various studies define a migrant as "highly skilled" if they have "at least one year of tertiary education" (i.e. Kerr, Kerr, Ozden, & Parsons, 2016: 7). The advantage with these data are that they are often gender disaggregated. Under these analyses, women's representation as skilled immigrants is growing as women are among the tertiary-educated. For instance, Arslan and collaborators (2016: 9) argue that "[o]ver the next decade, the share of tertiary educated rose faster for females than for males, and the gender balance of tertiary education became far more equal". Women may also be motivated to emigrate, not only due to their increased education but also gender discrimination in the country of origin (Docquier, Lowell, & Marfouk, 2009). Yet, analyses that present a rosy picture of women's improvement as skilled immigrants, fail to acknowledge that even with educational qualifications, women are often not acquiring skilled immigration visas because other dimensions of "skill" are important in the skills assessment processes of selecting governments. In an assessment of thirty-seven skilled visas, only one (the United States' Employment Second Preference) was found to focus mainly on educational qualifications (Boucher, 2016: 37). The other more common mechanisms for selecting skilled immigrants considered in this chapter hold more serious gender implications.

Occupational

"Skill" can also be defined according to occupational skill level. Such approaches to "skill" generally base their understanding upon benchmarks with the International Standard of Occupations (ISCO) 2008 classifications. The top three ISCO levels are often considered are 1) managers, social officials and legislators; 2) professionals and 3) technicians and associated professionals. Occupational classification lists are often used to identify who is considered "skilled" and who is not and this list, in its written form, can "mas[k] the social processes ... behind the prioritisation of some skills over other" (Webb, 2015: 271). These can include gendered processes of skills definition.

Much of the cross-national analysis of skilled immigration admissions has not disaggregated on gender grounds and there is very little or negligible discussion of the role of gender in such analyses (see for instance Parsons et al 2014; 2018 that overlook gender). Yet, analysis that have disaggregated according to gender, for instance Boucher's (2016: 120–121) study of Australian and Canadian occupational classifications of skilled immigration, demonstrates strong horizontal segregation—women are far more likely to enter through health and academic occupations, whereas men predominate in the IT sector. When occupational lists shifted towards occupations where women predominate in New Zealand in the mid-2000s, such as nursing, this also led to an increase in women's representation as independent skilled immigrants (Badkar et al., 2007: 130–131). Variation in women and men's occupational classifications suggests that as occupational lists are changed for skilled immigration selection, this in turn will affect gender composition of skilled immigration admissions.

Employer-Defined

Employer sponsored models are in fact the most common mode of skilled immigration selection (Boucher, 2016: 44). As the shift towards temporary modes of labour selection observed in recent years by the Organisation for Economic Cooperation and Development accelerates, it is also likely that this approach to skilled immigration selection will continue to grow (SOPEMI, 2019). Under employer-sponsored models, employers or their agents petition government for a visa for the migrant and generally, the migrant is tied to the employer for the period of their visa. These forms of selection are likely to disadvantage women. For instance, Reskin and McBrier (2000) demonstrate that "[m]ore informal recruitment practices are more likely to disadvantage women" as they rely upon existent informal networks that have been shown to advantage existing power structures, even in unintended or implicit ways. Further, employer-sponsored models may also heighten the risk of other forms of implied bias, whether they be ethnic or linguistic as it is ultimately the decision of the employer rather than a more transparent points test that is adjudicating selection of immigrants (Hawthorne, 2011: 157).

The discussion of employer-sponsored models brings us to the issue of softs skills. Soft skills play a greater role in employer-sponsored approaches than under other selection methods, especially where the employer has already met the individual. "Soft skills" can become a proxy for other attributes such as "demeanour, accent and style" (Warhurst & Nickson, 2009) and emotional attitude (Payne, 2009). A focus on physical appearance may also mean different things for women than for men. While there is a general positive bias to beauty in employment (Johnson, Podratz, Dipboye, & Gibbons, 2010), women's physical appearance is more heavily scrutinised in the workplace than men's and operates better in stereotypically female jobs but actually worsens job outcomes in leadership positions (Taledano, 2013: 695, citing Heilman and Saruwatari, 1979).[2] All of these aspects can play into perceptions of a person's "skills", especially at the high-skilled levels that we focus on here. Further, "soft skills" can include other things such as the capacity to build social networks or being in a position to take on lower level jobs (Webb, 2015: 276) which may be negotiated differentially for women than for men.

Extraordinary Talent

Most skilled immigration visa systems have a category of "skilled immigration" for those with "very high" or "extraordinary" levels of skill. This visa is often allocated to sportsmen and women, scientists, investors and artistic performers. The highest levels of skilled visas are also those with the lowest levels of women as principal applicants. For instance, in the United Kingdom's Tier 1 (business people, investors and those with exceptional talent) two-thirds are male and under the Tier 2 (highly skilled workers) 78% are male (Lim, 2019: 908). In other countries such as Austria, with its "Red White Red Card" for "Very Highly Qualified Migrants", workers are given higher numbers of points and more flexible visa conditions if they are employed in "mathematics, informatics, natural science of technology", all male-dominated occupations (Austrian government, 2019). In a recent philosophical piece on these patterns, Desiree Lim (2019: 911) argues that the policy design of exceptional talent visas is not intentional gender discrimination by government rather, negligent discrimination, as it fails to accommodate women's talent within definition of the "highly skilled". However, here perhaps there is hope that as advanced economies shift towards new areas of innovation and understandings of "knowledge", so too will definitions of "extordinary talent" change. As Walby (2011: 13) notes, if the "knowledge economy" includes areas such as service management, transport and financial mediation, it may also include more women in the remit of the highly skilled more than when there is a narrow focus upon IT-based occupations. Definitional choices on the part of government will be crucial in this regard.

[2]This assumption could itself be changing as more men enter into service-based work where appearance is valued for both genders and as conceptions of gender-norms in the workplace shift (Warhurst & Nickson, 2009: 400).

Skills Recognition and Labour Market Engagement

Aside from the normative questions of admission and its gendered effects is the issue of what happens to migrants once they are admitted into the host society and economy. Issues of whether their skills are acknowledged and what other obstacles they might face to labour market engagement are central issues here and have also been considered through a feminist lens. Devaluation of skills is a common experience among men and women as a result of the migration process. At times, give regulatory obstacles or accreditation, such obstacles mean that complete retraining is required (Wojczewski et al., 2015). The rates of devaluation do differ across visas and across countries, with some countries demonstrating low and/or improving rates of devaluation (i.e. Australia, see Hawthorne, 2015) compared with others, such as Canada where devaluation is more common (Creese et al., 2008: 278).

Yet, the implications of such skills devaluation through migration are different for women than for men. For women, migration can often lead to "re-domestication due to care or reproductive responsibilities" (see also Banerjee & Phan, 2015; see also Iredale, 2005, cited below). This trend towards "re-domestication" can be particularly the case for partners of skilled immigrants whose careers are often interrupted through the immigration process by domestic duties or entry into lower paid health and human services work (Webb, 2015) or through "survival jobs" (low-skilled, temporary work) in order to make ends meet (Banerjee & Phan, 2015: 348). One of the reasons for women's disproportionate "re-domestication" after migration is that it can be hard to find either paid or unpaid childcare (such as extended family) after migration and the choice of women to revert from paid to caring roles may be pragmatic or cultural (Phan, Banerjee, Deacon, & Taraky, 2015, 2072; Webb, 2015: 35–6) (see also Dodson, next chapter).

These patterns in turn can reinforce gendered patterns of male breadwinners and female care givers within migrant households (Banerjee & Phan, 2015). There is also evidence that admissions policy can play a role in deskilling—for instance, women who are illegally present in a host society may be more likely to be deskilled as they cannot have their skills and qualifications recognised given their illegal status (Ncube, Bahta, & Jordaan, 2019: 18). This phenomenon has been noted in developed Western immigration countries such as Australia, Canada Switzerland, but also authoritarian regimes, such as Qatar, where gendered institutional, cultural and employment norms can interact to affect female migrants and the recognition of their skills (Rodriguez & Scurry, 2019).

Other research finds that even when female migrants do not experience gendered obstacles in skills accreditation or rehiring, they may navigate gendered obstacles in the workplace that operate as an effective form of deskilling (Grigoleit-Richter, 2017: 2745). The experiences are generally worse for women with certain ethnic backgrounds and with certain accents (Grigoleit-Richter, 2017: 2745–6), although Western women can also experience deskilling. One such example is European female partners in Israel who have experienced considerable deskilling and redomestication following migration (Kranz 2019). That said, deskilling of female migrants is not a universal experience and some people actually experience upward mobility as a result of their migration (Charpentier and Quéniart 2017: 442). Further, the availability of childcare in the host society can act as a mitigating factor that can lower perceived feelings of negative settlement and improve outcomes for female migrants (Phan et al., 2015).

Even once issues of entry, skills accreditation and childcare are addressed, there is evidence that women face greater challenges in accessing employment than men. This may relate to the initial visa status of male applicants (evidence suggests that men, as primary applicants, are more likely to find work, especially if it is tied to their visa) or the better language abilities of men than women (O'Dwyer & Colic-Peisker, 2016). An evaluation of a four-week intensive professional transition course in Australia called the "Skilled Professional Migrant Programme" found that men were 1.7 times more likely than women at the end of the programme to find work (O'Dwyer & Colic-Peisker, 2016: 54). This study, although of a small sample, is important, because it suggests that early career training interventions already demonstrate gendered differences that can exacerbate over time.

The evidence of deskilling within countries of immigration is mirrored once female migrants return to their country of origin. Such deskilling is particularly an issue for migrants engaged in temporary or circular contract work, where they move frequently between sending and receiving nations, or in times of global economic recession. An assessment of return migration of Filipino migrants to the Philipines demonstrates that female migrants retrenched from overseas work can be unduly affected compared with male migrants. They can also be placed through Filipino government training programmes into highly feminised roles upon return, such as in nail palours, where their skills are not adequately recognised or develped (Spitzer & Piper, 2014: 1016–1018; see also Withers, 2019 for Sri Lanka). Interestingly, a distinctive focus upon female skilled return migrants as opposed to female migrants generally, is lacking in the scholarship.

Visa Privileges and Skill: Threshold Between Skill and Semi-Skill

The discussion around skills definition makes clear that the distinction between "skilled" and "unskilled" immigration is not bright line, rather contested and fungible. Categories of "unskilled" work have become "skilled" such as in the area of care work managers, or nursing, when sufficient labour shortages are in place. However, the differential rights and privileges that flow to "skilled" compared with "unskilled" migrant workers are considerable and given the disaggregation of workers along both gender and skill lines, raise gender discrimination concerns. As Martin Ruhs (2013) argues persuasively, in many countries there is a trade-off between the rights accrued to workers and the number of workers admitted. In essence, countries are likely to let in large numbers of temporary and unskilled workers but with fewer rights and a small number of skilled immigrants with more rights.

Where this analysis becomes relevant to gender is in the division between women's and men's representation across skilled and unskilled visa categories. Women are more likely to enter through low-skilled visa channels than high-skilled ones, particularly through care work, which aside from managerial roles, is generally classified as unskilled and where exploitation and abuse are more common (Stasiulis & Bakan, 2005). While recommendations have been made for the creation of skilled visas for carers (i.e. Hugo, 2009), these have not been adopted in policy settings. A review of 37 skilled visas demonstrated that only a few had clear pathways for care workers.[3] Disproportionately, migrants in care-related occupations, enter through unskilled visas (Boucher, 2016: 45–46).

Men predominate as principal skilled applicants in highly skilled channels, whereas women have high rates of entry as economic migrants in low-skilled channels and as undocumented workers (see for instance Boucher, 2016; Piper & Yamanka, 2008). This stratification along skill and gender lines creates differential rights for male and female economic migrants. As Boucher (2016: 146) argues: "Low-skilled pathways offer entry opportunities for female immigrants (and often racialized minorities from the Global South) that do not exist under a system skewed towards the highly skilled. On the other hand, low-skilled visas also limit the rights of such workers and lead to stratifications across immigrant groups along gender and source country lines".

Many low-skilled visas offer no access to permanent residency, welfare provision, health benefits and rights of family reunification that exist under skilled visas (Ruhs, 2013). Access to highly skilled status therefore offers access to a bundle of other benefits that assist with economic and social integration of migrants over the longer term, whereas denial of such benefits can have the opposite effect and reinforce precariousness.[4] Further, there is evidence that once in a low-skilled job, there is limited mobility for immigrant women into more skilled occupations with better pay. In a study undertaken across France, Germany, the Netherlands, Italy, Spain and the United Kingdom over seven years, only the United Kingdom and the Netherlands show some evidence of upward mobility for migrant women (Ballarino & Panichella, 2017: 133).

[3]Denmark for child care workers and Australia for child care managers. This states that law in 2011, see Boucher 2016 for further analysis.

[4]Fay Faraday (2016) provides an impressive overview of the interacting effects for a bundle of low-skilled visas in Canada.

Conclusion—A Marketised Model of Immigration Selection

Globally, in both democracies and non-democracies, governments are moving towards temporary labour regimes, with a focus on both low-skilled and high-skilled streams of entry. With structural ageing and diminishing fertility rates in many countries, it is clear that labour immigration for care work is playing and will continue to play an escalating role into the future. At the same time, skilled labour is needed to meet skills shortages in agile and changing knowledge economies. The contemporary picture of migration selection can be likened to a "market model" of immigration where we observe higher rates of temporary immigration, high labour intake compared with other immigration categories and reduced opportunities for naturalisation (Boucher & Gest, 2018). Consistent with other areas of social and labour policy that are becoming marketised (i.e. Meagher & Goodwin, 2015), employers and markets plan a heightened role in this model. Within this model, attention to equality concerns by government are arguably lessened where they cannot be subsumed within broader market logics. This chapter has demonstrated that many of the gender effects of policy are ignored in skilled immigration policy design, or considered only to the extent that they do not conflict with broader policy goals.

Acknowledgements: Some aspects of this chapter draw upon and extend my previous thinking as articulated in Boucher (2016). I am grateful to Eda Gunaydin for research assistance, funded through the University of Sydney SOAR Fellowship.

References

Anderson, D. J., Binder, M., & Krause, K. (2003). The motherhood wage penalty revisited: Experience, heterogeneity, work, effort and work-schedule flexibility. *Industrial and Labor Relations Review, 56*(2), 273–294.

Arslan, C., Dumont, J.-C., Kone, Z., Özden, C., Parsons, C., & Xenogiani, T. (2016) (International Migration to the OECD in the 21st Century). *Knomad Working Article*, Washington D.C, https://www.knomad.org/sites/default/files/2017-04/KNOMAD%20WP%2016%20Migration%20to%20OECD_final_0.pdf.

Austrian government. (2019). *Very highly qualified workers*. Retrieved from Vienna https://www.migration.gv.at/en/types-of-immigration/permanent-immigration/very-highly-qualified-workers/.

Badkar, J., Callister, P., Krishnan, V., Didham, R., & Bedford, R. (2007). Gender, mobility and migration in New Zealand: A case study of Asian Migration. *Social Policy Journal of New Zealand, 32* (November 2007), 126–154.

Ballarino, G., & Panichella, N. (2017). The occupational integration of migrant women in Western European labour markets. *Acta Sociologica, 61*(2), 126–142.

Banerjee, R., & Phan, M. B. (2015). Do tied movers get tied down? The occupational displacement of dependent applicant immigrants in Canada. *Journal of International Migration and Integration, 16*(2), 333–353. https://doi.org/10.1007/s12134-014-0341-9.

Becker, G. (1964). *Human capital: A theoretical and empirical analysis with special reference to education*. Chicago: University of Chicago Press.

Boucher, A. (2010). Gender mainstreaming in skilled immigration policy: From Beijing 1995 to the Canadian immigration and refugee protection act (2002). In A. Neville (Ed.), *Human Rights and Social Policy: A Comparative Analysis of Values and Citizenship in OECD countries* (pp. 174–200). Cheltenham: Edward Elgar.

Boucher, A. (2016). *Gender, migration and the global race for talent*. Manchester: Manchester University Press.

Boucher, A. (2019a). *Assessing gender effects in international labour mobility.* Paper presented at Sharing data: Sharing Solutions: APEC labour Mobility Statistics Forum Programme, Valparaíso, Chile.

Boucher, A. (2019b). How 'skill' definition affects the diversity of skilled immigration policies. *Journal of Ethnic and Migration Studies*(https://doi.org/10.1080/1369183X.2018.1561063).

Boucher, A., & Gest, J. (2018). *Crossroads: Comparative immigration regimes in a World of demographic change*. New York Cambridge University Press.

Boyd, M. (1997). Migration policy, female dependency, and family membership: Canada and Germany. In P. Evans & G. Werkle (Eds.), *Women and the Canadian Welfare State* (pp. 142–169). Toronto: Toronto University Press.

Charpentier, M., & Quéniart, A. (2017). Aging experiences of older immigrant women in Québec (Canada): From deskilling to liberation. *Journal of Women & Aging, 29*(5), 437–447. https://doi.org/10.1080/08952841.2016.1213111.

CIC, C. I. C. (2015). FW Canada, express entry: Minister Alexander Announces New Details Retrieved from https://www.canadianimmigration.net/news-articles/express-entry-minister-alexander-announces-new-details/.

Creese, G., Dyck, I., & McLaren, A. T. (2008). The 'Flexible' immigrant? Human capital discourse, the family household and labour market strategies. *Journal of International Migration and Integration / Revue de L'integration et de La Migration Internationale, 9*(3), 269–288. https://doi.org/10.1007/s12134-008-0061-0.

Czaika, M., & Parsons, C. (2016). *The gravity of high-skilled migration policies. In Knomad Working Article*. Washington D.C., World Bank, https://www.knomad.org/sites/default/files/2017-04/KNOMAD%20Working%20Paper%2013%20HighSkilledMigration_0.pdf.

Dauvergne, C. (2000). Citizenship, migration laws and women: Gendering permanent residency statistics. *Melbourne University Law Review, 24*, 280–309.

Dobson, B., & Crush, J. (2004). A report on gender discrimination in South Africa's 2002 immigration act: Masculinizing the migrant. *Feminist Review, 77*(1), 96–119.

Docquier, F., Lowell, B. L., & Marfouk, A. (2009). A gendered assessment of highly skilled emigration. *Population and Development Review, 35*(2), 297–322.

DHA, D. o. H. A. (2019). Australia's 2019–2020 Migration Program, Discussion Paper Retrieved from https://www.homeaffairs.gov.au/reports-and-pubs/files/discussion-paper-australias-2019-20-migration.pdf.

Faraday, F. (2016). *Decent WOrk or entrenched exploitation for migrant workers,* . Metcalfe Foundation Toronto, https://metcalffoundation.com/publication/canadas-choice/.

Fincher, R. (1997). Gender, age, and ethnicity in immigration for an Australian nation. *Environment and Planning a, 29*(2), 217–236.

Fincher, R., Foster, L., Giles, W., & Preston, V. (1994). Gender and migration policy. In H. Adelman, A. Borowski, M. Burstein, & L. Foster (Eds.), *Immigration and refugee policy: Australia and Canada compared* (pp. 149–184). Toronto, Buffalo: University of Toronto Press.

Folbre, N. (2012). The political economy of human capital. *Review of Radical Political Economics, 44*(3), 281–292.

Grigoleit-Richter, G. (2017). Highly skilled and highly mobile? Examining gendered and ethnicised labour market conditions for migrant women in STEM-professions in Germany. *Journal of Ethnic and Migration Studies, 43*(16), 2738–2755. https://doi.org/10.1080/1369183X.2017.1314597.

Hawthorne, L. (2011). *Competing for skills: Migration policies and trends in New Zealand and Australia*. Department of Labour, New Zealand/Department of Immigration and Citizenship, Canberra, https://www.homeaffairs.gov.au/research-and-stats/files/migration-policies-trends-fullreport.pdf.

Hawthorne, L. (2015). The impact of skilled migration on foreign qualification recognition reform in Australia. *Canadian Public Policy, 41*(Supplement 1), S173–S187 https://doi.org/10.3138/cpp.2015-027.

Heilman, M. E., & Saruwatari, L. R. (1979). When beauty is beastly: The effects of appearance and sex on evaluations of job applicants for managerial and non-managerial jobs. *Organisational Behaviour and Human Performance, 23*(3), 360–372.

Hugo, G. (2009). Care worker migration, Australia and development. *Population, Space and Place, 15*(2), 189–203.

ILO. (2019). A quantum leap for gender equality: For a Better Future of Work for All, ILO.

International Standard Classification of Occupations. (2008). *Structure, group definitions and correspondence tables*. Geneva: ILO.

Iredale, R. (2005). Gender, immigration policies and accreditation: Valuing the skills of professional women migrants. *Geoforum, 36*(2), 155–166.

ISCO. (2008). International Standard Classification of Occupations, ILO, available at https://www.ilo.org/public/english/bureau/stat/isco/.

Johnson, S., Podratz, K. E., Dipboye, R. L., & Gibbons, E. (2010). Physical attractiveness biases in ratings of employment suitability: Tracking down the "beauty is beastly " effect. *Journal of Social Psychology, 150*(3), 301–318.

Kahn, J. R., Garcia-Manglano, J., & Bianchi, S. M. (2014). The motherhood penalty at midlife: Long-term effects of children on women's careers. *Journal of Marriage and Family, 76*(1), 56–72.

Kerr, S. P., Kerr, W., Ozden, C., & Parsons, C. (2016). *High-skilled migration and agglomeration*. National Bureau of Economic Research: Cambridge, Massachusetts.

Kofman, E. (2014). Towards a gendered evaluation of (Highly) skilled immigration policies in Europe. *International Migration, 52*(3), 116–128.

Kranz, D. (2019). The Global North goes to the Global North Minus? Intersections of the integration of highly skilled, Non-Jewish female partner and Spousal migrants from the Global North in Israel. *International Migration, 57*(3), 192–207. https://doi.org/10.1111/imig.12574.

Lim, D. (2019). The indirect gender discrimination of skill-selective immigration policies. *Critical Review of International Social and Political Philosophy, 22*(7), 906–928.

Meagher, G., & Goodwin, S. (2015). *Markets, rights and power in Australian social policy*. Sydney: Sydney University Press.

Ncube, A., Yonas T. Bahta, & A. J. Jordaan. (2019). "Job market perceptions of African migrant women in South Africa as an initial and long-term coping and adaptation mechanism." Journal of International Migration and Integration September, z.

Nishikawa, M. (2011). (Re)defining care workers as knowledge workers. *Gender, Work and Organisation, 18*(1), 113–136.

O'Dwyer, M., & Colic-Peisker, V. (2016). Facilitating the professional transition of migrants in Australia: Does gender matter? *Australian Journal of Social Issues, 51*(1), 47–66. https://doi.org/10.1002/j.1839-4655.2016.tb00364.x.

Parsons, C. R., Rojon, S., Rose, L., & Samanani, F. (2018). High skilled migration through the lens of policy. *Migration Studies, 0* (0), 1–28.

Parsons, C. R., Rojon, S., Samanani, F., & Wettach, L. (2014). Conceptualising international high-skilled migration. *Working paper* (104), https://www.migrationinstitute.org/publications/conceptualising-international-high-skilled-migration.

Payne, J. (2009). Emotional labour and skill: A reappraisal. *Gender, Work and Organisation, 16*(3), 348–367.

Phan, M., Banerjee, R., Deacon, L., & Taraky, H. (2015). Family dynamics and the integration of professional immigrants in Canada. *Journal of ethnic and migration studies, 41*(13), 2061–2080. https://doi.org/10.1080/1369183X.2015.1045461.

Phillips, A., & Taylor, B. (1980). Sex and skill: Notes towards a feminist economics. *Feminist Review, 6*(1), 79–88.

Piper, N. (2008). International migration and gendered axes of stratification—Introduction. In N. Piper (Ed.), *New Perspectives on Gender and Migration—Rights, Entitlements and Livelihoods* (pp. 1–18). London: Routledge.

Piper, N., & Yamanka, K. (2008). Feminist migration in East and Southeast Asia and the securing of livelihoods. In N. Piper (Ed.), *new Perspectives on Gender and Migration: Livelihoods, Rights and Entitlements* (pp. 159–188). London: Routledge.

Reskin, B., & McBrier, D. (2000). Why not ascription? organisations' employment of male and female managers. *American Sociological Review, 65*(2), 210–233.

Rodriguez, J. K., & Scurry, Tracy. (2019). Female and foreign: An intersectional exploration of the experiences of skilled migrant women in Qatar. *Gender, Work & Organization, 26*(4), 480–500.

Ruhs, M. (2013). *The price of rights: Regulating international labor migration* New Jersey: Princeton University Press.

SOPEMI. (2019). *International Migration Outlook: Annual Report*, OECD: Paris, https://www.oecd.org/migration/international-migration-outlook-1999124x.htm.

Spitzer, D., & Piper, N. (2014). Retrenched and returned: Filipino migrant workers during times of crisis. *Sociology, 48*(5), 1007–1023.

Stasiulis, D., & Bakan, A. B. (2005). *Negotiating citizenship: Migrant, women in Canada and the global system*. Toronto: Toronto University Press.

Steinberg, R. J. (1990). Social construction of skill: Gender, power, and comparable worth. *Work and Occupations, 17*(4), 449–482.

Stewart, K. (2014). Employment trajectories and later employment outcomes for mothers in the British Household Panel Survey: An analysis by skill level. *Journal of Social Policy, 43*(1), 87–108.

Toledano, E. (2013). The looking-glass ceiling: Appearance-based discrimination in the workplace. *Cardozo Journal of Law and Gender, 19*, 683–714.

Walby, S. (2011). Is the knowledge society gendered? *Gender, Work and Organisation, 18*(11), 1–29.

Warhurst, C., & Nickson, D. (2009). "Who's got the look'? Emotional, aesthetic and sexualised labour in interactive services. *Gender, Work and Organisation, 16*(3), 385–404.

Webb, S. (2015). 'It's who you know not what': Migrants' encounters with regimes of skills as misrecognition. *Studies in Continuing Education, 37*(3), 267–285. https://doi.org/10.1080/0158037X.2015.1007938.

Weinberger, C. J. (2011). In search of the glass ceiling: Gender and earnings growth among US college graduates in the 1990s. *Industrial and Labor Relations Review, 64* (5)949–980.

Withers, M. (2019). *Sri Lanka's remittance economy: A multiscalar analysis of migration-underdevelopment*. London; New York: Routledge.

Wojczewski, S., Pentz, S., Blacklock, C., Hoffmann, K., Peersman, W., Nkomazana, O., & Kutalek, R. (2015). African female physicians and nurses in the global care Chain: Qualitative explorations from five destination countries. *PloS One, 10*(6), 1–20, e0129464. https://doi.org/10.1371/journal.pone.0129464.

Zlotnik, H. (1995). Migration and the family: The female perspective. *Asian and Pacific Migration Journal, 4*(2–3), 253–271.

Anna Boucher is a Associate Professor in Public Policy and Political Science at the University of Sydney. Her books Gender, Migration and the Global Race for Talent (Manchester University Press, 2016) and Crossroads: Immigration Regimes in a World of Demographic Change (Cambridge University Press, 2018) cover a wide array of immigration, ethnicity and gender issues.

CHAPTER 13

Gender and Gender Relations in Skilled Migration: More Than a Matter of Brains

Belinda Dodson

Abstract Gender aspects of skilled migration remain under-researched, despite significant advances in conceptualisation and understanding. The "brain drain" debate has gained nuance with emergence of alternative interpretations such as brain circulation, brain gain and brain waste, along with recognition of the contributions skilled migrants make to their countries of origin. Work examining gendered processes and experiences of skilled migration demonstrates its embeddedness in social relations and structures. Gender-informed analysis also raises questions of what "count" as skills and how skills themselves are socially constructed in gender terms. Yet much skilled immigration research remains framed in terms of skills, education, human capital, employment and labour markets. This chapter argues not only that gender has been neglected in research on skilled migration, but that the very distinction between skilled and unskilled migration reproduces existing gender and geographic inequalities.

Introduction

Immigration policies, especially those in countries in the global North, increasingly adopt models based on importing skilled labour to make up for skills shortages and ageing populations (Boucher, 2016; Czaika, 2018; Iredale, 2015). In parallel with these trends, research on skilled migration has expanded in volume and scope. Much of this research focuses on immigration,

B. Dodson (✉)
University of Western Ontario, London, ON, Canada
e-mail: bdodson@uwo.ca

C. Mora and N. Piper (eds.), *The Palgrave Handbook of Gender and Migration*, https://doi.org/10.1007/978-3-030-63347-9_13

for example on national immigration policies and the settlement and integration of skilled immigrants (Boucher, 2016; Cerna, 2014; Kaushik & Drolet, 2018; Triadafilopoulos, 2013). Other work considers countries of origin and the drivers and impacts of skilled emigration (Gibson & McKenzie, 2011; Jenkins, 2016; Skeldon, 2009). The "brain drain" narrative has been challenged and reformulated: from earlier concern about a unidirectional flow from developing to developed countries, and its assumed deleterious effects on countries of origin, there is now greater attention to mixed and potentially positive effects of skilled emigration on source countries—for example by providing incentives for education, or through migrant remittances, skills transfer and diaspora investment (Brinkerhoff, 2008; De Haas, 2005, 2010; Docquier & Rapoport, 2012; Gibson & McKenzie, 2011; Skeldon, 2020). Terms such as brain gain, brain waste, brain abuse, brain return and brain circulation have been added to the skilled migration lexicon (Banerjee, Verma, & Zhang, 2018; Bauder, 2003; Iredale, 2015; Saxenian, 2005).

Despite this expanding body of research, gender aspects of skilled migration remain under-researched. In a 2009 article titled "A Gendered Assessment of Highly Skilled Emigration", Docquier, Lowell and Marfouk observed that gender had received insufficient attention. In her 2016 book *Gender, Migration and the Global Race for Talent*, Boucher pointed to only limited improvement, as did Czaika in his 2018 edited volume *High-Skilled Migration: Drivers and Policies*. An extensive body of research on gender and migration, meanwhile, has focused on "less-skilled", feminised occupations such as domestic labour or care work and drawn attention to gendered constructions of what count as skills in research and policy (Bastia & Piper, 2019; Benería, Deere, & Kabeer, 2012; Kofman, 2013). Work on gender and skilled migration highlights the importance of gender constructs, gender relations and gendered social structures in understanding the drivers and outcomes of migration—whether deemed skilled or unskilled—and reveals the gender-uneven effects of migration policies (Bailey & Mulder, 2017; Boucher, 2016; Kofman, 2013; Kofman & Raghuram, 2005, 2006; Liu-Farrer, Yeoh, & Baas, 2020).

The first section of this chapter outlines patterns and trends in the gender composition of skilled migration, along with the gender implications of related policies. This is followed by a section examining evolution in conceptual framing of "brain drain" (or skills flow). The chapter then outlines three themes in understanding skilled migration from a gender perspective: (1) the gendering of skills and occupations, (2) gendered family and life-course factors, and (3) gender considerations in migration-development linkages. A brief conclusion draws together the chapter's main arguments and presents pointers for future research. The chapter argues not only that gender has received insufficient attention in skilled migration research, but that the very distinction between "skilled" and "unskilled" migration sustains existing gender and geographic inequalities.

Gender and Skilled Migration: Trends and Policies

Quantitative data, where available, reveal patterns and trends in the gender composition of skilled migration. For OECD countries, the female proportion of tertiary-educated immigrant residents was already close to 50 per cent by 2000 (Docquier, Lowell, & Marfouk, 2009), with subsequent increase to a slight majority. For many of the source countries of these immigrants, the proportion of highly educated women who emigrate is significantly higher than the equivalent proportion of highly educated men (Docquier et al., 2009; Docquier & Rapoport, 2012; Gibson & McKenzie, 2011; IOM & OECD, 2014). Docquier et al. (2009) presented rates of *emigration* to OECD countries from 195 source countries, broken down by gender and education level. They found overall that "women's highly skilled emigration rate is 17 percent above men's" (p. 316), albeit with considerable geographic variation. A later OECD report found the gender gap persisting over the period 2000/1–2015/16: "while the emigration rates of low-educated men and women are similar, highly-educated women born in non-OECD countries are significantly more likely to emigrate towards the OECD area than their male counterparts" (d'Aiglepierre et al., 2020: 10). Ultimately, however, quantitative male/female comparisons reveal little about the gendered social norms and family dynamics that structure migration, or about the policies and labour markets that produce particular skilled migration flows.

Paradoxically, higher female rates of skilled emigration and proportions of skilled immigrants do not mean more women than men entering through formal skilled immigration channels. Migration policies are expressed and defined in terms of education, qualifications, occupations, experiences or wage levels, as if these did not have gender already embedded within them. Boucher (2016) presents detailed case studies of Australian and Canadian immigration policy from 1988 to 2013. Among the factors she identifies as producing de facto gender discrimination are unequal access to education and formal qualifications, the gendering of skills and occupations, the role of family dynamics and life-course events, and gender pay gaps. Policies based on formal qualifications, particular occupations (e.g. engineering or IT), years of employment or wage thresholds are all likely to discriminate against women (see also Boucher, in this volume).

Canada, for example, has long practised a points-based immigration system using education, skills and work experience, along with other criteria such as age and language. Among principal applicants in the overall Economic Class of admissions, the proportion who were female in 2015, 2016 and 2017 was 43, 42 and 44%, respectively (IRCC, 2018). Within the Economic Class, Canada introduced two Express Entry migration streams in 2015: the Federal Skilled Worker (FSW) Program and Federal Skilled Trades (FST) Program. Of those admitted under FSW and FST in 2017, only 36% of principal applicants were female. The IRCC 2018 report notes the "historically gendered nature of labour market sectors that attract economic immigrants, such as engineering and information technology—occupations that are predominantly filled by men" (p. 16). A separate category of immigration to Canada was for caregivers, where the proportion of 2017 admissions who were female was 94%. A Business Immigrants category, by contrast, was 78% male. Women again dominate in the Family Class, where the female proportion was 57% in 2015, 58% in 2016 and 57% in 2017. Similarly-gendered patterns, with women under-represented in economic migration streams and over-represented in family streams, can be observed in other destination countries and regions (Boucher, 2016; Kofman, 2014).

These distinctions and patterns demand deeper examination. As outlined in the sections below, a multidisciplinary body of research examines gendered processes, impacts and experiences of skilled migration, demonstrating its embeddedness in social relations and structures (e.g. Bailey & Mulder, 2017; Kofman & Raghuram, 2005; Liu-Farrer et al., 2020). What also persists, however, is research and policy that represent skills flow as the movement of skills-bearing individuals or human capital, detached from family dynamics and social contexts and often lacking even simple disaggregation on the basis of sex.

Theorising Skilled Migration: From "Brain Drain" to "Brain Gain"?

Skilled emigration is sometimes represented in terms of "brain drain". Originally coined to describe the migration of British academics and scientists to North America in the 1960s, "brain drain" has come to refer mainly to migration of the highly educated from the global South to the global North, and more recently to South-South migration and various circular flows (Chikanda & Crush, 2014; Iredale, 2015; Skeldon, 2009, 2020). The medical field in particular is portrayed as experiencing skills losses as doctors, nurses and other health professionals migrate (Kalipeni, Semu, & Mbilizi, 2012; Raghuram, 2009a). Research re-examining the "brain drain" narrative provides evidence of positive as well as negative effects on skilled migrants' countries of origin, although this remains the subject of considerable debate. Along with geographic and conceptual broadening, there has been growing—if still insufficient—attention to gender.

Some research points to skilled emigration as potentially increasing, not decreasing, the supply of skills in migrants' countries of origin. Aspiration to future emigration can lead to more people pursuing higher education—although this can lead to distortion in the type of education pursued, and does not lead inevitably to economic or welfare gains (Docquier & Rapoport, 2012; Gibson & McKenzie, 2011). Whether emigration prospects lead to a domestic "brain gain" depends on the availability of education and the structure of national labour markets. Gendered patterns of education and occupation can be either reinforced or reshaped by potential migration. In the Philippines, Indonesia and India, for example, there has been an emphasis on nursing training in response to demand in the global labour market (Mavroudi & Nagel, 2016; Ortiga, 2018). Nursing has been an occupation in which women have predominated, but recent research by Margaret Walton-Roberts (2019) shows an increase in Indian men's interest in nursing related to prospects for migration, and a rise in the status of nursing as an occupation.

Other accounts that reconsider "brain drain" narratives emphasise emigrants' contributions to their countries of origin through remittances (discussed further in a section below). Remittances can be important for poverty alleviation and financial flows, but any wider development impact depends on structural economic, social and political factors (Brønden, 2012; De Haas, 2005). In terms of "brain gain", remittances are often invested in children's education (Gibson & McKenzie, 2011). Studies show varied gender outcomes in different migration settings, but many indicate positive effects for children's education overall, and some show particular improvement for girls. As Bouoiyour and Miftah (2016) conclude in their study of Morocco, remittances there and in other developing countries "could lead to higher investments in schooling for girls and a reduction in gender inequalities in access to schooling" (p. 278).

"Brain circulation" through transnational and return migration is attracting increasing research attention (Dickerson & Ozden, 2017; Iredale, 2015; Saxenian, 2005). This can constitute a different form of "brain gain", for example the return of international students to their home countries, or returned skilled migrants starting businesses or taking up employment after years abroad gaining experience and professional networks. The Indian high-tech industry, in which Indians in the diaspora and returnees to India from the US and elsewhere have played a catalysing role, is a commonly cited example, although one very much male dominated (Bhatt, 2018; Chacko, 2007). In Ghana, Wong (2014) examined intersecting gender, family, cultural and class dynamics in highly skilled migrants' return experiences. She found considerable diversity in migration decisions and outcomes, with some women able to navigate return in ways that challenge patriarchal norms. She concludes that "highly skilled women are equally vital to effective development as men who choose to return" (pp. 454–455), yet have been neglected in research and policy.

Countering optimistic, "win-win" readings of "brain gain" or "brain circulation" is a body of research examining "brain waste" or "brain abuse", describing immigrants employed in occupations that do not match their level of skill and education (Banerjee et al., 2018; Bauder, 2003). Census and survey data show unemployment or underemployment of skilled immigrants to varying extent in different geographical contexts (Lo, Li, & Yu, 2019; Lu & Hou, 2019; Rajendran, Ng, Sears, & Ayub, 2020), and there are myriad examples in qualitative research. In addition to racialised immigrant experiences, immigrant women often experience additional obstacles and intersecting disadvantage in securing employment commensurate with their skills and qualifications (Fossland, 2013; Liversage, 2009; Man, 2004; Meares, 2010; Purkayastha, 2005).

Within changing realities and evolving theorisation of skilled migration and "brain drain", gender-based analysis remains limited. This is perhaps inevitable in conceptual framings based on skilled individuals, human capital and economic outcomes, where superficial gender neutrality can obscure gendered processes and experiences. Applying a gender lens deepens and critiques both "brain drain" and alternative, more positive accounts of skilled emigration (Raghuram, 2009a). Gender-informed analysis of migration also raises questions about how skills themselves are socially constructed in gender terms.

The Gendering of Skill, Occupation and Employment

Determining what "counts" as skills is far from neutral, and is further complicated by migration (Boucher, 2019; Kofman, 2014; Liu-Farrer et al., 2020; Nowicka, 2014). A convenient but problematic definition is to equate skills with higher education. That approach is common in quantitative reporting and analysis of skilled migration, such as those based on OECD data on skilled migrants (Docquier et al., 2009; IOM & OECD, 2014). Formal educational qualifications are important in many countries' skilled immigration policies. Other definitions, and some immigration policies, are based on occupational categories and rankings. As Liu-Farrer et al. observe, such technical criteria "are rarely correlated with actual practices because migrants' ability to utilise skills depends on complex social and political negotiations" (2020: 3). Designation of certain qualifications and occupations on a hierarchy of skill levels reflects social constructions and differential valorisation of forms of knowledge and labour, including on the basis of gender, with "embrained" skills valued more highly than those that are "embodied" (Kofman & Raghuram, 2005, 2015). Women's association with reproductive labour and the domestic sphere has contributed to their marginalisation in skilled migration policy and research (Kofman & Raghuram, 2006).

Skilled women may not only be less likely to gain admission through skilled immigration channels, but can also face particular challenges in finding skills-commensurate employment. For male and female immigrants, foreign credential recognition can be problematic, leading to deskilling and underemployment (Bauder, 2003; Man, 2004). This can fall disproportionately on female-dominated professions such as teaching and nursing, which require formal accreditation by state regulatory bodies (Kofman & Raghuram, 2015). Employers, too, discriminate on the basis of intersections of gender, racialisation and migrant identity, with immigrant women often experiencing cumulative disadvantage in accessing employment matching their field and level of training (Liu-Farrer et al., 2020; Purkayastha, 2005; Riano, 2012). Further, as discussed in the section below, balancing family and work can present obstacles to skilled immigrant women's employment and career, depending on intra-family negotiations about household reproductive and care labour and which spouse's career is given precedence.

Literature on gender and migration has focused on women, and on feminised occupations. Kofman (2013) described the two "emblematic migratory figures" (p. 579) in Europe as the skilled migrant in "heavily male scientific, IT, financial and management occupations" (p. 591) and the "female domestic and, increasingly, care worker" (p. 580)—despite large numbers of foreign-born women employed in high-skilled occupations (and men in occupations considered less-skilled). Research that redresses the balance pays attention to immigrant women in occupations such as medicine, engineering and IT (Grigoleit-Richter, 2017; Iredale, 2005; Raghuram, 2004, 2008), and to how migrants' deskilling, commonly associated with women, encompasses job-skills mismatch among male migrants (Banerjee et al., 2018). This can include men working not just in lower-paid jobs but in jobs associated with "women's work" (Batnitzky, McDowell, & Dyer, 2009; Datta et al., 2009; McGregor, 2007). There is acknowledgement that the "trailing spouse" can be male, and attention to skilled female migrants as independent or primary-applicant immigrants (Banerjee & Phan, 2015; Cangià, 2018; Ressia et al., 2017). Research has also highlighted how people who enter as spousal or family immigrants, who are more likely to be women, may independently have met criteria for admission as skilled immigrants—but whose particular circumstances channelled them into a family or spousal admission category, often with negative consequences for their employment and career (Bhatt, 2018; Boucher, 2016; Elrick & Lightman, 2018; Iredale, 2005; Riano, 2012).

Examining intersections of gender, migration, skills, occupation and employment shows the limitations of individualised, human capital approaches to understanding skilled migration. Skills and occupations are embedded in gendered social norms and stereotypes, involving gendered subjectivities and requiring (re)negotiation of roles and identities in new locations. Skilled migrants are also not merely individuals but members of families and often couples, whose linked lives and intertwined life-courses shape migration decisions and experiences.

Life-Course and Family in Shaping Skilled Migration

Family and life-course factors are fundamental in shaping the migration process. This is as true for skilled migration as for migration that takes place specifically through family migration channels. Official migration policies sort people into economic and family streams and primary and accompanying migrants. In addition to, and in intersection with, the domains of immigration policy and labour markets, skilled migration is undertaken and experienced within gendered personal and family relationships. These are embedded in social and cultural systems with gender-based expectations of work versus family roles and responsibilities.

Marriage, family and life-course factors are important in skilled migrants' professional and migration decisions, before and after migrating (Bailey & Mulder, 2017; Kou & Bailey, 2017; Liversage, 2009; Riano, 2012; Shinozaki,

2014). As Bailey and Mulder wrote in their introduction to a special issue on gender and life-course in highly skilled migration, "there is a need to broaden our lens to understand the life-course choices the migrants make (entry into partnership, having a child, becoming home owners and retirement plans) which then shape their future migration decisions" (2017: 2691). People might choose to marry or delay marriage, to have children or delay having children, based on migration plans and trajectories. Migrants adapt their educational and career strategies based on a combination of family considerations with migration and settlement or return intentions. These "choices" are made within, and constrained by, formal and informal institutions.

Social norms and expectations around gender, marriage, family and work produce variously gendered outcomes in different countries and communities, and can be either transformed or reinforced through migration (Ho, 2006; Kou & Bailey, 2017; Yeoh & Willis, 2005). A rich diversity of case studies of particular groups, locations and migration forms defies any universalising narrative of women's disadvantage or "wives' sacrifice" in skilled migration. Yet gender norms and family obligations can constrain skilled migrant women's professional and career advancement. Raghuram (2004) noted for the high-skilled IT sector how expectations of career flexibility and geographical (hyper)mobility present particular obstacles to women IT specialists, especially those with children. In their study of Indian skilled immigrants in the Netherlands and UK, Kou and Bailey (2017) show how migration allows women to escape patriarchal norms to pursue education or a career, but also how skilled immigrant women "experience difficulties in combining work and family that could cause them to leave the labour market or to move to a part-time schedule" (p. 178). Gendered family roles and obligations that may constrain female immigrants' careers are, however, precisely what supports the economic and social integration of their partners and families. In their five-year longitudinal study of immigrants from a number of different countries now living in Vancouver, Canada, Creese, Dyck, and McLaren (2011) demonstrated immigrant women's "generation and deployment of cultural, social and emotional capital as important dimensions of support in their husbands' search for jobs commensurate to their training" (p. 153). Bhatt (2018) and Walsh (2018) describe gendered practices and negotiations around home, family and career in skilled transnational communities: Bhatt for the "high-tech housewives" married to Indian men in the IT sector in the US and India; Walsh for expatriates in the "globalising city" of Dubai.

Research on parenting and family brings human and emotional aspects of migration to the surface (Gilmartin & Migge, 2016; Ryan, 2008). Many forms of temporary or transient skilled migration produce transnational families, requiring performance and navigation of parenting, caring and intimate relationships across distance (Parreñas, 2008; Walsh, 2018). Even where nuclear families move as a unit, immigrants experience loneliness and isolation as they navigate new social environments, and this can take gendered form. A study of immigrant mothers in Ireland by Gilmartin and Migge (2016), as but one example, shows how separation from previous social and family networks acted to prevent women from establishing a sense of identity and belonging outside of motherhood and the domestic sphere. As these and other studies show, subjectivities, affect and emotion are integral to international and transnational migration (Boccagni & Baldassar, 2015; Conradson & McKay, 2007). Examining emotional and family dimensions of migration also shows how family is "embedded in the broader political economy of global work, such that the making of family is far from a private concern" (Walsh, 2018: 126).

Gender, Migration and Development

Skilled migration research has tended to focus geographically on migrants' countries of destination and employment, whether as place of temporary residence or of longer-term relocation and settlement. Research revising ideas around "brain drain", by contrast, has focused on the potential development impact of emigrants to their countries of origin. Narratives of brain drain or brain gain are themselves rooted in conceptualisations of development and of the much-debated relationship or "nexus" between development and migration (Brønden, 2012; De Haas, 2010; Gamlen, 2014; Glick Schiller, 2012; Hickey, 2016; Raghuram, 2009b). That broader debate is beyond the scope of this chapter, but feminist scholars have made significant contributions. Their work has critiqued both the marginalisation of gender in migration-development debates and the neoliberal economic framing of mainstream migration-development linkages (Bastia, 2013; Bastia & Haagsman, 2020; Benería et al., 2012; Piper, 2009; Pearson & Sweetman, 2019; Silvey, 2009). It draws attention to the spatial politics of migration and its governance, in which the "mobile governable subject of migration–development ... is both required to move in order to strategise their human capital, but also to act morally for the collective good of a distant place/community" (Raghuram, 2009b: 110).

Although the relationship between migration and development involves much more than remittances, these have been a focus of research and policy. Gender-informed analyses show how attempts by governments and development organisations to "harness" remittances for development in migrants' countries of origin serve to instrumentalise migrants as development agents, often deploying gendered myths and stereotypes (Bastia & Piper, 2019; Davids & Van Driel, 2015; Kunz, 2015). These portray women as more reliable and selfless remitters; and remittances being spent more "responsibly", for example on health and education rather than consumption, when women are the primary decision-makers. Critiques of these myths draw on migrants' practices and experiences to challenge binaries such as sender and receiver, developed and developing countries and skilled or unskilled migrants.

Migrant women's and men's family remittances have attracted extensive research attention, reflecting the significance and meaning of remittances in migrants' own practices and priorities. While acknowledging categorisation of migrants as "skilled", "less skilled" or "unskilled" as inherently problematic, some studies have examined the intersection of gender, skills and remittances. In a quantitative study of highly educated migrants from the Philippines, McDonald and Valenzuela (2017) examined how mismatch between skills and employment affected remittances, finding that skills mismatch decreased the amount of men's remittances while not seeming to affect women's. Skilled migrant women in Riano's 2012 study in Switzerland related how sending remittances impedes their own career advancement, through pressure to accept employment below their qualifications and by reducing financial resources for further education and training.

Similarly, Wong's 2006 study of immigrant Ghanaian women in Toronto, Canada, showed how "remittance expectations and obligations, coupled with their own material investments in Ghana, can impinge upon their economic and social mobility in Toronto" (p. 357). Wong highlights the tensions and contradictions that women navigate in sustaining transnational lives, juggling the interests of nuclear family in Canada, other family members in Ghana, and their own interests and ambitions. A study of female Indian migrants in Australia by Singh (2019) examined gender and generational contestation around remittances, as income-earning daughters-in-law challenge patriarchal norms that prioritise their husbands' parents as remittance recipients. These examples show the significance of gender in remittance flows, but in contextualised, socially and culturally embedded ways, examining how gendered subjectivities and intra-family relations are reshaped through migration and remittance practices.

Insights from the gender, migration and development literature enrich understanding of skilled migration. Centring gender and gender relations shows the significance of social and family factors in migration, and migrants as more than human capital to be harnessed for the economic ends of receiving or sending states. Rather than the national framing of much migration and development discourse, gender-centred and feminist approaches highlight migrants' priorities and aspirations for themselves and their families. Questioning the skilled/unskilled categorisation reveals how development gains from migration are attributable to people across a range of education, migration status and occupation, often in precarious and exploitative employment, sustaining social, economic and family relations across transnational space.

Conclusion

Fifteen years ago, in a special issue of *Geoforum* on "Gender and skilled migrants: into and beyond the workplace" (2005), Kofman and Raghuram set out the contours of a research agenda. First, they urged both recognition of women as skilled migrants, and examination of processes hindering migrant women's entry into skilled employment. Second, looking "beyond the workplace", they called for greater attention to family, social and cultural dimensions of skilled migration—approaches that had hitherto been applied largely to the study of less-skilled migration. Today, research placing gender at the centre of skilled migration can be found across multiple disciplines and interdisciplinary fields. Work adopting a gender-focused approach demonstrates the gendering of skills, occupations, labour markets and related migration streams; the importance of life-course and family factors in gendered experiences of skilled migration; and intersections of gender with nationality, racialisation, class and migrant status. It exposes de facto gender discrimination in skilled migration policy and governance, and draws attention to the limitations of narrowly economic framings of skilled migration and its impacts.

Three methodological (or even paradigmatic) trends, already signalled by Kofman and Raghuram (2005) and evident in work outlined in this chapter, suggest productive directions for research on gender, skills and migration. First is to avoid the "methodological sexism" (Dumitru, 2014: 203) through which research on gender and migration has sometimes been over-focused on feminised professions and occupations. Further research on women in masculinised, high-skilled occupations and professions, and men in "women's work", offers only partial rebalancing. Beyond that, continued attention to the gendering of skills, occupations, labour markets and related mobilities can shed light on changes and continuities, whether in skilled or less-skilled migration, and in economic, family or other migration channels. This can build understanding of the social construction and context-dependence of skills, and how gender, migration and skill are co-constituted.

Second is moving beyond methodological individualism. Quantifying proportions of male and female skilled migrants can be important in identifying patterns and trends, and comparing different outcomes for men and women can expose discrimination in immigration regimes and employment practices. Understanding those patterns, trends and practices, and how they are produced and experienced, requires situating individual migrants in family and wider social networks, and considering linked life-courses involving negotiation and contestation of roles, responsibilities, identities and mobilities. It also requires attention to structural gender discrimination in both source and destination countries, including inequality in access to education and skilled employment.

A third emerging shift is geographical, or geopolitical. Reified designation of countries and regions as developed or developing, North or South, influences how migrants and their skills are valued and governed. There has been growing recognition of the scale and significance of South-South and intra-regional flows (Bastia & Piper, 2019), along with calls for "recentering the South in studies of migration" (Fiddian-Qasmiyeh, Berg, & Waters, 2020), while simultaneously challenging those very categories. Literature on gender and skilled migration remains dominated by South-North flows, and there is certainly a need for more South-South research. But moving beyond the binary of North and South, and the methodological nationalism that characterises much migration research (Anderson, 2019), is an important part of any future research agenda.

Finally, as Liu-Farrer et al. (2020: 11) caution: "It is worthwhile to bear in mind that the categories of 'highly skilled professionals', 'skilled labour' and 'guest workers' are but visa categories, not actual people. A human-centred approach to migration might reveal that not only do we need to rethink the way nation-states categorise migrants, but also how we employ these concepts in our own studies".

References

Anderson, B. (2019). New directions in migration studies: Towards methodological de-nationalism. *Comparative Migration Studies, 7*(1), 1–13. https://doi.org/10.1186/s40878-019-0140-8.

Bailey, A., & Mulder, C. H. (2017). Highly skilled migration between the Global North and South: Gender, life courses and institutions. *Journal of Ethnic and Migration Studies, 43*(16), 2689–2703. https://doi.org/10.1080/1369183x.2017.1314594.

Banerjee, R., Verma, A., & Zhang, T. (2018). Brain gain or brain waste? Horizontal, vertical, and full job-education mismatch and wage progression among skilled immigrant men in Canada. *International Migration Review, 53*(3), 646–670. https://doi.org/10.1177/0197918318774501.

Banerjee, R., & Phan, M. B. (2015). Do tied movers get tied down? The occupational displacement of dependent applicant immigrants in Canada. *Journal of International Migration and Integration, 16*(2), 333–353. https://doi.org/10.1007/s12134-014-0341-9.

Bastia, T. (2013). The migration-development nexus: Current challenges and future research agendas. *Geography Compass, 7*(7), 464–477. https://doi.org/10.1111/gec3.12055.

Bastia, T., & Haagsman, K. (2020). Gender, migration and development. In T. Bastia & R. Skeldon (Eds.), *Handbook of migration and development* (pp. 103–113). London: Routledge.

Bastia, T., & Piper, N. (2019). Women migrants in the global economy: A global overview (and regional perspectives). *Gender & Development, 27*(1), 15–30. https://doi.org/10.1080/13552074.2019.1570734.

Batnitzky, A., McDowell, L., & Dyer, S. (2009). Flexible and strategic masculinities: The working lives and gendered identities of male migrants in London. *Journal of Ethnic and Migration Studies, 35*(8), 1275–1293. https://doi.org/10.1080/13691830903123088.

Bauder, H. (2003). "Brain abuse", or the devaluation of immigrant labour in Canada. *Antipode, 35*(4), 699–717.

Benería, L., Deere, C. D., & Kabeer, N. (2012). Gender and international migration: Globalization, development, and governance. *Feminist Economics, 18*(2), 1–33. https://doi.org/10.1080/13545701.2012.688998.

Bhatt, A. (2018). *High-tech housewives: Indian IT workers, gendered labor, and transmigration*. Seattle, WA: University of Washington Press.

Boccagni, P., & Baldassar, L. (2015). Emotions on the move: Mapping the emergent field of emotion and migration. *Emotion, Space and Society, 16*, 73–80. https://doi.org/10.1016/j.emospa.2015.06.009.

Boucher, A. (2016). *Gender, migration and the global race for talent*. Manchester: Manchester University Press.

Boucher, A. (2019). How 'skill' definition affects the diversity of skilled immigration policies. *Journal of Ethnic and Migration Studies*, *46*(12), 2533–2550. https://doi.org/10.1080/1369183x.2018.1561063.

Bouoiyour, J., & Miftah, A. (2016). The impact of remittances on children's human capital accumulation: Evidence from Morocco. *Journal of International Development, 28*(2), 266–280. https://doi.org/10.1002/jid.3147.

Brinkerhoff, J. (Ed.). (2008). *Diasporas and development: Exploring the potential*. New York: Lynne Reinner.

Brønden, B. M. (Ed.). (2012). Migration and development buzz? Rethinking the migration-development nexus and policies. Special issue of *International Migration, 50*(3), 1–97. https://doi.org/10.1111/j.1468-2435.2012.00752.x.

Cangià, F. (2018). Precarity, imagination, and the mobile life of the 'trailing spouse'. *Ethos, 46*(1), 8–26. https://doi.org/10.1111/etho.12195.

Cerna, L. (2014). Attracting high-skilled immigrants: Policies in comparative perspective. *International Migration, 52*(3), 69–84. https://doi.org/10.1111/imig.12158.

Chacko, E. (2007). From brain drain to brain gain: Reverse migration to Bangalore and Hyderabad, India's globalizing high tech cities. *GeoJournal, 68,* 131–140. https://doi.org/10.1007/s10708-007-9078-8.

Chikanda, A., & Crush, J. (2014). Diasporas of the South. In R. Anich, J. Crush, S. Melde, & J. Oucho (Eds.), *A new perspective on human mobility in the South* (pp. 65–88). Dordrecht, Netherlands: Springer.

Conradson, D., & McKay, D. (2007). Translocal subjectivities: Mobility, connection, emotion. *Mobilities, 2*(2), 167–174. https://doi.org/10.1080/17450100701381524.

Creese, G., Dyck, I., & McLaren, A. T. (2011). The problem of 'human capital': Gender, place and immigrant household strategies of reskilling in Vancouver. In A. Kraler, E. Kofman, M. Kohli, & C. Schmoll (Eds.), *Gender, generations and the family in international migration* (pp. 141–162). Amsterdam: Amsterdam University Press.

Czaika, M. (Ed.). (2018). *High-skilled migration: Drivers and policies*. Oxford: Oxford University Press.

d'Aiglepierre, R., David, A., Levionnois, C., Spielvogel, G., Tuccio, M., & Vickstrom, E. (2020). *A global profile of emigrants to OECD countries: Younger and more skilled migrants from more diverse countries* (OECD Social, Employment and Migration Working Papers No. 239). Paris.

Datta, K., McIlwaine, C., Herbert, J., Evans, Y., May, J., & Wills, J. (2009). Men on the move: Narratives of migration and work among low-paid migrant men in London. *Social and Cultural Geography, 10*(8), 853–873. https://doi.org/10.1080/14649360903305809.

Davids, T., & Van Driel, F. (2015). Framing gender in remittances. In T. Van Naerssen, L. Smith, T. Davids, & M. H. Marchand (Eds.), *Women, gender, remittances and development in the global South* (pp. 229–242). Farnham, UK and Burlington, VT: Ashgate.

De Haas, H. (2005). International migration, remittances and development: Myths and facts. *Third World Quarterly, 26*(8), 1269–1284. https://doi.org/10.1080/01436590500336757.

De Haas, H. (2010). Migration and development: A theoretical perspective. *International Migration Review, 44*, 227–264. https://doi.org/10.1111/j.1747-7379.2009.00804.x.

Dickerson, S., & Ozden, C. (2017). Diaspora engagement and return migration policies. In A. Triandafyllidou (Ed.), *Handbook of migration and globalisation* (pp. 206–225). Cheltenham, UK: Edward Elgar.

Docquier, F., Lowell, B. L., & Marfouk, A. (2009). A gendered assessment of highly skilled emigration. *Population and Development Review, 35*(2), 297–321. https://doi.org/10.1111/j.1728-4457.2009.00277.x.

Docquier, F., & Rapoport, H. (2012). Globalization, brain drain, and development. *Journal of Economic Literature, 50*(3), 681–730. https://doi.org/10.1257/jel.50.3.681.

Dumitru, S. (2014). From "brain drain" to "care drain": Women's labor migration and methodological sexism. *Women's Studies International Forum, 47 (Part B)*, 203–212. https://doi.org/10.1016/j.wsif.2014.06.006.

Elrick, J., & Lightman, N. (2018). Sorting or shaping? The gendered economic outcomes of immigration policy in Canada. *International Migration Review, 50*(2), 352–384.

Fiddian-Qasmiyeh, E., Berg, M. L., & Waters, J. (2020). Editorial: Recentering the South in studies of migration. *Migration and Society, 3*(1), vii–x. https://doi.org/10.3167/arms.2020.030101.

Fossland, T. (2013). Crossing borders—Getting work: Skilled migrants' gendered labour market participation in Norway. *Norwegian Journal of Geography, 67*(5), 276–283. https://doi.org/10.1080/00291951.2013.847854.

Gamlen, A. (2014). The new migration-and-development pessimism. *Progress in Human Geography, 38*(4), 581–597. https://doi.org/10.1177/0309132513512544.

Gibson, J., & McKenzie, D. (2011). Eight questions about brain drain. *Journal of Economic Perspectives, 25*(3), 107–128. https://doi.org/10.1257/jep.25.3.107.

Gilmartin, M., & Migge, B. (2016). Migrant mothers and the geographies of belonging. *Gender, Place & Culture, 23*(2), 147–161. https://doi.org/10.1080/0966369x.2014.991700.

Glick Schiller, N. (2012). Unravelling the migration and development web: Research and policy implications. *International Migration, 50*(3), 92–97. https://doi.org/10.1111/j.1468-2435.2012.00757.x.

Grigoleit-Richter, G. (2017). Highly skilled and highly mobile? Examining gendered and ethnicised labour market conditions for migrant women in STEM-professions in Germany. *Journal of Ethnic and Migration Studies, 43*(16), 2738–2755. https://doi.org/10.1080/1369183x.2017.1314597.

Hickey, M. (2016). Modernisation, migration, and mobilisation: Relinking internal and international migrations in the 'migration and development nexus'. *Population, Space and Place, 22*(7), 681–692. https://doi.org/10.1002/psp.1952.

Ho, C. (2006). Migration as feminisation? Chinese women's experiences of work and family in Australia. *Journal of Ethnic and Migration Studies, 32*(3), 497–514. https://doi.org/10.1080/13691830600555053.

Immigration, Refugees and Citizenship Canada (IRCC). (2018). *Annual report to Parliament on immigration*. Ottawa: IRCC. https://www.canada.ca/en/immigration-refugees-citizenship/corporate/publications-manuals/annual-report-parliament-immigration-2018/report.html.

International Organization for Migration (IOM) & Organization for Economic Cooperation and Development (OECD). (2014). *Harnessing knowledge on the migration of highly skilled women*. Geneva: IOM. https://publications.iom.int/es/books/harnessing-knowledge-migration-highly-skilled-women.

Iredale, R. (2005). Gender, immigration policies and accreditation: Valuing the skills of professional women migrants. *Geoforum, 36*(2), 155–166. https://doi.org/10.1016/j.geoforum.2004.04.002.

Iredale, R. (2015). High-skilled migration. In F. D. Bean & S. K. Brown (Eds.), *Encyclopedia of migration*. Dordrecht, Netherlands: Springer.

Jenkins, R. (2016). Brain drain. *British Journal of Psychology International, 13*(3), 53–55. https://doi.org/10.1192/s2056474000001215.

Kalipeni, E., Semu, L. L., & Mbilizi, M. A. (2012). The brain drain of health care professionals from sub-Saharan Africa: A geographic perspective. *Progress in Development Studies, 12*(2–3), 153–171. https://doi.org/10.1177/146499341101200305.

Kaushik, V., & Drolet, J. (2018). Settlement and integration needs of skilled immigrants in Canada. *Social Sciences, 7*(5), 76. https://doi.org/10.3390/socsci7050076.

Kofman, E. (2013). Gendered labour migrations in Europe and emblematic migratory figures. *Journal of Ethnic and Migration Studies, 39*(4), 579–600. https://doi.org/10.1080/1369183x.2013.745234.

Kofman, E. (2014). Towards a gendered evaluation of (highly) skilled immigration policies in Europe. *International Migration, 52*(3), 116–128. https://doi.org/10.1111/imig.12121.

Kofman, E., & Raghuram, P. (2005). Gender and skilled migrants: Into and beyond the work place. *Geoforum, 36*(2), 149–154.

Kofman, E., & Raghuram, P. (2006). Gender and global labour migrations: Incorporating skilled workers. *Antipode, 38*(2), 282–303. https://doi.org/10.1111/j.1467-8330.2006.00580.x.

Kofman, E., & Raghuram, P. (2015). *Gendered migrations and global social reproduction*. Basingstoke, UK: Palgrave Macmillan.

Kou, A., & Bailey, A. (2017). 'Some people expect women should always be dependent': Indian women's experiences as highly skilled migrants. *Geoforum, 85*, 178–186.

Kunz, R. (2015). 'Moneymaker and mother from afar': The power of gender myths. In T. Van Naerssen, L. Smith, T. Davids & M. H. Marchand (Eds.), *Women, gender, remittances and development in the global South* (pp. 207–228). Farnham, UK and Burlington, VT: Ashgate.

Liu-Farrer, G., Yeoh, B. S., & Baas, M. (2020). Social construction of skill: An analytical approach toward the question of skill in cross-border labour mobilities. *Journal of Ethnic and Migration Studies (Latest Articles)*. https://doi.org/10.1080/1369183x.2020.1731983.

Liversage, A. (2009). Vital conjunctures, shifting horizons: High-skilled female immigrants looking for work. *Work, Employment & Society, 23*(1), 120–141. https://doi.org/10.1177/0950017008099781.

Lo, L., Li, W., & Yu, W. (2019). Highly-skilled migration from China and India to Canada and the United States. *International Migration, 57*(3), 317–333. https://doi.org/10.1111/imig.12388.

Lu, Y., & Hou, F. (2019). *Over-education among university-educated immigrants in Canada and the United States*. Ottawa: Statistics Canada, Analytical Studies Branch Research Paper Series.

Man, G. (2004). Gender, work and migration: Deskilling Chinese immigrant women in Canada. *Women's Studies International Forum, 27*(2), 135–148.

Mavroudi, E., & Nagel, C. (2016). *Global migration: Patterns, processes, and politics*. Abingdon, UK and New York, NY: Routledge.

McDonald, J. T., & Valenzuela, M. R. (2017). How does skills mismatch affect remittances? A study of Filipino migrant workers. *Contemporary Economic Policy, 35*(1), 216–231.

McGregor, J. (2007). 'Joining the BBC (British Bottom Cleaners)': Zimbabwean migrants and the UK care industry. *Journal of Ethnic and Migration Studies, 33*(5), 801–824. https://doi.org/10.1080/13691830701359249.

Meares, C. (2010). A fine balance: Women, work and skilled migration. *Women's Studies International Forum, 33*(5), 473–481. https://doi.org/10.1016/j.wsif.2010.06.001.

Nowicka, M. (2014). Migrating skills, skilled migrants and migration skills: The influence of contexts on the validation of migrants' skills. *Migration Letters, 11*(2), 171–186.

Ortiga, Y. (2018). *Emigration, employability and higher education in the Philippines*. Abingdon, UK and New York: Routledge.

Parreñas, R. S. (2008). Transnational fathering: Gendered conflicts, distant disciplining and emotional gaps. *Journal of Ethnic and Migration Studies, 34*(7), 1057–1072. https://doi.org/10.1080/13691830802230356.

Pearson, R., & Sweetman, C. (2019). Introduction: Gender, development, and migrants in a global economy. *Gender & Development, 27*(1), 1–13. https://doi.org/10.1080/13552074.2019.1588587.

Piper, N. (2009). The complex interconnections of the migration–development nexus: A social perspective. *Population, Space and Place, 15*(2), 93–101. https://doi.org/10.1002/psp.535.

Purkayastha, B. (2005). Skilled migration and cumulative disadvantage: The case of highly qualified Asian Indian immigrant women in the U.S. *Geoforum, 36*(2), 181–196. https://doi.org/10.1016/j.geoforum.2003.11.006.

Raghuram, P. (2004). Migration, gender, and the IT sector: Intersecting debates. *Women's Studies International Forum, 27*(2), 163–176. https://doi.org/10.1016/j.wsif.2004.06.006.

Raghuram, P. (2008). Migrant women in male-dominated sectors of the labour market: A research agenda. *Population, Space and Place, 14*(1), 43–57. https://doi.org/10.1002/psp.472.

Raghuram, P. (2009a). Situating women in the brain drain discourse: Discursive challenges and opportunities. In H. Stalford, S. Currie & S. Velluti, S. (Eds.), *Gender and migration in 21st century Europe* (pp. 85–106). Aldershot: Ashgate.

Raghuram, P. (2009b). Which migration, what development? Unsettling the edifice of migration and development. *Population, Space and Place, 15*(2), 103–117. https://doi.org/10.1002/psp.536.

Rajendran, D., Ng, E. S., Sears, G., & Ayub, N. (2020). Determinants of migrant career success: A study of recent skilled migrants in Australia. *International Migration, 58*(2), 30–51. https://doi.org/10.1111/imig.12586.

Ressia, S., Strachan, G., Bailey, J., Broadbent, K., Strachan, G., & Healy, G. (2017). Gender and migration: The experiences of skilled professional women. In K. Broadbent, G. Strachan & G. Healy (Eds.), *Gender and the professions: International and contemporary perspectives* (pp. 105–116). New York: Routledge. https://doi.org/10.4324/9781315563954-7.

Riano, Y. (2012). The invisibility of family in studies of skilled migration and brain drain. *Diversities, 14*(1), 26–44.

Ryan, L. (2008). Navigating the emotional terrain of families "here" and "there": Women, migration and the management of emotions. *Journal of Intercultural Studies, 29*(3), 299–313. https://doi.org/10.1080/07256860802169238.

Saxenian, A. (2005). From brain drain to brain circulation: Transnational communities and regional upgrading in India and China. *Studies in Comparative International Development, 40*(2), 35–61.

Shinozaki, K. (2014). Career strategies and spatial mobility among skilled migrants in Germany: The role of gender in the work-family interaction. *Tijdschrift voor Economische en Sociale Geografie, 105*(5), 526–541. https://doi.org/10.1111/tesg.12111.

Silvey, R. (2009). Development and geography: Anxious times, anemic geographies, and migration. *Progress in Human Geography, 33*(4), 507–515. https://doi.org/10.1177/0309132508096351.

Singh, S. (2019). The daughter-in-law questions remittances: Changes in the gender of remittances among Indian migrants to Australia. *Global Networks, 19*(2), 197–217. https://doi.org/10.1111/glob.12215.

Skeldon, R. (2009). Of skilled migration, brain drains and policy responses. *International Migration, 47,* 3–29. https://doi.org/10.1111/j.1468-2435.2008.00484.x.

Skeldon, R. (2020). Skilled migration. In T. Bastia & R. Skeldon (Eds.), *Routledge handbook of migration and development* (pp. 136–145). Abingdon, UK and New York, NY: Routledge.

Triadafilopoulos, T. (Ed.). (2013). *Wanted and welcome? Policies for highly skilled immigrants in comparative perspective*. New York: Springer.

Walsh, K. (2018). *Transnational geographies of the heart: Intimate subjectivities in a globalising city*. Chichester, Sussex and Hoboken, NJ: Wiley.

Walton-Roberts, M. (2019). Asymmetrical therapeutic mobilities: Masculine advantage in nurse migration from India. *Mobilities, 14*(1), 20–37. https://doi.org/10.1080/17450101.2018.1544404.

Wong, M. (2006). The gendered politics of remittances in Ghanaian transnational families. *Economic Geography, 82*(4), 355–381. https://doi.org/10.1111/j.1944-8287.2006.tb00321.x.

Wong, M. (2014). Navigating return: The gendered geographies of skilled return migration to Ghana. *Global Networks, 14*(4), 438–457. https://doi.org/10.1111/glob.12041.

Yeoh, B., & Willis, K. (2005). Singaporeans in China: Transnational women elites and the negotiation of gendered identities. *Geoforum, 36*(2), 211–222. https://doi.org/10.1016/j.geoforum.2003.07.004.

Belinda Dodson's research examines intersections of gender and migration, with a regional focus on Southern Africa. She has degrees from the University of KwaZulu-Natal in South Africa and a Ph.D. in Geography from Cambridge University in England. She was a faculty member at the University of Cape Town in South Africa in the 1990s and joined the faculty at the University of Western Ontario in Canada in 2002. Her work has been published in edited volumes and in scholarly journals including *Africa Today*, *Agenda*, *Canadian Journal of African Studies*, *Canadian Geographer*, *Feminist Review*, *Gender Place and Culture*, *Geoforum*, *Health and Place*, and *Migration and Development*.

CHAPTER 14

Gender and International Student Migration

Parvati Raghuram and Gunjan Sondhi

Abstract Students have been important constituents of migratory flows but until recently they were seen as temporary short-term migrants and were therefore excluded from mainstream migration debates, including those on gender and migration. However, the rapid rise in international student migrants in recent years has led to a spate of research on this group. Yet, and despite the fact that women form a growing part of international student flows, little gender analysis has been conducted. This chapter begins by setting out the framings of parallel conversations in gender and migration and student migration. It then focuses on what the two bodies of work can learn from each other. The next section explores how and where gender matters in the student migration trajectory—from decision-making about migration to the migration experience and finally the post-study transitions. In doing so it points to the limited work exploring these issues and highlights areas for future research. The chapter concludes by broadening out these issues and interrogating gender, international and student in debates on gender and international student migration.

P. Raghuram (✉) · G. Sondhi
The Open University, Milton Keynes, UK
e-mail: Parvati.Raghuram@open.ac.uk

G. Sondhi
e-mail: gunjan.sondhi@open.ac.uk

C. Mora and N. Piper (eds.), *The Palgrave Handbook of Gender and Migration*, https://doi.org/10.1007/978-3-030-63347-9_14

Introduction

International students, like highly skilled migrants, are inherent to globalisation. Their international study is both an outcome of driving globalising aspirations among young people and also results in skilling up young people who can become potential architects of globalisation and use their knowledge to drive global economic growth. While international students have long been part of global mobility, it is only the recent rapid growth of student migration which has led to increasing attention to their presence within public and policy discourses around migration, education and labour markets. Perhaps as a result, the shift to examining migration through a gendered lens, which began in the late 1980s in wider migration research, has been much less evident in research on international student migration. Feminist scholarship has long made visible the mutually constitutive relationship between gender and migration—how gender shapes access and process of migration, while being shaped through the migration process and structures—but these relationalities are less discussed in international student migration literature. Yet, the international mobility of students too is deeply cut by gender throughout the migration process as we will go on to see (also see Bilecen & Mol, 2017).

This chapter explores the history of research on international student migration and compares this with the gendered contours of other forms of migration research globally. It then explores what the literature on gender and migration more broadly can contribute to that on student mobility but also what gendered migration research can gain from international student migration research. The chapter then proceeds to explore where and when gender affects student mobilities, providing examples of current findings but also pointing to gaps in existing knowledge. Finally, it sets out an agenda for future research in this field of gendered student mobility.

A Conversation That Never Happened?

Undoubtedly, the growing numbers and proportions of students travelling abroad to study have garnered widespread attention. In 2017, there were nearly 4 million international students studying just within the OECD countries (OECD, 2019). The most popular countries of destination remain the United States (984,898), United Kingdom (435,734) and Australia (258,873). Student migrants are lauded as ideal migrants, part of the skills habitus but yet usually short-term and hence discardable if their skills are not seen as necessary to the economy of the hosting country. Moreover, they are also central to bolstering higher education institutions through the valuable financial resources they contribute in terms of fees and through their expenditure, fostering local economies in university towns. This money is usually removed from poor and middle income sending countries and inputted into economics of developed receiving countries, creating a significant flow of negative remittance (Raghuram & Sondhi, 2017). Finally, their knowledge is important in so called knowledge economies where growth is driven by and through the knowledge industries, whose labour power students feed.

Along with their numerical growth in migration terms, their significance for higher education institutions and for the knowledge economy has led to an explosion of research on international student mobility. This work has spanned the breadth of disciplines from tourism research focusing on the mobility of youth for personal development and adventure (Brown, 2009; Michael, Armstrong, & King, 2004), to cost–benefit analyses within economics and development studies (Kim & Sondhi, 2015; Trilokekar & Kizilbash, 2013) which have focused on brain drain/gain/circulation as their calculative framework (Lee & Kim, 2010; Song, 1997; Song & McCarthy, 2018). The drivers for study abroad (King & Sondhi, 2018; Yang, 2018), experiences of international education (Abukhattala, 2013), living (Lulle & Buzinska, 2017), everyday subjectivities of migrants (O'Connor, 2019; Sondhi & King, 2017), post-study intentions (Kelly, Bennett, Giridharan, & Rosenwax, 2018; Lu, Zong, & Schissel, 2009) and transitions to employment and resettlement (Yoon & Kim, 2018, 2019) within both countries of study and those of origin have all been scrutinised.

Yet, it is their short-term stay, i.e. that students usually remain for three or four years that seems to mark their presence in migration research. Some students are seen as more transient because they engage in credit mobility rather than degree mobility, staying abroad for less than one year. They are not included in the discussion in this chapter. However, even those who undertake the longer graduate and postgraduate programmes are seen as transient and hence their inclusion into migration figures is often questioned.

This treatment of students as transient also led to their relatively late inclusion into broader discussions of migration. After the second world war, student mobility was fostered as part of soft diplomacy especially by North American (Fulbright scholarships) and Russian governments and by US private institutions (Ford Foundation) attempting to gain authority in an increasingly polarising world. These scholarship programmes were used to generate brain flows and were primarily offered for masters and doctoral level students but fee-paying students were also very significant in some contexts where the postcolonial state was unable to meet the educational aspirations of the new elite ruling classes (for Portugal and the Lusophony, see França, Alves, & Padilla, 2018). These mobilities were not always welcomed as they were seen to engender neocolonialism within the newly independent nations at best (for France, see Rice, 2013), or at worst to lead to the large scale removal of the educated from countries that needed them but despite these controversies they led to and shaped student flows. The scale of analysis in these debates was the nation with gains and losses theorised as modernisation or dependency and calibrated through the nation, i.e. brain gain or brain drain, respectively (Raghuram, 2013). Irrespective of the outcome of the mobility, the primary movable commodity in much of this analysis was knowledge, and its effects were seen as national gains and losses. The individuals who moved were rarely part of these discussions, nor their social lives, thus putting questions of identity, including gender identity, outside the frame of analysis.

The large-scale growth in international student mobility as well as questions about whether all these students did go back, i.e. were they more than student tourists, were they actually long-term migrants, brought students squarely into the migration literature. Spearheaded by long-standing migration researchers like Allan Findlay, they argued that some of the research issues facing migrants were also pertinent to student migrants (Li, Findlay, Jowett, & Skeldon, 1996).

Yet, during the same period gender issues rose and eventually became a very significant part of migration analysis and theory (Morokvasic, 1984; Pedraza, 1991). Initially much of this work focused on recognising the presence of women in the overwhelmingly male-centred narratives of labour migration (Phizacklea, 1983). Recognising the role of female family migrants in labour market streams and their agency in and beyond the workplace became central to the emerging gender and migration literature, especially in Europe and North America (Sassen-Koob, 1984). In some countries of the global South, the New International Division of Labour recognised women's roles in factory work and as part of the evolution of capitalist relations. However, most of that literature looked at the lives of internal migrants although here too paid domestic work remained an important focus (Radcliffe, 1990). They were clearly acquiring skills but not an education and their primary purpose was to earn money and support families through waged work.

Research also highlighted how gender relations had changed due to migration—some women were empowered, but not entirely; the uneasy relationship between patriarchy and capitalism continued to haunt women. Gender identities and complex family relations that arbitrated and interfered with migration however increasingly came to the focus and women and men mattered to migration, differently. More recent work on migration and gender has largely focused on the socially reproductive sectors of the labour market, retreating to a large extent from the productive sectors and thus only hesitantly entering discussions of women's contributions to skilled sectors of the labour market (and see Kofman & Raghuram, 2006, 2015 for a detailed discussion). In doing so, researchers have retained interest in the nation and class but often exploring the lives of migrants from working classes and poorer countries (Parreñas, 2015).

The large and growing literature on student migration has, on the other hand, been influenced by early theorisations of international students as privileged bearers of cultural capital (Waters, 2006). Despite recent interventions that highlight inequalities among students (Waters & Leung, 2012) and that not all international students are necessarily privileged the overwhelming assumption has been that international students are all upper class. A cursory glance at these two bodies of work may well lead to the erroneous assumption that all women migrants are working class while all students are upper class.

Class divides the two literatures but in doing so it also relegates gender to an issue of concern to the working classes alone.

So, what is gained by bringing these two bodies of work together? First, it is clear that one of the reasons why the student migrant numbers have risen is not because of an equivalent growth in the upper classes but rather because middle and lower classes are increasingly engaging in student migration. The rich class analysis which includes theorisations of contradictory class mobility, i.e. one can gain money through migration but that this may be obtained through working in lower class jobs and losing class status is very apparent in the gender and migration literature but is this also relevant among student migrants? The insights from longitudinal studies that trace the long-term outcomes of female migrants have not been replicated in the literature on students, so we know little about gendered variations in class mobility. Empirically, it is important to trace what students share with each other irrespective of gender and where the commonalities are based on gender. For example, how do patriarchal formations in sending and receiving countries influence women migrants' lives irrespective of mode of entry? Racism too continues to haunt all migrants—increasingly. In the context of anti-migrant sentiments, certain groups are racialised but this racialisation is particularly gendered—through dress, through the spaces that women are encouraged or permitted to occupy. Besides, study forms only one part of a portfolio of activities in which students engage as researchers increasingly; students also often work, have families and inhabit the same spaces as other migrant (King & Raghuram, 2013). However, the opportunities that students receive both within and outside the classroom may be unique to them as they upskill.

Another factor to consider is that not all students are the same; there are stark divides between universities even within the same country as universities themselves play games of distinction. This is increasingly so as international study expands beyond the global North countries to include educational hubs in many parts of the global south. How does class influence and shape student lives bringing some students lives closer to that of other male students in less prestigious universities and distinguishing them from more wealthy and privileged students? Moreover, what are the age profiles of these different groups of students; gendered expectations and performances are very age sensitive both independently but also in how they relate to the life course. Women with unequal responsibilities for childcare may have very different study experiences than men of the same age, but older women may find study is enabled because these responsibilities are complete, while men may have career-related responsibilities that foreclose these different more flexible international study options. The care chains literature has been widely deployed in female migration debates but has not been extended to students, as they are always perceived to be young and footloose. This is, however, not always the case, especially as students from poorer countries are involved in mobility. Thus, class, race, age and gender operate intersectionally to fracture and to unite migrant experiences, and both theoretical and empirical questions around these intersections are better exposed by relating the strengths of different bodies of research.

In the next section we outline current research questions on gendered student migration. Given the dearth of literature in the field we explore the student migration process through a gendered lens and suggest how and where gender might matter. In doing so we highlight the role of biographical temporalities, i.e. the migration process as lived and how that offers important insights into gender in student migration.

Gender in International Student Migration—How Gender Matters

Gender matters in shaping the drivers of migration, the experience and the outcomes and in doing so it involves a number of actors and sites. In this section, we explore how and when gender matters in student migration.

Pre-migration Factors

Student migration is driven by a range of factors such as the desire for world class education, better economic prospects and social mobility. However, just as in the broader gendered migration literature, underlying these gender-muted narratives of drivers of mobility are complex negotiations within the family, between parents and their differential expectations from their sons and daughters. For instance, in the Indian case, men's mobility (King & Sondhi, 2018) requires as much negotiation as women's due to the expected role of the son to care for ageing parents. The uncertainty surrounding the period of time abroad for studies, and the associated costs requires that men undertake set of negotiations with their family members to enable their mobility (Sondhi, 2013). By contrast, women's negotiations occur around the issues of safety and security in a new and unfamiliar place away from familiar home spaces (Sondhi & King, 2017).

The national-level policies of the sending country also impact the drivers and the negotiations that occur across gender and generations. While the family continues to remain the site of negotiation, the need for negotiation is shaped by the differential context of the sending country. For instance, the flows of international students from China are increasingly showing gender parity, and even a feminisation of the flows. The effect of liberalisation of China's mobility policies and programmes (Yang, 2018) along with the growth of daughter-only households due to the impact of the one-child policy (Kajanus, 2015) has meant that both the opportunities and the need to develop daughters' social and cultural capital through migration has been enhanced. Thus, the changing nature of the family works alongside national factors to influence who migrates as a student.

Migrant Experiences

Education research has long identified the classroom as a gendered site. The raced and gendered composition of a classroom leads to the exclusion and at time withdrawal of some minority groups. In STEM-related programmes such as engineering which are male dominated, migrant and non-migrant women face continual challenges, but these are further exacerbated as they cut across issues such as race. Hence, for international students where there is a gender balance within HEIs in STEM enrolment (Raghuram, Herman, Ruiz Ben, & Sondhi, 2017), such as those from India and China, a male-dominated classroom setting, with predominantly male lecturers and tutors who struggle to accept the presence of female students can be challenging (Griffith & Main, 2019). Students who struggle with these differently gender scripted spaces are blamed for differences in pedagogical and learning styles (Liu, 2011). The deficit is individual rather than structural, i.e. the differences may lie in how gendered racialised bodies are recognised and treated within the classroom and without.

Outside the classroom, social interaction with other students makes visible the intersections of racism and the gendered relations of power within the country of destination. For instance, O'Connor (2019) highlights how Muslim women studying at an Irish university feel a lower sense of belonging in part because they do not feel comfortable socialising in pubs. By contrast, men find it easier to engage in social interactions through hegemonic masculine sporting activities such as football and cricket. Additionally, as Sondhi (2019) points out in her research on Indian students in Canada, the expected gendered and often heteronormative performances in the majority of higher education institutions lead to the exclusion of those whose gender subjectivities are not in alignment with these hegemonic performances.

Students from Asian countries also report racism and social exclusion outside the classroom, i.e. in the cities where they are often located. No doubt African students too experience this racism (Beoku-Betts, 2004; Hanassab, 2006), but empirical literature on African international students remains relatively scarce. Racism often appears in the form of verbal abuse by 'established residents' of the city towards various groups (Collins, 2010); in more subtle actions through the expected gendered performances from the 'host society' (Ichimoto, 2004; Park, 2010); or violent forms of physical assault such as in the case of Indian students in Australia (Baas, 2010). These are further exacerbated among refugee youth, who may be considered as 'international' students due to their non-national status in the country of asylum but face discrimination on account of their refugee status and affiliations (Lee & Sehoole, 2019).

The gendered implications of immigration status on international study, i.e. the differences between young refugee women and men students, has, however, not been examined. Women migrants appear to encounter double discrimination based on their formal documentary status, and their racialised and gendered performances, most notably Muslim women wearing hijabs (Li & Zizzi, 2018). The experiences of exclusion produce different responses. Some women assert their agency by withdrawing from spaces in which they do not feel included and find other inclusive spaces instead (Li & Zizzi, 2018; O'Connor, 2019). Other students chose to continue to assert their often unintelligible gender performances as a way to continue to disrupt the hegemonic gender frames (Sondhi & King, 2017). Still others choose to adapt their gender performances to fit better into the hegemonic gender frames while continuing to disrupt other norms such as race and sexuality (Sondhi, 2013, 2019). In every practice of asserting their chosen gender frames and performances, international students particularly women, question and disrupt the normative patriarchal frames of the host countries.

Post-study

The intentions and decision to 'return' or stay have been the core of post-study transition literature (Kim & Sondhi, 2019). This temporal dimension of student mobility has, however, been accessed primarily through intentions of current students or by interviewing students who transitioned into work in the country of destination. The actual experiences of international students who returned remain underexplored.

Better job opportunities is often cited as a reason to stay but the 'better' sometimes refers to pay commensurate to their qualification and skills (Waters, 2009) but at other times, students might take lower pay, lower skills jobs in order to benefit from being in more wealthy countries (Baas, 2010). They therefore exhibit the same contradictory class mobility as seen among lower skilled migrant women in sectors such as domestic work although it has not been theorised as such, or compared to that of women migrants in other sectors (for insights on this issue, see Boucher and Dodson in this volume). Besides, international work experience may itself be a highly valued commodity enabling future career progression in a third country and in the country of origin but these issues have not been interrogated through a gender lens. What are the labour market outcomes for male and female international students who stay? Are these outcomes gender-differentiated? What are their experiences of accessing the labour market? What is the impact of visa/migration regimes on their ability to stay and work in the country of study? These are crucial questions to ask as post-study permit/visa requirements, in conjunction with lack of social networks is cited as the key reasons for reasons that students consider leaving the country of destination.

A majority of international students return to their country of origin upon completion of their studies. In the UK nearly 60% of international students return, with only a quarter studying in the UK upon completion of their education, and the remaining moving to a third country (UUKI, 2019). Some research points to experiences of reverse culture shock (Gaw, 2000) among labour migrants but the experiences of returning students remain understudied. Student's return and labour market experiences has received a little more attention. One finding is that securing an international world-class degree no longer guarantees easy access to better job opportunities for students who return (Brooks & Waters, 2011; Hao, Wen, & Welch, 2016). This failure of recognition of credentials and the institution is a result of the increasing competition among internationally mobile and educated applicants, their career-level at the time of application, and the relevance of their credentials and skills acquired in country of study to the labour market of the country of return. Hence, an examination of labour market outcomes in the country of origin raises questions about the portability of education and the knowledge acquired through international education (Collins, Ho, Ishikawa, & Ma, 2017). And when intersected with social locations such as gender and race, the accumulated social and cultural capital is insufficient to overcome the barriers in the local labour market. However, despite the importance of gendered implications, this too has remained largely unexamined.

Post-study return may not be permanent and are gender varied as Sondhi and King's research (2017) of students who moved back to India from Canada shows. They point to how women's return is more 'temporary' and part of a longer mobile trajectory than men's because the men they interviewed had to perform their role in the Indian patrifocal family where the responsibility of care of ageing parents falls on sons and their wives. This responsibility was made more acute by the limited welfare provision offered by the state and civil society. Women who return also helped with family responsibilities, but they also faced barriers in the labour market leading them to withdraw from work or move to a third country. For instance, (Holloway, O'Hara, & Pimlott-Wilson, 2012) found that accumulated social capital through study abroad was insufficient to overcome systemic gender biases within the labour market in Kazakhstan. Even where women accessed work, the lack of local social networks could affect their career development as (Moskal, 2018) found. Hence, sending countries' labour market structures can potentially lead to a gendered pattern of brain drain—where the drain in this case is due to failure of the labour market to value and incorporate women's knowledges and skills. However, there is limited gender sensitive research on this topic.

Conclusion

This chapter has explored the interplay between gender and migration in the student mobility literature and pointed to how the student journey raises important gendered questions, many of which remain unanswered. It has deployed two temporal registers—a historical overview of why gender and student migration research did not intersect and why this matters, followed by a more biographical take which explores student migration through the pre-, migration and post-migration experiences of students. By way of conclusion the chapter sets out a canvas of larger questions that too need to be thought through in future research.

First, *gender* does not mean the same thing in every country. It interpolates with study and mobility as well as the surrounding infrastructures such as family, social networks and financing. Women and men have differential access and rights over these infrastructures. For instance, how do women and men pay for their study and do they use different resources (bank loans, family support and scholarship)? These will also vary over time and across classes as various studies have shown us. What we do not yet know, however, is how is gender operationalised and performed differentially due to migration and due to study.

Secondly, the definition of the international *student* needs more thought. In focusing on those on study visas as the basis for research the experiences of family migrants, refugees and other groups who also engage in study is often lost. Upskilling by those not engaged in formal study can help to bridge the class divide usually seen in student migration research but also help to question what is meant by education and knowledge. Also, what is studied seems to matter. Globally women tend to be over-represented in the arts and social sciences but there are pockets where women play an important role in male-dominated education sectors. Similarly, a large number of students travel for religious studies but currently they fall outside the purview of current student migration research. Expanding study to include all types of study will expand our understandings of how gender plays out in knowledge constructions (Raghuram, 2013).

Thirdly, the *international* in international student migration requires a lot more interrogation. Currently most of our knowledge and categories adopt a Northern receiving country perspective because that is where most researchers catch up with these students. Postcolonial approaches to international study have pointed to the uneven geographies of migration and to the constitutive role that migrants have paid in producing knowledge (Madge, Raghuram, & Noxolo, 2015) but the new centres of growth are imbricated in other projects. For instance, the new diplomacies associated with the Belt and Road Initiative and the attendant student flows require new theoretical understandings. Their gendered implications are also likely to vary. The growth in regional flows too do not mimic existing migration corridors but rather are based on the nature and extent of marketisation of education. Thus Philippines, a major sender of migrant workers is emerging as a receiver of migrant students, marketing their migration past as the basis for skilling up people in the sectors in which they have a history of sending migrants. Nursing schools which teach Filipina migrant nurses now become the nurseries for other migrant groups too to follow the path of the Filipinas (Ortiga, 2018). Clearly gendered migration in other sectors and student migration are closely connected. Finally, students have been at the forefront of the decolonisation of knowledge (Booysen, 2016) expressing their agency with regard to who studies, what and how but this agency must also recognise that all decolonial projects must have a gendered dimension and must also include international students (Raghuram, Breines, & Gunter, 2020). After all international boundaries, educational infrastructures and gendered expectations have all been colonial constructs. The need to think through the presence of gendered international students in a gender unequal world is now pressing.

References

Abukhattala, I. (2013). What Arab students say about their linguistic and educational experiences in Canadian universities. *International Education Studies, 6*(8), 31–37.

Baas, M. (2010). *Imagined mobility: Migration and transnationalism among Indian students in Australia*. London: Anthem Press.

Beoku-Betts, J. A. (2004). African women pursuing graduate studies in the sciences: Racism, gender bias, and third world marginality. *NWSA Journal, 16,* 116–135.

Bilecen, B., & Mol, C. V. (2017). Introduction: International academic mobility and inequalities. *Journal of Ethnic and Migration Studies, 43*(8), 1241–1255. https://doi.org/10.1080/1369183X.2017.1300225.

Booysen, S. (2016). *Fees must fall*. Available at: http://witspress.co.za/catalogue/fees-must-fall/. Accessed 2 Nov 2018.

Brooks, R., & Waters, J. (2011). *Student mobilities, migration and the internationalization of higher education*. Basingstoke: Palgrave Macmillan.

Brown, L. (2009). The transformative power of the international sojourn: An ethnographic study of the international student experience. *Annals of Tourism Research, 36*(3), 502–521.

Collins, F. L. (2010). International students as urban agents: International education and urban transformation in Auckland, New Zealand. *Geoforum, 41,* 940–950.

Collins, F. L., Ho, K. C., Ishikawa, M., & Ma, A.-H. S. (2017). International student mobility and after-study lives: The portability and prospects of overseas education in Asia. *Population, Space and Place, 23*(4), e2029. https://doi.org/10.1002/psp.2029.

França, T., Alves, E., & Padilla, B. (2018). Portuguese policies fostering international student mobility: A colonial legacy or a new strategy? *Globalisation, Societies and Education, 16*(3), 325–338. Routledge. https://doi.org/10.1080/14767724.2018.1457431.

Gaw, K. F. (2000). Reverse culture shock in students returning from overseas. *International Journal of Intercultural Relations, 24,* 83–104.

Griffith, A. L., & Main, J. B. (2019). First impressions in the classroom: How do class characteristics affect student grades and majors? *Economics of Education Review, 69,* 125–137. https://doi.org/10.1016/j.econedurev.2019.02.001.

Hanassab, S. (2006). Diversity, international students, and perceived discrimination: Implications for educators and counselors. *Journal of Studies in International Education, 10,* 157–172.

Hao, J., Wen, W., & Welch, A. (2016). When sojourners return: Employment opportunities and challenges facing high-skilled Chinese returnees. *Asian and Pacific Migration Journal, 25*(1), 22–40. https://doi.org/10.1177/0117196815621806.

Holloway, S. L., O'Hara, S. L., & Pimlott-Wilson, H. (2012). Educational mobility and the gendered geography of cultural capital: The case of international student flows between Central Asia and the UK. *Environment and Planning a: Economy and Space, 44*(9), 2278–2294. https://doi.org/10.1068/a44655.

Ichimoto, T. (2004). Ambivalent 'selves' in transition: A case study of Japanese women studying in Australian universities. *Journal of Intercultural Studies, 25*, 247–269.

Kajanus, A. (2015). Migrant journeys. In A. Kajanus (Ed.), *Chinese student migration, gender and family*. Palgrave Studies on Chinese Education in a Global Perspective (pp. 73–99). London: Palgrave Macmillan UK. https://doi.org/10.1057/9781137509109_4.

Kelly, A., Bennett, D., Giridharan, B., & Rosenwax, L. (2018). Study experiences and the post-study intentions of female international undergraduate students. *Asia Pacific Journal of Education, 38*(2), 202–213. https://doi.org/10.1080/02188791.2018.1460256.

Kim, A., & Sondhi, G. (2015). Bridging the literature on education migration. *Population Change and Lifecourse Strategic Knowledge Cluster Discussion Paper Series/ Un Réseau stratégique de connaissances Changements de population et parcours de vie Document de travail, 3*(1). Available at: http://ir.lib.uwo.ca/pclc/vol3/iss1/7.

Kim, A., & Sondhi, G. (2019). Explaining international student mobility to Canada: A review. In *Outward and upward mobilities: International students in Canada, their families, and structuring institutions* (pp. 56–76). Toronto: University of Toronto Press.

King, R., & Raghuram, P. (2013). International student migration: Mapping the field and new research agendas. *Population, Space and Place, 19*(2), 127–137. https://doi.org/10.1002/psp.1746.

King, R., & Sondhi, G. (2018). International student migration: A comparison of UK and Indian students' motivations for studying abroad. *Globalisation, Societies and Education, 16*(2), 176–191. https://doi.org/10.1080/14767724.2017.1405244.

Kofman, E., & Raghuram, P. (2006). Gender and global labour migrations: Incorporating skilled workers. *Antipode, 38*(2), 282–303. https://doi.org/10.1111/j.1467-8330.2006.00580.x.

Kofman, E., & Raghuram, P. (2015). *Gendered migrations and global social reproduction*. Basingstoke: Springer.

Lee, J. J., & Kim, D. (2010). Brain gain or brain circulation? U.S. doctoral recipients returning to South Korea. *Higher Education, 59*(5), 627–643. https://doi.org/10.1007/s10734-009-9270-5.

Lee, J. J., & Sehoole, C. (2019). International students seeking political stability and safety in South Africa. *Higher Education Policy*. https://doi.org/10.1057/s41307-019-00171-z.

Li, F. L. N., Findlay, A., Jowett, A. J., & Skeldon, R. (1996). Migrating to learn and learning to migrate: A study of the experiences and intentions of international student migrants. *International Journal of Population Geography, 2*(1), 51–67. https://doi.org/10.1002/(SICI)1099-1220(199603)2:1%3c51:AID-IJPG17%3e3.0.CO;2-B.

Li, S., & Zizzi, S. (2018). A case study of international students' social adjustment, friendship development, and physical activity. *Journal of International Students, 8*(1), 389–408. https://doi.org/10.32674/jis.v8i1.171.

Liu, L. (2011). An international graduate student's ESL learning experience beyond the classroom. *TESL Canada Journal, 29*(1), 77–92.

Lu, Y., Zong, L., & Schissel, B. (2009). To stay or return: Migration intentions of students from people's republic of China in Saskatchewan, Canada. *Journal of International Migration and Integration, 10*(3), 283–310. https://doi.org/10.1007/s12134-009-0103-2.

Lulle, A., & Buzinska, L. (2017). Between a 'student abroad' and 'being from Latvia': Inequalities of access, prestige, and foreign-earned cultural capital. *Journal of Ethnic and Migration Studies, 43*(8), 1362–1378. https://doi.org/10.1080/1369183X.2017.1300336.

Madge, C., Raghuram, P., & Noxolo, P. (2015). Conceptualizing international education: From international student to international study. *Progress in Human Geography, 39*(6), 681–701. https://doi.org/10.1177/0309132514526442

Michael, I., Armstrong, A., & King, B. (2004). The travel behaviour of international students: The relationship between studying abroad and their choice of tourist destinations. *Journal of Vacation Marketing, 10*(1), 57–66.

Morokvasic, M. (1984). Women in migration. *The International Migration Review, 18*, 882–1382.

Moskal, M. (2018). Gendered differences in international graduates' mobility, identity and career development. *Social and Cultural Geography, 21*(3), 421–440. https://doi.org/10.1080/14649365.2018.1499039.

O'Connor, S. (2019). The intersectional geographies of international students in Ireland: Connecting spaces of encounter and belonging. *Gender, Place & Culture, 27*(20), 1460–1480. https://doi.org/10.1080/0966369X.2019.1693342.

OECD. (2019). Education at a glance: Share of international students enrolled by field of education. *OECD Education Statistics (Database)*. https://doi.org/10.1787/e86f4692-en.

Ortiga, Y. Y. (2018). Constructing a global education hub: The unlikely case of Manila. *Discourse: Studies in the Cultural Politics of Education, 39*(5), 767–781. Routledge. https://doi.org/10.1080/01596306.2018.1448703.

Park, H. (2010). The stranger that is welcomed: Female foreign students from Asia, the English language industry, and the ambivalence of 'Asia rising' in British Columbia, Canada. *Gender, Place and Culture, 17*, 337–355.

Parreñas, R. (2015). *Servants of globalization: Migration and domestic work* (2nd ed.). Stanford: Stanford University Press.

Pedraza, S. (1991). Women and migration: The social consequences of gender. *Annual Review of Sociology, 17*(1), 303–325. https://doi.org/10.1146/annurev.so.17.080191.001511.

Phizacklea, A. (Ed.). (1983). *One way ticket: Migration and female labour*. London and Boston: Routledge.

Radcliffe, S. A. (1990). Ethnicity, patriarchy, and incorporation into the nation: Female migrants as domestic servants in Peru. *Environment and Planning D: Society and Space, 8*(4), 379–393. SAGE Publications Ltd STM. https://doi.org/10.1068/d080379.

Raghuram, P. (2013). Theorising the spaces of student migration. *Population, Space and Place, 19*(2), 138–154. https://doi.org/10.1002/psp.1747.

Raghuram, P., & Sondhi, G. (2017). *Student migration and development*. Statement for the Fourth Informal Thematic Session towards the UN global compact for safe orderly and regular migration. Available at: https://refugeesmigrants.un.org/sites/default/files/ts4_student_migration_and_development.pdf. Accessed 19 Feb 2020.

Raghuram, P., Herman, C., Ruiz Ben E., & Sondhi, G. (2017). *Women and IT Scorecard—India, A survey of 55 firms*. Milton Keynes: The Open University. Available at: https://doi.org/10.13140/RG.2.2.10118.98882. Accessed 4 Mar 2017.

Raghuram, P., Breines, M. R., & Gunter, A. (2020). Beyond #FeesMustFall: International students, fees and everyday agency in the era of decolonisation. *Geoforum, 109*, 95–105. https://doi.org/10.1016/j.geoforum.2020.01.002.

Rice, L. (2013). Between empire and nation: Francophone West African students and decolonization. *Atlantic Studies, 10*(1), 131–147. Routledge. https://doi.org/10.1080/14788810.2013.764106.

Sassen-Koob, S. (1984). Notes on the incorporation of third world women into wage-labor through immigration and off-shore production. *The International Migration Review, 18*(4), 1144–1167 [Center for Migration Studies of New York, Inc., Wiley]. https://doi.org/10.2307/2546076.

Sondhi, G. (2013). Indian international students in Toronto: Exploring young men resisting their family's expectations. *South Asian Diaspora, 5*(2), 223–235. https://doi.org/10.1080/19438192.2013.750957.

Sondhi, G. (2019). 'Settlers' meeting the 'settled': International students encountering the South Asian Diaspora in Ontario. In *Outward and upward mobilities: International students in Canada, their families, and structuring institutions* (pp. 159–176). Toronto: University of Toronto Press.

Sondhi, G., & King, R. (2017). Gendering international student migration: An Indian case-study. *Journal of Ethnic and Migration Studies, 43*(8), 1308–1324. https://doi.org/10.1080/1369183X.2017.1300288.

Song, H. (1997). From brain drain to reverse brain drain: Three decades of Korean experience. *Science Technology & Society, 2*(2), 317–345. https://doi.org/10.1177/097172189700200206.

Song, X., & McCarthy, G. (2018). Governing Asian international students: The policy and practice of essentialising 'critical thinking'. *Globalisation, Societies and Education, 16*(3), 353–365. https://doi.org/10.1080/14767724.2017.1413978.

Trilokekar, R. D., & Kizilbash, Z. (2013). IMAGINE: Canada as a leader in international education. How can Canada benefit from the Australian experience? *Canadian Journal of Higher Education, 43*(2), 1–26.

UUKI. (2019). *International Graduate Outcomes 2019*. Universities UK International. Available at: https://www.universitiesuk.ac.uk/International/Documents/international_graduate_outcomes.pdf.

Waters, J., & Leung, M. (2012). Young people and the reproduction of disadvantage through transnational higher education in Hong Kong. *Sociological Research Online, 17*(3), 1–8. https://doi.org/10.5153/sro.2499.

Waters, J. L. (2006). Geographies of cultural capital: Education, international migration and family strategies between Hong Kong and Canada. *Transactions of the Institute of British Geographers, 31*(2), 179–192. https://doi.org/10.1111/j.1475-5661.2006.00202.x.

Waters, J. L. (2009). Transnational geographies of academic distinction: The role of social capital in the recognition and evaluation of 'overseas' credentials. *Globalisation, Societies and Education, 7*(2), 113–129.

Yang, P. (2018). Understanding youth educational mobilities in Asia: A comparison of Chinese 'Foreign Talent' students in Singapore and Indian MBBS students in China. *Journal of Intercultural Studies, 39*(6), 722–738. https://doi.org/10.1080/07256868.2018.1533534.

Yoon, H., & Kim, H. (2018). Career–family strategies and spatial mobility among South Korean women scholars. *Asian and Pacific Migration Journal, 27*(1), 101–120. https://doi.org/10.1177/0117196818760863.

Yoon, H., & Kim, H. (2019). Seeking a sense of belonging: The exclusion of female doctorate holders in South Korea and the US. *Gender, Place & Culture, 26*(5), 638–658. https://doi.org/10.1080/0966369X.2018.1555141.

Parvati Raghuram is Professor in Geography and Migration at the Open University. She has published widely on retheorising migration of international students and skilled migrants. She is currently leading a grant on decolonising peace education in Africa. She has co-authored *Gender, Migration and Social Reproduction* (Palgrave), *The Practice of Cultural Studies* (Sage), *Gender and International Migration in Europe* (Routledge) and co-edited *South Asian women in the diaspora* (Berg) and *Tracing Indian diaspora: Contexts, Memories, Representations* (Sage). She has written for policy audiences having co-authored research papers for a number of think-tanks. She co-edits the journal *South Asian Diaspora*, *Geographical Journal* and the Palgrave Pivot series *Mobility and Politics* with Martin Geiger and William Walters both at Ottawa.

Gunjan Sondhi is a Lecturer in Geography at the Open University. Gunjan's work focuses on the intersection of highly skilled mobility, gender and class. She has worked internationally on projects exploring the experiences of highly skilled migrants in Singapore, India, UK and Canada. Gunjan's work has appeared in Journal of Ethnic and Migration Studies, Globalisation, Societies and Education, and the annual publication India Migration Report 2018.

CHAPTER 15

The Promises of Migrant Entrepreneurship: A Kaleidoscopic Exploration

Denise L. Spitzer

Abstract There are approximately one billion entrepreneurs—self-employed workers or business developers—around the world, among them are millions of migrants. Successful migrant entrepreneurship is a central pillar of the migration and development nexus—the neoliberal prescription for the alleviation of poverty wherein out-migration is thought to allow for the accumulation of financial and human capital that can be deployed in the service of businesses to lift families and communities out of poverty. But is the promise of migrant entrepreneurship realised? Are there differences in the experiences and outcomes of returned or diasporic migrant entrepreneurs from disparate gendered and cultural backgrounds and the contexts from or in which they function? In this chapter, I use a kaleidoscopic lens to examine migrant entrepreneurship in both diasporic and home country contexts. To date, there is little evidence grounded in migrant experience in support of the migration and development paradigm. There are some instances of personal and material success associated with migrant entrepreneurship. I contend that research that deploys an intersectional lens is needed to account for the myriad variations of, and contexts in which, migrant entrepreneurs operate. These data generated will help inform and problematise the assertions of the migration and development paradigm.

D. L. Spitzer (✉)
School of Public Health, University of Alberta, Edmonton, AB, Canada
e-mail: spitzer@ualberta.ca

C. Mora and N. Piper (eds.), *The Palgrave Handbook of Gender and Migration*, https://doi.org/10.1007/978-3-030-63347-9_15

Introduction

There are approximately one billion entrepreneurs—self-employed workers or business developers—around the world (Naudé, Siegel, & Marchand, 2017). Among them are millions of internal, international, and returned migrants who rely on their own labour or that of their enterprises' employees to sustain themselves, their families, and (potentially) their communities. Successful migrant entrepreneurship is regarded as a central pillar of the migration and development nexus—the neoliberal prescription for the alleviation of poverty wherein out-migration is thought to allow for the accumulation of financial and human capital that can be deployed in the service of businesses to lift families and communities out of poverty (Spitzer, 2016).

But is the promise of migrant entrepreneurship realised? Does migration actually lead to upward mobility through the auspices of entrepreneurship? Furthermore, are there differences in the experiences and outcomes of returned or diasporic migrant entrepreneurs from disparate gendered and cultural backgrounds and the contexts from or in which they operate? In an effort to address these questions, I explore our current state of knowledge of migrant entrepreneurship. Notably, this exploration does not purport to be comprehensive. The inherent vastness and diversity of this issue and the relatively nascent emergence of empirical research grounded in migrants' experiences as opposed to population-based economic data is scant and scattered (Naudé et al., 2015; Spitzer, 2016). While acknowledging the lacunae that exist in our knowledge to date, in this chapter I synthesise available recent information in an effort to generate a base-line intersectional analysis of migrant entrepreneurship upon which further materials and interpretations can be incorporated.

To this end, I begin with a brief overview of the migration and development paradigm and its focus on remittances and migrant entrepreneurship. Next, I review the empirical literature on diasporic (host country) and returned migrant entrepreneurship, most of which highlights the intersections of a limited combination of major identifiers—notably gender, ethnicity, and place, *inter alia*. As the configurations of these markers differ, I refer to this exploration as kaleidoscopic, as we glimpse different images of focal groups, like the light refracted through the glass fragments of the instrument, with each turn—the images are compelling, but ultimately limited by the lens and the nature of the objects used to produce what we see. Importantly, this approach often lacks the dynamism, holism, and focus on power and oppression that is central to intersectionality. I conclude with a discussion as to how the experiences of migrant entrepreneurs rooted in diverse constellations of social location and migrant entrepreneurship detailed in the current literature disrupt the dominant discourse offered by proponents of the migration and development paradigm. I argue that more research grounded in the lives of migrant entrepreneurs (and in the individuals and communities they impact), which is informed and analysed through an intersectional lens that considers a multiplicity of interacting social markers, is needed to generate a more robust evidence-base to challenge, modify, and/or support the migration and development nexus.

Context

Migration and Development

The restructuring of the world economy over the past 50 years that has contributed to the exploitation and under-development of countries in the global South is rooted in the growth of multinational corporations and the expansion of global commodity production chains and outsourcing to poorly waged workforces in the South in a system of precarious labour (Delgado Wise, Covarrubias, & Puentes, 2013; Spitzer, 2016). As profits flow to the North world, Southern residents have contended with a paucity of remunerative employment, the implications of land grabbing and environmental degradation due to resource extraction industries, and the effects of the neoliberal agenda marked by increased privatisation and a concomitant reduction on spending for the public good. Rather than compelling sharing of resources, investments in local infrastructure, and a more equitable distribution of profits, multilateral institutions such as the World Bank, the International Monetary Fund (IMF), the Organization for Economic Cooperation and Development (OECD), and numerous governments North and South, have promoted migration as a development strategy that is meant to effect a triple win for sending countries, host countries, and migrants themselves (Cross, 2015; Piper, 2009; Raghuram, 2009). As de Haas opines: "'neo-optimism' on migration and development reflects neoliberal views in which migrants and markets, not states, become responsible for bringing about development" (2012, p. 10). Thus, as rural to urban and international migration have become survival strategies for many households, migrant workers are increasingly tasked with enhancing the economic prosperity of not only their families, but also their local communities, and countries (Anwar & Chan, 2016; Bakker, 2015; Démurger & Xu, 2011; Spitzer & Piper, 2014). Remittances, particularly those that invested in business, are the reputed mechanism through which poverty alleviation is to occur (see also Kunz & Maisenbacher, this volume).

Remittances and Migrant Entrepreneurship

Although a fully accurate total of migrant remittances is difficult to ascertain as funds transferred outside of formal banking channels are not readily captured in the data, Naudé et al., (2015) estimate the amount to total approximately $550 billion US, more than double the funds directed towards international development aid. Notably, the calculation of nation-level remittance data is configured for political and policy reasons that bolster the migration and development nexus (Bakker, 2015). These widely circulating numbers reinforce the discursively constructed image of the industrious yet frugal migrant from the South toiling for the benefit of their families at home and gesture towards the potential success of the migration as development paradigm (Page & Mercer, 2012). Evidence, however, troubles both the impact of the financial flows and the rosy portrayal of migrant workers and their families.

In coherence with trickle-down economics and the additional embrace of an individualist, bootstraps mentality, migrant remittances are presumed to enhance the economic well-being of their home countries through productive investments that create employment opportunities and generate wealth for their local communities (de Haas, 2012). Studies, however, indicate that most remittances are used by recipients for daily expenses such as food, housing and utilities, medical care, and school fees (McCormick & Wahba, 2004; Naudé, Siegel, & Marchand, 2015, 2017; Spitzer, 2016). Notably, migrants may be chastised in sending government discourse as irrational or self-centred for directing funds to family needs rather than entrepreneurial activities (de Haas, 2012; Page & Mercer, 2012). The distribution of remittances, however, may also contribute to greater intra-community inequality between and among migrant and non-migrant households depending on the amount of remittances received and whether these funds supplement or replace recipient household income (Anwar & Chan, 2016; Page & Mercer, 2012). The deployment of remittances also varies across place. In their cross-national review of remittance uptake, Naudé et al., (2017) found that migrant remittances had little impact on entrepreneurship in the Dominican Republic, the Philippines, and Ecuador; however, these monies were leveraged for small businesses in Mexico and El Salvador, particularly by rural women (see also Kunz and Maisenbacher, in this volume).

Migrant entrepreneurial activities—and expectations thereof—can, however, be fraught. While diasporic migrants are often heralded as innovative self-starters, in just over half of the 25 OECD countries are migrants more entrepreneurial than their native-born counterparts (Naudé et al., 2017). Moreover, with the exception of businesses that benefit from transnational connections such as export activities, migrant enterprises are no more successful than those of non-migrants (Naudé et al., 2017). Return migrants are, however, thought to have garnered sufficient funds and managerial and organisational skills to initiate entrepreneurial activities in their home countries and indeed some studies suggest that returning migrants appear likely to embrace self-employment (Démurger & Xu, 2011; Wahba & Zenou, 2012).

Entrepreneurial endeavours of return migrants, however, are not necessarily successful. Failures are apparently commonplace among businesses initiated by return migrants; moreover, those that remain do not necessarily generate additional employment (Naudé et al., 2015; Naudé et al., 2017; Spitzer 2016). Loss of social networks that are required to facilitate business development, problematic access to adequate financial backing, local policies that are not conducive to entrepreneurship, mismatch between the transfer skills acquired elsewhere to local conditions, and potential hostility from non-migrant community members are among the factors cited for these thwarted ambitions (Naudé et al., 2017; Spitzer, 2016).

Despite evidence to suggest that migrant entrepreneurship is not always sustainable or remunerative, sending countries appear committed to this agenda, often highlighting exemplary cases to bolster this perspective (Piper, 2009). As one Philippine authority from the National Reintegration Centre for OFWs (Overseas Filipino Workers) shared: “We tell them (OFWs) that now all Filipinos can be an entrepreneur at once. All of us can be an entrepreneur, but not all of us can be a successful entrepreneur” (Spitzer & Piper, 2014, p. 1016). Indeed, governments from both sending and host countries have initiated programmes to encourage return migrant entrepreneurship. For instance, Raghuram (2009) found that migrants from Senegal and Mali may avail themselves of financial support from the French government to return to their homelands to engage in entrepreneurial activities.

The Philippines, one of the major source countries for migrant labour, has a well-established structure for the facilitation of out-migration and reintegration of return migrants, much of which focuses on remittance sending and entrepreneurship (Rodriguez, 2010; Spitzer & Piper, 2014). Mandatory pre-departure orientation programmes for Filipino workers venturing overseas emphasise the importance of enacting their commitment to family and to their homeland by sending remittances, which requires industry and frugality to maximise the funds they are able to send, and to encourage investment in entrepreneurial activities even while working in a host country. NRCO also provides financial literacy and entrepreneurship training for families of migrants to place them on a similar trajectory (Spitzer, 2016; Spitzer & Piper, 2014). Return migrants may avail themselves of small business loans upon completion of entrepreneurship training. An evaluation of an earlier iteration of the programme revealed that 70% of these businesses failed (Weekley, 2006). Insufficient financial and material inputs, poor access to buyers, or being expected to succeed in an over-saturated market have been cited by former programme participants as reasons for the demise of their enterprises (Spitzer, 2016; Spitzer & Piper, 2014). Requests for loan repayment, however, does not cease when businesses fail, and many in this predicament face competing demands for their limited (if not shrinking) resources from family members, from government agencies, and often from loan sharks from whom they may have borrowed funds to pay for recruitment/employment placement fees for their overseas employment. Under these circumstances, repayments to government agencies may be given lesser priority. Resultantly, some government programme officials are inclined to mistrust loan recipients whom they blame for their inability to realise the government's entrepreneurial vision (Spitzer & Piper, 2014). Drawing from the Philippine experience, Indonesia, another major source country of migrant labour, has also established reintegration programmes for return migrants that highlight entrepreneurship that similarly offer gender-based training opportunities and financial literacy education (Anwar & Chan, 2016).

Theorising Migrant Entrepreneurship

The uptake, success, and failure of migrant entrepreneurship may be conceptualised in economic, cultural, social, and political terms; however, most perspectives may be allocated to schools of thought that contextualise migrant entrepreneurship as informed primarily by either culture or by structural issues (Brettell & Alstatt, 2007; Hagos, Izak, & Scott, 2019; Munkejord, 2017). Advocates for the former maintain that some cultures value and encourage entrepreneurship while others presumably do not. Those who favour a structural lens point to the opportunities and systemic challenges that migrant entrepreneurs may encounter as they attempt to establish their businesses. From this perspective, migrants may be either thrust into the role of entrepreneur due to barriers entering the labour market—and thus entrepreneurship is a form of excommunication from more stable employment—or conversely, they may be attracted to the possibilities afforded by self-employment and these ventures due to favourable business environments, which can be regarded as liberatory (Brettell & Alstatt, 2007; Munkejord, 2017). As Munkejord (2017) notes, both approaches turn on the structural context of the local economy as either attracting entrepreneurs or repelling wage labourers.

This dichotomous categorisation has been problematised by Kloosertman and Rath's concept of mixed embeddedness (2001, 2003 in Brettell & Alstatt, 2007) which takes into account the environment of the host country, although this framework may well be applied to return migrants resettling in familiar settings (Azmat & Fujimoto, 2016; Brettell & Alstatt, 2007; Munkejord, 2017). Specifically, mixed embeddedness draws attention to how social, political, economic, and policy environments influence—in both formal and informal settings—the potential population of migrant entrepreneurs and the opportunities afforded them at a particular time and place, and as well as the determinations regarding whether adequate returns on investment are available (Azmat & Fujimoto, 2016; Brettell & Alstatt, 2007; Munkejord, 2017). Moving forward from the mixed embeddedness framework, Brush et al. (2009 in Munkejord, 2017) incorporate proximate family embeddedness as in household role, motherhood, and family context as well as the meso- and macro-level phenomena, referring specifically to cultural norms, policies, legislation, and space, along with access to the 3Ms—market, money, and management (Munkejord, 2017). Others suggest that biographical embeddedness of entrepreneurship is also a vital consideration (Brettell & Alstatt, 2007). In this regard, the gendered social construction of entrepreneurship as a masculine risk-taking subject can be called in, and in this context, migrant women who take up the mantle are commonly perceived as engaging in acts of liberation, innovation, and independence that lead to upward mobility (Barragan, Erogul, & Essers, 2018).

Migrant Entrepreneurship Through an Intersectional Lens

Intersectionality highlights the dynamic and mutually constituted identities that situate individuals and communities in the social landscape where their position shapes exposures to oppressions and privileges (Spitzer, Torres, Zwi, Khalema, & Palaganas, 2019). In other words, intersectionality allows us "to speak about simultaneity and mutual co-constitution of different categories of social differentiation and to emphasise the specificity of the experiences shaped by these interactions" (Lutz, 2016, p. 18). More fluid in conceptualisation than the notion of embeddedness and moving beyond a focus on gender or gender in concert with racialised status and socioeconomic class, intersectionality opens up the historical, interactional, and agentic considerations of how individuals and groups of individuals respond and are responded to across dimensions of time, space, and place. While the constellation of factors that cohere to produce intersectionality can be vast, the current literature on migrant entrepreneurship that highlights any of these dimensions appears to primarily consider only a limited number of the following social markers in combination: gender, place (rural–urban or North–South), socioeconomic class, ethnicity or racialised status, and diasporic or returned migrant. In the upcoming section, I share some of the recent literature that addresses at least two of these axes of social stratification.

The Promise of Migrant Entrepreneurship

Diasporic Migrant Entrepreneurship

Diasporic migrant entrepreneurs often deploy their own ethnic identity and linkages situated in context of their host country environment (Azmat & Fujimoto, 2016). Hagos et al., (2019) study of male Eritrean entrepreneurs in Britain illuminates how ethnicity and racialised and socioeconomic statuses interact. Often excluded from the labour market due to racism and poor English language skills, Eritrean migrants also faced challenges accessing financial capital and the institutional supports that might in actuality be available to them. Moreover, they were concerned about their own entrepreneurial skills and were sometimes hesitant to seek out assistance from British agencies due to a lack of confidence in their ability to communicate. Taken together with the structural barriers, many preferred to work within the transnational Eritrean community. In addition, their commitment to family in Eritrea engendered caution about taking up potentially risky ventures that might disrupt or reduce their financial contributions to kin (Hagos et al., 2019).

In their study of male migrant entrepreneurs in the US state of Texas, Brettell and Alstatt (2007) informants from disparate home countries drew extensively from personal and familial resources to launch their enterprises, which they instigated as a means of demonstrating their agency and effecting a sense of independence. Furthermore, national origins figured prominently in their entrepreneurial trajectory. For example, study participants from Latin America tended to hew more closely to businesses with familial connections and with which they had some experience whereas Nigerian and Indian migrant entrepreneurs chose different enterprises. Fifty-per cent noted that their target clientele were members of their own ethnocultural community. Some migrant communities appear to favour particular occupations and that choice is reflected in trends that see some concentration of ethnocultural minority entrepreneurs in specific business categories. For example, approximately 60% of mid-sized motels and hotels in the USA are owned by South Asian Americans and between 40% and 50% of nail salons are operated by Vietnamese Americans (Brettell & Alstatt, 2007).

Like their male migrant entrepreneur counterparts in the US study, South Asian migrant women entrepreneurs in Melbourne, Australia assert that they were drawn to the path of entrepreneurship, which they attributed to their integration into mainstream Australian society (Azmat & Fujimoto 2016). Their journey, however, was not without its struggles. As women are generally tasked with caregiving and other household responsibilities and (as previously mentioned) entrepreneurship is often gendered male in the dominant discourse and public imagination, migrant women—particularly those from the global South where policies and procedures may differ dramatically from the Australian context—face a multiplicity of challenges. In addition to the lack of familiarisation with government and industry resources and regulations, and a paucity of social capital that they could leverage in aid of their business start-ups, South Asian migrant women entrepreneurs were compelled to manage familial responsibilities alongside their working lives. Importantly, most viewed their commitment to family and their families' happiness and well-being as their main priority. This attitude often engendered a supportive familial environment and some benefited from financial and in-kind contributions such as help with household duties to aid their business endeavours. Additionally, some participants noted that gender role expectations were shifting and their male partners were increasingly taking some responsibility for sharing household tasks. Some, however, still faced opposition from family and some opprobrium from some other members of the South Asian community (Azmat & Fujimoto, 2016).

Gendered cultural norms inform migrant women's capacity to engage in entrepreneurial activities and while they be seen as intervening in unhelpful ways, these norms, values, and relationships may also be harnessed in service of their activities. For example, Moroccan and Turkish migrant women in the Netherlands deployed intersecting identity markers of ethnicity and religion (Islam) to negotiate their family roles and relationships while furthering their business aspirations. (Barragan et al., 2018; Essers & Benschop, 2009). Some women who contended with opposition from their families grounded their arguments to be able to pursue their business endeavours by interpreting holy scripture or by invoking Islamic feminist thought, while others eluded familial control by aligning themselves more decidedly with mainstream Dutch society

(Barragan et al., 2018; Essers & Benschop, 2009). Eastern European migrant women entrepreneurs in the UK who were engaged in the transnational expansion of family businesses also used distanciation from patriarchal families to assert their independence by demonstrating their entrepreneurial competence, which ultimately benefited both the business and the family (Barragan et al., 2018; Vershinina, Rodgers, McAdam, & Clinton, 2019).

Place can also figure prominently in the opportunities afforded migrant workers. With increasing rural and feminised migration, migrant women in Northern Norway have become increasingly engaged in entrepreneurship due to the paucity of suitable employment, a desire for more flexibility, control, and independence in their working lives, and as a means to continue residing in the region (Munkejord, 2017). Both familial support and the notion that they were contributing to the social and economic well-being of the community also figured into their decision to pursue small business and self-employment and led to their ultimate success (Munkejord, 2017). In contradiction with the migration and development mantra, some countries delimit the possibilities of migrant entrepreneurship. For example, migrant workers in Hong Kong—a major site of temporary foreign worker migration—are prohibited from working outside the parameters of their work visa. While there are numerous organisations teaching financial literacy and entrepreneurial skills to migrant workers, especially foreign domestic helpers, they are prohibited from launching a small business initiative to supplement income or remittances could lead to deportation while in the special administrative region.

Return Migrant Entrepreneurship

As noted, return migrants are presumed to accumulate both financial resources and an array of skills that can be transferred into productive activities in their home countries thereby creating employment opportunities and fostering economic development (Lianos & Psieridis, 2009; Spitzer, 2016). This assertion presupposes that transnational migrant workers have the opportunity to not only save money, but to garner the requisite skills abroad to launch a business; however, not all skills acquired in a host country are fungible. Some research suggests that return migrants are more likely than non-migrants to be self-employed; however, the positive impact of these entrepreneurial activities on the greater community is not well-established (Naudé et al., 2015; Wahba & Zenou, 2012). Numerous researchers working in a wide array of countries attribute this phenomenon less to the lack of financial wherewithal or even skills acquired elsewhere, but more so to the loss of social capital and social networks that are vital to launching such endeavours (Démurger & Xu, 2011; Naudé et al., 2015; Pauli & Osowska, 2019; Wahba & Zenou, 2012).

In some instances, the loss of social connections appears to enhance the social exclusion of returnees (Anwar & Chan, 2016). Yu's (2015) work in the Philippines found that although return migrants were viewed by members of the non-migrant community as more independent, they were also regarded as haughty, materialistic, and disrespectful, and as behaving in a reputedly Western manner, which some locals found disturbing. Return migrants found it difficult to enter the labour market as job opportunities were lacking in their home communities, were poorly remunerated, or only available to those under 24 years of age (Yu, 2015). Perhaps unsurprisingly, self-employment and launching a small business would be regarded as one of the few viable options; however, they had to contend with the lack of networks and poor rates of success with the government-sponsored entrepreneurship programmes. In the Philippines, some returnees are eligible for specialised livelihood training programmes that direct participants into specifically gendered activities; women may learn to become a pedicurist or how to make lipstick while men may be engaged in sausage-making or in developing cart-racing venues (Spitzer, 2016; Spitzer & Piper, 2014).

Reviewing the data on return migrant entrepreneurship in Bosnia, Bulgaria, Georgia, Kyrgyzstan, Romania, and Tajikistan, Lianos and Psieridis (2009) uncovered distinct trends that speak to the salience of gender, family embeddedness, and temporal aspects of migration. Generally speaking, marital status and the length of time spent working as a migrant worker influenced migrant women's decision to engage in entrepreneurship while financial issues and family size were important factors in men's decision-making. Moreover, their analysis revealed that self-employment was taken up primarily by male return migrants, single individuals, those with large families, and those who were similarly self-employed before and during migration. Those who established businesses were generally better educated migrants with professional credentials who had employed others in the past. The benefits of return migrant entrepreneurship to migrant-sending countries appears to be "true mainly for men" (Lianos & Psieridis, 2009, p. 177).

In addition to gender, place also configures return migrant entrepreneurship. In Egypt, over 40% of migrants returning to urban areas and about 50% of those settling in rural regions invest in housing; 10% invested in economic activities (McCormick & Wahba, 2004). Overall, more migrant savings were returned to urban areas, which could increase rural–urban inequality in the country (McCormick & Wahba, 2004). Even within rural regions, place-based differences can be found. Anwar and Chan's (2016) examination of how two rural Indonesian communities engaged in return migrant entrepreneurship reveals how the specificities of place may influence enterprise development and impact. In one community, an enterprise engaged kin that though thriving, shored up boundaries of social and economic inclusion and exclusion in the small town. Conversely, in another village, financial resources were leveraged by a group of male return migrants for a community economic development project that benefitted more inhabitants who were engaged in the decision-making and a communal problem-solving process. Notably, gender disparities persisted as men dominated positions of power and women earned less than their male counterparts (Anwar & Chan, 2016).

Urban migrants returning to their rural communities in China have had limited options for wage labour and with the other major option being farm work, many choose self-employment (Wu, Fu, Gu, & Shi, 2018). Most self-employed returnees had been skilled, waged labourers in urban centres. Many of the businesses initiated by return internal migrants are small enterprises, 60% of which operated by a sole proprietor (Démurger & Xu, 2011). The remaining operations primarily employ family members (Démurger & Xu, 2011); therefore, any economic benefits are concentrated in the hands of local kin. The success of return urban migrant entrepreneurial activities in rural China are also influenced by family wealth, political affiliation, education, and status as a skilled labourer (Ma, 2002; Wu, et al., 2018). Some returnees are able to mobilise local social capital and motivate local officials to change policies to support business development (Wu, et al., 2018). As the most economically disadvantaged in the community are less likely to migrate (Démurger & Xu, 2011), greater inequality is likely to be entrenched.

Conclusion

Numerous governments and multilateral institutions continue to insist that individuals can migrate their way out of poverty and uplift the economic and social well-being of their families, communities, and countries. Downloading the responsibility of development onto migrant workers rather than addressing global, regional, national, and local inequalities that are grounded in historic legacies and the on-going consequences of colonialism and neoliberal globalisation, and failing to support an agenda for decent work and a reduction in precarious labour, leaves migrants vulnerable to attack for potentially failing in this task. Despite the dominant discourse that heralds the contributions of entrepreneurs, which is emphasised in pre-departure orientation programmes for outward-bound migrants offered in some countries, diasporic migrant entrepreneurship is not always possible, successful, nor of benefit to sending countries. Moreover, the boundaries between choosing to become an entrepreneur or being thrust into the role due to structural issues are blurry. As the mixed embeddedness model suggests, there are multiple formal and informal influences at micro-, meso-, and macro-levels that configure the possibilities and pitfalls for entrepreneurial activities.

Returning to the three questions that anchor this chapter: firstly, has the promise of migrant entrepreneurship realised? Secondly, does migration actually lead to upward mobility through the auspices of entrepreneurship? Thirdly, are there differences in the experiences and outcomes of returned or diasporic migrant entrepreneurs from disparate gendered and cultural backgrounds and the contexts from or in which they operate? Despite economic data that suggests remittance flows play a critical role in the economies of sending countries and the portrayal of select case studies that are used to shore up the migration and development position, there is little evidence that remittances are invested in business opportunities that benefit others. In contrast, some research indicates that remittances and migrant entrepreneurship leads to greater inequality and reduced social cohesion in some communities. When entrepreneurs succeed, benefits generally accrue to their own household and kin—although Anwar and Chan (2016) provide an example of pooled migrant resources being used to establish a community economic development project. The diversity of experiences, conditions, histories, and location and the paucity of published research that attempts to embrace this dynamic complexity leaves my response to the final query open-ended. Overall, studies indicate that the most successful migrant entrepreneurs—diasporic, internal, and returned—are skilled workers and male. Diasporic migrants engaged in export and/or the expansion of family enterprises are also more likely to have a sustainable livelihood. Thus, migrants who already have elevated gendered status and are more likely to have already possessed some skills prior to departure, acquiring more as migrant workers, are those most apt to fulfil the promise of migrant entrepreneurship. Women migrant entrepreneurs have more success in business by disembedding themselves from family—albeit not completely. Often prioritising family happiness and harmony, women negotiate and strategise to enable them to honour family and enterprise. Importantly, familial embeddedness is not only a consideration for those gendered female. Reflecting the construct of the successful entrepreneur as a masculine, assertive, and confident risk-taker, Eritrean male migrant entrepreneurs in the UK were labelled risk averse, lacking in entrepreneurial skills, and in essence feminised, as they factored in the need to conserve financial resources in support of their families at home (Hagos et al., 2019).

The concepts of mixed embeddedness and the 5 M framework are illuminating; however, the majority of the literature captures only a limited number of determinants and fails to consider not only the historical and contemporary socio-political, economic, gendered situatedness of the migrant's life story, but also how these phenomena interpenetrate. Intersectionality as both a lens and analytical framework allows us to explore the multiple social identifiers (socioeconomic class, gender, ethnicity, migration status, place including rural/urban status, dis/ability, sexuality, political affiliation, *inter alia*) that position us in the social landscape and that changes over time and place. Deploying intersectionality is essential to our examinations of this highly diverse global phenomenon and to our interventions into the proclamations offered by migration and development proponents. Future research should not only be intersectional, it needs to address various forms of mobilities in this context, that is internal as well as international and returned migrant workers, and be inclusive of, if not emerging from, the experiences of migrant entrepreneurs and their families.

In conclusion, while there are some examples of successful migrant entrepreneurship, there is little evidence to support the assertion that these activities will lead families and countries out of poverty. More research that deploys an intersectional lens is needed to account for the myriad variations of, and contexts in which, migrant entrepreneurs operate in order to inform and problematise the migration and development paradigm.

References

Anwar, R., & Chan, C. (2016). Contrasting return migrant entrepreneurship experiences in Javanese villages. *International Migration, 54*(4), 150–163. https://doi.org/10.1111/imig.12232.

Azmat, F., & Fujimoto, Y. (2016). Family embeddedness and entrepreneurship experience: A study of Indian Migrant Women Entrepreneurs in Australia. *Entrepreneurship & Regional Development, 28*(9–10), 630–656. https://doi.org/10.1080/08985626.2016.1208279.

Bakker, M. (2015). Discursive representations and policy mobility: How migrant remittances became a 'development tool'. *Global Networks, 15*(1), 21–42.

Barragan, S., Erogul, M., & Essers, C. (2018). 'Strategic (Dis)obedience': Female entrepreneurs reflecting on and acting upon patriarchal practice. *Gender, Work and Organization, 25*(5), 575–592.

Brettell, C., & Alstatt, K. (2007). The agency of immigrant entrepreneurs: Biographies of the self-employed in ethnic and occupational niches of the urban labor market. *Journal of Anthropological Research, 63*(3), 383–397.

Brush, C. G., de Bruin, A., & Welter, F. (2009). A Gender-Aware Framework for Women's Entrepreneurship. *International Journal of Gender and Entrepreneurship, 1*, 8–24.

Cross, H. (2015). Finance, development, and remittances: Extending the Scale of accumulation in migrant labour regimes. *Globalizations, 12*(3), 305–321.

de Haas, H. (2012). The migration and development pendulum: A Critical view on research and policy. *International Migration, 50*(3), 8–25.

Delgado Wise, R., Covarrubias, H., & Puentes, R. (2013). Reframing the debate on migration, development and human rights. *Population, Space and Place, 19*(4), 430–443.

Démurger, S., & Xu, H. (2011). Return migrants: The rise of new entrepreneurs in rural China. *World Development, 39*(10), 1847–1861.

Essers, C., & Benschop, Y. (2009). Muslim businesswomen doing boundary work: The negotiation of Islam, gender and ethnicity within entrepreneurial contexts. *Human Relations, 62*(3), 403–423.

Hagos, S., Izak, M., & Scott, J. (2019). Objective Institutionalized barriers and subjective performance factors of new migrant entrepreneurs. *International Journal of Entrepreneurial Behavior & Research, 25*(5), 842–858.

Kloosertman, R., & Rath, J. (2001). Immigrant Entrepreneurs in Advanced Economies: Mixed Embeddedness Further Explored. *Journal of Ethnic and Migration Studies 27*, 189–201.

Kloosertman, R., & Rath, J. (2003). *Immigrant Entrepreneurs: Venturing Abroad in the Age of Globalization*. Oxford: Berg.

Lianos, T., & Psieridis, A. (2009). On the occupational choices of return migrants. *Entrepreneurship & Regional Development, 21*(2), 155–181. https://doi.org/10.1080/08985620802176187

Lutz, H. (2016). Framing intersectionality: An introduction. In H. Lutz, M. Vivar, & L. Supik (Eds.), *Framing Intersectionality: Debates on a Multi-Faceted Concept in Gender Studies* (pp. 17–42). London: Routledge Press.

Ma, Z. (2002). Social-capital mobilization and income returns to entrepreneurship: The case of return migration in rural China. *Environment and Planning a, 34*(10), 1763–1784.

McCormick, B., & Wahba, J. (2004). *Return international migration and geographical inequality: The Case of Egypt*. Helsinki, Finland., UNU-WIDER: 28 pages.

Munkejord, M. C. (2017). Immigrant entrepreneurship contextualised: Becoming a female migrant entrepreneur in rural Norway. *Journal of Enterprising Communities: People and Places in the Global Economy, 11*(2), 258–276. https://doi.org/10.1108/JEC-05-2015-0029.

Naudé, W., Siegel, M., & Marchand, K. (2015). *Migration, entrepreneurship and development: A critical review* (IZA Discussion Paper No. 9284). Retrieved from Bonn: https://papers.ssrn.com/sol3/papers.cfm?abstract_id=2655324.

Naudé, W., Siegel, M., & Marchand, K. (2017). Migration, entrepreneurship and development: Critical questions. *IZA Journal of Migration, 6*(5), 1–16. https://doi.org/10.1186/s40176-016-0077-8.

Page, B., & Mercer, C. (2012). Why do people do stuff? Reconceptualizing remittance behaviour in diaspora-development research and policy. *Progress in Development Studies, 12*(1), 1–18.

Pauli, U., & Osowska, R. (2019). Building entrepreneurial potential abroad—Exploring return migrant experience. *International Journal of Entrepreneurial Behavior & Research, 25*(5), 936–954.

Piper, N. (2009). Guest editorial: The complex interconnections of the migration and development nexus: A social perspective. *Population, Space and Place, 15*(2), 93–101.

Raghuram, P. (2009). Which migration? What development? Unsettling the edifice of migration and development. *Population, Space and Place, 15*(2), 103–112.

Rodriguez, R. M. (2010). *Migrants for export: How the Philippines state brokers labor to the World*. Minneapolis, MN: University of Minnesota Press.

Spitzer, D. L. (2016). Return migrant entrepreneurship and the migration and development agenda: A Focus on Filipino and Indonesian migrant workers. *Migration, Mobility, & Displacement, 2*(2), 24–39.

Spitzer, D. L., & Piper, N. (2014). Retrenched and returned: Filipino migrant workers during times of crisis. *Sociology, 48*(5), 1007–1023.

Spitzer, D. L., Torres, S., Zwi, A., Khalema, E., & Palaganas, E. (2019). Towards inclusive migrant health care. *BMJ, 366,* l4256. https://doi.org/10.1136/bmj.l4256.

Vershinina, N., Rodgers, P., McAdam, M., & Clinton, E. (2019). Transnational migrant entrepreneurship, gender and family business. *Global Networks, 19*(2), 238–260. https://doi.org/10.1186/s40176-016-0077-8.

Wahba, J., & Zenou, Y. (2012). Out of sight, out of mind: Migration, entrepreneurship and social capital. *Regional Science and Urban Economics, 42*(5), 890–913.

Weekley, K. (2006). From wage labourers to investors? Filipina migrant domestic workers and popular capitalism. In K. Hewison & K. Young (Eds.), *Transnational Migration and Work in Asia* (pp. 193–112). London: Routledge Press.

Wu, C., Fu, Q., Gu, J., & Shi, Z. (2018). Does migration pay off? Returnees, family background, and self-employment in rural China. *The China Review, 18*(1), 59–77.

Yu, X. (2015). The sociocultural effects of returnee overseas Filipino workers in the Philippines. *Norsk Geografisk Tidsskrift-Norwegian Journal of Geography, 69*(1), 47–58.

Denise L. Spitzer, PhD is a Professor in the School of Public Health at the University of Alberta and an Adjunct Professor in the Institute of Feminist and Gender Studies at the University of Ottawa, where, from 2005 to 2015, she was the Canada Research Chair in Gender, Migration and Health and a Principal Scientist in the Institute of Population Health. In addition to undergraduate studies in Biology, Chinese Language, and Music, she holds a Master's degree and doctorate in Anthropology from the University of Alberta, Canada. Professor Spitzer has published in journals such as *The BMJ, Gender & Society, Medical Anthropology Quarterly,* and the *Canadian Journal of Public Health*. Her edited collection, *Engendering Migrant Health: Canadian Perspectives*, published by the University of Toronto Press in 2011, was recognized by the Women's and Gender Studies Association in 2013 with a WGSRF Outstanding Scholarship citation.

CHAPTER 16

Neoliberal 'Flexibility' and the Discursive Incorporation of Migrant Labour in Public Eldercare in Finland

Sirpa Wrede, Lena Näre, Antero Olakivi, and Camilla Nordberg

Abstract Globally spreading neoliberal doctrines regarding the need to 'flexibilise' the allegedly bureaucratic welfare states have been shown to contribute to the 'flexibilisation' of care work, involving austerity measures and reliance on migrant labour. Focusing on public eldercare in Finland, this chapter adds to the literature a close intersectional examination of the exploitative 'coping management' that the neoliberal discourse underpins. This management style fails to challenge inadequate resources and instead extorts nurses, treating them as disposable routine workers. We argue that coping management is expressed through 'flexibility talk' that emphasises personal characteristics rather than formal skills, and that there is an inherent affinity between 'flexibility talk' and 'migrant flexibility talk'. We maintain that the re-articulation of care work as work requiring 'flexibility' and therefore particularly fitting for migrants has problematic gendered and racialised implications in the workplace. Our analysis draws on interviews with managers in public eldercare and observations from eldercare units in Finland. We show how the practices and discourses of flexibilisation help care managers to reconcile the practical and ethical dilemmas they face in ways that appear morally and locally sustainable to them. We have observed that care managers do not currently recognise migrant nurses as stakeholders worthy of equal recognition.

S. Wrede (✉) · L. Näre · A. Olakivi
Sociology, University of Helsinki, Helsinki, Finland
e-mail: sirpa.wrede@helsinki.fi

C. Nordberg
Åbo Akademi University, Turku, Finland

C. Mora and N. Piper (eds.), *The Palgrave Handbook of Gender and Migration*, https://doi.org/10.1007/978-3-030-63347-9_16

Introduction

This chapter examines the discursive incorporation of migrant nurses in the workforce of public eldercare in Finland in the context of neoliberal austerity. We analyse discourses and practices related to care work organisation as linked to the globally spreading neoliberal doctrines regarding the need to 'flexibilise' the allegedly bureaucratic welfare state that began to surface in Finnish politics in the late 1980s. 'Flexibility' is one of the key concepts of neoliberalism, and one with a Janus-faced quality. Eagleton-Pierce (2016: 77, 81) argues that while 'flexibility' used in the context of work carries normatively positive overtones, particularly regarding the capacities of the workers, but also suggests unequal relations of power that may be subtle but still compelling.

Following Fraser (2003), we perceive 'flexibility' as a globally spreading buzzword that prescribes neoliberal recipes both for the mode of how care systems are to be socially organised and for the process of self-constitution to which care managers and nurses are subjected. We maintain that, in the context of care organisations, 'flexibility' offers care managers a conceptually loose way to name the hegemonic expectations they commit to in ways that carry positive overtones. It helps them to appeal to ideas associated with change, competition and innovation without engaging with the unequal capacity and resources that nurses have to adjust to, and the demands posed on them in the workplace (see also Eagleton-Pierce, 2016: 81).

We examine Finnish public eldercare at a time when policymaking first utilised the global idea that migrant labour is not only readily available but 'flexible' (e.g. Nordberg, 2016). We perceive the resulting incorporation of 'flexible' migrant workers in the organisational fabric of Finnish public eldercare as an active process of reimagining, which helps to reconfigure recruitment, management practices and nurse dispositions in ways that lead to inequalities in the workforce and the workplace.

This research contributes to the existing literatures on neoliberal austerity and growing diversity in the care sector (Misra, Woodring, & Merz, 2006; Schwiter, Berndt, & Truong, 2018; Wrede & Näre, 2013). We demonstrate how the ethnocentric system of care work organisation in Finland succeeds in treating the emerging practices through which migrant care workers are segregated as not only neutral but positive. We identify these discursive practices as 'flexibility talk' that is powerful in that it helps to silence possible critiques of the exploitative features of care management. We demonstrate the continued usefulness of the concept of 'coping management' for the analysis of management practices at floor level. Davies (1992), who first described the 'coping' style of management, describes it as an introverted management style that, in line with the neoliberal 'flexibility' discourse, explains problems with the quality of care as arising from the deficiencies of nurses, which can be solved through better leadership.

In what follows, we first describe the institutional features underpinning the 'flexibility talk' that produces migrant care labour as 'flexible'. After presenting our study and analytical approach, we describe 'coping management' in the Finnish context with the aim of highlighting its unequalising dynamics. The fourth section takes a closer look at 'migrant flexibility talk' in the context of understaffing. We conclude by drawing attention to the potent capacity of the Janus-faced 'flexibility' discourse to hamper and silence criticisms of austerity. We also discuss the wider implications of 'flexibility talk' for the status of migrant labour in the care sector.

Neoliberalisation of Finnish Eldercare, the Ambiguous Role of Care Managers and the Incorporation of 'Flexible' Workforce

The history of care work organisation in Finland mirrors the history of the Finnish welfare state. Socio-democratic principles regarding equal access and equity underpinned the expansion of Finnish health and social services as they did in the other Nordic countries (Anttonen, 2002). Finland's social policies of the Nordic type underpinned government-imposed occupational development and regulation in care work (Henriksson, Wrede, & Burau, 2006). Even in the Nordic context, where state investment in women-dominated occupations was a wider trend, Finnish investments in professionalism in care work, including eldercare, were exceptional (Wrede, Henriksson, Høst, Johansson, & Dybbroe, 2008). Eldercare remained fragmented, however, reflecting its comparatively low status even in the social democratic welfare state, which also explains why formal rights to care remained minimal (Karsio & Anttonen, 2013). Nevertheless, the social reforms of the 1980s supported professionalisation and occupational development, primarily through specialisation, qualification requirements and associated formal training and improvement in terms of employment and conditions at work (Julkunen, 1991; Henriksson et al., 2006). The 1990s brought global competitiveness as a dominant perspective that legitimated retrenchment measures (Kettunen, 2008). For the care sector, this neoliberal turn implied deregulation of occupational qualifications and skill-mixes in the context of marketisation and flexibilisation (see Henriksson et al., 2006; Henriksson & Wrede, 2012; Wrede, 2008). The implications of the related workforce policies in the workplace require closer attention.

The modest scale of eldercare coverage did not protect it from austerity measures. On the contrary, concerns about the ageing population and the associated growth of eldercare expenditure legitimated cost containment (e.g. Demos Helsinki, 2010). Cost containment measures targeting alleged inefficiencies include support for for-profit services, resulting in the rapid growth of the private care market (Karsio & Anttonen, 2013; Szebehely & Meagher, 2018).

New policies include, in addition to budget measures, new approaches to municipal care governance. Finnish national policies introduced 'management by results' (MBR, also 'management by objectives') in the late 1990s, entailing new tools such as 'information management' and 'quality management' (Ministry of Health and Social Welfare, 2009: 31–36). Information management focuses on employee performance that is supervised by care managers (such as ward managers, head nurses, home care supervisors, etc.) working at the junction between the top-level officials who articulate organisational policies and the nurses who do actual hands-on tasks. Care managers assess the care needs of clients and cater for the use of personnel resources by organising work schedules and skill-mixes, recruiting, and so forth. The new discourse also tasks care managers with the role of coaching the workers, 'empower(ing) the workforce and spur(ring) workers to develop their occupational skills and the activities of their respective unit' (Ministry of Health and Social Welfare, 2009: 38).

Public sector MBR borrows from globally spreading approaches to corporate governance. Since the mid-2000s, the Finnish capital Helsinki has implemented a management model that treats the city as a corporation with a sector-based organisation with centralised support and financial services (Aarva, 2009: 25). The policies juxtapose social principles of care with an economic rationality that promotes cost efficiency as a key criterion of good practice (Henriksson & Wrede, 2012). At the structural level, the neoliberalisation of care policies also transformed care work by deregulating care work occupations and skill-mixes (Karsio & Anttonen, 2013).

The treatment of care as an aspect of economic competitiveness and the emphasis on 'flexible' workforce policies clash with earlier policies that aimed to socially define quality of care by supporting professionalisation (Henriksson et al., 2006). The neoliberal labour policies promote new workforce sources, the intensification of labour through managerial control and the recomposition of labour (Clarke, 2004: 91). Stakeholders of the Finnish care sector 'detected' the instrumental value of migrant nurses—both those recruited among already resident populations and those recruited from abroad—relatively late (Näre, 2013). Unlike in the other Nordic countries, where immigration began earlier and in larger numbers, the small-scale immigration to Finland resulted mainly from humanitarian measures, including the relocation of individuals with historical Finnish origins from the former Soviet Union. More global immigration grew in the mid-2000s, amounting to 6 per cent of the employed workforce in 2016 (Statistics Finland, 2019). The main countries of origin are currently Estonia, Russia and Iraq.

The capital region leads to the diversification of the care workforce. In 2013, during our data collection, 13% of practical nurses and 10% of nurses were of migrant background, while this percentage among head and ward nurses was below 1% (Statistics Finland, 2019), exemplifying the unequal inclusion of migrant workers. Helsinki continues to spearhead the trend for diversification in the capital region, with the proportion of migrants among practical nurses tripling between 2004 and 2016, from 5 to 15% (Statistics Finland, 2019). The proportion was higher in social care (17%) than health care (11%) in 2016. The biggest numbers of migrant background people are found at the lowest levels of the care work hierarchy, however, as 24% of care assistants in Helsinki belonged to that category in 2016, which is much higher than their proportion in Helsinki's total employed workforce, which was 11%.

The majority of the individuals with a migrant background who work in eldercare have not come to Finland as labour migrants. Rather, policymakers view young women entering secondary education and older women subjected to formal integration activities in particular as important labour pools in a context of decreasing interest in eldercare work. For migrants of diverse backgrounds who choose to settle in Finland, care work is one of the key areas in which integration paths are offered while other options tend to be scarce (Brunila et al., 2011). Despite the modest prospects for career advancement and expectations of 'flexibility', leaving care work is not an option for the majority of migrant-background staff as labour market conditions are not particularly hospitable, especially for non-white people. Persons from outside the European Economic Area whose residence permits often depend on their employment status face particular difficulties due to their dependence on their employers.

Since the mid-1990s, reorganisations of care work training have aimed to produce more 'flexible' nurses (Wrede, 2008). Deregulation of qualifications was parallelled by cost containment, privatisation and effectivation of care management, aiming at dismantling 'unnecessary bureaucratic practices' in eldercare. Efficiency in leadership entails measurable results (Aarva, 2009: 23–26). The national eldercare policies in the 2010s remained silent about the problems caused by policy-induced austerity and instead continued to disentangle quality deficits from resources and costs, and to reframe them as questions of management and administration (Hoppania, 2019). Treating care management as the solution to quality problems is an international trend. Davies (1992) argues regarding the English NHS, where similar reforms occurred much earlier, that reforms have brought about a management style that confirms the treatment of the female rank-and-file nurses as disposable labour. She terms this approach 'coping management' as it remains silent about organisational neglect and its linkage to high wastage and turnover of labour force. Similarly, the 'flexibility' discourse underpins a focus on the deficiencies of workers. The claims that better leadership solves quality problems produces legitimacy for compelling existing staff to manage both increased pressure and compromises over service goals (Davies, 1992: 238). We argue that 'coping managers' find particular appeal in the idea of migrant background nurses as a 'flexible' workforce, as it supports their silence on organisational neglect.

We maintain that care managers refrain from raising criticism of inadequate resources to their superiors as futile, and that the idealisation of the 'flexibility' of migrant nurses aids managers in reconciling the conflicting demands they face (Davies, 1992: 238). The 'flexibility' discourse that includes the idea of nurses as autonomous actors also allows care managers to promote nurses' voluntary choice and self-governance (Bolton, 2004; Dahl, 2009; Olakivi, 2018).

The Study and Analytical Approach

Our study focused on public eldercare as we wished to examine the position of migrant workers in the most strictly regulated context in Finland. We conducted ethnographically oriented research in a large municipal nursing home in Helsinki where one-third of the nursing staff at that time had migration histories. We combined participant observation in four dementia wards with one to two-hour-long, semi-structured interviews with personnel of the nursing home across the organisation. Here we focus on care managers (including ward nurses and head nurses) trained in nursing or gerontology, most having work experience in nursing but also formal training in management. The interview themes covered recruitment practices, the everyday organisation and management of care work, and the perceptions and visions of good care now and in the future. Our participant observations focused on the care routines in the wards, including both hands-on routines and documentation. We paid special attention to the work of the nurses with migrant backgrounds. Our field notes contain observations by four researchers, compiled from 85 visits to the organisations (occurring between April 2011 and October 2012). We use pseudonyms when citing our participants.

Our subsequent discussion of our results draws on qualitative research on public eldercare in the City of Helsinki, including interviews with care managers and participant observation at the wards of a large nursing home. In analysing our research interviews, we focus on utterances that we identify as 'flexibility talk'. This term refers to how research participants draw from the discursive resources that the globally spreading doctrine of neoliberal flexibility offer as a means to present their management style as not only legitimate but in line with the wider innovative politics of care. Furthermore, we identify a particular variant of 'flexibility talk' as 'migrant flexibility talk', through which care managers articulate the increasing reliance on migrant labour as a neutral or desirable feature of 'flexible' care work organisation.

Our conceptualisations draw from Davies (1992: 229), who in her classic work on the management of nursing in the UK's National Health System analysed the linkage between 'gender talk' and the devaluation of nursing as women's work. She highlighted the problematic features of this 'gender talk', which at the time suggested that nurses, because they are women, are highly committed to the activity of direct care but less capable of leading. Indeed, 'gender talk' underpinned the idea that it was female nurses' 'womanly' commitment as much as their technical skill that made them good nurses. Our identification of 'flexibility talk' and 'migrant flexibility talk' builds on our observations of similar gendered and racialised arguments regarding care work in general and migrant nurses in particular. The fact that 'flexibility talk' regarding care work is silent about both old and emergent inequalities is in itself not surprising. As Davies (1992) and others have shown, inequalities are nothing new in care work and Finland is no exception to the rule of low pay and terms and conditions that are inferior to many other types of work that require skill and commitment. Eldercare in particular lacks recognition as professional work both internationally and in Finland, making it a problematic field even in comparison to other types of 'women's care work'.

Coping Management and 'Flexibility Talk' in Finnish Eldercare

The flexibility discourse invites public organisations to constitute themselves as modern, efficient, agile and innovative. Failing to live up to these ideals leads to public accusations of employing routine, stagnant, inefficient and bureaucratic practices (du Gay, 1996). By delineating her organisation as a spearhead of the transformation of public eldercare, the head of our target nursing home, Theresa, positioned herself as an eager proponent of flexibility: 'Our vision is to be a kind of centre of innovation and excellence so that we always try to stay a bit ahead of the others. We are pioneers in quite a few things, especially in improving care work'.

The austere city budget and everyday problems of organising care work pose a challenge to this type of 'flexibility talk', however. The image of the nursing home as 'flexible' and innovative was more fragile in the talk of the floor-level care managers. The management hierarchy consisted of the head of the nursing home at the top, followed by two head nurses and ward nurses responsible for their own wards and the staff and residents in the wards. Ward nurses in particular were less prone to depict the nursing home as an agile and post-bureaucratic unit. One of them, Annika, drew attention to the top-down approach of implementing development projects, and requirements concerning care practices as well as budget frames. Another ward nurse, Emma, described her situation as being 'between a rock and a hard place': 'There are the demands from above (…) and there are the demands from below, and when they do not always match, it is challenging indeed'.

In our target nursing home, the ideals of the earlier mentioned, result-oriented management approach (MBR) translated into, among other things, quarterly budget reporting and reporting on the condition of the residents based on the international Residence Assessment Instrument system (RAI). We found that many managers perceived the RAI-programme as being expensive and difficult to use because of the poor translation from English to Finnish. As RAI is a trademarked programme originating from the USA and in global use, the City of Helsinki lacks the authority to adapt it to local purposes. Instead, the nurses were trained to understand its logic. While using RAI clearly provides many benefits and its primary and original use is clinical, related to assessing individual residents' strength and care needs regularly (e.g.

Hawes et al., 1997), in the context of Finnish MBR it soon became a key management tool to compare the care intensity in different wards in relation to the staffing levels. The austerity-imposed focus on keeping staffing levels to a minimum gives RAI standardisation power. Ward nurse Annika drew our attention to problematic uses of RAI in comparisons of the different wards:

> We had to finish our RAI-assessment again by the end of September and I happened to notice that for one resident the RAI-assessment was done incorrectly. And when the results go to the top management, they never think if they are wrong or not. And instead of coming here to the ward to ask the nurses and to listen to their opinions, all the conclusions are made based on this one RAI-measurement. It's really dangerous to draw conclusions about the care intensity of a ward, if the results are not accurate. (Annika 31.10.2011.)

MBR relies on comparisons to achieve cost-efficient and up-to-date care management, treating the data collected at wards as objective indicators. According to Annika, however, indicators hide the fact that there are always different human factors behind such indicators. For her, the real problem is the hierarchical practice of relying on RAI without listening to the opinions of front-level personnel.[1]

Measurement of care needs as the basis for workforce allocation is an example of flexibility management devices for care organisation, but, as we show, interpreting care needs is never straightforward. The translation problems related to the RAI programme are a good example of the complexities of measurement, as is the use of the measures in staff planning. It is hardly surprising, then, that care managers have mixed feelings about MBR. Annika, for instance, likens her work as ward nurse to coaching: she wants to 'create team spirit' and 'spur' her staff members to 'carry out high-quality care', ideals that are difficult to combine with the hierarchical logic of MBR.

Ward nurses at our target organisation were each responsible for their ward budgets, on which they reported to their superiors. In our observations, we saw ample evidence for how the ward budgets were set to very low, even unrealistic levels:

> Liisa [the ward nurse in C] is coming from a meeting and hurrying past me looking stressed. "We are exceeding our budgets," she tells me in passing. I respond that then it means the budget is too small. To which Liisa responds: "Well yes, it is the absolute minimum." (Field notes 28.4.2011.)

[1]Another globally used indicator in the MBR of care institutions and hospitals is Case Mix Index (CMI), a relative value referring to the patients' condition and care needs. CMI is used to determine the allocation of resources to care. In the care work organisation we studied, the CMI is calculated for each ward and, in theory, the higher the index, the better the staffing levels should be. According to our interviews, the CMI is something that is closely followed by the top management.

The responsibility to maintain the budget caps creates stress in the daily management work of ward nurses. Liisa's comment further illustrates how difficult it may be for care managers to express overt criticism of their employers. In the above, she reformulates the suggested interpretation of the budget being too small, stating that the budget is at the absolute minimum, not yet (clearly) too small.

A budget set to an unrealistic level is a ruthless, omnipresent care management condition that affects every choice made on the wards. For instance, ward nurses regularly have to make decisions regarding whether there is money to hire substitutes:

> I go to the office to write down my field notes and the ward nurse Emma comes into mark the work shift lists that one nurse is sick today and tomorrow and that the physiotherapist is on sick leave until November. Emma is wondering whether they will get a substitute for the physiotherapist. I ask why they wouldn't. "Well, because we have this budget." (Field notes 10.10.2011.)

The austerity measures regarding personnel are not a passing phenomenon. The nurse positions of our target nursing home were cut by nearly 30% in 2013, soon after we completed our fieldwork. The job descriptions of practical nurses were expanded to include doing laundry, serving meals and other house-keeping tasks. A third means of cost containment, also regularly practiced when we were in the field, was limiting the use of temporary personnel when nurses were absent due to illness.

Inadequate personnel resources constrain the daily rhythms of care, creating a constant feeling of haste. The most time-consuming care routines, such as helping residents from and into their beds and showering and feeding residents in need of assistance were done early in the day or given little time. We observed that the time used to feeding bed-ridden residents was limited to five to ten minutes (Field notes, 4.5.2011). Even the personal hygiene needs of residents fitted poorly in the care plans: helping in the shower or heating the sauna 'destroyed the whole day', as one of the nurses stated (Field notes, 20.4.2011).

At the level of discourse, neoliberal doctrines task nurses with the responsibility to provide personalised care, but what we witnessed was blatant disregard on the part of the City of Helsinki to provide the resources needed for them to fulfill such goals. The silence of the 'flexibility' discourse regarding this neglect trickled down in the organisation to those care managers who worked at the floor level. Our data shows that this Janus-faced approach to care management offers them only a poor defence from failing to fulfil expectations of quality care, as the 'flexibility' discourse does not leave space to discuss its lack of realism regarding the use of personnel resources. Instead, care managers found ways to keep up what could be characterised as a façade of 'coping'.

The next section looks at how the availability of 'flexible' migrant nurses helps 'coping managers' to imagine a future in which coping with austerity can be seen as tenable, despite the constant pressures to 'flexibilise' care.

'Migrant Flexibility Talk' in the Context of Inadequate Staffing

In the context of coping management, recruitment and adequate staffing levels emerged as specific challenges for our target organisations. The managers interviewed described recruitment as the most challenging part of their work; filling vacant positions with qualified staff was especially difficult. High turnover and recruitment problems are in line with the exploitative dynamics of coping management (Davies, 1992: 239). A key example of the pressure directed at workers in our field was named by managers and nurses as 'minimum level of staff' (*minimimiehitys*). The expression of 'minimum level', repeatedly mentioned as a common situation in the wards, maintains an impression of a level of staffing that is not (clearly) below the minimum and thus impossible to cope with, yet far from ideal. In our fieldwork, however, the 'minimum level' often came across as inadequate staffing, considering the demands of the wards.

The consequences of constraining working conditions are clearly visible in Finland, as an increasing number of educated nurses either plan to quit their jobs in eldercare (Kröger et al., 2018: 70–78) or have already done so (Ailasmaa, 2015). The possibility of measures that would improve working conditions did not, however, come up in the interviews with care managers. Instead, they presented migrant workers and job seekers as an answer to the challenge of recruitment. As Niina (a head nurse) noted, 'Soon we'll have a real [issue]. Where are we getting employees from? In my view, people of migrant backgrounds are one of those resources. (…) Immigrants are, you might almost say, indispensable for us so that we can keep the wheels rolling in eldercare'.

Care managers often discussed migrant labour not only as the most realistic labour pool but one that offered care workers with better attitudes than local workers. Marja (a quality planner) expressed a view she shared with the other care managers regarding the quality potential of migrant care workers:

> I wish there would be more immigration (…) because we really do not have enough of our own staff. And I don't think it will have a negative influence [on the quality of care] if they have good language skills. (…) [Migrants] coming from many countries, for instance from Asian countries, have a really lovely attitude towards the elderly; they have a much more respectful attitude than we have.

Marja distinguishes migrants from 'our own' staff, invoking—perhaps inadvertently—an instrumental approach to the diversification of care work. She addresses an unexpressed but potential criticism over language skills, after which she describes the increased recruitment of migrants as a way of better serving the care recipients. Finnish language skills were commonly discussed as a key requirement for migrant-background nurses and the nursing home did offer Finnish language courses. However, nurses were themselves responsible for arranging the time for participation, which due to constant staff shortages was often impossible. Moreover, Marja's statement evokes a stereotype of Asian workers, but she refers to migrant nurses more generally. The ward nurses we talked to similarly constructed cultural difference by referring to migrant nurses' 'deeper' respect towards old people, discussed as something that nurses from 'different cultures' acquire from their 'mothers' milk'.

Previous research has highlighted how racialised stereotypes of migrant workers increase their marginalisation and create new hierarchies (Dahle & Seeberg, 2013). The 'migrant flexibility talk', which presents migrant workers' skills and qualities as extra-professional, treats respectful attitudes as an ethnic trait that is related to 'culture' rather than recognising it as a professional skill to be valued. Like 'gender talk' about care work as 'women's work', 'migrant flexibility talk' confirms the devaluation of hands-on care work and the migrant nurses who do that work (see Davies, 1992). Ethnic and racial stereotypes can also perform other kinds of work in care managers' social worlds. By evoking ethnic stereotypes as cultural resources, care managers can tell themselves—and other stakeholders—morally and affectively appealing stories about the neoliberal flexibilisation of care work. In these stories, the neoliberal flexibilisation of care work serves all actors in tandem, as discussed in the following account by Annika, a ward nurse:

> I've noted that if they come from these African countries, like many of our care workers do, or those who work here as temps, somehow you can see a kind of respect towards the work [in them]. They do it extremely well and with high quality, and they accept any assignments whatsoever. Somehow, they don't make a fuss about anything. [...] They are used to working more, and they are always ready to be flexible [...] They've come here to work for the residents.

Annika describes migrant workers, African workers especially, as 'flexible'. Instead of considering structural factors that might explain why non-white temporary staff in particular may be compelled to accept any kind of assignment, she describes the migrant workers as themselves agential: 'They are used to working more', 'they are always ready to be flexible' and 'they've come here to work for the residents'—that is, not for the managers. By drawing on the above stereotypes and by presenting flexibility, motivation and commitment as *inherent* and *authentic* qualities of migrant workers, managers can present the diversification and flexibilisation of care work as a win-win tendency that does not seem to exploit anyone or violate anyone's interests (Olakivi, 2018). This image of migrant diversification of work and labour as neutral, even desirable, efficiently silences concerns related to inequalities in the workforce.

Whether migrant nurses find these interpretations of self-governance (rather than exploitation) convincing, is of course another question (Olakivi, 2017). Indeed, challenging these interpretations of authentic self-governance is a central way for migrant workers to display hidden hierarchies and criticise their unequal position.

The depictions of migrant employees and job-seekers as authentic and self-governing nurses draw from discourses that treat the numerically female-dominated care work labour force as disposable labour and from colonialist and racist discourses regarding international recruitment (Henriksson et al., 2006; Näre & Nordberg, 2016). Importantly, 'migrant flexibility talk' resonates with deep-seated historical stereotypes of care work as a 'labour of love' that mitigates the importance of formal training in care work (Finch & Groves, 1983; Henriksson & Wrede, 2012). These stereotypes help care managers to tell themselves and other stakeholders a story in which the quality of care can be improved regardless of the harsh conditions and the hazardous consequences of MBR as long as nurses have the proper ethics and ability to practice self-governance.

Conclusions

While the imperative of global competitiveness that underpins the discourse of neoliberal 'flexibility' may remain unuttered in workplace interactions, it serves to legitimate a meta-level organisational design that not only enforces austerity measures on the public sector but underpins a set of measures aiming at the 'flexibilisation' of care work. This chapter shows how care managers are adjusting care work organisation to the new values of public policy, working under the discipline of austerity budgeting. We argue that care managers exhibit exploitative coping management, familiar from previous studies of gendered inequalities in care work.

Care managers struggle with the realities of care work in eldercare in that it defies standardisation and micromanagement (Bolton, 2004). Leaving austere budgets unchallenged leaves pressuring the staff their only option, which, in turn, results in problems such as low morale, burnout, high turnover, low self-esteem and infighting (Davies, 1992: 239). 'Migrant flexibility talk' suggests that non-migrant local staff may indeed be more rigid and therefore prone to exhibit such 'symptoms' of coping management. In this context of staff shortages and high turnover, migrant labour holds, at least at the level of discourse, a specific promise in that it is self-evidently understood as more 'flexible' than other labour. We have observed that migrant nurses were expected to be less likely to question inadequate working conditions. This expectation of 'flexibility' on the part of precarious staff members is hardly neutral in the context of unequal labour market dynamics.

Like the 'gender talk' that devalued the technical skills of nurses (Davies, 1992), the racialised talk about migrant nurse 'flexibility' distinguishes the caring 'them' from uncaring 'us'. It praises the commitment of migrant-background nurses but does not recognise it as professional skill, a misrecognition that confirms the role of migrant-background nurses as a buffer workforce. In effect, this discourse devalues migrant nurses as skilled workers in ways that limit their inclusion and deny them equal opportunities.

Austerity further strengthens the silence around the intersecting problems of the inclusion of migrant nurses in care work organisations. Understaffing efficiently hampers their access to the language training otherwise deemed necessary to their ability to provide quality care. Migrant-background women in particular lack opportunities in the Finnish labour market. Our analysis shows that, having gained access to a job in the care sector, they not only remain vulnerable to exploitation but that their access to language training and their other needs regarding occupational development tend to be overlooked. Access to equal integration would require adequate time allowed for tasks that pose more demands on them as non-native speakers, including the daily task of documentation. Language skills are not only required for career advancement, limited language skills tend to freeze nurses in a 'novice' position due to needing support in routine tasks (Vartiainen, 2019).

Increasing gendered and racialised inequalities loom if the professional integration of migrant nurses is not afforded more attention. This issue is pressing in that it is likely that the proportion of migrant workers will increase rapidly in a situation of waning traditional labour pools. If neoliberal 'flexibility talk' continues to dominate, it is likely that the future of eldercare is bleak and that exploitation of primarily female migrant-background workers is just one, albeit crucial, type of inequality.

Instead of pinning the blame on care managers, who are a comparatively low-paid group of women professionals, we hope we have shown that neoliberal austerity measures place them between a rock and a hard place. Care managers have not authored the reforms they implement, and they lack adequate resources to ensure quality care. Even though the often hardworking 'coping managers' can themselves be seen as being exploited by policymakers who enforce austerity, we cannot ignore the fact that care management has an ethical side that should be awarded more attention. Our study shows that, rather than focusing on the well-being of their staff, care managers tend to prioritise making ends meet, even when faced with constant understaffing. This is understandable in a situation where, first, it is likely that they would have little chance of successfully lobbying for more resources, and, second, an enthusiastic and appealing management discourse posits that problems in care should be solved by skilful management instead of investment in resources. There is ample evidence from other countries regarding similar neoliberal developments in the care sector. Previous research has demonstrated how the neoliberal 'flexibility' discourse invites all parties to accept a specific version of efficient and customer-oriented care, helping care managers to bridge the expectations of the different stakeholders (Bolton, 2004; Dahl, 2009).

What our work adds is our close intersectional examination of 'flexibility talk' and 'migrant flexibility talk' and, in particular, the inherent affinity between the two. We have shown how flexibilisation can be carried out through practices and discourses that help care managers to reconcile the practical and ethical dilemmas they face in ways that—to them—appear morally and locally sustainable. We have observed that care managers do not currently recognise migrant nurses as stakeholders worthy of equal recognition. Like the care managers whom Davies (1992) studied, who used 'gender talk' to legitimise their devaluation of female nurses as workers, the coping managers we studied inadvertently used 'migrant flexibility talk' to legitimise their misrecognition of migrant nurses. Care managers referred to migrant workers as a solution to their own management dilemmas, perceiving them as ideal carers but only for the purpose of filling the floor-level void. As a result, the dead-end status of the migrant-background workforce is becoming cemented as a potent source of inequality in the care sector.

References

Aarva, K. (2009). *Hoivan ja hoidon lähijohtaminen*. Tampere: University of Tampere.

Ailasmaa, R. (2015). *Terveys- ja sosiaalipalvelujen henkilöstö 2013*. Helsinki: National Institute for Health and Welfare.

Anttonen, A. (2002). Universalism and social policy: A nordic-feminist revaluation. *Nora: Nordic Journal of Women's Studies, 10*(2): 71–80.

Bolton, S. C. (2004). A simple matter of control? NHS hospital nurses and new management. *Journal of Management Studies, 41*(2), 317–333.

Brunila, K., Kurki, T., Lahelma, E., Lehtonen, J., Mietola, R., & Palmu, T. (2011). Multiple transitions: Educational policies and young people's post-compulsory choices. *Scandinavian Journal of Educational Research, 55*(3), 307–324.

Clarke, J. (2004). *Changing welfare, changing states: New directions in social policy*. London: Sage.

Dahl, H. M. (2009). New public management, care and struggles about recognition. *Critical Social Policy, 29*(4), 634–654.

Dahle, R., & Seeberg, M. L. (2013). "Does she speak Norwegian?" Ethnic dimensions of hierarchy in Norwegian health care workplaces. *Nordic Journal of Migration Research, 3*(2), 82–90.

Davies, C. (1992) Gender, history and management style in nursing: Towards a theoretical synthesis. In M. Savage & A. Witz (Eds.), *Gender and bureaucracy* (pp. 229–252). Oxford: Blackwell Publishers. *The Sociological Review, 39*, 229–252.

Demos Helsinki (2010). *Metropolin hyvinvointi*. Helsinki: Demos Helsinki.

Eagleton-Pierce, M. (2016). *Neoliberalism: The key concepts*. Abingdon: Routledge.

du Gay, P. (1996). Organizing identity: Entrepreneurial governance and public management. In S. Hall & P. du Gay (Eds.), *Questions of cultural identity* (pp. 151–169). London: Sage.

Finch, J., & Groves, D. (Eds.). (1983). *A Labour of love: Women, work and caring*. London: Routledge.

Fraser, N. (2003). From discipline to flexibilization? Rereading Foucault in the shadow of globalization. *Constellations, 10*(2), 160–171.

Hawes, C., Morris, J., Phillips, C., Fries, B., Murphy, K., & Mor, V. (1997). Development of the nursing home resident assessment instrument in the USA. *Age and Ageing, 26*(S2), 19–25.

Henriksson, L., Wrede, S., & Burau, V. (2006). Understanding professional projects in welfare service work: Revival of old professionalism? *Gender, Work & Organization, 13*(2), 174–192.

Henriksson, L. & Wrede, S. (2012). The making of medico-managerial care work culture in public home care for the elderly. In C. Ceci, K. Björnsdóttir & M. E. Purkis (Eds.), *Perspectives on care at home for older people* (pp. 171–185). New York, NY: Routledge.

Hoppania, H.-K. (2019). Politicisation, engagement, depoliticisation–the neoliberal politics of care. *Critical Social Policy, 39*(2), 229–247.

Julkunen, R. (1991). Hoiva Ja Professionalismi. *Sosiologia, 28*(2), 75–83.

Karsio, O., & Anttonen, A. (2013). Marketisation of eldercare in Finland: Legal frames, outsourcing practices and the rapid growth of for-profit services. In G. Meagher & M. Szebehely (Eds.), *Marketisation in nordic eldercare: A research report on legislation, oversight, extent and consequences* (pp. 85–126). Stockholm: Stockholm University.

Kettunen, P. (2008). *Globalisaatio ja kansallinen me: kansallisen katseen historiallinen kritiikki*. Tampere: Vastapaino.

Kröger, T., Van Aerschot, L., & Mathew Puthenparambil, J. (2018). *Hoivatyö muutoksessa: Suomalainen vanhustyö pohjoismaisessa vertailussa*. Jyväskylä: University of Jyväskylä.

Ministry of Health and Social Welfare. (2009). *Johtamisella vaikuttavuutta ja vetovoimaa hoitotyöhön. Toimintaohjelma 2009–2011*. [Sosiaali- ja terveysministeriön julkaisuja 2009:18.]. Ministry of Health and Social Welfare.

Misra, J., Woodring, J., & Merz, S. (2006). The globalization of care work: Neoliberal economic restructuring and migration policy. *Globalizations, 3*(3), 317–332.

Näre, L. (2013). Ideal workers and suspects: Employers' politics of recognition and the migrant division of labour in Finland. *Nordic Journal of Migration Research, 3*(2), 72–81.

Näre, L., & Nordberg, C. (2016). Neoliberal postcolonialism in the media: Constructing Filipino nurse subjects in Finland. *European Journal of Cultural Studies, 19*(1), 16–32.

Nordberg, C. (2016). Outsourcing equality: Migrant care worker imaginary in Finnish media. *Nordic Journal of Working Life Studies, 6*(3), 101–118.

Olakivi, A. (2017). Unmasking the enterprising nurse: Migrant care workers and the discursive mobilisation of productive professionals. *Sociology of Health & Illness, 39*(3), 428–442.

Olakivi, A. (2018). *The relational construction of occupational agency: Performing professional and enterprising selves in diversifying care work*. Helsinki: University of Helsinki.

Schwiter, K., Berndt, C., & Truong, J. (2018). Neoliberal austerity and the marketisation of elderly care. *Social and Cultural Geography, 19*(3), 379–399.

Statistics Finland. (2019). *Labour statistics on migrant and foreign-born employees available on request*. http://www.stat.fi.

Szebehely, M. & Meagher, G. (2018). Nordic eldercare—Weak universalism becoming weaker? *Journal of European Social Policy, 28*(3), 294–308.

Vartiainen, P. (2019). *Filippiiniläisten sairaanhoitajien polut Suomeen: Tutkimus oppimisesta ja työyhteisöintegraatiosta kansainvälisen rekrytoinnin kontekstissa*. Tampere: Tampere University Press.

Wrede, S. (2008). Educating generalists: Flexibility and identity in Auxiliary nursing in Finland. In E. Kuhlmann & M. Saks (Eds.), *Rethinking professional governance: International directions in health care* (pp. 127–140). Bristol: Policy Press.

Wrede, S., Henriksson, L., Høst, H., Johansson, S., & Dybbroe, B. (Eds.). (2008). *Care work in crisis: Reclaiming the Nordic ethos of care*. Lund: Studentlitteratur.

Wrede, S., & Näre, L. (2013). Glocalising care in the Nordic countries. *Nordic Journal of Migration Research, 3*(2), 57–62.

Sirpa Wrede , D. Soc.Sc., is a Professor of Sociology at the University of Helsinki. Her research interests deal with the dynamics of inequality in the social organisation of social service professionalism and social citizenship. Her current work focuses on ageing and social ties in the context of migration and on professional groups in the context of neoliberal globalisation. She leads a research team within the Centre of Excellence in Research on Ageing and Care (funded by the Academy of Finland, #312310), of which she is Vice Director.

Lena Näre (DPhil, PhD, Docent) is tenure-track Associate Professor of Sociology at the University of Helsinki. Her research interests include the study of migration and asylum, work and employment, care and transnationalism, intersectionality and ethnographic methods. She is currently the PI of a four-year Kone Foundation funded project *Struggles over Home and Citizenship—Neighbourhood Solidarity as a Response to the Asylum 'Crisis' (Neighbourhood Solidarities)* (2018–2021). She is the editor-in-chief (with S. Bendixsen) of the *Nordic Journal of Migration Research*.

Antero Olakivi , D. Soc.Sc., is a postdoctoral researcher at the University of Helsinki, in the Centre of Excellence in Research on Ageing and Care (funded by the Academy of Finland, #312310). His PhD dissertation, *The Relational Construction of Occupational Agency: Performing Professional and Enterprising Selves in Diversifying Care Work* (University of Helsinki, 2018), examined the political construction and silencing of racial and migration-based inequalities in eldercare work in Finland. He has published articles in journals such as *Sociology of Health and Illness* and *Current Sociology*.

Camilla Nordberg , D. Soc.Sc., is a senior lecturer in social policy at Åbo Akademi University. Her research has revolved around the migration welfare-state nexus, focusing on issues of citizenship, power, political discourse and street-level bureaucratic encounters. She is currently the project leader of the Academy of Finland funded research project *Orderingthe migrant family: power asymmetries and citizenisation in restructuring welfare bureaucracies* (#334686).

Part IV

Gendered Migration as Site of Governance and Development

CHAPTER 17

Gendering the Global Governance of Migration

Rianne Mahon

Abstract This chapter will examine the discourses and practices that key international organisations (IOs) have developed to understand and to govern the 'feminisation of migration'. International organisations (IOs) have made an important contribution to the gendering of contemporary international debates on migration by developing gendered representations of the issues and engaging in various practices designed to provide solutions. Yet just as 'gender equality and women's empowerment' remain a contested norm, so too do IOs offer competing versions. For some, this is limited to gendering the 'migration management' discourse. IOs more closely linked to feminist networks however offer more critical rights-centred alternatives.

Introduction

While the governance of migration long remained state-centric, over the last two decades there have been increasing efforts to develop mechanisms at the global and regional scales as rising flows of migrants[1] made it clear that greater inter-state cooperation and the involvement of a wider range of actors was necessary. The growing interest in transnational migration governance coincided with increased awareness of the 'feminisation' of migration'—i.e. that

[1]The number of international migrants increased form 82 million in 1970 to 244 million in 2017 (Betts & Kainz, 2017: 5).

R. Mahon (✉)
Carleton University, Ottawa, ON, Canada
e-mail: prmahon@rogers.com

C. Mora and N. Piper (eds.), *The Palgrave Handbook of Gender and Migration*, https://doi.org/10.1007/978-3-030-63347-9_17

women not only accounted for nearly half of migrants, but also were migrating on their own account as workers, as unequal development constraining possibilities at home coincided with rising demand for care work in wealthier countries. The constitution of a global 'gender equality and women's empowerment' norm, in turn, helped to generate awareness of the specific challenges that women migrants face. With noted exceptions,[2] however, the growing literature on global governance of migration has remained largely gender blind.

This chapter briefly reviews the literature on global governance of migration. Here it will be argued that a feminist approach has much in common with critics of the dominant 'migration management' project but also raises new questions and claims. For feminist scholars, an important question is how 'gender equality and women's empowerment' is translated (or mis-translated) in key international organisations. The second section examines the gendering of the managed migration project, both the 'triple win' and 'women as victims' variants. Here the focus is on the World Bank and the International Organisation for Migration (IOM). The third section turns to examine feminist, rights-centred alternatives. While there is some discussion of the International Labour Organisation (ILO) as one of the main organisations concerned with the rights of migrants as workers, the main focus is on two UN agencies—INSTRAW and UNIFEM,[3] now merged into UN Women—which have fought to inject a feminist understanding.

On the Global Governance of Migration

Grugel and Piper (2007) have done an excellent review of various conceptions of global governance, which I will not attempt to reproduce here. I do, however, adopt their way of thinking about global governance as 'project(s)'. That is, 'projects of global governance are essentially strategic interventions or policies that aim to manage the stresses of global politics and globalization through global rule-making. They may be the work of politicians, global bureaucrats, international institutions, states or other policy actors such as NGOs.... They take different institutional forms and have different normative underpinnings; different actors are caught up within them and different kinds of resources are exchanged within them' (2007: 11). This approach serves to highlight their aspirational character while also opening up the possibility of competing projects.

Three distinct approaches to analysing global governance of migration projects can be discerned: those that accept the aims and claims of the 'managed migration' project; those that take a critical perspective and work to

[2] Piper (2006), Basok and Piper (2012), Benería, Deere, and Kabeer (2013), Hennebry, Williams, and Walton-Roberts (2016)

[3] INSTRAW stands for International Research and Training Institute for the Advancement of Women; UNIFEM stand for the UN Development Fund for Women.

illuminate alternative, rights-based projects; and feminist research that shows how the 'gender equality and women's empowerment' project has been (differentially) taken up in the gendering of migration governance.

Pécoud provides a good description of the managed migration project:

> The core elements of this framework…include: (1) the description of migration as a 'global' issue that must be 'globally' addressed through international cooperation; (2) a pragmatic appreciation of migration as a normal process that should benefit sending and receiving countries and migrants alike ('triple-win'); (3) the aspiration to 'well-managed' or 'orderly' migration flows (as opposed to unauthorised migration); (4) the linkages between migration and other policy fields (like development or climate change); and (5) an adherence to universal principles, including (to some extent) human rights, but also free market…. This framework pervades the rhetoric of most IOs…and has become dominant at the intergovernmental level. (2018: 1632)

Those favouring the managed migration project often provide a useful description of the accelerated efforts to establish an institutional and normative framework for this project, culminating in the 2018 Global Compacts for Migration and on Refugees. However, this research largely takes for granted the good intentions behind the managed migration project that underestimates tensions between this project and alternatives and fails to challenge the underlying power structures that drive migration. For example, Betts and Kainz define global governance as 'the norms and organizational structures that regulate and facilitate states' and other actors' responses to migration. Its primary purpose is to ensure that states work collectively in ways that make them better able to fulfil their obligations than they would be acting alone' (2017: 1).[4] While few would disagree with the first part of this definition, the second reveals the lack of critical assessment of the migration management project. This literature is also largely silent on gender, beyond perhaps noting that women constitute a significant share of the flow of migrants. To be sure, the authors recognise tensions between (largely Southern) countries of origin and (largely Northern) countries of destination but argue that two decades of discussion have brought to the fore their common interest in managing migration.

Critics of the managed migration project see it developing out of, and working to reinforce, a power imbalance between states (receiving and sending) and between powerful corporate interests seeking cheap labour and those who speak for migrant workers' rights. For example, Pécoud argues that the '"migration management" agenda would aspire at governing the migration flows stemming from global capitalism, underdevelopment, or inequalities, without addressing these underlying factors' (2018: 1030). In addition to critiquing the migration management project, an important component of

[4]Another good example is Martin and Weerasignhe's chapter in the 2018 *World Migration Report*. See also Newland (2010).

this research agenda has been making visible the existence of rights-based project alternatives thereto. Likic-Broboric and Schierup contrast what they see as the Bretton Woods institutions' project to the rights-based project of UN agencies like the ILO (2018). Others would put greater emphasis on the role played by trade unions and migrants' rights organisations in advocating a rights-based agenda (Grugel & Piper, 2011; Piper & Rother, 2012; Rother, 2013). For them, while UN agencies can play an important role, it is pressure 'from below' that is needed to challenge the root causes. Such a project poses a transformational challenge to the existing system as it 'involves not only (legal) recognition and (political) representation but also (material) redistribution in the sense that better opportunities "here" and "there" would result in greater gains from migration, plus would provide more choice as regards the decision to migrate in the first place' (Piper, 2015).

A feminist perspective on the global governance of migration brings into view additional ILO Conventions[5]; governance mechanisms—e.g. Convention on the Elimination of all Forms of Discrimination Against Women (CEDAW), especially General Resolution 26; and 'friendly' agencies such as INSTRAW and UNIFEM (Benería et al., 2013; Hennebry et al., 2016; Piper, 2006). To understand the gendering of migration governance projects, it is also important to reflect on their intersection with another global project, 'gender equality and women's empowerment', which emerged from the two decades of UN Conferences on Women and the associated international feminist activism to which the conferences gave voice. As a result, 'the UN feminist activists have formed a unique apparatus of international governance that has made possible remarkable changes in gender regimes since the mid-twentieth century' (Caglar et al., 2013: 2). At the same time, feminist discourses have often been translated into terms consistent with the existing projects of international organisations in ways that neutralise and distort the egalitarian implications (Bedford, 2009; Caglar et al., 2013; Mahon, 2018, 2020). In other words, those responsible for integrating gender into an international organisation ('femocrats') are often constrained to translate it into terms that fit the organisation's dominant discourse. This can result in significant distortions as Bedford (2009) has shown for the World Bank. There may be more scope for advancing a feminist interpretation however if the femocrats find allies within the organisation. It is especially important to develop and maintain links to feminist epistemological and activist communities (Eyben, 2010; Mahon, 2018).

As we shall see, international organisations centrally involved in the migration management project, like the World Bank and the IOM, translate gender equality in ways that undercut the transformative thrust of the feminist project (see also Dannecker & Piper, this volume), while a gendered version of the

[5] In addition to ILO #189 discussed by Fish, these would include ILO #100 on equal remuneration and #111 Convention concerning discrimination in employment and occupation ad #181 on Private Employment Agencies.

migration management project has also been picked up by some of the UN agencies. While the ILO has been more effective at incorporating gender into its decent work for migrant workers project, the most transformative interpretation is to be found in the work of INSTRAW and UNIFEM, to be examined in the third section.

Gendering the Managed Migration Project

The gendered version of the 'win-win' managed migration project sees migrant women as escaping patriarchal domination while women in countries of destination benefit from low cost care labour performed by migrants. There is also a darker version of this discourse more often found in UN agencies that try to combine migration management with a concern for migrants' rights. This version depicts women migrants as vulnerable and unwitting victims (of trafficking or exploitation as domestic servants) and/or highlights the costs borne by those 'left behind'.

The World Bank's engagement with the gender project as part of its embrace of the broader post-Washington Consensus has been well described by Bedford (2009). Two key elements of the Banks' interpretation of gender reflected in its approach to migration are: (1) that women need to be included in labour markets for development to succeed 'even as the documents footnote or fudge inconvenient data showing the opposite' and (2) that structural adjustment policies had undermined men's capacity to be breadwinners in the South, leading them to adopt destructive behaviours (Bedford, 2009: 13).The Bank only began to recognise the importance of migration in 2004, following Ratha's work demonstrating the monetary value of remittances.[6] Thus from the start its overarching concern was with how migration- and, in particular, remittances—could best contribute to development.

The Bank's interpretation of gender is reflected in its 'win-win' understanding of migration as 'good not just for migrants, but also for the communities they leave behind and for the countries that welcome them' (World Bank, 2018: xi). Its first report on gender and migration, *The International Migration of Women* (Morrison, Schiff, & Sjöblom, 2008) drew very selectively on feminist work, notably Ehrenreich and Hochschild's classic (2003) volume. It noted that women spend in ways that yield better child outcomes—education, nutrition and reduced child labour—and that men are more likely to (mis)spend the money remitted by their wives (Morrison et al., 2008: 189).[7] It also cited the Hochschild volume in support of the thesis that Southern women use migration to escape abusive relationships with their

[6] DeParte, Jason (2008 March 17) 'The World Banker and his Cash Return Home', *New York Times*. https://www.nytimes.com/2008/03/17/world/asia/17remit.html.

[7] Note: this claim is often repeated in gendered international migration narratives thus demonising non-hegemonic masculinities from the South—an image contested inter alia by the important research of Lam and Yeoh (2016) on Filipino fathers.

husbands (Morrison et al., 2008: 186, 191–192). Migrant women were also lauded for boosting productivity in the North by enabling 'the more educated native-born women in the destination countries to enter the labor force' (Morrison et al., 2008: 198). However, the report also failed to note Ehrenreich and Hochschild's points that demand for migrant care workers is linked to 'the marked failure of First World governments to meet the needs created by its women's entry into the workforce' (2003: 8); that Northern women hire women migrant care workers 'as mistress and maid, employer and employee, across a great divide of privilege and opportunity' (Ehrenreich & Hochschild, 2003: 11); and that male unemployment in the South was often the result of Bank-supported structural adjustment policies (Ehrenreich and Hochschild: 11).

The darker side of the gendered migration project depicts women, especially when undocumented, as hapless victims needing protection, often placing them alongside other 'vulnerable' categories like children and the elderly. This variant is more often found in the discourse of certain UN agencies and the IOM as they try to reconcile their rights-orientation with the migration management project. One way to do so is to simultaneously support the rights of migrants (women and men) and destination states' interest in controlling access to suit their labour market needs and security concerns, by focusing on the issue of trafficking. Trafficking fits the managed migration narrative as it is a particularly pernicious form of 'irregular' migration that exploits migrants while undermining state capacities to control migration. Trafficking is also very much associated with women (and girls).

The IOM has focused much of its gender work on trafficking. A report commissioned by the IOM's gender unit *Working to Prevent and Address Violence Against Women Migrant Workers* (IOM, 2009) nicely illustrates the uneasy union of gender equality concerns and those of 'migration management'. The project combined training of Royal Thai police and Immigration Detention Centre officials to identify victims of trafficking with a one-day training session for women detainees on 'safe migration and trafficking, legal channels to enter and work in Thailand and basic rights of migrant workers' as well as assisted 'voluntary' return (IOM, 2009: 29). While the first part involved capacity-building to enable the Thai state to better control undocumented migrants, the latter reflects the use of 'softer' instruments to recruit migrant women as active participants in their own management. Feminist scholars are particularly critical of such narratives (Hennebry et al., 2016: 22).

The women as victim theme also appears in the discourse of the UN Population Fund (UNFPA). For the UNFPA, 'All migrants are vulnerable to abuse and exploitation, but female migrants are particularly at risk. Women and girls account for 71% of all human trafficking victims, according to a 2016 report by the UN Office on Drugs and Crime' and, 'when abuses occur, many migrant women and girls lack the resources, support systems and knowledge to seek help' (UNFPA, 2018). It did recognise that 'protection of human rights and access to legal and health services, including reproductive health services, are especially important when it comes to addressing the challenges faced by migrant women' (UNFPA, 2005: 3). Nonetheless it fell back on maternalist stereotypes: although remittances are seen to play a positive role in promoting development and poverty reduction, 'there are significant social costs of migration that are not always obvious: children without mothers, husbands without spouses, elderly relatives without anyone who will care for them. The migration of mothers often results in children dropping out of school or finding themselves in situations of neglect or abuse' (UNFPA, 2005: 2–3).

This representation of the negative impact of the missing mother ignores the fact that father's absence can also have a harmful effect (UNRISD, 2005: 109). It also assumes the nuclear (and heteronormative) family form yet, as Raghuram notes, 'in many countries the nuclear family may not be the norm and extended families may be commonplace... Many women may thus be involved in care distribution which means that the migration of one woman does not have as great an impact on care' (2012: 164). This is not to negate the impact of migrant absences on those left behind, including the elderly. A more gender-responsive solution would be to 'ensure the widest protection of the family unity of migrants, facilitating family reunification, and preventing arbitrary or unlawful interference with the right of migrants to the enjoyment of private and family life' (Global Migration Group, 2016: 40) but the migration management project embraces temporary migration and is silent on the right to family reunification.

Feminist Rights-Centred Alternatives

The four UN conferences on women (1975-1995), which helped to foster transnational links among women's organisations, put the gender equality and women's empowerment project on the global agenda. As Caglar et al. note, the UN and its affiliated agencies 'have played a catalytic role in focusing feminist strategizing internationally, and they have been instrumental in advancing the status of women worldwide' (2013: 2). The Third UN Conference on Women (1985) was the first to make a connection between migration and gender equality, drawing attention to the double discrimination migrant women face as women and as migrants. However, it was only in the late 1990s that the gender equality project began to be linked to the emergent migrant rights project. UN agencies, especially those specifically charged with advancing feminist goals, played a key role but they were not alone. At times femocrats within other IOs, such as the IOM, have also sought to advance a feminist, rights-based global migration governance project.

In fact, the IOM offers a nice example of both how having a feminist in a key position can open the way to the articulation of a more gender egalitarian ideas and, at the same time, how organisational structure can place limits on its translation into organisational practice. IOM's version of the migration management project places it somewhere between the rights-oriented discourse of ILO and the neoliberal version articulated by the World Bank

(Basok & Piper, 2012: 37; Georgi, 2012). The introduction of a rights-element into its discourse helped create space for the promotion of gender equality and women's empowerment within the IOM. In 1995, Council adopted 'equal opportunity and treatment of men and women' as one of its guiding principles and appointed a gender issues coordinator. Overall responsibility was subsequently transferred to the then-Deputy Director General, Ndioro Ndiaye, an African feminist. Ndiaye strongly supported the gender unit, which was able to commission research by leading feminist scholars such as Piper (2009). Ndiaye herself articulated a strong rights-based approach in which women migrants were seen as agents in the '*avant garde* of change'. For her, it was important not only to provide migrant women with access to recourse against injustices but also to help them develop collective agency (Ndiaye, 2006: 18).

Nevertheless, the IOM was and remains a highly decentralised IO, reliant on the capacity of its field units to secure project funding. As Geiger and Pécoud note, 'this particular style of management … confers a strong influence to key 'donors' and to field offices, which are responsible for attracting financial support' (2014: 872). In their work, the field offices attempt to balance concern for the socio-economic well-being of migrants with the IOM's role in supporting the wealthier states' efforts to control migration. Thus, even when Ndiaye was in office, the gender unit had little impact on the IOM's field operations. Of the first five 'model projects' sponsored by its Working Group on Gender Issues, three dealt with trafficking (Mahon, 2020: 61). A subsequent report noted the organisation's tendency to conflate gender-sensitive programming and anti-trafficking activities (IOM CRP/22, 2006: 2, 23).

The ILO is seen as champion of a workers' rights perspective, including those of migrant workers. Yet initially it focused on male migrant workers. While some conventions dealt with equal rights for women workers, Convention #97 (1949), Migration for Employment, which sought to extend to migrants social and collective bargaining rights, and Convention 143 on equal opportunities and treatment of migrant workers (1975) clearly envisioned the migrant worker as a male breadwinner.[8] ILO 189 on Decent Work for Domestic Workers, discussed by Fish (this volume), marked an important break with this assumption that came about as a result of cooperation between insiders and those advocating from outside. The way for change was paved by the appointment of Juan Somavía as Director General, who 'encouraged an emphasis on women's labour, the informal economy and migration', in the process strengthening the position of the unit dealing with gender (Boris & Fish, 2014: 427–428). The director of that unit, Manuela Tomei, in turn would help frame the issue in terms of the ILO's new 'decent work' discourse. Pressure from outside came from the International Domestic

[8]Both conventions use the language of the worker/he. For example, ILO Convention #97 article 8 refers to 'a migrant for employment who has been admitted on a permanent basis and the members of his family who have been authorised to accompany him'.

Workers Network and its allies. The ILO's ongoing involvement in promoting the rights of migrant domestic workers at the regional and national scales has subsequently strengthened its connections with migrant workers' organisations (Mahon, 2020).

The ILO has also commissioned studies that reflect a gender equality, rights-based argument. Thus, *Tricked and Trapped: Human Trafficking in the Middle East* counters victims of trafficking discourse of the migration management project. While acknowledging that 'victims of trafficking usually have limited financial resources, incur debt and are poorly educated', the report goes on to note that 'many are resilient and courageous women and men, who are aware of the possible risks of exploitation but, impelled by the lack of viable job opportunities at home and the pressing needs of their families, have nevertheless made their individual decisions to travel abroad in search of work' (Haraff-Tavel & Nasri, 2013: 14). In other words, the women are agents who are making choices, albeit not in circumstances of their choosing.

This raises the question of creating conditions for meaningful choice as to whether to migrate and for a better range of options on the labour market if one does. This would entail recognising the connection between women's migration and what is now called 'the care economy'. As early as 2006, Ndiaye had presciently located the kinds of jobs migrant women usually do firmly within the care economy:

> We know that migrant women find themselves in certain employment sectors which unfortunately rely solely on female workers: jobs as "caregiver" in the widest sense, such as domestic service, care of the elderly, the sick and children; jobs requiring patience, an eye for detail and repetition; and sexual work… These are also the sectors where one can find the worst pay, irregular situations, marginalization and a lack of social protection. (Ndiaye, 2006: 17)

The ILO has also come to recognise the care economy's importance to achieving decent work for migrant women workers (ILO, 2018).

The ILO has thus learned to integrate gender equality into its decent work project in a way that poses a potential challenge to the conditions underpinning the dominant migration project. A more consistently feminist challenge has come, however, from the UN agencies dedicated to women's equality—UNIFEM, INSTRAW and UN Women.

The first INSTRAW document to deal with migration, *The Migration of Women: Methodological Issues in the Measurement of Internal and International Migration* (1994), was critical of women's relative invisibility in the data on migration, where they were primarily seen as dependents. Migration only became central to INSTRAW's project, however in 2003, when gender, remittances and migration were identified as one its priority areas. *Crossing Borders* and *The Development Potential of Remittances* (2006) developed a theoretical framework to illuminate the link between remittances and gender-responsive local development, explore the role and nature of global care chains and identify women migrants' rights at all stages of the migration process. Here remittances are seen as more than 'just periodic financial transfers; they are the result of complex processes of negotiation within households that are immersed in an intricate network of relations between the Diaspora and countries of origin' (Ramirez, Dominquez, & Morais, 2005: 31). *Crossing Borders* also highlighted the way that the Bank's structural adjustment policies have led to the destruction of men's jobs, forcing men and women to turn to migration as a survival strategy. It went on to caution against 'ethnocentric views that consider that destinations societies offer women from the developing world opportunities for their personal "liberation"' (Ramirez et al., 2005: 36). Women migrants' subordinate insertion in countries of destination is shaped by class, gender and race-ethnicity structures (Ramirez et al., 2005: 23). These studies in turn laid the basis for the creation of a virtual community on gender and migration, which boasted over one thousand members by 2010, when INSTRAW was absorbed into UN Women.

Of particular interest is INSTRAW's training manual, *Gender on the Move* (Petrozziello, 2013). The manual reflects INSTRAW's feminist philosophy, which focused on enhancing women migrants' organisational capacity in a process where the trainer simply functions as a facilitator, drawing on and developing the participants' own lived experience. It was designed to encourage critical reflection on gender—understood as a social construct, as intersectional (race-ethnicity, class and other relations of difference), as involving masculinities as well as femininities; and migration practices as shot through with power relations at the micro (household), meso (national) and macro (global) scales. As it noted, gender analysis of migration 'should be conducted not only at the household level, but also at the community, institutional, national and transnational levels, taking into account the diversity of men and women and the ways in which gender identities are constructed and reconstructed throughout the migration process' (Petrozziello, 2013: 28).

Unlike INSTRAW, UNIFEM never identified migration as central to its overall programme. However, UNIFEM offices in Asia became aware of the feminisation of migration flows within as well as out of the region and in 2001 adopted an Asia-Pacific and Arab States Regional Programme on empowering women migrant workers 'to claim their rights and entitlements' (UNIFEM, n.d.: 1). This led its regional and local offices to engage with migrants' organisations, national governments and regional organisations like ASEAN (Association of Southeast Asian Nations).

UNIFEM's strengths were its links with women's CSOs as well as governmental and intergovernmental agencies at both national and regional scales, which enabled it to play an advocacy role. UNIFEM thus contributed to the gendering of ASEAN's Declaration on Promoting and Protecting the Rights of Migrant Workers.[9] It also used its position to highlight the gender

[9]All of what follows draws on *Policy and Programme Work on International Migration* UNIFEM (n.d.).

dimensions of migration at the 62nd meeting of the Economic and Social Commission for Asia and the Pacific, which formed part of the regional preparations for the first High Level Dialogue on Migration and Development. UNIFEM also supported the participation of the migrants' organisations at the first Global Forum on Migration and Development (GFMD), organising events on women migrant workers. It played an even more important role in getting women migrant workers' concerns inserted into the GFMD meeting in the Philippines in 2008. Working with the National Commission on the Role of Filipino Women, other UN agencies, the Lola Grande Foundation and Migrant Forum Asia, organised the International Conference on Gender, Migration and Development: Seizing the Opportunities and Upholding Rights, which produced the Manila Call for Action that was subsequently incorporated into key documents and statements at the GFMD.

UNIFEM contributed to the adoption of CEDAW general recommendation 26 which, as Hennebry et al. argue 'is the most comprehensive international instrument within the normative framework relation to women migrant workers specifically' (2016: 69). To do so, UNIFEM convened regional and global multi-stakeholder meetings and provided technical assistance in the drafting of the recommendations. UNIFEM also worked to incorporate the concerns of women migrants into the CEDAW Review process, a practice that CEDAW has continued. UNIFEM also engaged in capacity-building for Asian governments and NGOs on how to use CEDAW to address women migrants' rights. Currently much of the work done by the Asia-Pacific regional offices of UN Women focuses on women migrant workers. It has conducted research highlighting the governance gaps within ASEAN concerning the migrant women workers' rights sharing the results with migrants' organisations such as the Migrant Forum in Asia and the Mekong Migration Network.

In 2010 UNIFEM and INSTRAW were merged into a new entity, the UN Entity for Gender Equality and the Empowerment of Women or UN Women. Since then UN Women has continued to support the rights of migrant women workers in various ways and at other scales. Globally, it pushed for a gendered perspective on migration at the Global Migration Group (GMG), which brought together all UN agencies with an involvement in migration. It convenes the Women in Migration Caucus at the annual Council on the Status of Women sessions and the Women's Caucus of the Global Forum on Migration and Development. Both events create important spaces for women migrants' organisations. As chair of the GMG in 2016, it sought to emphasise links between Sustainable Development Goal 5 on gender equality and targets in other goals dealing with migrants.

More importantly, UN Women played a 'pivotal role' in making space for civil society organisations and feminist academics during the negotiations around the Compacts for Migration and on Refugees (Hennebry & Petrozziello, 2019: 131). In particular, UN Women acted as the secretariat for the Expert Working Group for Women's Human Rights in the Compact for Migration. The Groups' first joint statement underscored the need to 'protect, promote, respect and fulfil the human rights of women and girls, including those who identify as lesbian, bisexual or transgender, at all stages of migration' (Expert Working Group, 2018: 1–2). In addition to pushing for the right to family reunification, the Group sought to challenge the current system by including the right to permanent migration and to regularisation for the undocumented (Hennebry & Petrozziello, 2019: 129).

Conclusions

The managed migration project is thus contested not only by migrants' organisations and trade unions 'from below', but also by key UN agencies. UN Women and its predecessors have played a key role in making gender equality part of this counter-project. How successful have they been in making their gendered, rights-based project hegemonic? The results of the Global Compact processes are sobering. Although the Expert Group and its allies managed to insert some important 'gender-responsive' language, 'safe, orderly and regular migration', the mantra of the migration management project, remains the dominant approach to migration, one which continues to limit women migrants to temporary and irregular pathways to low paid, precarious jobs in wealthier countries while leaving their families behind. Although the New York Declaration promised access to sexual and reproductive health care services, this is too absent from both Compacts.

Of particular concern is the structure established to follow up and review the results of the Compacts. Although the UN Migration Network, which replaces the Global Migration Group, includes some agencies committed to the migrants' rights projects, UN Women is not a member. As Hennebry and Petrozziello argue, 'without the involvement of UN Women, the Expert Working Group for Women's Human Rights…and other civil society actors directly involved in each working group, it is unclear how the capacity-building mechanism that is envisioned to support Member States in implementation…will be realized in a way that is gender-responsive' (2019: 133) (see also Hennebry, in this volume). One important area for future research then is what strategies UN Women and migrants' organisations will adopt in response both at the global and regional scales. Will UN Women enlist the ILO's support to gain entry to these negotiations? Will migrants' organisations create alternative spaces for challenging the status quo?

More broadly, in this period marked by the resurgence of right-wing, anti-immigrant populisms in the North and the South, the rights of immigrants are in danger of being sacrificed to the darker, disciplinary side of the managed migration project. How are migrants' rights organisations' countering anti-immigrant populism within key regions (e.g. Europe) and states (e.g. the US)? In what ways might UN Women and the ILO support their struggles? What role will the IOM, now a 'related' organisation of the UN that has been given a central role in implementing the Compact for Migrants. Its 2015 Migration Governance Framework committed it to adhere to international standards and the fulfilment of migrants' rights and to advancing the socio-economic well-being of migrants (Martin & Weerasinghe, 2018: 134). Will this lead it to side with the rights-oriented UN agencies or will its dependence on donors from destination states take precedence?

References

Basok, T., & Piper, N. (2012). Management vs rights: Women's migration and global governance in Latin America and the Caribbean. *Feminist Economics, 18*(2), 35–61.

Bedford, K. (2009). *Developing partnerships: Gender, sexuality and the reformed World Bank*. Minneapolis and London: University of Minnesota Press.

Benería, L., Deere, C., & Kabeer, N. (2013). Gender and international migration: globalization, development and governance. In L. Oro & N. Ribas-Maleos (Eds.), *The international handbook on gender, migration and transnationalism* (pp. 45–68) Cheltenham: Edward Elgar.

Betts, A., & Kainz, K. (2017). *The history of global migration governance* (Refugee Studies Centre, University of Oxford Working Paper Series 122).

Boris, E., & Fish, J. (2014). "Slaves no more": Making global standards for domestic workers. *Feminist Studies, 40*(2), 411–443.

Caglar et al. (2013). *Feminist strategies in international governance*. Abingdon: Routledge.

DeParte, J. (2008, March 17). The World Banker and his cash return home *New York Times*. https://www.nytimes.com/2008/03/17/world/asia/17remit.html.

Ehrenreich, B., & Hochschild, A. (Eds.). (2003). *Global women: Nannies, maids and sex workers in the new economy*. New Work: Holt.

Expert Working Group. (2018). *Statement by the Expert Working Group for addressing women's human rights in the global compact for safe, orderly and regular migration, calling for a gender-responsive compact*. https://www.ohchr.org/Documents/HRBodies/CERD/UNWomen_EWG_FirstJointStatement_January2018.pdf.

Eyben, R. (2010). Subversively accommodating feminist bureaucrats and gender mainstreaming. *IDS Bulletin, 41*(2), 54–61.

Geiger, M., & Pécoud, A. (2014). International organizations and the politics of migration. *Journal of Ethnic and Migration Studies, 40*(6), 865–887.

Georgi, F. (2012). For the benefit of some: The International Organization for Migration and its Global Migration Management. In M. Geiger & A. Pécoud (Eds.), *The politics of International Migration Management*. Basingstoke and New York: Palgrave-Macmillan.

Global Migration Group. (2016). *Principles and guidelines, supported by practical guidance, in the human rights protection of migrants in vulnerable situations within large or mixed movements*. https://www.ohchr.org/EN/Issues/Migration/Pages/HRC37.aspx.

Grugel, J., & Piper, N. (2007). *Critical perspectives on global governance: Rights and regulation in governing regimes*. London and New York: Routledge.

Grugel, J., & Piper, N. (2011). Global governance, economic migration and the difficulties of social activism. *International Sociology, 26*(4), 435–454.

Haraff-Tavel, H., & Nasri, A. (2013). *Tricked and trapped: Human trafficking in the Middle East*. Geneva: ILO. https://www.ilo.org/beirut/publications/WCMS_211214/lang–en/index.htm.

Hennebry, J., & Petrozziello, A. (2019). Closing the gap? Gender and the global compacts for migration and refugees. *International Migration, 57*(6), 115–138.

Hennebry, J., Williams, K., & Walton-Roberts, M. (2016). *Women working worldwide: A situational analysis of women migrant workers*. New York: UN Women.

ILO. (2018). *Care work and care jobs: For the future of domestic work*. Geneva: ILO. https://www.ilo.org/global/topics/care-economy/care-for-fow/lang–en/index.htm.

INSTRAW. (1994). *The migration of women: Methodological issues in the measurement of internal and international migration*. Santo Domingo: INSTRAW.

INSTRAW. (2006). *The development potential of remittances*. Santo Domingo: INSTRAW.

IOM. (2009). *Working to prevent and address violence against women migrant workers*. Geneva: IOM. https://publications.iom.int/es/books/working-prevent-and-address-violence-against-women-migrant-workers.

IOM CRP/22. (2006). *Evaluation of gender mainstreaming policy and strategy in IOM*. Prepared by the Office of the Inspector General for the 92nd Session of Council Geneva, 2 November–1 December. https://governingbodies.iom.int/system/files/jahia/webdav/shared/shared/mainsite/about_iom/en/council/92/CRP-22.pdf.

Lam, T., & Yeoh, S. (2016). Migrant mothers, left-behind fathers: The negotiation of gender subjectivities in Indonesia and the Philippines. *Gender Place Culture, 25*(1), 104–117.

Likic-Brboric, B., & Schierup, C. (2018). Labour rights as human rights? Trajectories in the global governance of migration. In C. U. Schierup, R. Munck, B. Likic-Brboric, & A. Neergaard (Eds.), *Migration, precarity ad global governance: Challenges and opportunities for labour* (pp. 224–244). London: OUP.

Mahon, R. (2018). Shaping the way international organizations "see" gender equality: The OECD and ECLAC. In S. Shaver (Ed.), *Handbook on gender and social policy* (pp. 267–286). Cheltenham: Edward Elgar.

Mahon, R. (2020). Gendering migration management. In M. Geiger & A. Pécoud (Eds.), *The International Organisation for Migration: The new "UN Migration Agency' in critical perspective* (pp. 53–74). Palgrave: Macmillan.

Martin, S., & Weerasinghe, S. (2018). *Global migration governance: Existing architecture and recent developments* (World Migration Report). Chapter 5. https://publications.iom.int/books/world-migration-report-2018-chapter-5-global-migration-governance-existing-architecture-and.

Morrison, A., Schiff, M., & Sjöblom, M. (Eds.), (2008). *The international migration of women*. Washington, DC: IBRD/World Bank. http://documents.worldbank.org/curated/en/673051468328175051/The-international-migration-of-women.

Ndiaye, N. (2006). *Overview: Situation and contribution of migrant women*. Female Migrants: Bridging the Gaps Throughout the Life Cycle: Selected Papers of the UNFPA-IOM Expert Group Meeting. New York: UNFPA.

Newland, K. (2010). The governance of international migration: Mechanisms, processes, and institutions. *Global Governance, 16,* 331–343.

Pécoud, A. (2018). What do we know about the International Organization For Migration? *Journal of Ethnic and Migration Studies, 44*(10), 1621–1638.

Petrozziello, A. (2013). *Gender on the move: Working on the migration-development nexus from a gender perspective*. Santo Domingo, DR: UN Women. https://www.unwomen.org/en/digital-library/publications/2013/12/gender-on-the-move.

Piper, N. (2006). Gendering the politics of migration. *Migration Review, 40*(1), 133–164.

Piper, N. (2009). 'Overview' and 'Conclusion' to *Gender and Labour Migration in Asia*. Geneva: IOM. https://publications.iom.int/system/files/pdf/gender_and_labour_migration_asia.pdf.

Piper, N. (2015). Democratising migration from the bottom up: The rise of the Global Migrants Rights Movement. *Globalization, S, 12*(5), 788–802.

Piper, N., & Rother, S. (2012). Lets argue about migration: advancing a right(s) discourse via communicative opportunities. *Third World Quarterly, 33*(9), 1735–1750.

Raghuram, P. (2012). Global care, local configurations—Challenges to conceptualizations of care. *Global Networks, 12*(2), 155–174.

Ramirez, C., Dominquez, M.G., & Morais, J (2005). *Crossing borders: Remittances, gender and development*. Santo Domingo, DR: UN-INSTRAW. https://reliefweb.int/report/world/crossing-borders-remittances-gender-and-development.

Rother, S. (2013). A tale of two tactics: Civil society and competing visions of global migration governance from below. In M. Geiger & A. Pécoud (Eds.), *Disciplining the transnational mobility of people* (pp. 41–62). Palgrave-Macmillan: Basingstoke.

UNFPA. (2005). *Policy and programme of work on International Migration of United Nations Population Fund*. https://www.un.org/en/development/desa/population/events/pdf/7/P09_UNFPA.pdf.

UNFPA. (2018). *Five reasons why migration is a feminist issue*. https://www.unfpa.org/news/five-reasons-migration-feminist-issue.

UNIFEM. (n.d.). *Policy and programme work on International Migration by the UN Development Fund for Women*. https://www.un.org/en/development/desa/population/events/pdf/7/P06_UNIFEM.pdf.

UNRISD. (2005). *Gender equality: Striving for justice in an unequal world*. Geneva: UNRISD. http://www.unrisd.org/80256B3C005BCCF9/search/1FF4AC64C1894EAAC1256FA3005E7201.

World Bank. (2018). *Moving for prosperity: Global migration and labor markets*. Washington, DC: World Bank. https://www.worldbank.org/en/research/publication/moving-for-prosperity.

Rianne Mahon is a distinguished research professor in the Sociology and Anthropology Department at Carleton. She has published on industrial policy, labour market restructuring, childcare politics, and the redesign of welfare regimes at the local, national and global scales. Mahon has co-edited numerous books including *Achieving the Social Development Goals: Global Governance Challenges* (with S. Horton and S Dalby), and co-authored *Advanced Introduction to Social Policy (*with Daniel Béland) Her current work focuses on the 'gendering' of global governance, with a particular focus on migrant domestic workers.

CHAPTER 18

On the Gendered Structures and Outcomes of Interstate Bilateral Labour Agreements as Migration Governance Instruments

Jenna Hennebry and K. C. Hari

Abstract As states look to expand "regular pathways" for migration, as evidenced in the Global Compact for Safe, Orderly and Regular Migration (UN, 2018), many are turning to managed temporary labour migration schemes under bilateral labour agreements (BLA), rather than permanent immigration pathways. For states, such schemes represent a viable (and manageable) "orderly and regular" migration pathway, which serves the dual purpose of regulating migration and addressing labour demands. Yet, such instruments remain largely gender-blind and channel women into gendered occupations—such as caregivers, cleaners, service/sales clerks and entertainers—occupations that are typically low skilled, low waged, with high levels of precarity and low levels of social protection—often in sectors without consistent collective bargaining rights (e.g. agriculture, care work). Further, few BLAs address rights or social protection. This chapter weighs the caveats and concerns, as well as the opportunities and innovations, in these and other approaches to governing labour migration.

J. Hennebry (✉) · K. C. Hari
Wilfrid Laurier University, Waterloo, ON, Canada
e-mail: Jhennebry@wlu.ca

K. C. Hari
e-mail: hkc@balsillieschool.ca

C. Mora and N. Piper (eds.), *The Palgrave Handbook of Gender and Migration*, https://doi.org/10.1007/978-3-030-63347-9_18

Introduction

As states look to expand "regular pathways" for migration, as evidenced in the Global Compact for Safe, Orderly and Regular Migration (UN, 2018), many are turning to managed labour migration schemes under bilateral labour agreements (BLAs), rather than permanent immigration pathways. For states, such schemes represent a viable (and manageable) "orderly and regular" migration pathway, which serves a dual purpose of regulating migration and addressing labour demands. At a time when the feminisation of labour migration is a global trend, unfolding at different speeds and intensities regionally, labour migration (particularly of "low-skilled" workers) tends to be highly gendered, with male migrants concentrated in construction, agricultural and manufacturing sectors and female migrants concentrated in "feminised" service occupations such as domestic work (see Shutes, in this volume). Yet, such bilateral instruments remain largely gender-blind in significant ways. In the first place, the BLAs do not seem to pay heed to the gendered challenges of women migrant workers and refrain from providing specific provisions, consequently producing gendered outcomes. Simultaneously, they contribute to perpetuating the gendered occupations by channelling women into jobs such as caregivers, cleaners, service/salesclerks and entertainers, which are typically low skilled, low waged and involve high levels of precarity, often in sectors without consistent collective bargaining rights. Further, very few address rights or social protection despite considerable efforts by countries such as the Philippines to use BLAs to address rights violations for citizens abroad. During the entire processes when BLAs are conceived, negotiated and implemented, they largely ignore gender sensitivities and thus fail to address specific issues faced by women migrant workers.

Interstate bilateral labour migration arrangements vary in names (e.g. BLAs, Memoranda of Understanding (MOUs), Framework Agreements, Protocols and Cooperation Agreements) as well as in forms and in appellations. Though some of such bilateral arrangements are legally binding, most are non-binding and whose enforcement depends on the signing states' good faith and political will. Given the variations in the bilateral labour instruments, we use the term "bilateral labour agreement" generically to refer to both formal and legally binding bilateral agreements and the legally non-binding MOUs, which set out broad frameworks of cooperation to address common concerns, as well as to other arrangements, including those between specific government agencies (Table 18.1).

As the table shows, BLAs are diverse, however they provide some common clauses pertaining to recruitment procedures and criteria for migrant workers, obligation of sending countries to screen migrant workers and the responsibility of host countries to give legal protections to the workers while the provisions for overseeing recruiters and employers and institutional mechanisms for administering the agreement (e.g. Chilton & Posner, 2017). Although the general provisions of the BLAs apply to women migrant workers

Table 18.1 Variations in bilateral instruments across regions

Type	*Asia*	*%*	*Africa*	*%*	*Europe and Americas*	*%*	*Total*	*%*
Memorandum of Understanding (MOU)	45	69.2	1	3.1	2	3.7	48	31.8
Memorandum of Agreement	1	1.5	0	0.0	2	3.7	3	2.0
Bilateral Agreement (BA)	10	15.4	23	71.9	43	79.6	76	50.3
Inter-Agency Understanding (IAU)	6	9.2	0	0.0	0	0.0	6	4.0
Framework Agreement	0	0.0	7	21.9	1	1.9	8	5.3
Protocol	3	4.6	1	3.1	6	11.1	10	6.6
Total	65	100	32	100	54	100	151	100

Source ILO (2015)

as well, most such instruments have failed to pay attention to specific challenges and vulnerabilities faced by women migrant workers. Even within research communities, the gender-blind structures of such BLAs across regions and different sectors, and the gendered outcomes they engender, have received scant attention. Given a wide range of bilateral labour migration governance instruments across regions and sectors, building a knowledge base on existing practices is of critical significance not only to further research but also assess and de-gender their structures and outcomes.

In this chapter, we examine the ways in which the BLAs as interstate instruments of labour migration governance are structurally gendered and thus perpetuate gender-based migration flows and employment uptake. As we analyse the gendered intent, structures and outcomes of the BLAs as migration governance instruments, we draw upon practices globally while using the specific examples of Nepal and the Philippines. The rationale behind choosing these countries is that the two countries have a great deal of similarities as well as significant differences in terms of the content and structures of the BLAs and the ways in which such instruments are implemented to address issues of women migrant workers. Further, the Philippines has signed numerous bilateral agreements with migrant receiving countries and is widely recognised as "a model of migration governance" while Nepal's labour migration governance, in contrast, is often deemed to have gaps and lapses, pertaining especially to women migrant workers in the Gulf countries.

In addition to basing our analyses on available literature on bilateral labour migration governance as well as on media outputs and reports produced by international organisations, we carry out a content analysis of some of BLAs signed by Nepal and the Philippines with migrant receiving countries. The chapter proceeds as follows. First, we provide a brief overview of the BLAs as instruments of migration governance in the global and historical context from a gender perspective. Second, we reflect on the ways in which the BLAs as labour migration governance instruments are structurally gendered, specifically focusing on the gender-blind spots and gendered outcomes. Third, we weigh in the caveats and concerns in the BLAs from a gender lens, as well as opportunities and innovations that they can provide as instruments of labour migration governance.

Governing Labour Migration Through Bilateral Labour Agreements

Labour migration is an increasingly complex and dynamic global phenomenon. Overall, three levels of international migration governance currently exist in terms of governing interstate mechanisms: bilateral, regional and international. And, in terms of processes, there are three modes in which global migration governance operates: national, international and transnational. Gamlen and Marsh argue that in the recent times there has been a "proliferation in bilateral, regional and informal multilateral structures" and this can be taken as a "growing recognition of the limitations of unilateralism" on the part of the nation-states (2011). The growing inclination of states towards bilateral instruments shows that the nation-states are reluctant "to relinquish sovereignty over migration decision-making" (Gamlen & Marsh, 2011).

Among a range of governance instruments and approaches employed by states in the management of international labour migration, BLAs have become increasingly popular migration governance instruments. Through BLAs, states aim to address simultaneously both shortages of labour demands in migrant receiving countries and unemployment problems in countries of origin. BLAs refer to cooperation frameworks between countries of origin and destination to ensure that labour migration is regulated according to agreed principles and procedures. Bilateral agreements thus control the conditions under which countries of origin send temporary migrant workers to host countries, in addition to delineating detailed specific responsibilities of both migrant sending and receiving countries. A bilateral agreement is a legal instrument that reflects the interests of states of origin and states of employment while protecting the rights of migrant workers to ensure decent working conditions, equitable compensation, non-discrimination, legal redress and access to justice. According to UN treaties, BLAs are legally binding, in the same way as a treaty holds states accountable under the law.[1] MOUs are usually non-binding frameworks which may be more general in content, referring to broad principles of cooperation, concepts of mutual understanding, goals and plans shared by the states. In the context of labour migration governance, host countries tend to sign BLAs to secure assistance from origin countries in screening migrant workers so as to control the flow of migrants by preventing terrorists, criminals and other undesirable people (Plotnikova, 2011).

Although BLAs became increasingly popular globally following the 1990s, different countries in Europe had used them as institutional mechanisms of labour migration in the wake of the World War II and oil-price shocks in the mid-1970s. Looking at BLAs in historical context, they have evolved in two distinct stages: the "first generation" BLAs in the late 1940s focused mainly on labour recruitment for reconstruction of post-war European economies (Wickramasekara, 2015). One distinctive feature of early agreements was the key role played by states in organising and supervising recruitment, employment and return of migrant workers (ILO, 2015). From 1945 to 1973, BLAs were used mostly by European countries, which sought labour from eastern and southern Europe and North Africa, for reconstruction. From 1974 to 1989, relatively few new BLAs were negotiated, perhaps because the 1970s were a period of economic stagnation (Chilton & Posner, 2017) (Fig. 18.1).

Along with the end of the Cold War and the emergence of a new era of globalisation in the 1990s and early 2000s, there was an upsurge in ratification of BLAs. During the period, countries in the Middle East and Asia emerged as significant signatories of such instruments both as migrant sending and receiving countries. Unlike the first generation of BLAs were concluded during a period where states organised labour migration, the second generation BLAs operate in a context where various non-state actors such as profit-oriented private recruitment agencies as well as various civil society actors working at subnational, national and transnational levels have become significant part of labour migration governance in major migratory corridors.

Further, in the more recent years, the legally non-building bilateral arrangements in forms of MOUs have become more preferable to migrant receiving countries, since they are easier to negotiate and implement than the more formal legally binding agreements while providing unobstructed access to labour to meet the changing economic and labour market needs (ILO, 2015). Similarly, since such instruments are signed by states, they tend to be state-centric in that the interests of the states and particularly those of the migrant hosting countries are preponderant while marginalising provisions for protecting labour and human rights of individual migrant workers. As per an ILO study (2015), about 70% of the 144 labour bilateral agreements in

[1]As per the Vienna Convention on the Law of Treaties Signed at Vienna (23 May 1969), a treaty is: "an international agreement concluded between States in written form and governed by international law, whether embodied in a single instrument or in two or more related instruments and whatever its particular designation".

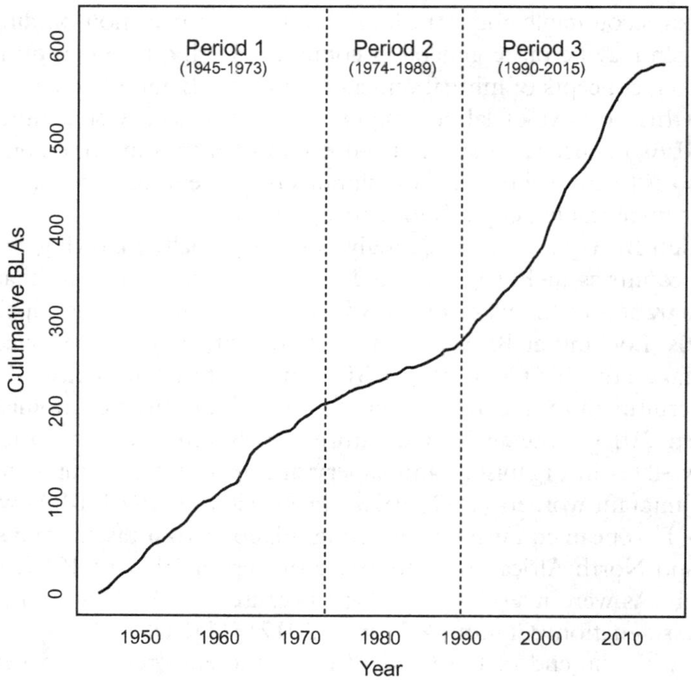

Fig. 18.1 Cumulative number of BLAs globally (*Source* Adam S. Chilton & Eric Posner, "Why Countries Sign Bilateral Labor Agreements," Coase-Sandor Working Paper Series in Law and Economics, No. 807 [2017] [Reprinted with permission])

Asia were MOUs that simply outlined broad goals and cooperative frameworks of mutual understanding. In addition to the legally non-binding nature of such MOUs, they provide feeble clauses for the protections of migrant workers' rights and hardly make any room for gender-sensitive provisions and overseeing mechanisms for addressing women migrants' specific challenges.

The issue of enhancing better protections for vulnerable groups of migrant population through such legally non-binding bilateral structures is highly questionable. Almost 70% of the BLAs failed to provide social security to migrant workers such as health benefits (Wickramasekara, 2015). Further, since the BLAs are negotiated between states, the asymmetric political and economic as well as bargaining power that exists between the origin and destination countries pushes the sending states on the receiving end. Unless states remain committed to upholding the human and labour rights of migrant workers and negotiate legally binding and rights-based BLAs over non-binding bilateral instruments, migrant workers and their rights continue to be marginalised and vulnerabilised. The legally non-binding flimsy frameworks lack enforceability while allowing states to use bilateral labour arrangements simply as tools to promote a profit-centred migration management system, solely driven by the host states' needs for flexible, cheap labour while easing unemployment burdens for sending states, without ensuring protections of migrants' labour and human rights (Schierup, Likic-Broboric, Delgado, & Teksöz, 2019).

Concerning the governance of global migration, Gamlen and Marsh (2011) identify three different modes: national, international and transnational. Governance in the national mode, since the nation-state is the most important regulator, places the state at the centre to control migration through borders, safeguarding access to territory, nation, citizenship and rights. The international mode entails the formal and informal cooperation between nation-states to regulate migration issues that include bilateral, regional and multilateral governance. The transnational mode of migration governance involves "a neoliberal assessment of the role of states in regulating market mechanisms such as migration" (2011, xxiv). Informed by "the new migration and development optimism", the transnational mode promotes temporary and circular labour migration schemes, producing 'triple wins' for sending and receiving states, as well as the migrants themselves (Gamlen & Marsh, 2011, p. xxiv), since such a model of governance has its 'conceptual roots in the new migration and development optimism' (Gamlen & Marsh, 2011, xxiv). For example, the Philippine government has acclimatised the role of the state in the changing global market economy to maximise gains from the 'triple wins'.

In the Philippines' case, civil society organisations have been actively working for the recognition and protection of the rights and welfare of overseas Filipino workers through direct services, organising education and advocacy at all levels: local, national, regional and international (Grugel & Piper, 2007). For example, the 2013 Saudi Arabia-Philippines BLA includes protective clauses recognised by international bodies in protection of the rights, security and welfare of the migrant domestic workers who are not covered by Saudi Arabia's national labour laws. Although the agreement is touted for providing protections for migrant domestic workers, it does not address social security issues, exclusion of the employer/sponsor in legal measures against the recruitment offices, companies or agencies for violations of any applicable laws, minimum and maximum age limits and freedom of observance of religion by the domestic worker. We argue that the three modes of labour migration governance that Gamlen and Marsh elucidate are not and cannot be treated as separate governance processes—they rather overlap and are interconnected structurally and if BLAs fail to ensure and monitor a proactive involvement and roles of the other non-state actors, merely signing bilateral instruments will fail to implement and protect migrant workers.

Gender-Blind Spots in Bilateral Labour Agreements

Out of the total of 164 million international migrant workers globally, 68 million are women migrants (IOM, 2020). Although the geographies of the global migrant population have massively transformed with large numbers of women migrating for employment purposes, the BLAs are highly gendered and fail to respond to women migrants' specific experiences and challenges, producing gendered outcomes. Even if BLAs exist, very few of them provide social security mostly to migrant workers in formal sectors, dominated by male migrants while those in the informal economy remain unprotected (Van Panhuys, Aoul, & Binette, 2017) and it is women migrants who suffer because most women work in irregular and informal areas. Aside from multiple ways in which women migrant workers have benefitted from transnational labour migration, those working in informal, low-skill sectors suffer from significant decent work deficits, including violations of the fundamental human and labour rights in the destination countries. Most women migrant workers globally are thus the most disadvantaged "precariat" victimised by informal labour, wage cuts, temporariness, uncertainty and risks (Schierup, Ålund, & Likić-Brborić, 2015). In particular, in the context of women migrant workers working in informal and fragmented sectors, especially those working as migrant domestic workers whose employment is excluded even from domestic labour laws, not having gender-sensitive and gender-responsive BLAs with effective overseeing mechanisms produces deleterious gendered outcomes.

Migration at the conceptual level has increasingly been understood as a gendered process; however, a lot has yet to be done in making this understanding percolate indeed to various migration governance actors, especially the state actors, who play a crucial role in the governance of labour migration. Very few such bilateral agreements address the specific concerns and challenges of women migrants. Despite the feminisation of labour migration, the lack of gender-sensitivity is a major gap in the BLAs. Right from the beginning when BLAs are conceived until they are negotiated and implemented on the ground, they largely ignore gender issues and lack gender-sensitive monitoring mechanisms. While treating BLAs as broad frameworks, it is critical to supplement them by negotiating for the introduction of concrete measures and mechanisms in several areas (Wickramasekara, 2012).

More importantly, women migrant workers working in informal sectors and not even covered by national labour laws in the destination countries are confronted with much more heightened precarities than others. For example, women migrant domestic workers, domestic drivers, gardeners and agricultural workers are among the most vulnerable categories of labourers. BLAs should offer protections to such most vulnerable groups of population who are at greater risks of being abused and exploited in the destination countries. However, many interstate bilateral arrangements do not conform to international norms and good practices on protection of migrant rights. Most of such instruments which exist as MOUs merely provide lip service stating that appropriate steps will be taken to address problems of non-covered workers without specifying any concrete measures (Wickramasekara, 2012). For example, in Asia, most BLAs are geared towards managing irregular migration and promoting orderly labour movements while meeting the demands of labour markets. At the same time, such instruments provide vague clauses that aim to promote cultural-political ties and exchanges; however, they are indeed intended to reduce unemployment pressures in origin countries while supplying cheap labour to the migrant receiving countries.

A significant omission applicable to agreements across all regions relates to the absence of almost any reference to women migrant workers or other categories of migrant workers, such as low-skilled migrants, who are at particular risk of abuses. In other cases, such BLAs maintain a complete silence on addressing specific issues and challenges of women migrant workers, despite the rapidly increasing feminisation of migration processes in the less developed countries of the Global South. For instance, Nepal has so far signed nine MOUs with the migrant receiving countries, including six from the Middle East (e.g. Qatar, Bahrain, Jordan and the UAE) while the Nepali government claims to be in the process of negotiating such agreements with more countries at a time when an estimated 2.5 million Nepali women are currently working as migrant workers, mostly in the Gulf countries (Nepal Labour Migration Report, 2020). In almost all the labour agreements that Nepal has signed with the destination countries in the Gulf, women migrant workers almost remain invisible and the agreement is completely oblivious to the cases of exploitation and abuse faced by migrant domestic workers in the Gulf countries. For example, the MOU between Nepal and the United Arab Emirates does not provide any clauses to protect women migrant workers' human and labour rights while an emphasis is placed on enhancing 'the existing friendly relations between the two countries by developing cooperation in the field of manpower' and 'recognizing the benefit to be derived by both countries from close cooperation in the field of manpower' (Government of Nepal 2007). The lack of a gender lens and gender-sensitivity in such bilateral agreements is a major gap, which reflects that such governance mechanisms are emerging as the other sites where women migrants are invisible and their voices silenced.

More recently, the number of sector-specific BLAs, has been on the rise, particularly pertaining to women migrant workers and those working in precarious sectors such as domestic work. Yet, such instruments remain largely gender-blind and channel women into gendered occupations—such as caregivers, cleaners, service/sales clerks and entertainers—occupations that are typically low skilled, low waged, with high levels of precarity and low levels of social protection—often in sectors without consistent collective bargaining rights (e.g. agriculture, care work). In the context of the ILO Domestic Workers Convention, 2011 (No.189) and the establishment of the International Domestic Workers Federation (IDWF) in 2013 (Lim, 2014). If such agreements are not negotiated with enough caution, they help to further entrench gender stereotypes and gendered outcomes. Though bilateral labour agreements, though they carry the potential to protect women migrant workers, if they are structured solely on neoliberal ideologies, they end up further perpetuating the 'international division of reproductive labour' (Parrenas, 2001, 61). Instead of 'streamlining of migration governance into business-friendly migration management and a geopolitical securitisation agenda' (Likić-Brborić, 2018, 762), there is a need for overhauling the governance system by adopting a gender-sensitive, rights-based approach to migration governance.

En-gendering Bilateral Labour Agreements

During the entire migration journey, most often women migrant workers are confronted with gender-specific disadvantages, barriers and vulnerabilities. From migration though, women can have empowering experiences; however, they are vulnerable in many gendered ways (e.g. sexual harassment, flexible and informal work, etc.) (OSCE, 2009). The acknowledgement of such a dual process in migration and women-specific vulnerabilities requires that BLAs should not only explicitly provide provisions to address those challenges but also be structured and implemented in a gender-sensitive manner. However, since migration is a multi-scalar and multipronged process, where multiple actors play key roles, BLAs alone fall short of addressing women migrant workers' gender-specific challenges and a whole of society approach, actively involving various civil society partners, is of paramount significance. Since BLAs are interstate instruments, they can, however, institutionalise and implement the process while providing legal protections and support to women migrant workers in particular. Similarly, since BLAs are negotiated between two states and also given the asymmetric political and economic powers between sending and receiving countries and the bargaining power and bargaining capacity, not all women migrant workers can be protected in similar ways. To what extent BLAs provide protections to women migrants are thus different in differing contexts.

It is crucial to understand the importance of including a gender dimension in BLAs, as well as consider the inclusion of specific gender-sensitive measures throughout the design and implementation of such agreements. The contents of BLAs should reflect the norms and protections provided by international human rights instruments and relevant international labour standards. For example, the "ILO Multilateral Framework on Labour Migration: Non-binding principles and guidelines for a rights-based approach to labour migration" is a compendium of principles and guidelines on labour migration (ILO, 2006). Guideline 4.5 of the ILO Multilateral Framework on Labour Migration calls on states to ensure that 'labour migration policies are gender-sensitive and address the problems and particular abuses women often face in the migration process' Similarly, gender-responsive governance is highlighted in the Committee on the Elimination of Discrimination against Women (CEDAW) General Recommendation No. 26. General Recommendation No. 26 provides a framework for developing gender-responsive migration policies to eliminate discrimination in line with State obligations under CEDAW. The Recommendation highlights the formulation of evidence-based, gender-responsive and human rights-based migration policies as a core common responsibility. BLAs should specially provide protective provisions and mechanism for those women migrant workers who work in informal sectors such as domestic work which is not even included in destination countries' national labours laws as in the context of the Gulf Cooperation Council countries.

Simply, the negotiation of BLAs between states does not suffice to provide protections to women migrants. Even when women migrant workers are specifically mentioned in the BLAs, they are not complete on their own; they just constitute a crucial part of the rest of other governance actors, structures and apparatuses, since labour migration governance is multi-levelled, multi-layered and multi-dimensional. In summary, while BLAs are important mechanisms in the overall governance of labour migration, so too are the active roles and involvement of all other actors and stakeholders in the implementation of the interstate bilateral instruments. Any legal provisions and institutional practices that do not synchronise with the provisions provided by the BLAs should be reformed in such a way that they do not hinder the effective implementation of the BLAs. For instance, in destination countries where national labour laws do not cover certain groups of migrant workers such as domestic work in the Gulf countries, there is a need for their reforms, or the BLAs should explicitly delineate the ways in which the sector-specific challenges will be addressed and women migrants' protections ensured. Similarly, migrant sending countries should take legal and institutional measures such as regulation of recruitment agencies, national labour migration policies, pre-departure orientation programmes, consular support and migrant resource centres to reinforce, supplement and thus implement the BLA provisions effectively. If such bilateral arrangements dedicated to such vulnerable sectors and groups of women migrants are not structured, monitored and implemented without sufficient gender-sensitivity, they invariably entail the danger of challenging women into certain gendered forms of employment and thus further perpetuate gender discriminations and entrench the gendered outcomes in contrast to providing protections.

Further, a gender-responsive and gender-sensitive approach to BLAs is not a once-and-for-all phenomenon. During the entire cycles of BLAs from the drafting, implementation to their evaluation and revisions, it is critical to have women migrants' concerns and voices heard and have their specific challenges and issues addressed. A gender-responsive approach to labour migration governance and to BLAs should embrace and integrate the norms and principles of gender equality, rights and empowerment into all the processes and cycles of the bilateral agreements beginning with the drafting of BLA until their evaluation and revisions as shown in the figure below. BLAs can benefit female migrant workers by providing human and labour rights protections for women migrants if they are gender-responsive. However, the ILO report (2015) revealed that the BLAs as they currently exist exhibit many shortcomings in terms of their design, content, monitoring, implementation and impact (ILO, 2015). When such bilateral instruments are negotiated, states should do so with a broad-based consultative process with all stakeholders, including migrant workers, members of their families, civil society and trade unions, in the development, implementation and evaluation of BLAs (ILO, 2015) (Fig. 18.2).

During the drafting process, the bilateral agreement should involve all relevant state line agencies as well as the broad spectrum of stakeholders including women migrant workers' representatives and social partners including gender advisers with expertise on labour migration. In the negotiation process, instead of a state-centric neoliberal approach to BLAs that preponderates the state interests and simply focuses on controlling migration flows and securitisation, the broad-based consultative process can help focus on the rights-based BLA that provides protections provisions and mechanisms. Only such a rights-based gender-responsive approach can mitigate the repercussions that the dominance of state interests in BLAs can have on women migrants' social protections as well such as the provision of portability of social entitlements. Since most women migrants work in informal, flexible and precarious sectors, they tend to be deprived of obtaining social security benefits and pension claims and providing such a social protection in the BLA can protect and empower women migrants, especially those employed in the so-called "gig" economy.

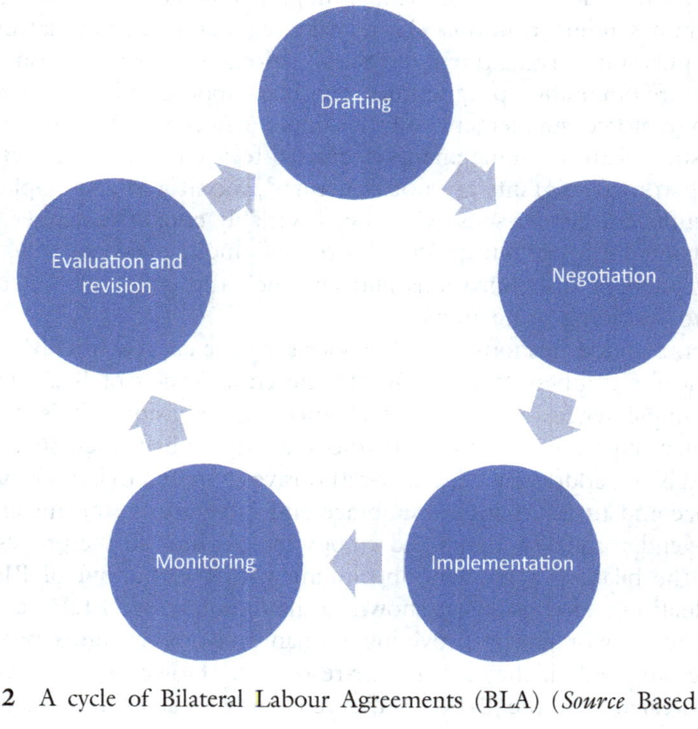

Fig. 18.2 A cycle of Bilateral Labour Agreements (BLA) (*Source* Based on ILO, 2015)

Labour migration has indeed grown into a burgeoning industry of its own that includes informal intermediaries, unauthorised subagents and others operating within the illegal spaces of the informal economy. Temporary migration is thus 'buttressed' by an extensive web of labour market private recruitment agencies and brokers (Wickramasekara, 2005). BLAs should provide gender-sensitive monitoring mechanisms to ensure that their effective implementation and gender impact assessments. In order for BLAs to be gender-friendly, states that enter into BLAs build institutions and procedures that can transparently monitor and evaluate the implementation and compliance with normative frameworks and report results to all stakeholders for policy review, development and renewal or termination. States should also ensure the collection and dissemination of gender-disaggregated data in order to take into account the specific vulnerabilities inherent in female-dominated sectors. Further, both migrant sending and receiving states should synchronise their national laws and institutional processes. The new trend for development of dedicated agreements concerning domestic workers should be promoted given that this group has high vulnerability which cannot be addressed in a general agreement when they are not even covered by labour laws of destination countries.

Conclusion

Along with the feminisation of labour migration globally, states are increasingly using BLAs as instruments of labour migration governance. However, such instruments have largely ignored the specific vulnerabilities and challenges faced by women migrant workers, especially those in the informal sectors. In order for BLAs to be gender-sensitive and transform the gendered structures of the BLAs, states should have political will and commitment to negotiate agreements in ways that uphold the international human and labour rights norms, instead of resorting to BLAs as neoliberal tools to serve state-centric interests. In the context of the Domestic Workers Convention, 2011 (No. 189) and the establishment of the International Domestic Workers Federation (2013) along with other civil society activism, the number of sector-specific bilateral agreements that focus on women migrant workers is also increasing. BLAs should refer to normative frameworks such as ILO R86 and the Multilateral Framework on Labour Migration and cover the whole spectrum of migration, including recruitment, pre-departure, arrival in the countries of destination, return and reintegration. Such dedicated BLAs can be instrumental to enhancing the protections of women migrant workers if they are devised in gender-responsive ways during the entire cycle of the bilateral agreement. However, in the absence of ample caution, such instruments can potentially have adverse impacts on women migrants by channelling them to certain "feminised" and informal employment sectors such as domestic, service and entertainment sectors.

Additionally, there is a need for a gender-responsive strategies integrated during the entire cycle of BLAs from their drafting through implementation, monitoring, evaluation and revisions. Given the multi-dimensional and multi-layered governance of labour migration involving various state and non-state actors at different levels and scales, BLAs which are negotiated between state parties are crucial, but they should be considered as part of the overall governance of labour migration. States should ensure, throughout the entire cycles of BLAs, a broad-based consultation and engagement with civil society, linkages and interactions between different state agencies, women migrant workers' organisations and women migrants themselves. Further, it is critically important to create legal and social protective provisions for sectors not covered by national labour laws, such as agriculture and domestic services.

References

Chilton, A., & Posner, E. (2017). Why countries sign bilateral labor agreements. *The Journal of Legal Studies, 47*(S1), S45–S88. Available at: https://www.journals.uchicago.edu/doi/abs/10.1086/694456?journalCode=jls.

Gamlen, A., & Marsh, K. (2011). *Migration and global governance* (International library of studies on migration; 12). Cheltenham; Northampton, MA: Edward Elgar.

Government of Nepal. (2007). Memorandum of Understanding Between the Government of Nepal and the Government of United Arab Emirates in the Field of Manpower. Available at: https://www.ceslam.org/uploads/backup/GoN_2007_MoUBetweenNepalAndUAE.pdf

Grugel, J., & Piper, N. (2007). *Critical perspectives on global governance: Rights and regulation in governing regimes*. London: Routledge.

ILO. (2006). *ILO multilateral framework on labour migration: Non-binding principles and guidelines for a rights-based approach to labour migration*. Geneva. Guidelines 3.4, 5.3 and 9.9. Available at: https://www.ilo.org/wcmsp5/groups/public/—asia/—ro-bangkok/documents/publication/wcms_146243.pdf.

ILO. (2015). *Bilateral agreements and memoranda of understanding on migration of low skilled workers: A review*. Geneva: International Labour Office. Available at: https://www.ilo.org/wcmsp5/groups/public/—ed_protect/—protrav/—migrant/documents/publication/wcms_385582.pdf.

IOM. (2020). *World migration report 2020*. Geneva: International Organization for Migration. ISSN:1561-5502. Available at: https://publications.iom.int/system/files/pdf/wmr_2020.pdf.

Likić-Brborić, B. (2018). Global migration governance, civil society and the paradoxes of sustainability. *Migration, Civil Society and Global Governance, 15*(6), 762–778.

Lim, Lin Lean. (2014, December 1–2). *ILO—KNOMAD Thematic Working Group 3 on Low-skilled Labour Migration Technical Workshop on Review of Bilateral Agreements1 on Low-skilled Labour Migration*. Kathmandu, Nepal. Available at: https://www.knomad.org/sites/default/files/2017-03/KNOMAD%20TWG3%20BLA%20workshop%20summary.pdf.

Nepal Labour Migration Report. (2020). *Ministry of labour, employment and social security*. Kathmandu. Available at: https://moless.gov.np/wp-content/uploads/2020/03/Migration-Report-2020-English.pdf.

OSCE. (2009). *Guide on gender-sensitive labour migration policies*. Vienna: Organization for Security and Cooperation in Europe (OSCE). Available at: https://www.osce.org/secretariat/37228.

Parrenas, R. (2001). *Servants of globalization: Women, migration and domestic work*. Stanford, CA: Stanford University Press.

Plotnikova, E. (2011). *Recruiting foreign nurses for the UK: The role of bilateral labour agreements*. Thesis, The University of Edinburgh.

Schierup, C., Ålund, A., & Likić-Brborić, B. (2015). Migration, precarization and the democratic deficit in global governance. *International Migration, 53*(3), 50–63.

Schierup, C.-U., Likic-Broboric, B., Delgado, R., & Toksöz, G. (2019). Migration, civil society and global governance: An introduction. In C.-U. Schierup, B. Likic-Broboric, R. Delgado, & G. Toksöz (Eds.), *Migration, civil society and global governance* (pp. 3–15). London and New York: Routledge.

UN. (2018). *Global compact for safe, orderly and regular migration*. Available at: https://refugeesmigrants.un.org/sites/default/files/180713_agreed_outcome_global_compact_for_migration.pdf.

Van Panhuys, C., Aoul, S. K., & Binette, G. (2017). *Migrant access to social protection under bilateral labour agreements a review of 120 countries and nine bilateral arrangements* (ILO Working Paper 994955792602676). ILO. Available at: https://ideas.repec.org/p/ilo/ilowps/994955792602676.html.

Wickramasekara, P. (2005). *Rights of migrant workers in Asia: Any light at the end of the tunnel?* (International Migration Papers No. 75). Geneva: International Migration Programme, International Labour Office. Available at: http://www.ilo.org/public/english/protection/migrant/download/imp/imp75e.pdf.

Wickramasekara, P. (2012). *Something is better than nothing: Enhancing the protection of Indian migrant workers through bilateral agreements and memoranda of understanding: Migrant forum in Asia, Manila*. Available at: http://ssrn.com/abstract=2032136.

Wickramasekara, P. (2015). *Bilateral agreements and memoranda of understanding on migration of low skilled workers: A review*. Available at: http://dx.doi.org/10.2139/ssrn.2636289.

Jenna Hennebry Ph.D. is an Associate Professor with the Balsillie School of International Affairs at Wilfrid Laurier University in Waterloo, Canada. Hennebry's research expertise includes labour migration governance, gender and migration, migrant worker health and human rights.

KC Hari is a doctoral candidate in Global Governance at the Balsillie School of International Affairs, Wilfrid Laurier University, Waterloo, Canada. His research interests are in the intersections of globalization, migration and governance, particularly in the South Asian context. His Ph.D. dissertation examines labour migration governance with a focus on Nepali women migrant workers in the Gulf countries.

CHAPTER 19

Revisiting the Migration–Development Nexus Debate Through the Prism of Gender, Politics and Agency

Petra Dannecker and Nicola Piper

Abstract Academic and policy circles have debated the link between migration and development for decades, and thanks to the 'rise' in feminist critiques, there has been greater gender awareness and attention being paid to the need for a gender analysis. Yet, as we will argue, governmental and international organisational actors involved in the migration–development nexus debate at the global level have not taken up feminist critiques of gender specific power structures and intersectional forms of agency in a sufficiently comprehensive manner, demonstrating the persisting divide between practice and research. A more nuanced understanding of migrant agency at the intersection of key debates around migrant rights, care and development—which have tended to take place in siloes—is required. Such understanding would also have the potential to inform the design of policy interventions that would result in better gender outcomes.

P. Dannecker (✉)
University Vienna, Vienna, Austria
e-mail: petra.dannecker@univie.ac.at

N. Piper
Queen Mary University of London, London, UK
e-mail: n.piper@qmul.ac.uk

C. Mora and N. Piper (eds.), *The Palgrave Handbook of Gender and Migration*, https://doi.org/10.1007/978-3-030-63347-9_19

Introduction

The relationship between migration and development (also referred to as the 'migration-development nexus') has been discussed in academic and policy circles in various forms and shapes since the 1960s, undergoing a period of dormant interest, while experiencing a revival in attention as of the early 2000s when a flurry of activities and actions promoting knowledge on and interest in this nexus occurred on the part of the United Nations and donor agencies (see Mahon, in this volume). Thanks to the rising influence of feminist critiques to the fields of both migration and development studies, gender has been increasingly integrated into these debates, not only as a topic for policy intervention but also as an analytical lens and as a basis for theory building. However, several aspects of the migration–development nexus debate have not considered a gender analysis of agency in its economic and socio-political manifestations. In our view, this has to do with the tendency of key debates on gendered aspects of migrants' human and labour rights, care and development not having gelled sufficiently enough—if at all—which is due to those debates having largely taken place in siloes. We would argue that better integration of such debates is required in order to move the discussion forward and arrive at more effective policy interventions which would benefit migrants in general and female migrants in specific in a more wholesome manner.

The mainstream debate and policies around the migration and development nexus is too narrowly focused on economic aspects of development, with too little attention being paid to social development of which care is one vital (and highly gendered) element. As far as agency is concerned, the main image created of migrant workers in general and female migrants in particular, is that of heroes or agents of development which serves the political need to emphasise the valuable contributions they are making to society and why their migration is 'good' for development. What is de-emphasised are the limitations placed on migrant's economic agency as the result of intersectional forms of precarity (Piper, Rosewarne, & Withers, 2017) as well as migrants' own development visions and ambitions which differ by gender (Dannecker, 2009) and affect migrants' political agency.

In this chapter our aim is to show that governmental and international organisational actors involved in the migration–development nexus debates at the global level have not taken up feminist critiques of gender specific power structures and intersectional forms of agency in a sufficient and comprehensive manner with the effect of perpetuating the practice-research divide for largely political reasons. We start this endeavour by providing a 'historical' background to the migration–development nexus debate and its gendered critiques. The section thereafter focuses in more detail on the missing linkages between care, migration and development, to take this to the realm of political activism in the section that follows. In so doing, we raise the issue of how feminist critiques are translated into political demands made by civil society actors; and whether collective activism and global policy making in fact converge on bypassing the need for structural change to address gender differences and outcomes.

The Lure of the (Labour)Migration–Development Nexus: A Gendered Critique

The debate and policy implications of the interdependent processes of cross-border movements of people and 'development' have undergone various changes over time. Up to the 1990s migration was defined primarily as a reaction to 'underdevelopment' in the so-called sending countries and economic growth defined as development. It was argued however, that neither remittances nor the returning of 'skilled' migrants could initiate 'development' processes because of the persisting 'traditional' structures in the societies of origin. This changed in the 1990s when the positive aspects of migration and the role migrants have in the development of their 'home' countries and communities were celebrated in academic circles and by international development organisations, donors and national governments. This was the result, as Faist (2008) and de Haas (2010) argue, of a paradigm shift in social theory away from emphasising structure in a deterministic manner to foregrounding agency. The advent of transnational perspectives in migration theory (Withers & Piper, forthcoming) along with the construction of communities and civil society organisations as important development actors and a renewed emphasis on bottom-up processes as the main development principle (so-called 'rights-based' approach) reflect this shift. The understanding of development as first and foremost an economic process was, however, not buried entirely as the main solutions presented by practitioners—such as self-employment or micro-credit programmes and projects, relying heavily on the market, reveal (Koehler, 2015).

Migrants and certain types of migrant organisations were especially targeted as 'development agents' since the publication of the *Global Development Finance Report* by the World Bank in 2003 (Faist, 2008). The collating of global remittance data led to a booming interest by academics, donors and policymakers in ascertaining the potential development impact of migratory flows and remittances (Lindley, 2010). The new hype around the issue of remittance was accompanied by debates, policies and strategies on how governments and international organisations could cultivate the positive relationship between migration and development. Tapping into overseas communities, under the discursive appropriation of 'diasporas' by the development sector, became an essential aspect of such cultivation efforts (Gamlen, 2014: 582) (incidentally, research into diaspora involvement in development has been largely gender neutral). Since then empirical studies that observe positive effects of migration on development have appeared in greater numbers, unlike the few critical studies pointing to the negative relations between migration and development (de Haas, 2012). The fewer, more sceptical studies range from critiquing the neoliberal orientation with its strong focus on individual-level agency thereby neglecting structural macro aspects hampering more broad-based agency (Withers, 2019), to research arguing that migrants' own visions and ideas about development are rarely considered (Dannecker, 2009; Piper & Rother, 2014; Raghuram, 2009). In doing so, such critical inquiries question who this 'development' is benefitting, thus raising the issue of unequal power relations of which gender is constituent.

The debates and policies around the migration and development nexus reveal that the different experiences of migrant women, and the fact that migrant women are not a homogenous group, are considered neither in the mainstream discussions in academia nor politics. This is especially surprising given the fact that since the 1950s different frameworks and approaches in the interdisciplinary field of 'gender and development' have been discussed and a variety of concepts and tools developed. Starting with the 'Women in Development' (WID) framework in the 1970s criticising the existing development theories and practices by arguing that modernisation (read 'development') was impacting women in an unfavourable manner, and demanded as a solution the better integration of women into the economic system. In 1980s, out of the criticisms of the earlier WID concept, the Gender and Development (GAD) concept was introduced to approach the matter in terms of the 'social relations of gender', viewing women as agents of change rather than as passive recipients of development efforts, thus placing a strong emphasis on women's emancipation. The important strategy and instrument of the GAD concept was and still is 'gender mainstreaming', which aims at increasing gender awareness in all areas and all levels of public life (Razavi & Miller, 1995). In the 1980s empowerment as a relatively new approach in women and development discourse was put forward and, in contrast to other policy approaches discussed above, emerged from feminist writings and grassroots organisational experience of 'Third World' women (Moser, 1993: 74), in particular DAWN (Development Alternatives with Women for a New Era), a network of activists, researchers and policymakers from the Global South (Sen & Crown for DAWN, 1988: 9). The empowerment approach points to the existing structures in society as sources of women's subordination, deepened further by socio-cultural norms. This approach further highlights that women's experiences are also determined by other factors such as class, nation, ethnicity and race, thereby criticising Western feminism and development actors for its exclusive focus on gender as a singular axis of inequality. Since then empowerment as a concept has been intensively debated by feminist theorists. Different theoretical conceptualisations have been put forward as well as strategies and indicators to measure empowerment (Cornwall, 2016; Kabeer, 2001, 2008; Wierenga, 1994). In the field of migration studies, migration processes have also begun to be analysed from gender and feminist perspectives since the 1980s, albeit from varying vantage points. Starting from recovering women's participation in, and experiences of, primary migration in the 1980s (Grasmuck & Pessar, 1991; Morokvasic, 1984), extending to gendered perspectives inherent to international migration as a whole (Carling, 2005; Hondagneu-Sotelo & Cranford, 2006; Kofman, 2004), to studies revealing that gender (as an analytical category) is central to all phases of migration, is played out at multiple sites (Mahler & Pessar, 2006; Piper, 2007) and mediated by other socially constructed categories (Lutz, 2010).

The wide range of feminist contributions that have been made to the fields of development and migration studies are, however, largely absent from the mainstream academic and political debates on the nexus between migration and development, as are the different experiences of migrant women, and their heterogeneity due to differences in class, race, sexuality or culture. Those differences are hardly reflected in the policies accompanying the recent 'discovery' of migrants as development actors and bottom-up processes as a contributing factor to development. This does not imply, however, that gender was not promoted in development discourses or policies; nor that empowerment was not defined as an important development target or part of female migration processes. In the last decades, however, policies and strategies became more technical in nature, at times based on empty slogans or buzzwords (Cornwall & Eade, 2011; Cornwall, Harrison, & Whitehead, 2008; Smyth, 2007), leading critics to suspect that the actual objective is to target women as the untapped resource for economic development.

This can be explained by the fact that mainstream migration scholars' understanding of development seems to be stuck in modernistic development thinking, the prevalence of a narrowly defined neoliberal economic paradigm and the understanding of migration a result of 'underdevelopment'. Despite the feminist critiques of development as a form of economic reductionism or the result of postcolonial legacies, neither such thinking nor concepts like human development referring to people's resources beyond material and technocratic aspects filtered into the mainstream migration–development nexus policy and research debate. The renewed enthusiasm concerning migration and development, as we argue, has not translated into differentiated conceptualisations about development as a multidimensional, complex and gendered process with different meanings attached for the different actors involved. The reasons for this persistent refusal to incorporate more nuanced understandings of development are largely of political nature, as we argue. This is so because in meanstream policy discourse, development is primarily related to remittances and their economic impact in the less developed migrant countries or communities of origin.

In relation to the migration–development nexus debate the main interest of international organisations, national governments of sending countries and donors regarding female migrants can therefore be summarised under the heading of 'what can migrant women do for development'. Similarly to mainstream discussions in the field of development studies and development policies, the inclusion of migrant women tends, thus, to be resource-oriented and technocratic (Cornwall & Rivas, 2015; Narayanaswamy, 2016). This is revealed by the fact that since the millennium, scholars have identified the 'feminisation of migration' as a core dimension of international migration and globalisation.[1] It was only in the 1990s that global data on female migrants began to be gathered and published by the UN expert group on 'International Migration Policies and the Status of Female Migrations' (UN, 1995) for example. In 2005 UN-INSTRAW published its first conceptual framework on gender, remittances and development (Ramírez, Domínguez, & Morais, 2005) which can be read as a response to the discovery of 'female migrants' and the emergence of the 'feminisation of migration' discourse.[2] Since then migrant women as development agents are at least rhetorically present in publications and strategic planning by international organisations and national governments. It is especially as remitters that female migrants have entered the sphere of international and national policy and research debates. The final Report of the Global Commission on International Migration (2005, 26), for example, states that female migrants not only send back remittances in higher amounts and more regularly but also use the remittances more efficiently for local and family-oriented 'development' processes compared to their male counterparts. Such a statement reproduces an image and representations of women as having particular qualifications well suited for development programmes, namely their supposed innate characteristics as: responsible and caring mother, docile and hard-working person, the prudent entrepreneur or the reliable credit taker (see Cornwall et al., 2008; White, 1997).

The kind of thinking emanating on the basis of these assumptions, therefore, was to '...develop gender-sensitive strategies that aim to increase the development potential of remittances' (IOM, n.d.: 6), like 'financial literacy' projects, business training (IOM, ibid., 6)[3] or more generally to support the '...inclusion of gender considerations in international, regional and bilateral policy dialogues on remittances and development' (IOM, ibid., 6). UN Women as well as the ILO have also initiated and supported many research projects and a variety of initiatives on gender, migration and development, especially but not solely to do with financial remittances. Thus, female migrants have entered the discourses and politics regarding the migration and development nexus as devoted remitters, sacrificing themselves for their families' and communities' wellbeing—and ultimately for the economic development of their 'home' countries. The gendered context in which the 'remittance economy' operates, the purpose or the social and political meaning of remittances, as well as the gendered interpretation of remittances, however, continue to be 'blind spots', both in scientific debates and in development

[1]Although for more than four centuries, women have been part of global migration flows (Donato & Gabaccia, 2015).

[2]The argument about 'feminisation of migration' was not meant to imply that women did not migrate before, but rather that migratory patterns in particular regions and the patterns of migration have changed such that more women migrated as sole income earners (Piper, 2008; Zlotnik, 2003).

[3]https://www.iom.int/sites/default/files/about-iom/Gender-migration-remittances-infosheet.pdf.

practice as are the development visions of the female migrants.[4] Thus, the debates and policies in the framework of the migration–development nexus reflect hegemonic discourse that views development as primarily an economic process. Feminist agendas are co-opted (de Jong & Kimm, 2017) also in the area of migration and development.

The strategies and initiatives reveal that channelling and appropriating female migrant's 'potential' or 'agency' as a resource for economic development is indicative of the neoliberal project of development and concomitant migration policy design, thereby ignoring feminist critiques of the single-axis category of the 'female migrant' (Bastia, 2014), or studies taking an intersectional perspective (Bastia, 2012; Lutz, Herrera Vivar, & Supik, 2011; Manalansan, 2006) to analyse the differences and relations of power between migrants in general and female migrants (and non-migrant women) in specific. This is also revealed in the area of care as the following section will show.

Care, Development and Migration

Reproductive labour has been an important dimension of women's economic and human development (Razavi, 2012). On the one hand, it has shaped women's access to and treatment in the labour market and the education sector; and on the other hand, it has led to (or deepened) socio-economic inequalities and discrimination. Importantly, reproductive labour has increasingly become a key focal issue in the scholarly work on gender and migration, with care considered an important dimension of reproductive labour, especially from a social policy point of view (Yeates, 2009). A contentious aspect of reproductive labour is its unpaid nature. This is acknowledged by Goal 5 of the SDGs in relation to women's achievement of their development potential in socio-economic terms by providing empowerment and autonomy. Women are assumed to be natural providers of unpaid work such as home-making, child rearing and looking after the sick and the elderly. This has obvious implications for their participation in formal labour markets and their ability to achieve their full potential. Merely distinguishing between paid and unpaid work, however, does not recognise women's contribution to development, a claim supported by the ethnographic record in the global care chain literature (Ehrenreich & Hochshild, 2004). When reproductive labour crosses borders and is outsourced to migrant women, women from the resource richer parts of the world can achieve their social and economic aspirations, which reveals that not only gender but also class and racial divisions are deeply implicated in social reproduction.

Many analysts of reproductive and care labour argue that women providing reproductive labour of this nature engage in global processes of exchanging, buying, selling, objectifying and consuming commodified forms of intimacy.

[4]For a critical reflection see Kunz (2008) and Van Naerssen, Smith, and Marchand (2015).

Constable (2009) borrows the latter notion from Marx, defining commodification as a 'process of assigning market value to goods or services that previously existed outside of the market'. As global care chain scholars point out, this commodification process has had a profound effect on women from the Global South, notably in their treatment as disposable and objectified labour (Ehrenreich & Hochshild, 2004). Although this strand of feminist critique usefully exposes market structures and mechanisms that define and constrain the role of women, it neglects the role of the state in condoning, developing and reinforcing the globalised private exchange of care services (Yeates, 2009).

Others argue the commodification of affect has been the product of state policies devising class-, gender-, and ethnicity-specific responses. This strand of feminist critique actively engages with state discourses. For example, Pei-Chia Lan (2008) identifies Foucauldian-style control of sexuality and the body in state policies on gender and migration. She explores what she labels the 'care deficit' and 'bride deficit' in her study of the reproductive labour provided by migrant domestic workers and marriage migrants in Taiwan. While the care deficit is, in part, triggered by middle-class women pursuing tertiary education, professional careers, and 'modern' life-styles, she attributes the bride deficit to the rising number of single women in Taiwan. Compared to the past, significantly more Taiwanese women of marriageable age are now pursuing university degrees, participating in the labour market, building careers and enjoying financial independence. They combine a concern for the high costs of child rearing with a desire for an equal partnership. These factors are a bane for Taiwanese men who are already challenged in their search for a partner, given Taiwan's skewed gender ratio. Consequently, working-class men have been drawn to the reproductive labour offered by marriage migrants, leading some scholars to suggest marriage migration is a working-class response to the reproductive crisis in Taiwan and Singapore (Yeoh, Chee, & Baey, 2013).

The Dilemma of (Re-)Productive Agency

When gender is understood in a relational sense, the expanded agency of one woman can lead to another woman's confines, revealing the heterogeneity of the category 'women' as well as the necessity of an intersectional analysis. But this is rarely taken up in debates on the migration–development nexus where a specific understanding of gender dominates which ignores the process and complexity of individual and collective agency and its context specific nature, by dismissing how key axes of differentiation—class, ethnicity or race—impact on agency (Bastia, 2014).

By contrast, the issue of difference and power relations between female migrants and non-migrant women, or between migrant women from different nationalities or classes, have been well demonstrated in the case of migration's link to social reproduction and the growing care economy, one of *the* key sectors providing employment for migrant women. Women's migration often results in care deficits or care gaps which are in turn typically filled by (mostly female) members of their families. Parreñas (2000) has captured these dynamics in her three-tier conceptualisation of the global division of reproductive labour with which she distinguishes between the following groups of women: (1) middle-class women in receiving countries; (2) migrant domestic workers and (3) women from the Global South who are too poor to migrate or lack the necessary human and social capital. The migrant domestic worker as a result is ranked in the middle of this division of labour and the care deficit created by her own out-migration might be filled by the poor woman who cannot migrate (in paid form) or by the eldest daughter or another female relative (in unpaid or paid form; Hochschild, 2000).

To return to the structural aspects of gendered migration and development, Chin's study (1998) on foreign domestic work has situated the import of such workers within the 'politics of development' of the destination country, in this case Malaysia, and the expansion of the middle class, resulting in more educated women entering the labour market, leaving a care gap to be filled. Without state provision of such services, this gap has been filled by foreign domestic workers (Chin, 1998; also on Singapore, see Teo & Piper, 2009; on Japan, see Onuki, 2009) based on the construction of cultural stereotypes, such as that of Philippine care workers being better care workers than women from other countries. Increased competition between the various sending countries for a slice of the overseas market has led to a situation in which not only gender but also the nationality and level of development of sending regions intersect in creating a hierarchy of female migrants as a marketable product (Dannecker, 2014). In this regard, Chin's work echoes approaches by other social scientists who have linked foreign domestic worker issues to larger questions of global economic restructuring and the feminisation of labour migration (Sassen, 2000). The issue of agency has, however, been side-lined especially in such structural analyses.

What does this mean in terms of agency in relation to development? One message is that agency does not only involve individual but is also about collective forms. One place where this is evident is rising collective political activism by or on behalf of migrant women, such as domestic workers.

Politicising the Gendered Labour Migration–Development Nexus: Rights, Transnationalism and New Protest Strategies

Our discussion so far has demonstrated how the dominant development discourse operationalises migrants as 'agents' for development despite the contradictory mismatch between the economic benefits (such as migrant remittances) at the macro level versus social and political changes that are prompted at the micro level. By contrast, on the ground, migrants' desire for development is embedded in a complex socio-political reality that is shaped by the process of migration. Yet, popularisation of 'migrant-led' development models and mainstream imaginations of migrant agency in development fall short in incorporating the lived reality of migration as often differently experienced by men and women (Dannecker, 2009). Migrants' ability to enhance their developmental agency, we argue, hinges crucially upon their ability to act collectively.

Global Governance of Gendered Migration and Development

At the global level, organisations have not only shaped the discourses and policy prescriptions on this specific nexus since the late 1990s but also initiated programmes and strategies targeting and promoting migration in gender neutral ways so that gender effects are largely invisible. Given the feminisation of migration due to the channelling of women into highly gendered forms of labour, epitomised by the global care economy, women's choices are heavily circumscribed and as a result, so are the empowering effects of migration that would lead to enhanced personal agency. This focus on economic aspects of development has instead led to an analytical bias in favour of structural issues to do with the desire to address root causes of migration in countries of origin, thereby moving away from agency (or a specific definition thereof) (Collins, 2015).

As a result, female migrants are not as prominently targeted or mentioned in current policies or initiatives in relation to enhancing their agency. Instead, a growing tendency can be observed towards discussing and funding initiatives and programmes that focus on border control and return migration, the key elements of migration management. With temporary or circular migration being seen as the ideal regulatory framework for the realisation of the triple win mantra, return migration is one of the inevitable outcomes of 'safe, regular and orderly migration'. What is, however, given less attention is the exact nature of income generating activities, or decent work as per the ILO parlance, in source as well as destination countries; nor the way in which return migration reaffirms gender norms and can lead to the stigmatisation of women workers and crises of masculinity among under- or unemployed men (Withers, 2019). In other words, resources are predominantly spent on border management and security, and to a lesser extent on broader mobility, economic justice via internationally agreed upon labour standards or social development in the broadest sense (Hujo & Piper, 2010; Lavenex & Kunz, 2008).

At the same time, the significance of, and dynamics involved in, South-South migration have begun to gain a modicum of attention, with a few studies having appeared which highlight its growing importance and magnitude (e.g. Hujo & Piper, 2010; Ratha & Shaw, 2007). Research on South-South labour migration has shed light on the particular dynamics of temporary circulation, under the heading of transnational migration, intra-regional streams or cross-border circulation. That migration to the Global North requires different human and social capital, is a well established fact.

In addition, the majority of migrants moving intra-regionally are categorised as low-wage, low-skilled. It has been observed that migrants who travel to other Southern countries enjoy much lower increases in income and are more likely to be irregular, and thus subject to greater risks of exploitation (Ratha & Shaw, 2007). This heightened awareness and attention being given to South-South migration in relation to gender has so far led to the focus being often not so much on development as discussed above, nor on agency.

Rather, what we can observe in this context is the tendency towards the construction of 'Third World' women as the result of victimisation discourses which find an easy entry point into discussions on low-skilled women migrating to middle-income countries or countries not classified as western liberal democracies (such as the Gulf States). There they carry out low paid types of work, often located in the informal economy and/or involving tasks socio-legally unrecognised as work (as e.g. the case of sex work, or domestic work, at least up to 2011 when ILO Convention 189 was adopted), or at the extreme end of the 'labour' spectrum: i.e. tasks classified as forced or trafficked labour. Important is what is seen as a solution: often it is the banning of women's migration, rather than tackling lacking or constraining regulatory frameworks.

Yet, women's migration is also being celebrated as a form of empowerment and as a contributing factor to development, thus signifying considerable agency—but of a certain kind in a specific context. The celebratory aspects are usually tied to the idea that migration allows women (in the Global South) to escape from narrowly defined gender regimes, gender roles and underdevelopment, without taking the heterogeneities and the multiplicity of social and structural factors that affect women in general and migrant women in particular into account (Mohanty, 1984; Withers, 2019). Moreover, the sustainability of such newly found freedoms under conditions of permanent temporariness as prevalent in intra-Asian and Asia-Middle Eastern migration corridors, enforced return migration after completion of temporary contracts or deportability, and migrant women's channelling into a narrow range of job categories by bilateral agreements (see Hennebry & Hari, this volume) is not fully considered in all its ramifications, nor the fact that migrant women not only leave gendered regimes in countries of origin but also enter gender regimes in countries of destination (Piper, 2008), requiring a gender analysis and policy intervention at all phases of the migration cycle.

Political Agency via Collective Activism

The debate on collective organisations as key vectors for driving economic development largely revolves around so-called hometown associations, or political diaspora organisations, business persons' networks and other less formalised community actions (Faist, 2008; Piper, 2010). There are two problems with such existing studies on migrant collectives: (1) they assume the existence of diaspora groups with immigrants who are settled permanently or semi-permanently in the destination countries and thus are mostly derived from the destination country perspective; and (2) they rarely if ever offer a gendered analysis in terms of representation of men's and women's interests or their roles in such organisations. What is left aside is the widespread experience if employer-tied temporary contract migrants—and the many women among them. Rather, conceptualisation of migrant-driven development typically involves imagining a migrant who lives in affluent, democratic countries in North America, Oceania and other parts of the so-called 'West'. Overlooked in such discourses are the many migrants who move within the Global South, or intra-regionally, a migratory flow which parallels if not exceeds South–North migration in its numeric significance (Hujo & Piper, 2010).

There are also meso level institutions which are involved in political activism aimed at bettering the situation of the many migrant workers around the world. Collective activism by female migrants has focused on better working conditions, extended labour rights and freedom from violence. There are limitations though as to the extent to which structural issues are being critiqued and fought against by migrant rights activists. In addition, certain organisational forms are more prominent than others: unions e.g. are somewhat absent from female-dominated sectors but even here new developments have taken place. The prime example for collective, cross-organisational activism involving unions, migrant rights and sectoral worker organisations in a new form of 'networks of labour' (Zajak, Egels Zanden, & Piper, 2017) is domestic work in relation to the success story of the adoption of ILO Convention 189 in 2011 (Fish, 2017; see also her chapter, this volume). The significance of its size and the high numbers of migrant women involved in the domestic work sector have given rise to collective organising by domestic workers and the formation of transnational advocacy networks (Ford & Piper, 2007), ultimately leading to their integration into complex 'networks of labour' (Piper et al., 2017; Zajak et al., 2017). This is thus a prime example where migrant agency is played out in a collective manner and with tangible outcomes as far as global norms are concerned as well as the formation of new collective organisations as vehicles for women's political agency.

The political opportunity provided by the ILO process around negotiating its Convention No. 189 partly explains the overwhelming attention domestic and care work has received over the last years. This has, however, led to the eclipsing of other types of work (one exception is the chapter by Pearson & Kusakabe, this volume) traditionally not associated as tightly with women as care and domestic work has been. In so doing, the experience of other groups of female workers, as economic and political actors, have been left on the sidelines and would deserve greater attention by scholars as well as policymakers alike.

In sum, as far as political activism at the global level is concerned, much of the 'insider' advocacy efforts have focused on finding meaningful participatory channels into global policy processes and fora to the detriment of spending more energy on developing political messages and strategies aimed at achieving structural change in the globally networked economy that would lead to more evenly distributed development outcomes and greater freedom of mobility (Piper & Rother, 2014). Such radical demands are voiced from the 'outside' by a smaller group of CSOs which have opted to stay away from participation in global processes and to organise side-events and engage in other strategies instead. As a result, however, they find themselves marginalised from the growing global migrant rights movement leaving their voices even less heard.

Conclusion

The main argument which emerges from our chapter relates to the mainstream political and academic discussions around the relationship between migration and development which have included gender and female migrants in a selective and narrow manner, ignoring the accumulated knowledge regarding gender in its complexity and as a relational concept. As a result, the rich, nuanced discussions at the intersection between care and development by feminist scholars highlighting the development dimension of 'care' and the care dimension of development have not been integrated into mainstream discussions. Secondly, the mainstream migration–development nexus debate has largely ignored the issue of political agency especially in the form of collective activism. In doing so, the importance of enhancing migrant women's collective political agency to allow them to self-organise, express their own visions and fight for their rights have failed to be addressed by 'development projects'.

What the mainstream academic and policy debates have in common is the tendency to bypass feminist debates entirely or to only partially incorporate feminist perspectives, with the result that the differences between the diverse group of 'migrant women', and thus the diverse expressions of migrant women's agency, remain invisible. The key reasons to explain such omission is the economic logic dominating the mainstream approaches to, and political interests in, the migration–development nexus. As a result, the power structures between North and South involved in the provision of and demand for care and social protection are not topicalised; neither is development being talked about in terms of 'whose' and 'what' development—and 'how' to achieve it.

References

Bastia, T. (2012). 'I am going, with or without you': Autonomy in Bolivian transnational migrations. *Gender, Place & Culture, 20*(2), 1–18.

Bastia, T. (2014). Intersectionality, migration and development. *Progress in Development Studies, 14*(3), 237–248.

Carling, J. (2005). *Gender dimensions of international migration* (Global Migration Perspectives No. 35). Global Commission on International Migration.

Chin, C. (1998). *In service and servitude: Foreign female domestic workers and the Malaysian "modernity" project*. New York: Columbia University Press.
Collins, P. (2015). Intersectionality's definitional dilemmas. *Annual Review of Sociology, 41*(1), 1–20.
Constable, N. (2009). The commodification of intimacy: Marriage, sex and reproductive labor. *Annual Review of Anthropology, 38*, 49–64.
Cornwall, A. (2016). Women's empowerment: What works? *Journal of International Development, 28*(3), 342–359.
Cornwall, A., Harrison, E., & Whitehead, A. (2008). *Gender myths & feminist fables*. Hoboken: Blackwell Publishing.
Cornwall, A., & Eade, D. (2011). *Deconstructing development discourse: Buzzwords and fuzzwords*. Rugby: Practical Action Publishing.
Cornwall, A., & Rivas, A. M. (2015). From gender equality and 'women's empowerment' to global justice: Reclaiming a transformative agenda for gender and development. *Third World Quarterly, 36*, 396–415.
Dannecker, P. (2009). Migrant visions of development: A gendered approach. *Population, Space and Place, 15*(2), 103–118.
Dannecker, P. (2014). Migration, gender and development: Trends, discourses and politics in Asia. In S. Poma Poma & K. Pühl (Eds.), *RLS papers—Perspectives on Asian migration: Transformation of gender relations and labour relations* (pp. 16–24). Berlin: Rosa Luxemburg Foundation.
De Haas, H. (2010). The internal dynamics of migration processes: A theoretical inquiry. *Journal of Ethnic and Migration Studies, 36*(10), 1587–1617.
De Haas, H. (2012). The migration and development pendulum: A critical view of research and policy. *International Migration, 50*(3), 8–61.
De Jong, S., & Kimm, S. (2017). The co-optation of feminisms: A research agenda. *International Feminist Journal of Politics, 19*(2), 185–200.
Donato, K. M., & Gabaccia, D. (2015). *Gender and international migration*. New York: Russel Sage Foundation.
Ehrenreich, B., & Hochshild, A. (Eds.). (2004). *Global woman: Nannies, maids, and sex workers in the new economy*. New York: Henry Holt and Company, LLC.
Faist, T. (2008). Migrants as transnational development agents: An inquiry into the newest round of the migration-development nexus. *Population, Space and Place, 14*(1), 21–42.
Fish, J. (2017). *Domestic workers of the world unite! A global movement for dignity and human rights*. New York: New York University Press.
Ford, M., & Piper, N. (2007). Southern sites of female agency: Informal regimes and female migrant labour resistance in East and Southeast Asia. In J. Dobson & L. Seabrooke (Eds.), *Everyday international political economy* (pp. 63–80). Cambridge: Cambridge University Press.
Gamlen, A. (2014). The new migration-and-development pessimism. *Progress in Human Geography, 38*(4), 581–597.
Global Commission on International Migration. (2005). *Migration in an interconnected world: New directions for action*. International Organization for Migration. https://www.iom.int/jahia/webdav/site/myjahiasite/shared/shared/mainsite/policy_and_research/gcim/GCIM_Report_Complete.pdf.
Grasmuck, S., & Pessar, P. (1991). *Between two islands: Dominican international migration*. Berkeley: University of California Press.

Hochschild, A. (2000). Global care chains and emotional surplus value. In W. Hutton & A. Giddens (Eds.), *On the edge: Living with global capitalism* (pp. 130–146). London: Jonathan Cape.

Hondagneu-Sotelo, P., & Cranford, C. (2006). Gender and migration. In J. S. Chafetz (Ed.), *Handbook of sociology of gender* (pp. 105–126). Berlin: Springer.

Hujo, K., & Piper, N. (Eds.) (2010). *South-south migration: Implications for social policy and development*. London: Palgrave Macmillan.

International Organization for Migration. (n.d.). *Gender, migration and remittances* [Infosheet]. IOM. Retrieved from https://www.iom.int/sites/default/files/about-iom/Gender-migration-remittances-infosheet.pdf.

Kabeer, N. (2001). Resources, agency, achievements: Reflections on the measurement of women's empowerment. In SIDA Studies, *Discussing women's empowerment: Theory and practice*. No. 3. Stockholm: SIDA (pp. 17–57).

Kabeer, N. (2008). *Paid work, women's empowerment and gender justice: critical pathways of social change* (Working Paper 3). Brighton, UK: Institute of Development Studies, Sussex University.

Koehler, G. (2015). Seven decades of 'development'. And what now? *Journal of International Development*, *27*(6), 733–751.

Kofman, E. (2004). Family-related migration: A critical review of European studies. *Journal of Ethnic and Migration Studies, 2*(1), 243–262.

Kunz, R. (2008). 'Remittances are beautiful'? Gender implications of the new global remittances trend. In R. Munck (Ed.), *Globalisation and migration: New issues, new politics* (pp. 160–180). London: Routledge.

Lan, P.-C. (2008). Migrant women's bodies as boundary makers: Reproductive crisis and sexual control in the frontiers of Taiwan. *Signs: Journal of Women in Culture and Society 33*(4), 833–861.

Lavenex, S., & Kunz, R. (2008). The migration-development nexus in EU external relations. *Journal of European Integration, 30*(3), 439–457.

Lindley, A. (2010). *The early morning phone call: Somali refugees' remittances*. New York: Berghahn Books.

Lutz, H. (2010). Gender in the migratory process. *Journal of Ethnic and Migration Studies, 36*(10), 1647–1663.

Lutz, H., Herrera Vivar, M. T., & Supik L. (2011). *Framing intersectionality: Debates on a multi-faceted concept in gender studies*. Farnham: Ashgate.

Mahler, S. J., & Pessar, P. (2006). Gender matters: Ethnographers bring gender from the periphery toward the core of migration studies. *International Migration Review, 40*(1), 26–43.

Manalansan, M. (2006). Queer intersections: sexuality and gender in migration studies. *International Migration Review, 40*(1), 224–249.

Mohanty, C. (1984). Under western eyes: Feminist scholarship and colonial discourses. *boundary, 12*(3), 333–358.

Morokvasic, M. (1984). Birds of passage are also women. *International Migration Review, 18*(4), 886–907.

Moser, C. (1993). *Gender planning and development: Theory, practice and training*. London and New York: Routledge.

Narayanaswamy, L. (2016). Whose feminism counts? Gender(ed) knowledge and professionalisation in development. *Third World Quarterly*, *37*(12), 2156–2175.

Onuki, H. (2009). Care, social (re)production and global labour migartion: Japan's 'Special Gift' toward 'Innately Gifted' Filipino Workers. *New Political Economy 14*(4), 489–516.

Parreñas, R. (2000). Migrant Filipina domestic workers and the international division of reproductive labour. *Gender & Society, 14*(4), 560–580.

Piper, N. (2008). Feminisation of migration and the social dimensions of development: The Asian case. *Third World Quarterly, 29*(7), 1287–1303.

Piper, N. (2010). Migration and social development: Organizational and political dimensions. In K. Hujo & N. Piper (Eds.), *South-South migration: Implications for social policy and development* (pp. 120–157). Basingstoke: Palgrave Macmillan.

Piper, N. (Ed.) (2007). *New perspectives on gender and migration—Rights, entitlements and livelihoods*. Abingdon: Routledge.

Piper, N., Rosewarne, S., & Withers, M. (2017). Migrant precarity: 'Networks of Labour' for a rights-based governance of migration. *Development and Change, 48*(5), 1089–1110.

Piper, N., & Rother, S. (2014). More than remittances—Resisting the global governance of migration. *Journal fuer Entwicklungspolitik/Austrian Journal for Development Studies*, *30*(1), 30–45.

Raghuram, P. (2009). Which migration, what development? Unsettling the edifice of migration and development. *Population, Place and Space, 15,* 103–118. https://doi.org/10.1002/psp.536.

Ramírez, C., Domínguez, M., & Morais, J. (2005). *Crossing borders. Remittances, gender and development*. United Nations International Research and Training Institute for the Advancement of Women.

Ratha, D., & Shaw, W. (2007). *South-south migration and remittances* (World Bank Working Papers). The World Bank. Available at http://elibrary.worldbank.org/doi/book/10.1596/978-0-8213-7072-8. Accessed 23 May 2016.

Razavi, S. (2012). *Seen, heard and counted: Rethinking care in development context.* Oxford: Blackwell.

Razavi, S., & Miller, C. (1995). *From WID to GAD: Conceptual shifts in the women and development discourse* (Occasional Paper 1). Geneva: United Nations Research Institute for Social Development.

Sassen, S. (2000). Women's burden: Counter-geographies of globalization and the feminization of survival. *Journal of International Affairs, 53*(2), 503–524.

Sen, G., & Crown, C. for DAWN (1988). *Development, crisis and alternative visions: Third world women's perspectives.* Oxford: Earthscan Publications.

Smyth, I. (2007). Talking of gender: Words and meanings in development organisations. *Development in Practice, 17*(4–5), 582–588.

Teo, Y., & Piper, N. (2009). Foreigners in our homes: Linking migration and family policies in Singapore. *Population, Space and Place, 15*(2), 147–159.

United Nations. (Ed.) (1995). *International migration policies and the status of female migrants*. Proceedings of the United Nations Expert Group Meeting on International Migration Policies and the Status of Female Migrants. New York: United Nations.

Van Naerssen, T., Smith L., & Marchand M. (2015). *Women, gender, remittances and development in the Global South*. Farnham: Ashgate.

Wierenga, S. (1994). Women's interests and empowerment: Gender planning reconsidered. *Development and Change, 25*(4), 829–848.

White, S. C. (1997). Men, masculinities, and the politics of development. *Gender & Development, 5*(2), 14–22.

Withers, M. (2019). *Sri Lanka's remittance economy: A multiscalar analysis of migration-underdevelopment*. London: Routledge.

Withers, M., & Piper, N. (forthcoming). Transnationalism and temporary labour migration. In: B. Y. S. Yeoh & F. Collins (Eds.), *Handbook on transnationalism*. Edward Elgar Publishing.

Yeates, N. (2009). Production for export: The role of the state in the development and operation of global care chains. *Population, Space and Place, 15*, 175–187.

Yeoh, B., Chee, H. L., & Baey, G. H. (2013). The place of Vietnamese marriage migrants in Singapore: Social reproduction, social 'problems' and social protection *Third World Quarterly, 34*(10), 1927–1941.

Zajak, S., Egels Zanden, N., & Piper, N. (2017). Networks of labour: Collective action across Asia and beyond. *Development and Change, 48*(5), 899–921.

Zlotnik, H. (2003). *The global dimensions of female migration*. Migration Information Source. Retrieved from http://www.migrationinformation.org/feature/display.cfm?ID=109.

Petra Dannecker is Professor of Development Sociology and Head of the Department of Development Studies at the University of Vienna. Before coming to Vienna she was assistant professor at the Department of Sociology at Bielefeld University, Germany and Senior Research Fellow at the German Development Institute in Bonn responsible for coordinating research and knowledge transfer between the Federal Ministry for Economic Cooperation and Development (BMZ) and scientific communities. She is working on issues related to development sociology, development politics, globalization and migration processes, transnationalism and gender and migration focusing regionally on South and Southeast Asia.

Nicola Piper , a political sociologist, is Professor of International Migration at the University of Sydney (and the Founding Director of the Sydney Asia Pacific Migration Centre). She is currently a British Academy Global Professor Fellow hosted by Queen Mary University of London's School of Law, to conduct research on global governance of labour migration and decent work for migrant workers (January 2019 to December 2022). Nicola is also partner investigator in the UKRI funded Global Challenge Research Hub *on South-South Migration and Inequality* as co-lead of the Gender Work Package (2019–2023). Her publications include the edited volumes *New Perspectives on Gender and Migration: Livelihoods, Rights, and Entitlements* (2008), *South–South Migration: Implications for Social Policy and Development* (with Katja Hujo, 2010) and the co-authored book *Critical Perspectives on Global Governance: Rights and Regulation in Governing Regimes* (with Jean Grugel, 2007). She is co-founder and co-editor of the Routledge book series *Asian Migration* and one of the chief editors of the journal *Global Social Policy*.

CHAPTER 20

Gender and Remittances

Rahel Kunz and Julia Maisenbacher

Abstract Since the migration and development promise has made its inroads into the debates of the global development community, remittances have become heralded as a key tool to promote development and reduce poverty. The gender dimensions of remittances have long been ignored but have come into focus with media and policy voices emphasising the empowerment potential of remittances for women. More recently, remittances have become linked to development financing through the international financial inclusion agenda, which has led to the financialisation of remittances (FOR). One key novelty of the FOR is its explicit emphasis on empowering women and reducing gender inequality through linking remittances to financial inclusion. This chapter focuses on the ways in which gender is implicated within the FOR by analysing its three main gender tropes: women sending and receiving remittances as financial customers and untapped resources, as entrepreneurs and as part of the 'remittance family'. We explore how these tropes form different subjectivities in deeply gendered, classed and racialised ways; discipline subjects through prescribed roles and norms and promote financial and economic forms of empowerment that silence other forms of empowerment.

R. Kunz (✉) · J. Maisenbacher
Université de Lausanne, Lausanne, Switzerland
e-mail: rahel.kunz@unil.ch

J. Maisenbacher
e-mail: Julia.Maisenbacher@unil.ch

C. Mora and N. Piper (eds.), *The Palgrave Handbook of Gender and Migration*, https://doi.org/10.1007/978-3-030-63347-9_20

Introduction

Since the migration and development promise has made its inroads into the debates of the global development community in the late 1990s, remittances—the money that migrant workers send to their families or communities of origin—have become heralded as a key tool to promote development and reduce poverty. Officially recorded remittance flows[1] to lower- and middle-income countries reached $529 billion in 2019 and constitute up to 35.2% of GDP in countries such as Tonga (World Bank & KNOMAD, 2019). An estimated 800 million people worldwide are supported by remittances sent by over 200 million migrants.[2] More recently, remittances have been increasingly linked to development financing through the global financial inclusion agenda, which has led to the financialisation of remittances (FOR) (Kunz, Maisenbacher, & Paudel, 2020). In the context of the FOR, remittances have become understood primarily in terms of financial logics, and various international, governmental, non-governmental, private and civil society actors promote their formalisation, digitalisation and securitisation. Remittances senders and receivers are targeted with financial education programmes and new financial services and products to promote the 'productive' investment of remittances, strengthen financial infrastructure and spur development.

Earlier forms of the remittances-based development agenda mostly ignored gender equality but implicitly drew on, and reproduced, gender dynamics (Kunz, 2011). One key novelty of the FOR is its explicit emphasis on empowering women and reducing gender inequality through linking remittances to financial inclusion. Yet, in the emerging literature analysing the implications and different forms of the FOR its gender dimension has not yet received much attention (Datta & Guermond, 2019; Guermond, 2019; Hudson, 2008; Zapata, 2018). Most studies dealing with the gender dimensions of remittances are primarily interested in 'typical female' sending and investment behaviour. They ask how women receive, send and invest remittances using (digital) financial services and whether that empowers them economically. For example, an influential and widely quoted study asks to what extent digital financial services allow remittance receiving women to move into business and out of agricultural production (Suri & Jack, 2016). In addition, a wide range of market studies analyse what type of products suit supposedly typical female sending behaviour and how remittance sending and receiving women can be provided with digital financial service products (see for example GSMA, 2013; Totolo & Gubbins, 2018; UNCDF, 2017).

[1]Official numbers do not include 'informal' remittances, thought to be substantial. Yet, the increase in official remittances figures is also influenced by the increased capacity to register such flows.

[2]https://refugeesmigrants.un.org/sites/default/files/stocktaking_ifad.pdf (all websites were accessed in March 2020).

Most of these studies take a behaviouralist approach, treating gender as a variable to investigate how it affects remitting behaviour in relation to development, poverty reduction or economic growth. Instead, in this chapter, we draw on feminist IPE insights to analyse the social dynamics underlying the FOR. We ask: In what ways is the FOR constituted by gendered, class-related and racialised power relations and what are its implications? We analyse three dominant gender tropes of the debates surrounding the FOR in order to elucidate their implications: women who send and receive remittances have been portrayed as financial customers and untapped resources, as entrepreneurs and as part of the 'remittance family' (see also Spitzer, this volume).

This chapter draws on the analysis of documents published after 2008 (around which year the FOR emerged), which deal with remittances, gender equality, financial inclusion, digitalisation and development financing. The corpus includes key documents by institutions involved in the FOR: international organisations, international financial institutions, private actors and civil society organisations. To complement the document analysis, we conducted twenty expert interviews between 2018 and 2019 with officials from international financial institutions, NGOs, UN agencies and private actors such as fintech start-ups.[3] To provide some illustrations from one particular context, we also draw on insights from fieldwork carried out in Kenya in 2019.[4]

The chapter begins with an account of how the FOR emerged out of the earlier remittances-based development model, the Global Remittances Trend (Kunz, 2011). We present its key novelties and analyse how gender is implicated within the FOR. Part two analyses the three main FOR gender tropes. We explore how they form different subjectivities in deeply gendered, classed and racialised ways; discipline subjects through prescribed roles and norms and promote financial and economic forms of empowerment that silence other forms of empowerment. The chapter concludes with some avenues for further feminist research on remittances.

Gender and the Financialisation of Remittances (FOR)

When the international development community became interested in remittances in the late 1990s with the emergence of the Global Remittances Trend (GRT), gender equality and women's empowerment were largely absent from the debates and most early academic, media and policy publications on remittances did not include references to women or gender. Interest in the gender

[3]Financial technology (Fintech) refers to technology that seeks to automate and transform the delivery and use of financial services, e.g. through specialised software and algorithms used on computers smartphones.

[4]We would like to thank our interview partners. All interviews have been anonymised for confidentiality. The fieldwork in Kenya was carried out by Julia Maisenbacher.

dimensions of remittances emerged in the early 2000s, notably with UN-INSTRAW's research and policy work on gender and remittances, and with the first official session on gender at the *International Forum on Remittances* in 2005 (Kunz, 2011). Since then, publications by international governmental and non-governmental organisations and private institutions on the gender dimensions of remittances have multiplied (Bourguignon, 2006; De Bruyn & Kuddus, 2005; Morrison, Schiff, & Sjöblom, 2007; United Nations Population Fund, 2006). In these early documents, two gender tropes dominated the debate (Kunz, 2011, 2018). First, women were understood mainly as passive remittance receivers, whereas men were referred to as (active) remitters, based on the binary gendered assumption that associates 'active' men with migration and 'passive' women with 'staying at home'. Linked to the so-called 'feminization of migration'[5] and growing empirical evidence of women's roles as remitters, their positive contributions to development received increased attention in the international policy discourse, which led to the second trope that portrays women as better remitters and heroines (Kunz & Schwenken, 2014). This trope asserts that women remit a higher percentage of their income than men, send remittances over longer time periods, channel remittances into health and education purposes rather than consumption and are more strongly motivated by altruism, in contrast to men who tend to be cast as more self-interested remitters. In this context, women migrants became the new 'development agents' par excellence (Lee & Piper, 2017) and their role as mothers and primary care takers was naturalised (Kunz, 2015).

Emerging after the global financial crisis in 2008 and against the backdrop of stagnating official development aid, the FOR evolved out of the GRT (Kunz, 2011). The FOR involves a wide range of international (financial) institutions—such as the World Bank, Inter-American Development Bank, G8/20, OECD, IFAD, IOM, ILO, UNCDF, UN Women—as well as private and some civil society actors. This group of institutions started to promote the financial inclusion of migrant workers as an 'innovative financing' mechanism for development (Ketkar & Ratha, 2009: 1). A high-level governmental initiative promoting the FOR, the Global Partnership on Financial Inclusion (GPFI) founded by the G20 in 2010, defines remittances as an 'important pathway to financial inclusion' (GPFI, 2016: 1). The FOR is characterised by the promotion of the integration of remittances into the formal financial system for 'productive investment', digitalisation and securitisation of remittances flows, and linking remittances to financial inclusion and financial education agendas.

One key novelty of the FOR, which has not received much attention in the literature so far, is its explicit focus on linking remittances and finance to promote the economic empowerment of women and gender equality. This

[5]This term has been used mostly to refer to the quantitative increase in women migrants as well as qualitative changes in migration flows due to the increase in women migrating independently.

is embedded in the broader global 'gender as smart economics' and financial inclusion agendas, which promote the mobilisation of women's labour power and their financial inclusion in order to decrease poverty and increase economic growth (G20, 2017: 11). The emphasis on women's empowerment and gender equality within the FOR manifests in various ways For example, FOR actors have launched a variety of initiatives that specifically target migrant women with financial inclusion, such as the United Nations Capital Development Fund (UNCDF) programme entitled Shaping Inclusive Financial Transformations (SHIFT). This programme encourages low-income people sending remittances to use formal financial services with a 'special focus on women' (UNCDF, 2017). Thereby, migrant women are encouraged to participate in financial education programmes and to use financial services in a 'productive' way.

Linked to this is the promotion of digitalisation and financial innovation related to remittances as a tool for women's empowerment and for closing the gender gap in the use of financial services (Kunz, Maisenbacher & Paudel, 2020). In comparison with the GRT, the FOR is intensely shaped and supported by financial corporations and Fintech start-ups. Philanthropy foundations, public private partnerships and financial corporations play a crucial role in assisting development agencies and governments in the developing world in designing policies that promote the financial inclusion of women sending and receiving remittances through the provision of gender-sensitive data and policy recommendations. In their understanding, 'poor women' are an untapped resource whose potential needs to be unlocked through financial services (GSMA, 2013). They assume that once 'poor women' are provided with suitable digital financial education, services and products they will be economically empowered and lifted out of poverty, as Claire Scharwatt, Director of Policy and Advocacy at GSMA, assures in a blog post entitled 'Mobile money: A product of choice for women to send and receive remittances' (Scharwatt, 2019).

A final element of the FOR focus on women's empowerment and gender equality is its emphasis on the family[6] (Kunz, Maisenbacher, & Paudel, 2020). In 2015, the Global Forum on Remittances, Investment and Development (GFRID) declared June 16 as the *International Family Remittance Day* (IFAD, 2018: 3), an initiative supported by the UN General Assembly (UN General Assembly, 2018). In 2019, an expert roundtable entitled 'Women's financial inclusion and the role of remittances' was organised during this day to discuss women's financial empowerment, development and 'the active role

[6]The family currently receives sustained attention within the international community, as illustrated in the Global Family Initiative (see http://www.globalfamilyinitiative.org or the 2019 title of UN Women's flagship report 'Progress of the World's Women 2019–2020: Families in a changing world'.

of migrant women in financial savings management'.[7] This FOR emphasis on women's empowerment and gender equality has led to the emergence of three main gender tropes.

Gender Tropes in the FOR

Existing studies on the links between remittances and finance tend to take a behaviouralist approach, treating gender as one variable among others shaping people's behaviour linked to remittances (López-Anuarbe, Cruz-Saco, & Park, 2016; Mahapatro, 2017; Paerregaard, 2015; Pfeiffer, Richter, Fletcher, & Taylor, 2008). Instead, in our analysis, we draw on feminist IPE to focus on the ways in which global agendas such as the FOR are constituted by gendered, classed and racialised power relations (Allon, 2014; Calkin, 2015; Elson, 2010; Griffin, 2013; Joseph, 2013; Montgomerie & Young, 2010; Pollard, 2013; Young, Elias, & Roberts, 2018). For this purpose, we analyse the gender tropes that dominate the debates surrounding the FOR. These tropes are narratives about how poverty can be reduced, empowerment can be achieved and economic growth must be promoted. They are not mutually exclusive, but complementary and co-existing.

These FOR gender tropes are embedded in two broader global agendas: the 'gender as smart economics' and the financial inclusion agenda. The 'gender as smart economics' agenda started before, and was reinforced after, the global financial economic crisis and is promoted most prominently by international financial institutions and private financial corporations, such as the World Bank and Goldman Sachs, but also by development agencies (Calkin, 2015; Chant & Sweetman, 2012). In this agenda, women are portrayed as rational and utility-maximising economic actors who—empowered through using financial services—move into business and lift themselves and their families, communities and countries out of poverty. The financial inclusion agenda aims to extend financial services to those traditionally excluded from the mainstream financial system, in order to bank the unbanked (Rankin, 2013; Weber, 2004). Thereby, the expansion and deepening of finance is understood as a way to lead the global economy out of recession while simultaneously addressing the problem of gender inequality. Financial inclusion is linked to broader global tendencies of the financialisation of everyday life, i.e. the increasing importance of finance in people's lives and the growing role of debt in financing social reproduction (Rankin, 2013: 549).

Women Senders and Receivers as Untapped Resource and Customers

A first gender trope in the FOR portrays women who send or receive remittances as untapped resources. Women are represented not only an 'untapped' resource in the sense that their labour power is supposedly not yet fully

[7] https://www.ifad.org/en/web/latest/event/asset/41195068.

exploited to promote development, they are also considered as an underserved market segment. This customer group is assumed to behave in particularly gendered ways, hence financial actors aim to develop products tailor-made for the supposedly typical sending and receiving behaviour of women—e.g. sending small amount of money, regularly and over a long period of time to be invested in reproductive necessities (e.g. food, school fees and housing/shelter).

This trope is promoted by various international financial and development institutions that consider the provision of financial products for remittance sending and receiving women as key to advance their financial inclusion and development more generally. For example, a UN Women research paper considers remittances as a main driver for financial inclusion since formal remittances institutions offer seemingly cheaper alternatives for remittances senders (Hennebry, Holliday, & Moniruzzaman, 2017: 28). In a report entitled *Remittances as a driver of women's financial inclusion in the Mekong region,* the UNCDF recommends to 'support customers, especially women, to adopt formal remittance channels and remittance-linked financial products' (UNCDF, 2017: 7). Another case in point are the initiatives of the Alliance for Financial Inclusion (AFI)—a network of central banks and financial regulatory institutions. Its creation was endorsed by the G20 and supported by the Gates Foundation (Gabor & Brooks, 2017: 1). In a guideline note entitled 'Innovative Cross-border remittances service', AFI presents a variety of policy measures that governments can adopt to promote the financial inclusion of women senders and receivers (AFI, 2018). Among these features research on 'sex-disaggregated data, which can help regulators identify necessary policy or regulatory actions and enable (financial service providers) FSPs to improve the development and marketing of products to women' (AFI, 2018: 9).

This trope also suggests that remittance-based financial inclusion should be linked to financial innovation and digitalisation. According to this trope, digital financial services are particularly beneficial for women since they are often excluded from traditional banking while the gender gap of mobile financial services users is low. Linking remittances sending and receiving activities to digitalisation and in particular mobile financial services is expected to empower women economically. For example, UN Women highlights the role of digital remittance transfers whose rise 'has also resulted in greater engagement with formal transfer processes, signifying a move toward greater financial inclusion' (Hennebry et al., 2017: 28). This development is considered as most beneficial for 'those with the lowest levels of financial inclusion, such as women, those in rural areas, and those in poverty' (Hennebry et al., 2017: 29). Another example is a recent FinAccess Survey (2019) in Kenya—often heralded as a role model in promoting the financial inclusion of remittances senders and receivers through technological innovation such as mobile money and digital credit. This Survey praises digital financial services as 'the optimal market-based solution in narrowing the gap in usage of financial services between male and female' (Central Bank of Kenya, 2019: 18). Much of the emphasis on mobile money derives from an influential article published in *Science* by Suri and Jack, which is widely quoted and acknowledged by policy-makers and financial institutions as the most compelling evidence that mobile money is reducing poverty and promoting the economic empowerment of women (Demirgüç-Kunt, Klapper, Singer, Ansar, & Hess, 2018; Rayiyan Kabir & Klugman, 2019; Scharwatt, 2019; Tiwari, Schaub, & Sultana, 2019).[8] The article investigates whether and how mobile money can lift remittance receiving households out of poverty. Based on a household survey and an analysis of the consumption patterns of digital financial services users, the authors suggest that using such services empowers remittance receiving women, as 'mobile money allows women to directly access remittances and or have more agency' (Suri & Jack, 2016: 1291). Others also suggest that digital financial services increase the likelihood that women receive remittances directly, which supposedly gives them 'much more control over household budgets and increases the likelihood of greater economic empowerment and decision-making' (Scharwatt, 2019).

In this trope, women are assumed to be more inclined to lack official documents necessary to access mobile banking services and formal remitting channels. Thus, the digital finance industry is lobbying for digital identity systems in order to promote gender equality. AFI recommends 'creating a digital identity system could also enhance the ability of women to access digital international remittance services in some markets, as research has found that women are less likely to have the official documents required to open mobile money accounts other financial services' (AFI, 2018: 16). It is also assumed that women tend to be less technically and financially literate than men and have less confidence in their ability to use mobile money and other digital financial services. Women, particularly in rural areas, are represented as having 'lower levels of literacy and education than men' and digital financial literacy is promoted as playing a key role 'to overcome women's risk adversity, improve ease of use and address their lack of trust in the banking system' (UNCDF, 2017: 34). Thus, for example, the GSMA has set up a special 'Connected Women Team' that developed the Gender Analysis and Identification Tool (GAIT) based on a machine learning algorithm to explore the usage behaviour of men and women in order to 'assist operators in reaching underserved female customers with relevant and tailored products and services' (GSMA, 2018: 3). UNCDF recommends that in order to 'facilitate the evolution of remittances into a portfolio of financial services, they (providers) will need to develop products and services appropriate to the needs of women migrants and women remittance receivers (…) Such financial services should speak to the current financial service preferences of senders and receivers' (UNCDF, 2015: 23). Thereby, women are also portrayed as particularly loyal and valuable customers that can provide stability of profits for financial actors. For

[8]For a more critical reading see for example (Bateman, Duvendack, & Loubere, 2019).

example, a recent report published by GSMA, VISA and the Bankable Frontier Association emphasises: 'Mobile operators can achieve scale and stability if they build and maintain the women's segment of the MFS (Mobile financial service) market (…) When their needs are met consistently, women can be very loyal and evangelizing customers, providing both scale and stability for core mobile financial services business' (GSMA, 2018: 1). Here, we can see how women's financial empowerment is no longer an objective in itself but becomes instrumentalised by financial institutions for the provision of financial stability and profit. This is embedded in broader current efforts to deepen financial markets through the targeting of 'poor women' and by broadening financial markets to include dimensions of social reproduction.

Women Senders and Receivers as Entrepreneurs

Building on the first trope, the second trope is based on the idea that increased access to remittance-linked financial services, and in particular digital services, will transform women sending or receiving remittances into entrepreneurs (see also Spitzer, this volume). This move into business is presented as the primary poverty reducing factor: 'financial inclusion helped them to graduate from subsistence agriculture and to reduce their reliance on multiple-part time occupations' (Suri & Jack, 2016: 1291). Most prominently, the above cited study by Suri and Jack reaches the conclusion that using digital financial services encourages remittance receiving women 'to switch into business or retail as their main occupation' (Suri & Jack, 2016: 1289). Female entrepreneurs are expected to invest digital credit in small scale enterprises or retail in the informal sector (Suri & Jack, 2016; Wandibba, Nangendo, & Mulemi, 2014). Their entrepreneurial activities are considered to automatically spur economic growth and reduce poverty in their family, community and country.

This entrepreneur trope assumes that women act as utility maximiser and rational economic actors who move into business as self-reliant and independent market participants, competing in a gender-neutral marketplace. This process is equated with empowerment and development. The only thing that could prevent women from benefiting from this economic empowerment is their supposed inclination to risk-adverse behaviour and their financial illiteracy which has to be overcome through financial education programmes and marketing strategies, as seen above. For example, AFI encourages financial service providers to consider outreach and marketing strategies targeted specifically at women and girls (AFI, 2018: 17). Such strategies can build the client base, increase women's confidence in their abilities and trust in providers and, if designed appropriately, can help women make more informed decisions about investing remittances.

This FOR trope echoes global development discourses regarding the entrepreneurial woman and the woman development agent. While the entrepreneurship literature traditionally focused mostly on male entrepreneurs, there is now an increasing interest in women entrepreneurs, and more specifically in women entrepreneurs as development agents (Calkin, 2015; Roberts, 2015; Roberts & Mir Zulfiqar, 2019; Roy, 2012). This global agenda is driven by a 'belief in the mutually reinforcing nature of women's entrepreneurship, economic growth, and empowerment' (Roberts & Mir Zulfiqar, 2019: 413). Mostly, the emphasis is on how to 'fix the problem' of women entrepreneurs, to bridge the entrepreneurial gap between men and women and to make women become like male entrepreneurs. Women's entrepreneurship is seen as a new development mantra and the entrepreneurial woman is considered the ideal subject of the 'smart economics' and the financial inclusion agenda. Thus, this FOR trope ties into the global preoccupation with women's entrepreneurship, that 'works to reaffirm and extend the neoliberal agenda of market deepening, not least through various 'financial 'inclusion' strategies' along gendered lines (Roberts & Mir Zulfiqar, 2019: 411). Yet, this trope problematically obscures the fact that most of the small scale and retail business run by women on the basis of short-term digital credit are in the informal sector that is characterised by high gender wage gaps and precarious working conditions. What is left out are structural issues that disadvantage women entrepreneurs in the first place and this approach also justifies neoliberal reforms (Weekley, 2006).

Women in the 'Remittances Family'

The third trope is based on the FOR emphasis regarding the importance of the family. This is illustrated in the creation of new terms, such as the 'remittance family' (IFAD, 2017a) or 'family remittances' (Inter-American Development Bank, 2010). It was endorsed with the proclamation in 2015 of the celebration of June 16 as the *International Family Remittance Day* (IFAD, 2018: 3), later supported by a UN General Assembly Resolution (UN General Assembly, 2018). It was also echoed in the slogan 'one family at a time', which has become popular within the FOR: 'Now is the time to fully engage the ambition of about 200 million migrants who send money home and bring to scale their remittances and savings to help reach the SDGs by 2030: One family at a time' (IFAD, 2017b, 12). This trope promotes the notion of the 'remittance family' as a new entity to be advanced in order to reduce poverty and promote economic growth. This is summarised as follows by an IFAD publication:

> By promoting financial inclusion, and moving towards account-to-account transactions, remittance families will be enabled to save and invest more. Private-sector companies targeting remittance families to become account holders, clients and investors will benefit their own bottom lines, while also contributing to local economies. (IFAD and Financing Facility for Remittances (FFR), 2018: 6)

Contrary to the first two tropes, this trope does not explicitly mention women's roles. Yet, implicitly, it presumes and normalises a model of the heteronormative transnational family where women play the role of the spouse in charge of social reproduction. Playing a key role in the remittance family, women's roles as caring mothers and loyal married women are considered increasingly important in promoting financial inclusion and acting as responsible agents of development.

The remittance family trope silences the plurality of women's relationships and life projects. It problematically reduces the family to the main characteristic of financial flows, defining family through money. The remittance family is portrayed as a collectivity in charge of decision-making around migration and remittances, implying that the money belongs to the family as a whole. As long shown by the feminist literature, this treatment of the (transnational) family as a black box is problematic because it glosses over gendered and other power relations within households (Bedford, 2009). Moreover, the invoking of positive emotive meaning associated with the family obscures the gendered economic, physical, psychological and emotional hardship and sometimes disintegration that the migration experience often brings (CARAM Asia, 2010; Kunz, 2011). This trope is also problematic because it draws on an ahistorical understanding of poverty and social reproduction and silences the underlying causes of poverty, reducing it to a lack of financial inclusion. Moreover, the strong focus on women as members of remittance families to promote development silences the responsibility of the state. This trope has to be analysed against the background of strict austerity programmes imposed on developing countries from the 1990s onwards. Decreased public investment in education, health care and childcare has increased the pressure on women to invest remittances in social reproduction.

Implications of the FOR Gender Tropes

From a feminist IPE perspective, the three gender tropes have a number of problematic implications. First of all, they tend to silence the social reproductive work that is often mostly done by women. All tropes are largely based on the assumption that women were 'untapped' or unoccupied before being empowered as entrepreneurs or customers. This problematically reproduces the productive–reproductive dichotomy and the masking of the reproductive sphere, as long criticised by feminist literature (Piper, 2011).

Secondly, these FOR gender tropes are based on an understanding of empowerment as economic and financial. Yet, it is questionable whether this empowerment increases the well-being of women since they might be confronted with a double burden of paid and unpaid work and the duty to manage severe indebtedness (Guérin, Morvant-Roux, & Villarreal, 2014). For example, ethnographic fieldwork from rural Kenya highlights that female entrepreneurs' income is often only supplementary, suggesting that men remain the key income-earners and women remain the primary caretakers responsible for the reproductive work on top of their business (Wandibba et al., 2014). Our own interviews in Kenya also reveal the multiple workloads and the burden of indebtedness that many women entrepreneurs face (Interview Fahamu, African grassroot NGO, Nairobi, February 2019). Presenting women's lack of financial inclusion and education as the main reason for their refusal to use digital finance to start businesses silences the strong impact of (lacking) public childcare service and social provisioning on the occupational choice of women (Mateo Diaz & Rodriguez-Chamussy, 2013). In addition, it also obscures the possibility that women choose not to become entrepreneurs or to stay outside formal financial mechanisms as an active form of resistance. The focus on financial and economic empowerment silences other forms of empowerment and can lead to the decline of support groups or initiatives for women's rights, delinking empowerment from broader social transformations (Roberts, 2015: 117).

Thirdly, FOR gender tropes reproduce particular class relations. It is problematic that women's entrepreneurial activities—based on the investment of debt-related financial services—are presented as innovation and emancipation, while the unequal relationships between capital owners and borrowers are obscured. Evidence suggests that the entrepreneurial activities that supposedly empower women often result in the need to cut back daily expenses to pay back loans and debts (Totolo & Gubbins, 2018). Moreover, existing research seems to suggest that women are especially vulnerable to debt cycles and over-indebtedness related to digital financial services (Hennebry et al., 2017: 26). Portraying women as customers and untapped resources and promoting the remittance family as an instrument for development fits with a general tendency to deepen financial markets through the targeting of 'poor women' and to enlarge financial markets to increasing dimensions of social reproduction. This is linked to a class project that sets ever more people and areas under the discipline of financial markets while enriching a small group of large financial corporations (Roberts, 2015).

Finally, FOR gender tropes reproduce racialised images of the 'poor Third World woman': there is a recurrence of racialised stereotypes that represent women in developing countries as traditional and without agency (Mohanty, 1984). As the literature shows, discourses of the good financial subject are racially encoded: they reconfigure 'longer standing narratives in which citizen 'fitness' is intertwined with racialized social hierarchies' (Meltzer, 2013: 649). Thereby, various 'character flaws' and 'traditional practices' of racialised population groups may be identified as obstacles to social and financial inclusion and modernity. According to Roy, the Third World woman victim in need of saving has been transformed into the 'heroic entrepreneur and selfless altruist', understood as the 'technosocial subject of millennial modernity', the 'good' financial subject who transforms risk into responsibility and poverty into enterprise (Roy, 2012: 143). This is echoed in the tropes that portray 'poor' women remittance senders and receivers as entrepreneurs and customers.

Often, the terms 'women', 'rural', 'poor', 'uneducated', 'remittance receiving' are conflated into what is understood as a marginalised 'underserved' group in need of financial education and inclusion. For example, the influential Global Findex report refers to women and low-income people as 'most likely to be financially inexperienced' (Demirgüç-Kunt et al., 2018: 90). Education is considered necessary to overcome women's reluctance to use financial services and take entrepreneurial risks, as if they had not been financially and entrepreneurially active before. As has been shown in other contexts, financial education programmes include 'gendered notions of morality and respectability' on how to become 'good financial citizens' (Loomis, 2018: 2). In this sense, financial education contributes to discipline behaviour, normalise debt, responsibilise individuals, thereby obscuring structural causes for poverty and invisibilising gendered and racialised discriminations in terms of access to finance.

The FOR gender tropes also contribute to silence the longstanding, well-developed and functional financial practices and experiences of women in many developing countries and discredit 'informal' financial mechanisms. Many societies have a long tradition of 'informal' lending and credit which is often dominated and controlled by women. In Kenya, for example, this is often referred to as 'table-banking' or *chama*, whereby women organise in small groups and put their earnings together to lend each other small amounts of money to be used for entrepreneurial activities. Such alternative forms of cooperative finance can be interpreted as resistance to the FOR. They are also promoted by some anti-debt activists as a way to fight increasing indebtedness of private households in developing countries (Interview with a representative of CADTM Kenya, Nairobi, February 2019).

Conclusion

In contrast to the GRT, the FOR is characterised by an explicit emphasis on empowering women and reducing gender inequality through linking remittances to financial inclusion and entrepreneurship. Our chapter has analysed three main FOR gender tropes: women sending and receiving remittances are portrayed as financial customers and untapped resources, entrepreneurs and as part of the 'remittances family'. These tropes echo broader global 'gender as smart economics' and financial inclusion agendas. As such, the FOR reproduces gendered power relations regarding social reproduction and neoliberal market deepening and broadening agendas. Moreover, our analysis of the gender dimensions of the FOR also shows how women are naturalised in their role as primary caretakers and responsibilised as agents of development (see also Mahon, in this volume). This echoes preoccupations by scholars who diagnose the current development regime as characterised by a 'feminisation of responsibility and obligation' (Chant, 2006: 206). Future feminist research related to remittances and financialisation could focus on the context-specific ways in which the FOR manifests. We know that earlier attempts to link remittances to development worked out in very different ways depending on the contexts and this is presumably also true for the FOR. Future research could also investigate in more detail the ways in which remittances and financialisation intersect with gendered, class-based and racialised dimensions of current development regimes. Finally, more attention could be paid to the voices of the people who are involved in remittance sending, receiving and investing to better understand the complex ways in which they experience, negotiate and resist these global policy agendas.

Acknowledgement This chapter draws on findings from a research project on the Financialisation of Remittances (https://www.unil.ch/crhim/en/home/menuinst/recherche/recherche-en-cours/the-financialisation-of-remittances.html). Funding by the Swiss National Science Foundation (10001A_172945) is gratefully acknowledged. We would like to thank Lekh Nath Paudel for his comments on an earlier version of the chapter, and the editors Claudia Mora and Nicola Piper for their patience and constructive comments.

References

AFI. (2018). *Innovative cross-border remittance services: Experiences from AFI member countries*. Guideline Note No. 30.

Allon, F. (2014). The feminisation of finance: Gender, labour and the limits of inclusion. *Australian Feminist Studies, 29*(79), 12–30.

Bateman, M., Duvendack, M., & Loubere, N. (2019). Is Fin-Tech the new panacea for poverty alleviation and local development? Contesting Suri and Jack's M-Pesa findings published in science. *Review of African Political Economy, 46*(161), 480–495.

Bedford, K. (2009). *Developing partnerships: Gender, sexuality, and the reformed World Bank*. Minnesota: University of Minnesota Press.

Bourguignon, F. (2006). *Women on the move: Magnitudes, trends and impacts of the international migration of women*. UN, New York, September 14. http://www.mtnforum.org/sites/default/files/publication/files/5212.pdf.

Calkin, S. (2015). "Tapping" women for post-crisis capitalism: Evidence from the 2012 World Development Report. *International Feminist Journal of Politics, 17*(4), 611–629.

CARAM Asia. (2010). *Remittances: Impact on migrant workers' quality of life*. https://www.ilo.org/dyn/migpractice/docs/141/Remittances.pdf.

Central Bank of Kenya. (2019). *2019 FinAccess Household Survey*. https://www.centralbank.go.ke/wp-content/uploads/2019/04/2019-FinAcces-Report.pdf.

Chant, S. (2006). Re-thinking the "feminization of poverty" in relation to aggregate gender indices. *Journal of Human Development, 7*(2), 201–220.

Chant, S., & Sweetman, C. (2012). Fixing women or fixing the world? "Smart economics", efficiency approaches, and gender equality in development. *Gender & Development, 20*(3), 517–529.

Datta, K., & Guermond, V. (2019). Financialization and/of migrant labor. In *The Routledge international handbook of financialization*. London: Routledge.

De Bruyn, T., & Kuddus, U. (2005, January). Dynamics of remittance utilization in Bangladesh. *IOM Migration Research Series* 18. https://hiva.kuleuven.be/resources/pdf/publicaties/R1065.pdf.

Demirgüç-Kunt, A., Klapper, L., Singer, D., Ansar, S., & Hess, J. (2018). *The Global Findex Database 2017: Measuring financial inclusion and the Fintech revolution.* Washington, DC: World Bank. https://globalfindex.worldbank.org/.

Elson, D. (2010). Gender and the global economic crisis in developing countries: A framework for analysis. *Gender and Development, 18*(2), 201–212.

G20. (2017). *G20 Leaders*. Hamburg: Declaration: Shaping an interconnected world.

Gabor, D., & Brooks, S. (2017). The digital revolution in financial inclusion: International development in the Fintech era. *New Political Economy, 22*(4), 423–436.

GPFI. (2016). *2016 update to leaders on progress towards the G20 remittance target.* https://www.gpfi.org/sites/gpfi/files/documents/2016%20Update%20to%20Leaders%20on%20Progress%20Towards%20the%20G20%20Remittance%20Target.pdf.

Griffin, P. (2013). Gendering global finance crisis, masculinity, and responsibility. *Men and Masculinities, 16*(1), 9–34.

GSMA. (2013). *Unlocking the potential: Women and mobile financial services in emerging markets*. https://www.gsma.com/mobilefordevelopment/wp-content/uploads/2013/02/GSMA-mWomen-Visa_Unlocking-the-Potential_Feb-2013.pdf.

GSMA. (2018). State of the industry report on mobile money.

Guérin, I., Morvant-Roux, S., & Villarreal, M. (Eds.). (2014). *Microfinance, debt and over-indebtedness: Juggling with money.* Oxford and New York: Routledge.

Guermond, V. (2019). Financialising migration? Remittances, algorithms and digital finance. *Developing Economics. A Critical Perspective on Development Economics* (blog). 2019. https://developingeconomics.org/2019/03/22/financialising-migration-money-algorithms-and-digital-finance/.

Hennebry, J., Holliday, J., & Moniruzzaman, M. (2017). *At what cost? Women migrant workers, remittances and development.* UN Women.

Hudson, D. (2008). Developing geographies of financialisation: Banking the poor and remittance securitisation. *Contemporary Politics, 14*(3), 315–333.

IFAD. (2017a). *International Family Remittance Day endorsements*. IFAD. https://www.ifad.org/documents/38714170/40313865/International+day+of+family+remittances.pdf/ec2d0b5f-5a6d-49fe-b2d5-870330991832.

IFAD. (2017b). *Sending money home: Contributing to the SDGs, one family at a time.* IFAD.

IFAD. (2018). *International Day of Family Remittances. June 16. Endorsements 2018.* http://www.un.org/en/events/family-remittances-day/assets/pdf/IDFR%20_Endorsements%202018%20Brochure_final-17072018.pdf.

IFAD., & Financing Facility for Remittances (FFR). (2018). *RemitSCOPE: Remittance markets and opportunities, Asia and the Pacific.* RemitScope.Org. IFAD. https://www.ifad.org/documents/38714170/40286666/2018-05-01_RemitSCOPE_web_final.pdf/1c454276-4fd3-4feb-a8d8-24b4ef9353cf.

Inter-American Development Bank. (2010). *Ten years of innovation in remittances lessons learned and models for the future.* New York: Inter-American Development Bank. https://publications.iadb.org/bitstream/handle/11319/3652/Ten%20Years%20of%20Innovation%20in%20Remittances%3A%20Lessons%20Learned%20and%20Models%20for%20the%20Future.pdf?sequence=1.

Joseph, M. (2013). Gender, entrepreneurial subjectivity, and pathologies of personal finance. *Social Politics: International Studies in Gender, State & Society, 20*(2), 242–273.

Ketkar, S., & Ratha, D. (2009). *Innovative financing for development*. Washington, DC: World Bank.

Kunz, R. (2011). *The political economy of global remittances: Gender, governmentality and neoliberalism*. RIPE Series in Global Political Economy. London: Taylor & Francis.

Kunz, R. (2015). "Moneymaker and mother from afar": The power of gender myths. In *Women, gender, remittances and development in the global south* (pp. 207–228). Farnham and Burlington: Ashgate. https://www.taylorfrancis.com/books/e/9781315546629/chapters/10.4324/9781315546629-22.

Kunz, R. (2018). Remittances in global political economy. In R. Elias, A. Roberts, (Eds.), *Handbook on the international political economy of gender* (pp. 265–280). Cheltenham, Northampton: Edward Elgar.

Kunz, R., Maisenbacher, J. & Paudel, L. N. (2020). The financialisation of remittances: Governing through emotions. *RIPE Review of International Political Economy*. https://doi.org/10.1080/09692290.2020.1785923

Kunz, R., & Schwenken, H. (2014). MigrantInnen Als HoffnungsträgerInnen in Der Entwicklungszusammenarbeit? Geschlechterspezifische Subjektivitäten Im Migration-Development Nexus. *Politische Vierteljahresschrift, 48*, 328–356.

Lee, S., & Piper, N. (2017). Migrant domestic workers as "agents" of development in Asia. *European Journal of East Asian Studies, 16*(2), 220–247.

Loomis, J. (2018). Rescaling and reframing poverty: Financial coaching and the pedagogical spaces of financial inclusion in Boston, Massachusetts. *Geoforum, 95*(October), 143–152.

López-Anuarbe, M., Cruz-Saco, M., & Park, Y. (2016). More than altruism: Cultural norms and remittances among Hispanics in the USA. *Journal of International Migration and Integration, 17*(2), 539–567.

Mahapatro, S. (2017). Why do they remit? Examining factors influencing migrant remittances in India. *Journal of Development Policy and Practice, 2*(2), 225–235.

Mateo Diaz, M., & Rodriguez-Chamussy, L. (2013). *Childcare and women's labor participation: Evidence for Latin America and the Caribbean*. Inter-American Development Bank. https://publications.iadb.org/publications/english/document/Childcare-and-Women-Labor-Participation-Evidence-for-Latin-America-and-the-Caribbean.pdf.

Meltzer, J. (2013). "Good Citizenship" and the promotion of personal savings accounts in Peru. *Citizenship Studies, 17*(5), 641–652.

Mohanty, C. T. (1984). Under Western eyes: Feminist scholarship and colonial discourses. *Boundary, 2*, 333–358.

Montgomerie, J., & Young, B. (2010, February). *Home is where the hardship is: Gender and wealth (dis)accumulation in the subprime boom* (CRESC Working Paper Series 79). 19.

Morrison, A., Schiff, M., & Sjöblom, M. (2007). *The international migration of women*. Washington, DC: World Bank.

Paerregaard, K. (2015). The resilience of migrant money: How gender, generation and class shape family remittances in Peruvian migration. *Global Networks, 15*(4), 503–518.

Pfeiffer, L., Richter, S. Fletcher, P. & Taylor, J. (2008). Gender in economic research on international migration and its impacts: A critical review. In A. R. Morrison, M. Schiff, & M. Sjöblom (Eds.), *The international migration of women* (pp. 11–49). The World Bank.

Piper, N. (2011). Toward a gendered political economy of migration. In N. Phillips (Ed.), *Migration in the global political economy*. International Political Economy Yearbook Series, v. 17 (pp. 61–82). Boulder, CO: Lynne Rienner Publishers.

Pollard, J. (2013). Gendering capital: financial crisis, financialization and (an agenda for) economic geography. *Progress in Human Geography, 37*(3), 403–423.

Rankin, K. (2013). A critical geography of poverty finance. *Third World Quarterly, 34*(4), 547–568.

Rayiyan K., & Klugman, J. (2019). *Women's financial inclusion in a digital world: How mobile phones can reduce gender gaps*. Washington, DC. http://giwps.georgetown.edu/wp-content/uploads/2019/12/Womens-Financial-Inclusion-in-a-Digital-World.pdf.

Roberts, A. (2015). Gender, financial deepening and the production of embodied finance: Towards a critical feminist analysis. *Global Society, 29*(1), 107–127.

Roberts, A., & Mir Zulfiqar, G. (2019). The political economy of women's entrepreneurship initiatives in Pakistan: Reflections on gender, class, and "development". *Review of International Political Economy, 26*(3), 410–435.

Roy, A. (2012). Subjects of risk: Technologies of gender in the making of millennial modernity. *Public Culture, 24*(1), 131–155.

Scharwatt, C. (2019). Mobiel money: A product of choice for women to send and receive remittances. *IFAD Blog* (blog). https://www.ifad org/en/web/latest/blog/asset/41410919.

Suri, T., & Jack, W. (2016). The long-run poverty and gender impacts of mobile money. *Science, 354*(6317), 1288–1292.

Tiwari, J., Schaub, E., & Sultana, N. (2019). Barriers to "last mile" financial inclusion: Cases from Northern Kenya. *Development in Practice, 29*(8), 988–1000.

Totolo, E., & Gubbins, P. (2018). *Digital credit in Kenya: Evidence from a demand-side surveys*. Nairobi: FSD Kenya. https://fsdkenya.org/publication/digital-credit-in-kenya-evidence-from-demand-side-surveys/.

UNCDF. (2015). *Women on the move. Harnessing the economic force of cross-border mobility in ASEAN*. SHIFT Report. Asia Pacific Regional Office: UNDP. https://www.uncdf.org/article/813/women-on-the-move-economic-force-cross-border-mobility-asean.

UNCDF. (2017). *Remittances as a driver of women's financial inclusion in the Mekong region*. UNCDF. https://uncdf-cdn.azureedge.net/media-manager/75322?sv=2016-05-31&sr=b&sig=1vr1i%2F08UYoN%2FqAK1%2FsVS%2BbHNTQ2SC2XiO4uiMIx%2FsU%3D&se=2018-09-06T06%3A53%3A10Z&sp=r.

UN General Assembly. (2018). *Resolution Adopted by the General Assembly on 12 June 2018*. In: New York: UN. https://undocs.org/A/RES/72/281.

United Nations Population Fund. (2006). *State of world population 2006: A passage to hope: Women and international migration*. New York: UNFPA.

Wandibba, S., Nangendo, S., & Mulemi, B. (2014). *Gender empowerment and access to financial services in Machakos Country, Eastern Kenya*. Institute for Money, Technology and Financial Inclusion, 16.

Weber, H. (2004). The "new economy" and social risk: Banking on the poor? *Review of International Political Economy, 11*(2), 356–386.

Weekley, K. (2006). From wage labourers to investors? Filipina migrant domestic workers and popular capitalism. In K. Hewison & K. Young (Eds.), *Transnational migration and work in Asia* (pp. 193–212). Oxford and New York: Routledge.
World Bank., & KNOMAD (2019). *Migration and development brief 31.* Washington, DC: World Bank. https://www.knomad.org/sites/default/files/2019-04/Migrationanddevelopmentbrief31.pdf.
Young, B., Elias, J., &Roberts, A. (2018). Financialization, unconventional monetary policy and gender inequality'. In *Handbook on the international political economy of gender* (pp. 241–251). Cheltenham: Edward Elgar.
Zapata, G. (2018). Transnational migration, remittances and the financialization of housing in Colombia. *Housing Studies, 33*(3), 343–360.

Rahel Kunz is a senior lecturer at the Institute of Political Studies of the University of Lausanne, Switzerland. Her research draws on feminist poststructuralist and postcolonial theories and focuses on migration and development, and the politics of knowledge production. Rahel is the author of *The Political Economy of Global Remittances: Gender, Governmentality and Neoliberalism* (Routledge 2011) and co-edited *Multilayered Migration Governance: The Promise of Partnership* (Routledge, 2011) with S. Lavenex and M. Panizzon.

Julia Maisenbacher is a post-doctoral researcher at the Institute of Political Studies of the University of Lausanne, Switzerland. Her main research interests are gender and labour relations, critical political economy, the transnational dimension of EU competition policy and the marketisation of corporate control in peripheral and emerging countries. She has recently published "*Transnational Capital Unbound? A Critical Institutionalist Perspective on the Marketisation of Corporate Control in Serbia and Turkey*" in *Competition and Change* 22(3), 293–312 (2018).

CHAPTER 21

Human Rights in Households: Gender and the Global Governance of Migrant Domestic Workers

Jennifer N. Fish

Abstract This chapter chronicles the mutual creation of an activist migrant worker movement and the production of a comprehensive global policy that recognized the labour and voices of domestic worker organizations worldwide. Data from direct research with the United Nations' International Labour Organization offers an analysis of the most substantial movement of migrant workers as policy informers. Key social protection debates elucidate larger global conditions that illuminate larger struggles to develop standards for migrant labour in private households. By positioning the household as a site of global governance and policy formation, this chapter looks at the role of migrants in creating human rights protections. This pivotal policy victory for domestic worker rights is then imagined in a larger arc of efforts to realize social justice in everyday conditions of labour power relations, and their meaning for transnational migrants.

> Freedom is not something that will be given to us on a silver platter, it has to be consistently fought and guarded, less it be taken away and diminished by tyrants whose views and vision are different from our own vision of freedom, equality and human rights. Marcelina Bautista, domestic worker activist, International Labour Conference, 2011

J. N. Fish (✉)
Old Dominion University, Norfolk, VA, USA
e-mail: jfish@odu.edu

C. Mora and N. Piper (eds.), *The Palgrave Handbook of Gender and Migration*, https://doi.org/10.1007/978-3-030-63347-9_21

Introduction

The International Labour Organization galvanized a movement for migrant rights through the attainment of the world's first international policy standard for domestic work. In 2010, those who have served as the "backbone of society" accessed the international house of labour regulation, altered its process of policymaking, and played a pivotal role in the passing of Convention 189 on Decent Work for Domestic Workers (Blackett, 2019; Fish, 2017). This moment marked an historic victory for women workers—both in terms of its outcome and the politics of inclusion that set this standard apart from any other ILO regulation. The realization of this policy captures a landmark first moment when "actual workers" took part in the terms of human rights policymaking within the ILO. As Myrtle Witbooi, President of the International Domestic Workers Federation put it, "this is something the ILO has never seen before."[1] At the same time, the attainment of Convention 189 captures the most comprehensive effort to bring human rights policy into the household realm. This chapter chronicles the mutual creation of an activist migrant worker movement and the production of a comprehensive global policy that recognized the labour and voices of domestic worker organizations worldwide. It highlights key social protection debates in relation to larger global conditions that elucidate the struggles to develop standards for migrant labour in private households. By positioning the household as a site of global governance and policy formation, this chapter looks at the role of migrants in creating protective policy, and the particular points of conversation that emerged in the negotiations of this first set of global standards.

The collective organization of migrants, both nationally and globally, created a new space for those most directly impacted by policy to engage in setting the terms of their own protections. I explore the politics of policy inclusion through key processes and meanings that emerged in this first global effort to assure standards for domestic workers. My reflections draw from ethnographic data collected from within the ILO policymaking process and fieldwork alongside the world's only international domestic workers' organization for the past nine years. With a focus on the key social welfare protection debates, I analyze points of resistance and possibility in adopting global standards for migrant women read as "most vulnerable" in the global economy. From this conversation, I examine implications for related human rights struggles worldwide.

[1]2010 Public speech at the International Labour Conference, Geneva, Switzerland.

The Gender of Migration in Decent Domestic Work

Juan Somavía, the Director General of the ILO from 1999 to 2012, hailed the integration of policy debates on household labour in the informal economy as a reflection of his focus on supporting "decent work" and "fair globalization."[2] With "diplomats and domestic workers" coming to the same table to determine protections for this heavily exploited sector, the largest international labour standards organization recognized the pervasive reliance on unprotected paid household labour in the upholding of the global economy. As delegates from 187 countries joined together to co-create this first convention, they embarked on an historic moment that symbolized recognition of a quintessentially feminized history of work, ensnared in a legacy of colonialism, racism, and slavery (Blackett 2019; Boris, 2019; Boris & Fish, 2014). To debate domestic work meant that governments, alongside labour unions and employers, collectively recognized the contemporary residuals of injustice, as well as an escalating scale of dependency in the increased transnational migration. Yet, the colossal task of regulating the household through international policy demanded state recognition of the value of household labour, and new flexibilities in the ILO system of policy formation.

Domestic work grounds the global economy (Anderson, 2000; Chang, 2001; Hondagneu-Sotelo, 2001; Parreñas, 2001), global justice (Basok & Piper, 2013; International Domestic Workers Federation, 2019), and global governance (Boris, 2019; Piper, 2017; Piper, Rosewarne, & Withers, 2017). As a central source of informal labour, the private household space has become an emblematic site of globalization. The ILO estimates approximately 100 million domestic workers worldwide, over 90% of whom are women and girls (International Labour Organization, 2011). Migrant women leave their own children to care for the families of their employers, often in countries far from their communities and languages of origin (Olievera, 2018). As a global gendered commodity, paid household labour has become the largest occupation for young women worldwide.[3] Dependence on this form of transnational exchange has fuelled gendered and racialized geopolitical relations around the trade of care work between sending and receiving countries.[4] As Cinzia Solari asserted, "nothing signals 'Third World' in the international arena like the

[2]For a comprehensive account of these guiding principles under Somavía's first year of leadership, see the International Labour Organization's 1999 "Report of the Director-General: Decent Work."

[3]For comprehensive analyses of the global statistics on domestic labour, see International Labour Office. 2013. "Domestic workers across the world: Global and regional statistics and the extent of legal protection." *International Labour Organization*. Retrieved April 15, 2013. (http://www.ilo.org/public/libdoc/ilo/2013/113B09_2_engl.pdf).

[4]For a comprehensive analysis of the state's participation in the export of domestic worker, see Christine Chin's (1998) account of the relationship between the Philippines and Malaysia see, *In Service and Servitude: Foreign Female Domestic Workers and the Malaysian "Modernity" Project*. (New York: Columbia University Press).

mass emigration of women to do domestic labour abroad."[5] Yet, human rights and labour protections have not travelled with domestic workers.

The complexities of setting standards for household work stem from core ideological assumptions about the devalued nature of domestic labour in the private household sphere, where deeply feminized, racialized, colonial practices persist. Women of colour, 90% of whom are from the developing world, comprise this sector, making it a conduit for the reproduction of sharp divides between "maids and madams" (Cock, 1989). In many senses, migrant domestic workers present one of the greatest challenges to policy's reach. Isolated in private households, often without citizenship benefits, domestic workers have struggled to access education, economic empowerment, and basic human rights protections. These structural barriers set a bar to not only protect workers in the private household, but to overcome the symbolism of this neo-colonial institution through the moral achievement of inscribed global policy. With the attainment of Convention 189, the focus on household labour simultaneously recognized the value of women's labour in this historic and symbolic global institution.

The Global Movement for Inclusion

> We don't want to be considered slaves. We are workers like any other worker.
> Myrtle Witbooi, ILC 2010 address

The inclusion of domestic workers in ILO policy became revolutionary not only because of its focus on this formerly excluded group. The process by which domestic worker representatives themselves entered the policymaking house of the ILO and took part in the creation of their own package of rights protections marked an all-time first for this largest transnational labour institution. Domestic worker organization leaders and representatives gained access to the ILO through international NGOs, global unions, and policy institutes. Within the formal tripartite system, where governments, labour leaders, and employer bodies are assured equal representation, domestic worker representatives overtook the fourth "observer" seat within the policy debates. Their presence carved a place for civil society to take part in international policymaking. At the same time, domestic worker representatives' observation and strategic participation in the policy debates instilled a consistent moral accountability that asked decision makers to look at those who would receive the direct impact of the protections that emerged from these debates.

The process of getting into the ILO, as the largest group of "actual workers," reflects a strategy of collective mobilization across international human rights organizations, as well as a dual emphasis on workers' and

[5]Excerpt from dialogue at the 2015 Sociology of Development Conference at Brown University, March 14, 2015.

women's rights. Domestic workers gained access through two ally organizations–Women in Informal Employment: Globalizing and Organizing (WIEGO), a global policy-research institute, and the International Union of Food, Agricultural, Hotel, Restaurant, Catering, Tobacco and Allied Workers' Association (IUF), the only global union to take on the cause of domestic work as part of its own mission.[6] These organizations afforded access to the International Labour Conference—the annual ILO gathering where policy is debated in a *tripartite* format[7] over a two-year process. Even though they were not allowed to speak formally, domestic workers fed their experiences into government delegates and the chair of the worker group to humanize policymaking and assure that those most directly impacted by the discussions carried a tangle and prominent place in the negotiation room.

To get to this table, domestic worker organizations drew from two years of transnational organizing to build capacity and align interests and causes across five continents. In 2008, the International Domestic Workers Network formed in Amsterdam, under the aspirations of linking causes and building a global movement for awareness and rights. Part NGO, part union in infancy, this organization's global collective mobilization around ILO Convention 189 became the largest coordinated social-political movement to address the economic gender disparity and neo-colonial expressions of paid household labour worldwide. With "one foot in the labour movement and one foot in the women's movement,"[8] the "decent work for domestic workers" campaign mobilized dual concerns for economic and gender equity through the prominence of UN-level attention. As Ai-Jen Poo, Director of the US National Domestic Workers Alliance contended, "movements matter."[9] The foundation of an international domestic worker organization linked the capacity of social movements to tangible policy negotiations at the transnational and symbolic UN level.

Just as the International Domestic Workers Network set up its structure and solidified its cause, the ILO announced the negotiations for Convention 189 would take place in the 2010–2011 International Labour Conferences. The possibility of gaining global policy protections centred the mission of the International Domestic Workers Network around a very tangible common claim to

[6]Dan Gallin spearheaded this movement, as the leader of the IUF. For a more comprehensive history of these leaders' roles see Fish (2017) *Domestic Workers of the World Unite!*

[7]The ILO is the only tripartite organization within the United Nations. This overarching ideology and practice assure that social dialogue, organizational practice and policymaking include worker and employer delegates from "peak" national associations and government representatives from all member states. The organizational composition, processes, and all of its outputs reflect this balance among the three parties.

[8]Marty Chen, founder of WIEGO, originally made this statement at the 2010 ILC meetings among domestic worker representatives.

[9]This insight is drawn from a personal conversation with Ai-Jen Poo at the 2011 ILC meetings in Geneva.

rights. The ILO served as a critical ground to galvanize national domestic worker movements around a common plight for global human rights and decent work. Bolstered by new strategies of activism and a transnational public presence, the International Domestic Workers Network (IDWN) embodied this sector as "decent," worthy of human rights, and deserving of global policy protections. By putting their voices on the table, in an organized activist form, the International Domestic Workers Network infiltrated the world's largest labour institution and demanded a space for the experiences of those most directly impacted by the existing informal migrant economy.

This collision of a global movement in formation and the sliver that opened the ILO potential to include domestic workers in the informal care economy manifested an historic first. On 16 June 2011, government, employer, and labour delegates to the International Labour Conference (ILC) voted nearly unanimously to adopt Convention 189, on Decent Work for Domestic Workers. As Luc Demaret, one of the ILO's internal agitators contended, this moment marked "a little step for the ILO, a bit step for humanity."[10] When the ILO house announced the final vote, IDWN leaders took over the Salle des Nations, the largest venue within the United Nations, by dropping a massive banner that read "Congratulations ILO, Now the work is up to governments, implement, ratify" (Fish, 2017; Fish, Crockett, & Ormiston, 2012). While their voices had been curtailed by the formality of the ILO process throughout the formal negotiations, with the announcement of the vote, domestic workers chanted, cheered, and sang as if they were in a street protest. "Up, up domestic workers, down, down with slavery" and "My mother was a kitchen girl, my father was a garden boy, that's why, I'm a unionist!" This moment encapsulated the intersection of global policymaking as a site of transnational women's coalition building and human rights activism within the halls of the ILO. The process catapulted a human rights movement by bringing clarity to domestic workers' next stage of global organizing in efforts to expand transnational activism and implement the actual protection terms debated in Geneva.

Formalizing Care Work: The Substance of Social Protections

Convention 189 set standards that universalized domestic work by embedding distinct fair labour practices—regardless of location. The rights-based reframing of "domestic work as real work" generated a series of debates—in the collective imagination of paid household labour, the places in need of protections, and the actual processes that assure basic human rights. Resistance to the formalization of domestic work came from employers who suggested that the sector could "never be regulated," holding to a notion that a domestic work convention would only be "on paper" and impossible to implement. Yet IDWN representatives emphatically promoted the notion of domestic work

[10] ILC Fieldnotes, 2010

as "real work," with the right to protections "just like any other work." Their presence—in their physical and rhetorical bodies—forced policy makers to look at the impact of the history of exclusion and the ultimate outcome of their decisions.

The initial terms outlined in Convention 189 reflect the prevailing frameworks so readily associated with domestic work's history of servitude, colonialism, and slavery. For example, the very need to put in place basic conditions of employment reflects the history of power relations that left domestic workers in live-in, unregulated work, without contracts, the assurance of paid maternity, holiday and sick leave, or retirement protections. Thus, the opening articles in Convention 189 require that "Each Member shall take measures to ensure that domestic workers are informed of their terms and conditions of employment in an appropriate, verifiable and easily understandable manner and preferably, where possible, through written contracts in accordance with national laws, regulations or collective agreements." To put in place a simple contract of employment requires a recognition of the value and need to regulate domestic work—in hours, accumulation of benefits, and long-term protections. Without such assurances in writing, prevailing power relations leave domestic workers vulnerable to exploitation and labour abuse. As one domestic worker leader contended, "we need contracts to assure our lives." Domestic workers came to the ILO negotiations with demands for the basic terms of employment provided to "all other workers." In the items negotiated for each particular article of the Convention, the virtual absence of prevailing protections becomes clear.

The foundation of the notion of policy protections for this sector hinged on core debates between this work as "just like any other" versus that which should be considered "special" because of its distinct nature in the private household. To regulate domestic work meant to place it in the same category as any other paid sector, bringing the household under the same umbrellas as labour in the public sphere. At the same time, domestic worker representatives and allies repeatedly brought up notions of the highly particular nature of the sector—where women are awoken in the middle of the night to care for children, hours are determined by daily care needs, and payment sometimes takes the form of used clothing and monthly food rations. To pass a policy for protections, domestic worker advocates walked a line between pushing for universal decency in domestic work "like any other" and a moral, emotional plea that asked policy makers to make up for the past abandonment of this sector and consider the women—often framed as a symbolic generalized "mother figure"—at the heart of these protections. "Think of your mother when you vote for this convention" and consider this work "just like any other" they demanded. These hybrid demands played out in every single term of the convention.[11] As Article 10 in the final document reveals, "normal hours of work, overtime compensation, periods of daily and weekly rest and paid annual leave in accordance with national laws, regulations or collective agreements, taking into account the special characteristics of domestic work." Even in the final language of the instrument, we see the fusion of "normal" and "special," reflecting a direct example of the particularity and complexity of this sector.

Crafting Migrant Rights and the Terms of Inclusion

Let us turn to three key examples in the text of the adopted Convention to discuss the core complexities in reaching a social welfare policy for this largest sector of migrant women workers worldwide. Of the 27 articles, these three capture some of the most heavily debated terms, as a symbolic reflection of the challenges of reaching human rights for this formerly forgotten sector.

Living In

Under the historical terms of servitude and slavery, domestic work has most often come with the expectations of live-in labour. This requires workers to often lose their privacy, leave their families, and work extended hours in a constant "on-call" labour relationship. Thus, the demand for human rights asked policy makers to separate labour from live-in terms of servitude. As ILO delegates debated a domestic work convention, the regulation of live-in labour became one of the most complex and heavily contested terms, where a history of servitude met an aspiration for universal human rights.

Article 9 of the convention addressed the nature of potential live-in work as follows:

Domestic workers

(a) Are free to reach agreement with their employer or potential employer on whether to reside in the household;
(b) Who reside in the household are not obliged to remain in the household or with household members during periods of daily and weekly rest or annual leave; and
(c) Are entitled to keep in their possession their travel and identity documents.

These three conditions speak to an entire history of exploitation. In the worst scenario, domestic workers lose access to their own identity documents, allowing for some of the most extreme forms of exploitation. Given this larger

[11] Eileen Boris originally named this notion of a hybrid set of demands for domestic workers. For more analyses, see *Making the Woman Worker: Precarious Labour and the Fight for Global Standards, 1919-2019*. Oxford: Oxford University Press.

context, domestic worker representatives demanded that the nature of such live-in and on-call requirements merits "real work," particularly given the "lack of freedom" they experienced, along with the constant invasion in their personal lives. Unlike any other job, live-in domestic labour presupposes the seeming limitless availability of women's work. It sets up a notion that women are constantly available, even when technically "off duty." As many domestic worker representatives suggested, "there is no off time," reminding the ILO house that "we are always on call" to the demands of employers.

These live-in conditions fed related policy protections on rest from work. Convention 189 assured that even when domestic workers live at their place of employment, they must be able to take protected rest time. As the policy states, "Weekly rest shall be at least 24 consecutive hours." Thus, a Thursday afternoon "maid's day" does not suffice for the minimum number of hours required to assure proper rest. Even though domestic workers often live at their place of work, this policy recognized the need for a period of break that would allow for adequate rest. Convention 189 also recognized the frequent practice of domestic workers' inability to determine and rely upon a regular day of rest—limiting their capacity to plan non-work activities, engage in meaningful social relationships, and enrich their lives beyond their place of work.

Among the most foundational forms of labour injustice, domestic workers' time is often so ambiguously defined that it creates a constant on-call scenario. This made counting hours of work difficult in the eyes of policy makers who sought to regulate live-in care labour. As one government delegate suggested, "surely when the worker gets up in the middle of the night to change the nappy, we cannot consider that hours of labour. That is simply the nature of the job."[12] The domestic worker representatives pushed back fervently on this notion, in efforts to make policy a tool to turn social power relations and change perceptions in ways that valued workers' time, throughout all terms of employment relations. The health and well-being of domestic workers came to the forefront of the dialogue, with a demand that interruption in sleep, perpetual on-call expectations, and extended night hours takes a tremendous toll on workers' physical rest and overall health. These debates proved to be one of the most contentious policy negotiations. To recognize on-call and live-in expectations, and to regulate these practices "just like any other," meant that domestic workers would be assured social and labour protections in line with those standard forms of employment in the public sphere. In the end, Convention 189's recognition of on-call labour requirements read as follows:

> Periods during which domestic workers are not free to dispose of their time as they please and remain at the disposal of the household in order to respond to possible calls shall be regarded as hours of work to the extent determined by

[12] Excerpt for policy negotiation ethnographic observations, 2010.

> national laws, regulations or collective agreements, or any other means consistent with national practice. (ILO C189, Domestic Workers Convention, Article 10.3)

As this text reflects, state practices become the tool to assure consistent implementation here. This requires that national regulations support the practices of recognizing on-call labour. This particular dimension of the policy points to the vital relationship between national laws and the more aspirational intentions of the Convention. Without compliance under the assurance of national laws, the notion of a transnational set of protections is rendered rather meaningless—a complexity that emerged throughout virtually all aspects of the debate.

The final policy terms on live-in labour expectations point to an underlying attempt to assure domestic workers' agency, such as in the notion of "choosing" to reside in the place of employment. Furthermore, these terms guarantee that domestic workers can easily navigate the geographies outside of their employers' residence. When domestic workers are employed in isolated suburbs, high-crime contexts, or countries where they do not enjoy citizenship, such ease and flexibility in so-called "off time" may be more illusionary than practical. Furthermore, the notion of "reaching agreement" with employers and potential employers assumes that the deeply embedded power relations can be overcome with the attainment and use of policy protections. When economic disparities, high levels of unemployment, and neo-colonial relations continue to construct the landscapes of migrant employment, the tools of protection intended in this policy are most often less accessible. Thus, through the lens of live-in regulation, we see that the victories of policy protection recognize the particular injustices fed through this sector. In practice, unequal power relations pose a serious challenge to implementation. Yet, the naming of the need to regulate domestic labour's live-in terms recognizes its distinct history and potential for extreme exploitation.

Kind vs. Fair Payments

Among the related forms of neo-colonial practices, domestic work's devalued nature has taken historical form in a lack of regulation of payment processes. Domestic workers themselves, throughout the negotiations, proclaimed, "we don't want to be paid in used clothing."[13] The notion of "payment in kind" became a defining source of debate, where workers demanded regular forms of payment that would assure standardized value for their labour. The live-in, historical conditions of servitude often limited domestic workers' pay to exchange relations of material goods, such as food, used clothing, and even toiletries. In the worst scenarios, domestic workers would receive rice, toilet paper, and soap as payment.

[13]Excerpt from International Domestic Workers Network meetings at the 2011 ILC.

In the negotiations of this particular article, domestic workers' testimonials conveyed a particular persuasive power in their bodily, affective, and moral dimensions. Domestic workers spoke of the limitations placed on their daily existence through the parsimonious food allocations they received in lieu of payment. When workers themselves addressed the rationing of basic needs for the body, such as food, clothing, and hygiene products, extreme potential for abuse and the deeply embedded power relations at the core of this profession surfaced. Only domestic worker representatives' testimonials could convey this dimension of suffering through the withholding of the most basic resources for living. These accounts generated a direct connection to demands for moral accountability to assure policy protections that would recognize the deeply colonial, racialized, gendered, and class-based exploitations embedded in this practice of in-kind payment.

To move domestic work out of its historical conditions of severe inequality and colonial relations, an international policy would need to redirect resource compensation to predominantly monetary compensation, as standard to any other form of paid labour. The final policy on payment reads:

> Domestic workers shall be paid directly in cash at regular intervals at least once a month. Unless provided for by national laws, regulations or collective agreements, payment may be made by bank transfer, bank cheque, postal cheque, money order or other lawful means of monetary payment, with the consent of the worker concerned. (C189, Article 12.1)

Yet even with vehement debate and a direct reflection of colonial conditions of labour, domestic workers could not win the assurance of exclusive monetary payment. Some forms of "in-kind" compensation would be allowed, provided their proportion is relevant to other sectors. In the formal negotiation "give and take," in-kind payment became a term that the workers' group conceded in order to assure the attainment of other core rights and standards. The final policy provided for in-kind payment as long as it is in line with related national standards and protective practices. As it read:

> National laws, regulations, collective agreements or arbitration awards may provide for the payment of a limited proportion of the remuneration of domestic workers in the form of payments in kind that are not less favourable than those generally applicable to other categories of workers, provided that measures are taken to ensure that such payments in kind are agreed to by the worker, are for the personal use and benefit of the worker, and that the monetary value attributed to them is fair and reasonable. (C189, Article 12.2)

This notion of equity rests in the determination of states. Ultimately, its implementation is assured only through the decision-making of employers, which leaves room for substantial exploitation. Who is to determine the extent to which in-kind payment is fair or reasonable? In the final form of the policy, domestic workers may not have the opportunity to exercise resistance to a determined distribution of payment, given the larger power relations that continue to govern this form of employment. At the same time, to name the need to regulate in-kind payment recognizes a history of exploitation, and severe practices of servitude that have benefitted employers at the cost of domestic workers' livelihoods and potential to move beyond the sector.

Recognizing Migration in Decent Domestic Work

> Members shall take measures to cooperate with each other to ensure the effective application of the provisions of this Convention to migrant domestic workers. Convention 189, Article 8.3

Convention 189 could only exist within a wider recognition of the role of migration in the history and growth of this transnational care economy. The terms of cooperation demanded that not only employers assure fair labour standards; the convention called states themselves to task in their practices of reliance upon third-party employment/recruitment agencies and bilateral agreements that supply and impact women workers as a core feature in the global economy. As Ip Fish Yu, the International Domestic Workers Federation Regional Coordinator for Asia, demanded in her speech to the entire ILC body, "We need sending and receive countries to cooperate to assure the protection of domestic workers" (Fish et al., 2012). This cooperation required states and third-party agencies to be responsible not only for the recruitment of domestic workers, yet also the assurance of migrants' return to their home countries—a source of frequent exploitation. As Article 4 reads, "Each Member shall specify, by means of laws, regulations or other measures, the conditions under which migrant domestic workers are entitled to repatriation on the expiry or termination of the employment contract for which they were recruited." Here, the testimonials of domestic workers themselves, and their direct experiences with the misleading recruitment processes, brought a human factor to the negotiations in ways that associated the creation of policy to the actual life courses of the women whose lives would be most directly impacted by this policy.

In the end, a global convention on domestic work became a way to recognize an important avenue to regulate migration in the transnational informal economy. As then Director General Juan Somavía contended, the instrument provides a gateway to a wider sense of responsibility to regulate informal labour, migration, and women's care work.[14] The ILO house provided a model of a higher moral standard for equity in labour, rather than simply

[14]These guiding perspectives emerged throughout most of Juan Somavía's speeches in both the 2010 and 2011 International Labour Conferences. For full transcripts of Juan Somavía's speeches, see the closing speech of the ILO Director-General Juan Somavía at the 100th International Labour Conference. https://www.ilo.org/global/about-the-ilo/multimedia/audio/WCMS_158195/lang--en/index.htm.

an instrument for operationalization of policy standards. These wider ideological and symbolic connections are made most evident in the direct reference to regulating international migration through the terms of domestic work employment. The final instrument dealt with migration as follows:

> National laws and regulations shall require that migrant domestic workers who are recruited in one country for domestic work in another receive a written job offer, or contract of employment that is enforceable in the country in which the work is to be performed, addressing the terms and conditions of employment referred to in Article 7, prior to crossing national borders for the purpose of taking up the domestic work to which the offer or contract applies. (C189, 8.1)

To take responsibility for recruitment agencies meant to recognize that the industry of global domestic work requires the inclusion of networks that operate far beyond the borders of nation states, where policy would be ratified. This point acknowledged that until international governance agencies address the translational flow of labour, policies directed to state domains alone will be partial and virtually ineffective in address the actual terms of global care labour. Like the management of monetary resources in the World Bank and International Monetary Organization, and the oversight of global goods in the World Trade Organization, the ILO would become the international governance agency with the most comprehensive attempt to regulate the human transnational flows of labour. This first policy on the terms of migrant care labour thus became the most comprehensive attempt to govern transnational human rights in informal labour flows.

A Labour Standard Instrument as a Human Rights Hook

> [Domestic workers] "... are challenging the tripartite sectors in the ILO, who are considered instruments of change to prove our value for freedom and human rights. How we respond to the challenge would reflect our own values on how we look at freedom and human rights." ILO official, 2011 ILC

Convention 189 put in place a wider aspiration that human rights can accompany migration. The implementation of Convention 189 brought different victories for each member of the tripartite dialogue. For domestic workers, as those most directly representative of labour, finally their lives and rights had been recognized by the most significant and largest international labour organization. As Peruvian domestic worker activist Ernestina Ochoa contended, "like all people, we're reclaiming our human rights."[15] For domestic worker organizations, in their first engagement with the ILO, this victory meant not

[15]As a leader of the International Domestic Workers Network, Ernestina Ochoa echoed this sentiment at a press conference during the 2011 International Labour Conference.

only inclusion in global human rights frameworks; the IDWN would be fortified in its mission and amplified in its capacity to grow a transnational migrant movement.

For employers, whose interests in human rights would constantly be countered by costs, the capacity to demonstrate a level of decency and investment in workers' rights brought not only legitimacy but a new face to the conveyers of capitalism. Paul MacKay, the 2011 employer representative, shared these perspectives with me during an interview immediately after the passing of Convention 189.

> And I think that, on that day, everyone was aware that these are human rights, and there is no longer any space in today's world to oppose this. Imagine how it would look to do that! The headlines would be "employers are opposed to…" and I think it would be so politically incorrect that it would be unimaginable. So, I really think the employers held a large role, they got back into the discussion in a constructive way. They expressed their concerns and the group of workers also had the sensitivity to understand their points in the same way as the governments did, and, in accordance with that, we all work for a common goal, for a common objective, which was arriving at a final agreement that would be acceptable for everyone, right? And we arrived at an instrument that might not be ideal, but it is undoubtedly without any historic doubt and we are very proud of that.

As we see in this reflection, the inability to deny human rights guided the establishment of this instrument. While migrant domestic workers may have been overlooked in a legacy of colonialism and its contemporary expressions in the global economy, the prevalence and expansive meaning of human rights could not be overlooked, and therefore became the hook in assuring mutual investment in this instrument, across all of the tripartite parties (see also Palmary, in this volume).

For governments, a global domestic worker convention meant contemplation of their capacity to ratify this set of global standards. In some cases, this meant an increased attention to existing national law frameworks. For others, these expectations would be out of reach. Yet the negotiation of Convention 189 meant that all governments would need to evaluate the larger human rights frameworks that emerged in these negotiations. As Vicky Kanyoka, IDWN's Africa Regional Coordinator contended in her 2010 public speech:

> Facing the economic crisis, some governments – including from the rich countries - are afraid of taking on further commitments. They feel they do not have the resources to organise such things as social security, maternity benefits, occupational health and safety, and regulation of employment agencies. But these are key human rights as well as labour issues.

Thus, even for countries, such as the US, who negotiated with no intention of ratification, the larger focus on human rights in the existing global labour system would hold them accountable to recognizing the relationship between labour and social protections in the existing transnational system. Furthermore, the ILO venue gave migrant domestic workers a chance to speak directly to governments in ways that would otherwise be completely inaccessible without this international policymaking physical and ideological forum.

The Convention 189 policy victory made moral standards concrete in the first global set of standards for domestic workers in the informal and migrant economy. As many analysts and activists have suggested, the real measure of implementation will only come through ratification and the extent to which domestic workers can actually access these protections. Yet this policymaking process served as a vital and symbolic "hook" for human rights worldwide. With Convention 189 in place, migrant organizations could call upon a global set of standards to hold governments and employers accountable through the use of human rights at the centre. As Lead ILO legal specialist Martin Oelz contended in our personal interview, this overarching link to human rights became a core feature of the effectiveness of this international policy:

> One of the strengths of the instrument is its relevance for our organization is its connection and relevance for the promotion of human rights, which makes it really a good instrument that's also being promoted by other organizations.

These organizations joined a social movement that promoted wider human rights through the plight of domestic workers worldwide. Thus, human rights frameworks got propped up widely in the making of Convention 189, as organizations with much wider causes engaged with domestic work as a tool to reach a higher call for global human rights. In the end, the domestic workers convention reinforced grounding moral commitments to human rights as an anchor for governments around ratification. Yet this framework also made a domestic workers' movement much more relative to the larger call for justice at the heart of transnational labour rights. The International Domestic Workers' Network moved to the first global union led by women, with these same human rights at the core of the organization's framework. Thus, the human rights centre of Convention 189 fed a moral movement, an activist identity, the formation of a global union, and the process that would transfer global ILO standards to state commitments through national legislation. While these measures mutually fortify each other, the recognition that domestic workers received in their call for human rights may be the most immeasurable and widely accessible outcome of this policy process. As IDWN President Myrtle Witbooi proclaimed, Convention 189 brought rights and visibility simultaneously. "We're no longer invisible. Now we're visible, and our rights are being recognized." As implementation unfolds, for domestic worker leaders to access this expression of human rights is the real victory of this historic moment. With Convention 189 in place, decades of calls that "domestic worker rights are human rights" gain ground in an international policy, guidelines for practice, and a moral guidepost for global governance. The arc of our next measure of real transformation, however, rests in analyses of the actual changes migrant domestic experience in their daily labour and livelihoods as a result of this international human rights policy victory.

References

Anderson, B. (2000). *Doing the dirty work? The global politics of domestic labor*. London, UK: Zed Books.

Basok, T., & Piper, N. (2013). Justice for migrants: Mobilizing a rights-based understanding of migration. In S. Ilcan (Ed.), *Mobilities, knowledge and social justice* (pp. 255–276). Montreal: McGill-Queen's University Press.

Blackett, A. (2019). *Everyday transgressions: Domestic workers' transnational challenge to international labor law*. Ithaca: Cornell University Press.

Boris, E. (2019). *Making the woman worker: Precarious labor and the fight for global standards, 1919–2019*. Oxford: Oxford University Press.

Boris, E., & Fish, J. (2014). "Slaves no more": Making global labor standards for domestic workers. *Feminist Studies, 40*(2), 411–443.

Chang, G. (2001). *Disposable domestics: Immigrant women in the global economy*. Cambridge, MA: South End Press.

Chin, C. (1998). *In service and servitude: Foreign female domestic workers and the Malaysian 'modernity' project*. New York, NY: Columbia University Press.

Cock, J. (1989). *Maids & madams: Domestic workers under apartheid* (2nd ed.). London, UK: Women's Press Limited.

Fish, J. (2017). *Domestic workers of the world unite! A global movement for dignity and human right*. New York: New York University Press.

Fish, J., Crockett, R., & Ormiston, R. (Directors). (2012). *C189: Conventional wisdom*. Norfolk, VA: Sisi Sojourner Productions.

Hondagneu-Sotelo, P. (2001). *Domestica: Immigrant workers cleaning and caring in the shadows of affluence*. Berkeley: University of California Press.

International Domestic Workers Federation. (2019). Five Years a Federation Congress Report.

International Labour Office. (2013). *Domestic workers across the world: Global and regional statistics and the extent of legal protection*. Geneva: International Labour Organization. http://www.ilo.org/public/libdoc/ilo/2013/113B09_2_engl.pdf.

International Labour Organization. (1999). *Report of the Director-General: Decent work*. https://www.ilo.org/public/english/standards/relm/ilc/ilc87/rep-i.htm.

International Labour Organization. (2011). *Convention 189: Domestic workers, convention*. Policy adopted June 16, 2011. For full article reference see: https://www.ilo.org/dyn/normlex/en/f?p=NORMLEXPUB:12100:0::NO::P12100_ILO_CODE:C189.

Kanyoka, V. (2010). *Statement on domestic workers rights*. Speech presented at the 2010 International Labour Conference, Geneva, Switzerland.

Olievera, G. (2018). *Motherhood across borders: Immigrants and their children in Mexico and New York*. New York: New York Univesity Press.

Parreñas, R. (2001). *Servants of globalization: Women, migration and domestic work*. Stanford, CA: Stanford University Press.

Piper, N. (2017). Global governance of labour migration: from "management" of migration to an integrated rights-based approach. In P. Drahos (Ed.), *Regulatory theory—Foundations and applications* (pp. 377–394). Canberra: ANU University Press.

Piper, N., Rosewarne, S., & Withers, M. (2017). Migrant precarity: 'Networks of Labour' for a rights-based governance of migration. *Development and Change, 48*(5), 1089–1110.

Solari, C. (2015). *Remittances as development? Gender, nation-building & Ukrainian Emigration*. Paper presented at the Sociology of Development Conference, Brown University, Providence, Rhode Island.

Jennifer N. Fish is Chair of the Department of Women's Studies at Old Dominion University. Over the past twenty years, she has worked with domestic worker organizations on the attainment of human rights under structures of national and global governance. Her most recent book, *Domestic Workers of the World Unite! A Global Movement for Dignity and Human Rights* (New York University 2017) chronicles the development of the International Domestic Workers Federation and the realization of the International Labour Organization's first global policy for household labor protections. She is a member of the Women in Informal Employment: Globalizing and Organizing (WIEGO) research policy institute and an active contributor to several initiatives in support of domestic worker rights.

Part V

Forced Migration, Gender-Based Violence and Conflict

CHAPTER 22

Gender, 'Refugee Women' and the Politics of Protection

Heaven Crawley

Abstract A substantial body of research has demonstrated the particular difficulties faced by women in securing access to protection under international refugee law. It is typically men who have been considered the principal agents of political resistance and therefore the (more) legitimate beneficiaries of protection resulting from associated persecution. Forms of violence and discrimination more commonly experienced by women have routinely been dismissed as 'private' and therefore falling outside the scope of international refugee law. This chapter argues that whilst all refugees have increasingly struggled to secure access to international protection, efforts to engender international refugee law have also been undermined by use of the category 'Refugee Women'. This category has been harnessed by policy makers, lawyers and civil society organisations to draw attention to the particular experiences of women fearing gender-based violence but has served to 'flatten out' the complex gendered experiences of those seeking protection, extrapolating their experiences from the socio-political contexts in which they occur. The result has been a form of 'exclusionary inclusion' in which women's experiences are effectively 'depoliticised': they are represented almost exclusively as 'victims' of male violence rather than activists and/or resisters of patriarchal norms.

H. Crawley (✉)
Centre for Trust, Peace and Social Relations (CTPSR), Coventry University, Coventry, England
e-mail: heaven.crawley@coventry.ac.uk

C. Mora and N. Piper (eds.), *The Palgrave Handbook of Gender and Migration*, https://doi.org/10.1007/978-3-030-63347-9_22

Introduction

A substantial body of research and policy analysis has shown that many women (and some men) face difficulties securing protection under international refugee law because their gendered experiences are not recognised as constituting persecution within the meaning of the UN Convention on the Status of Refugees ('Refugee Convention'). Moreover, gender is not one of the enumerated grounds on which a claim for international protection can be made (Campbell, 2001; Charlesworth, Chinkin, & Wright, 2001; Crawley, 1999, 2000; 2001; Edwards, 2010; Freedman, 2008; 2015; Greatbach, 1989; Indra, 1999; Macklin, 1995; Oswin, 2010; Spijkerboer, 2000). Although the Refugee Convention and its 1967 Protocol are ostensibly gender-neutral and do not distinguish between male and female refugees, the dominant interpretation of refugee law has evolved through an examination of male asylum applicants and their activities. It is men who have been considered the principal agents of political resistance and therefore the legitimate beneficiaries of protection resulting from persecution. Forms of violence and discrimination most commonly experienced by women—for example, rape and sexual violence, familial (domestic) violence, honour killings, female genital mutilation (FGM), forced marriage—are routinely dismissed as being 'private' and outside the scope of international refugee law.

In many respects, the failure to properly address issues of gender in claims for protection under international refugee law is a product of a more general failure of international refugee law to recognise social and economic rights due to its emphasis on individual targeting and the deprivation of civil and political rights. As Foster (2007) suggests, the Refugee Convention is capable of accommodating claims based on the deprivation of economic and social rights which may, themselves, be violated for political reasons. Nonetheless refugee-receiving states have been keen to maintain what they consider to be a clear differentiation between 'refugees' and 'economic migrants' despite the fact that these categories are often meaningless in the context of refugees' increasingly complex and multi-layered journeys and experiences (Crawley & Skleparis, 2017). But the failure to address issues of gender is also related to a larger criticism of human rights law and discourse, namely, that it privileges male-dominated 'public' activities over the activities of women, which take place largely in the so-called 'private' sphere. Modern international law, including refugee law, rests upon and reproduces various dichotomies between the public and private spheres: a distinction is made between matters of international 'public' concern and matters 'private' to states that are considered within their domestic jurisdiction, and in which the international community has no recognised legal interest. Moreover both international human rights law and, by extension, international refugee law, are based on an androcentric 'male-as-norm' model which privileges the family—often a source of intense and intimate violations of women's human rights—as 'the natural and fundamental group of society' (Charlesworth, 2005; Edwards, 2010). When gender becomes central to the analysis therefore, international refugee law dissolves into a normative struggle whose outcome is determined largely by power and which allows issues of particular concern to women to be either ignored or undermined.

In this context, feminist jurisprudence has emerged as a systematic critique of the practice and profession of law, with its central theme that the law is a thoroughly gendered system that marginalises women's interests and reinforces male domination. Feminist legal scholars have argued that such distinctions render violations of rights at the hands of individuals within the family and community largely invisible, with the result that what women do and what is done to them is often view as irrelevant (Campbell, 2001; Charlesworth et al., 2001; Crawley, 1999, 2001; Freedman, 2015; Oxford, 2014). Feminists have sought to mitigate this bias in the interpretation of the refugee definition by making women's experiences of persecution in the 'private' sphere more visible (Firth & Mauthe, 2013). They have highlighted the uses of rape and sexual violence as weapons of war, emphasised the particular structural dimensions that shape violence against women within the family and community, and drawn attention to oppressive cultural and ideological norms and practices that contribute to the harm experienced by women and which, they argue, should be identified as violations of women's human rights (Oswin, 2010).

Lawyers, women's organisations and refugee support groups have drawn on this feminist legal scholarship over the past thirty years, to highlight the experiences of women asylum seekers resulting in important changes to policy and practice that have legitimised the factual basis for gender-based asylum claims. These changes include the production of international guidelines on gender (UNHCR, 2002) and national guidance that specifically draws on international human rights law to make claims concerning women's right to protection and to assist decision-makers in understanding the importance of gender in asylum policies and procedures (Freedman, 2015). There have also been significant developments in case law across a number of jurisdictions which have resulted in important legal decisions which have derived in an understanding that the Refugee Convention can, and should, be interpreted to include domestic violence, female genital mutilation and other harms previously regarded as 'private' (Firth & Mauthe, 2013). In some contexts at least, there is evidence that resistance and ambivalence have been slowly inching towards recognition of gender-based asylum claims (Musalo, 2010).

The importance of these developments in creating opportunities for protection that might not otherwise exist is not in dispute. Nonetheless, there is growing concern that many of the earlier efforts to engender international refugee law have, over recent years, lost considerable ground. Whilst it is increasingly clear that gender issues are no longer in vogue within refugee studies (Hyndman, 2010), the project of engendering refugee law remains at best half done (Arbel, Dauvergne, & Millbank, 2014). Indeed it could be argued that the gains made during the 1990s and 2000s are increasingly being undone as seen, for example, in challenges to the recognition of domestic violence as the basis for securing asylum in the US (Anker, 2020). This chapter explores the reasons why.

The Politics of International Refugee Protection

The treatment of gendered refugee claims in law and practice cannot be understood outside of a broader analysis of the politics of asylum (Geddes, 2003; Gibney, 2004; Squire, 2009). Often presented as Europe's attempt to reconcile itself with the horrors of World War II, which led to the death and displacement of millions, the Refugee Convention quickly became a political tool for states to signal approval or disapproval of the ideology of geopolitical contexts from which asylum applicants originated. This means that different forms of asylum (and, in turn, the protection afforded to different types of harm) recede or come to dominate depending on historical or political necessity. Asylum may be a means of undermining one's enemies, augmenting a declining population or legitimating one regime over another: in the first few decades after World War II, the acceptance of refugees arriving from the communist bloc was a way to position the West as morally and ideologically superior (Crawley & Setrana, forthcoming). Although Western nations became less interested in championing the rights of refugees after the collapse of the Soviet Union in 1991 refugees have, nonetheless, remained a critical tool in global 'migration diplomacy' (Adamson & Tsourapas, 2019). Refugee policy may also be a means of constructing a country's sense of self and belonging (Squire, 2009).

At the same time there has, since the late 1990s in particular, been growing political concern about the scale of asylum flows to the countries of the Global North (Europe, North America and Australia) resulting in a series of legislative and policy changes to tighten access to asylum procedures. This is despite the fact that the vast majority of refugees—and estimated 85%—remain in the countries of the Global South (UNHCR, 2020). These political shifts have been associated with the externalisation of border controls, increasingly restrictive entry procedures, the use of detention, dispersal and deportation to reduce the number of applications and control those who arrive, and restrictions in the availability of welfare support for those awaiting a decision about their future. Such trends have created a 'race to the bottom' as refugee-receiving countries compete to create hostile political and policy environments that are perceived to deter would-be refugees from arriving. Although other forms of political, religious and ethnic violence are widespread, it has proved increasingly difficult to make the case that women fleeing patriarchal violence should be protected due to concerns about 'opening the floodgates' to potentially huge numbers of applicants (Freedman, 2008; Mullally, 2011).

One of the consequences of these increasingly restrictive political contexts' is the existence of widespread disparities in the way countries handle gendered asylum claims and a significant 'implementation gap' (Crawley & Lester, 2004;

Edwards, 2010; European Parliament, 2012; Freedman, 2015). None of the 41 countries surveyed by Crawley and Lester (2004) had officially adopted the UNHCR Gender Guidelines into their legislation or policy. Just two countries had introduced their own guidance on the assessment of gender-related asylum claims, although a further eight had included some gender-related points within their general asylum policy or guidelines. Where progress had been made at the policy level, implementation was found to be inconsistent. A more recent report on nine countries in Europe has also found significant inconsistencies in the interpretation of gender-based claims for protection despite the existence of European legislation including the Qualification Directive (European Parliament, 2012).

Many gender-based and gender-related asylum claims continue to be regarded as controversial, with female asylum applicants told that their experiences of sexual and domestic violence, female genital mutilation (FGM) and honour killings are insufficient violations of their human rights to warrant protection, or that they can return to their countries of origin and seek protection from a complicit state against the harms inflicted by members of their family or the wider community (see also O'Connell Davidson, in this volume). UNHCR's (2002) guidance, along with that produced by refugee-receiving countries themselves, is routinely ignored (see also Oxford, 2014). But the treatment of gender-based and gender-related claims for asylum is not just symptomatic of the broader politics of asylum: it is also a reflection of the ways in which the politics of gender are represented and understood. In particular, there is evidence that the category 'Refugee Women', harnessed by policy makers, lawyers and civil society organisations to draw attention to the particular experiences of women feared in gender-based violence, has resulted in a form of 'exclusionary inclusion' (Christensen & Siim, 2010; Freedman, 2009; Kneebone, 2005) that ultimately serves to undermine rather than enhance access to protection for those fearing persecution on the basis of gender (Campbell, 2001; Crawley, 2001; Freedman, 2008).

The Exclusionary Inclusion of 'Refugee Women'

Kneebone's (2005) concept of 'exclusionary inclusion' takes us some way in explaining, and making sense of, the ongoing difficulties experienced by women in securing protection against violations of their human rights. Writing with reference to Australia, Kneebone suggests that a major problem with the current dominant approach to gender-based asylum claims is that it constructs the 'Refugee Woman' according to her vulnerability in a patriarchal society in which women are subordinate to men as victims, or potential victims, of sexual or other violence directed towards them as women. This tendency toward 'protective' responses that assume women's vulnerability, dependency and need for 'special protections' has led to an over-emphasis on the construction of 'women' or specific groups of women as members of a Particular Social Group (PSG). This framing positions women primarily as 'victims' of male violence rather than situating gender-based violence within the broader political (and other) contexts within which such violence takes place. Membership of a PSG has now become the default ground for women's claims in many jurisdictions, even when one or more of the other grounds may be equally or more relevant (Edwards, 2010). Many of these cases are not legally binding or are factually specific, which means that they are of limited precedent value for others. Most importantly an analysis of gendered power relations in countries of origin is often absent.

Whilst the use of PSG has been an effective strategy for securing access to protection in *some* gender-based asylum claims (the treatment of claims based on domestic violence in the US provides perhaps the clearest example) therefore, the framing of 'Refugee Women' as vulnerable, dependent and in need of 'special protection' serves to 'flatten out' the complexity and diversity of women's experiences and, in turn, depoliticize the specific circumstances under which violence and other threats occur (McKinnon, 2010; Mullally, 2011). 'Refugee Women' are presented as uniformly poor, powerless and vulnerable victims of, for example, 'male oppression' or 'oppressive cultures, religions or traditions', while Western women are the reference point for modern, educated, sexually liberated womanhood. Any protection received is provided to them as 'victims' of human rights violations rather than as rights holders whose access has been negated or undermined by patriarchal structures and institutions (Firth & Mauthe, 2013).

This approach both reflects, and is reinforced by, a number of different factors. The first is the tendency to equate gender with 'sex', which prevails within human rights theory and advocacy, including that directed specifically towards international refugee law (Crawley, 1999). Despite the emancipatory possibilities opened up by the language of gender (for both women and men), it remains common for feminists and those advocating for women's rights in the refugee context to use gender as a synonym for 'women', which undermines the idea that gender is a social and relational category and threatens to reduce women, once again, to biology (Charlesworth, 2005). Otto (2013) makes a similar point, noting that whilst feminists use the term 'gender' to explore socially constructed differences between women and men that have been used to justify women's inequality, the term is more commonly used within institutional practice as a synonym for women's issues.

The second, contributory factor, is the role of the transnational institutions and the 'refugee sector' broadly defined, which, in the absence of political space in which complex discussions of gender and gendered power relations can gain traction, has often been complicit in constructing a particular 'script' or narrative that requires 'Refugee Women' to be represented in particular ways in order to be 'saved'. This has resulted in the development of a dominant narrative through which women's experiences are evaluated and understood, and which can serve to feminise or even infantilise those seeking protection (Arbel, 2013; Freedman, 2008; Oxford, 2014). Thus feminist scholarship and activism has inadvertently contributed to the process of essentialisation (Edwards, 2010; Freedman, 2015).

This script has real consequences 'on the ground', not least because some practitioners and advocates may push women towards claims that they believe will be more 'acceptable' in terms of their knowledge of the decision making process, for example, the case of a woman who said she would kill her husband if forced to return to him but whose account was charged to say that she feared that her husband would kill *her* (Berger, 2009). Oxford (2005: 32) argues that 'the gender regime of asylum provides few opportunities for women to voice outrage and desire for retribution toward their persecutors. Asylum seekers must therefore articulate an account of victimisation in order to make the claim that they have been persecuted. Intrinsic in the claim to persecution is a claim of victimisation; it is a fundamental dimension of the regime of asylum practices'. She suggests not only that the law fails to hear women's narratives of gender-related harms but claimants themselves censure, edit or redraft their narratives to fit within legal categories and expectations of how a 'Refugee Woman' should be. Over time, women may come to realise the power of stories of 'powerless and victimhood' and re-appropriate them in order to secure access to resources and rights (Freedman, 2015). Reduced to her victim identity, she may secure protection but in the process lose her sense of personhood (Firth & Mauthe, 2013).

Building on this theme, Bains (2004) and Johnson (2011) explore the gender narrative of refugee representation by the UNHCR and how it intersects with political (non-) agency. As Johnson (2011) suggests, how we imagine particular categories of people determines how we engage with them, who we accept, who is able to participate and on what terms. Both note that, whilst women are increasingly included in refugee regimes through policies, guidelines and other initiatives, they are included as a *broad and undifferentiated category* of victims, lacking agency and unable to determine their own futures. This positioning of women has played an important role in the discourses of victimisation and depoliticisation employed by UNHCR to reduce the threat perception of the refugees and generate support for humanitarian aid (Bains, 2004; Freedman, 2015).

Whilst this construction of the 'Refugee Woman' as victim has proved to be useful as a tool for the mobilisation of support behind humanitarian intervention and refugee work, it reproduces and reinforces the idea that women from the Global South are powerless and lack agency. This point is made most explicitly by Berger (2009) who, in her analysis of the *Matter of RA* (discussed above) and other Latina women seeking asylum in the United States, argues that in order to establish the merits of the case, advocates almost invariably position the countries from which women come as 'uncivilized' countries that fail to 'protect' women, in turn implying that refugee-receiving states are 'civilized' countries within which women are protected. She suggests that '[t]his discursive merry-go-round bolsters the paternalistic international hierarchy among countries and sets the stage for depoliticised solutions that ignore power inequalities' (Berger, 2009: 671–2).

The Politics of Gender

The concept of 'politics' is firmly embedded in the Refugee Status Determination (RSD) procedures undertaken across the countries of the Global North to determine whether an individual is 'deserving' of protection under international refugee law. Perhaps not surprisingly, the dominant conceptualisation on politics is deeply gendered. Indeed the assumption that women are unlikely to be actively involved in political activities typically associated with asylum claims, contributes to the decisions of advocates and practitioners to harness the category 'Refugee Women' in an effort to secure refugee protection for their clients. Critically, this fails to take into account the geopolitical contexts within which women's engagement in politics takes place and their resistance to deeply political (and politicised) patriarchal social and cultural norms (Crawley, 2000).

Gendered critiques of politics and political participation shed new light on the relationship between women and politics, both by pointing to the structural features of political life which have tended to exclude women from positions of power, and by revealing the history of women's involvement in political action (Christensen & Siim, 2010; Hyndman, 2010). In addition, they have challenged the tendency to separate the public world of politics and employment from the private sphere of family and interpersonal relations (Indra, 1999). Women are as vulnerable to political violence as their male counterparts, even though their political participation often takes place at a so-called 'low level' (Crawley, 2000, 2001). In many societies, indeed, the penalties for political participation and resistance are even more severe for women than for men, because of cultural and social norms that preclude women's involvement. For example, women who are imprisoned by the authorities run the risk of 'double punishment'. They are punished not only because they oppose the regime in some way, but also because they shun the traditional role of women by being politically active at all. As a result, they are often 'put back in their place' by prison guards or military men.

Moreover, violence against those who oppose political regimes is not confined to the public sphere, because politics and political resistance are not exclusive to the public sphere. Political violence by the state aims to disable opposition or resistance by intimidating a population and to forcibly 'depoliticize' it. Bringing violence into the 'private' sphere of the home and family, appears to be a particularly effective means of achieving this aim. As a result, even where women do not participate in formal politics, a woman nonetheless may be harmed as a means of intimidating, coercing or harming other family members, who hold dissenting political views or who engage in political activities which are disapproved of by the persecutor. In addition, women who have had little or no involvement in formal political institutions often take up all sorts of practical and innovative ways to exert pressure on the political scene in times of conflict. These forms of political participation by women often grow out of and subvert their gender roles as providers and nurturers; because women are seen as political innocents, they are able to use this immunity to take initiatives and responsibilities of a covert political nature.

It is also important to understand that gender itself can become politicised with consequences for both women and men, who are able, or refuse, to comply with gendered expectations regarding their identity, behaviours and relationships, especially sexual relationships and those within the family (Arbel et al., 2014; Indra, 1999). Many anti-colonial nationalist projects aim to recover or reinvent 'tradition' in order to develop a new nationalist consciousness (Yuval-Davis, 2011). Within this process, national difference is often constructed in cultural terms against the West and, because this difference is often located in the private sphere, in family and sex roles, women have been constructed as the bearers of an authentic/authenticated culture. In many parts of the world, women who do not live up to the moral or ethical standards imposed on them by their societies are imputed with a political opinion and as a consequence suffer cruel or inhuman treatment. Refusing to marry, having sexual relations outside marriage, providing unsatisfactory dowry or even wearing certain dress can result in persecution.

Perhaps one of the clearest examples of the way in which political participation and resistance is gendered can be seen in the politics of dress codes, of which the veil in some Islamic countries is a well-publicised example. Women's opposition to the imposition of dress codes during the process of Islamisation should be seen in the context of the political symbolism of the veil. The concept of 'honour' has become strongly associated with women's sexual behaviour in many contexts. Any actual or perceived violation of what are deemed to be 'appropriate' gender relations is viewed as defiling the honour of the woman and, in turn, the honour of the nation itself. Protecting the honour of the woman and the nation therefore gains political significance and will be enforced either directly through the state—as seen in legislated discrimination and laws regulating women's behaviour—or through a woman's family and community.

The implications of these processes for women can be seen in the experiences of those seeking asylum under the Refugee Convention. Women may be subjected to discriminatory treatment, which is enforced through law or through the imposition of social or religious norms which restrict their opportunities and rights. The consequences for women of failure or refusal to comply with social norms and mores regarding their behaviour can vary enormously. At one extreme, a woman may fear that she will be subjected to threats on her life if she is forced to return to her country of origin. On the other, even where a woman does not fear threats on her life, she is often forced to submit to measures to retain or re-establish her honour. The harm experienced or feared in these cases will often be marriage-related, including forced marriage, a form of abuse that is often not recognised. In many cases of women who refuse to agree to such arrangements, there will be a punishment inflicted for refusing to abide by discriminatory social mores, rather than the marriage itself, which will rise to the level of 'serious harm'. The repercussions for women of divorcing their husbands or entering into mixed marriages may be equally devastating.

Towards an Intersectional Analysis of Gender and Refugee Protection

Since the early 1990s, feminist legal scholars have drawn attention to the failures of international refugee law to adequately take account issues of gender in procedures for determining who is—and is not—entitled to protection. As was noted above, some of the difficulties in gendering international refugee law are due to the narrow and increasingly restrictive interpretation of the Refugee Convention, in the context of political and public concerns around migration and the growing emphasis on immigration controls. But they are also symptomatic of a more fundamental problem, namely, the ongoing failure of policy makers, lawyers and practitioners alike to fully comprehend the role of gender in shaping both women's experiences of persecution and the (mis)interpretation of these experiences in the application of international refugee law. Whilst there has been increased acknowledgment of gender in asylum policy and practice reflected, for example, in the publication of gender guidelines in a number of countries, the assessment of women's claims for international protection reflects a form of 'exclusionary inclusion' (Kneebone, 2005) which draws heavily on a script of vulnerability and victimhood. It has been argued in this chapter that this framing, which frequently harnesses the category 'Refugee Women' to draw attention to the particular experiences of women fearing gender-based violence, reinforces gendered stereotypes, potentially limiting rather than opening up opportunities for international protection. This is because it has served to 'flatten out' the complex gendered experiences of those seeking protection, extrapolating their experiences from the socio-political contexts in which they occur. In the process, women's experiences are effectively 'depoliticised': they are represented almost exclusively as 'victims' of male violence rather than activists and/or resisters of patriarchal norms.

None of which is intended to diminish the efforts of activist academics, lawyers, international and civil society organisations—often working in partnership—to secure protection for those seeking protection on the basis of gender. There have been important gains, often hard won, for individual women (and some men) for whom gender-based and gender-related persecution lies at the heart of the asylum claim (Anker, 2020; Crawley, 2001; Freedman, 2009, 2015; Musalo, 2010). But the gains are often limited to individuals, rooted in increasingly complex and convoluted PSG definitions which fail to acknowledge the deeply gendered geopolitical, social and cultural contexts within which violence against women is situated and the use of gender as a political tool by refugee-receiving States in the Global North to signal their cultural, religious and racial superiority (Clark, 2007). The gains in terms of embedding gender as a core concept underpinning inequalities in access to protection under the Refugee Convention remains as elusive as ever. As noted by Hennebry and Petrozziello (2019) the development of the Global Compacts on Refugees presented an important opportunity to continue enhancing the international framework for protecting the rights of women and men on the move but here, too, there has been a failure to grasp the political significance of gender. As has been seen in recent policy and practice, the primary mode of understanding gender in the Refugee Compact is in terms of the specific vulnerabilities and needs of women and girls. Gender is not viewed as a lens of analysis or power relations, much less a question of rights. Most significantly, there is no gender analysis of root causes of displacement or forced migration, missing the opportunity to recognise gender-based persecution.

So where do we go from here? One approach would be to move away from gender, which continues to be equated with 'women'. Ever since the renowned law scholar and critical race theorist Kimberlé Crenshaw (1989) introduced the term, the concept of intersectionality has become an increasingly important lens though which to understand the experiences of women, highlighting the ways in which gender relations are intertwined in complex ways with other forms of social inequality (Yuval-Davis, 2011). This concept goes beyond merely merging separate identities but considers the unique identity developed from an individual belonging to multiple categories simultaneously. It has been suggested that part of the reason for the lack of progress in relation to gender-based asylum claims has been the failure of lawyers and decision-makers to unpack 'gender', relying instead upon static, decontextualised and essentialising categories (Crawley, 2001; Freedman, 2009). It is surprising therefore that, with some notable recent exceptions (e.g. Davis, 2015; Hancock, 2015), there has been very little application of an intersectionality framework to the areas of international human rights and refugee law.

Understanding the *intersection* of gender with other dimensions of identity and the larger political projects of nation states is crucial to understanding both, the experiences of women in their countries of origin, and the ways in which these experiences are framed in the process of deciding whether or not they are 'deserving' of international protection. What is needed is a multi-level, contextual and dynamic understanding of intersectionality which is able to link the experiences of individual women seeking protection with the contexts within which these experiences take place. We need to develop new theoretical insights and approaches in relation to gender-based asylum claims, by bringing into dialogue three main areas of academic thinking: feminist legal studies, postcolonial feminism and intersectionality, each of which is situated within a broader understanding of the contemporary politics of asylum. With some notable exceptions (Clark, 2007; Hyndman, 2010; Oxford, 2005), these areas of academic investigation have generally been conducted in isolation, progressing within geographical and disciplinary silos. This has resulted in a striking disconnect between highly theoretically sophisticated work on gender, migration and identity, which does not always engage with the empirical realities of gender-based asylum claims on the one hand, and very detailed empirical (mostly legal) analysis which has not explicitly engaged with theoretical insights into core concepts, including gender and 'race', on the other. Bringing together these areas of academic endeavour has the potential to inform, shape and guide the next generation of research on gender-based asylum claims, challenge the legal framing and construction of 'Refugee Women' primarily as apolitical 'victims', disconnected from the geopolitical contexts within which their experiences occur.

References

Adamson, F. B., & Tsourapas, G. (2019). Migration diplomacy in world politics. *International Studies Perspectives, 20*(2), 113–128.

Anker, D. (2020). The history and future of gender asylum law and recognition of domestic violence as a basis for protection in the United States. *Human Rights Magazine, 45*(2) https://www.americanbar.org/groups/crsj/publications/human_rights_magazine_home/immigration/the-history-and-future-of-gender-asylum-law/.

Arbel, E. (2013). The culture of rights protection in Canadian refugee law: Examining domestic violence cases. *McGill Law Journal, 58*(3), 729–771.

Arbel, E., Dauvergne, C., & Millbank, J. (Eds.). (2014). *Gender in refugee law: From the margins to the centre*. Abington, Oxon: Routledge.

Bains, E. (2004). *Vulnerable bodies: Gender, the UN and the global refugee crisis,* Oxon: Routledge.

Berger, S. (2009). Production and reproduction of gender and sexuality in legal discourses of asylum in the United States. *Signs: Journal of Women in Culture and Society, 34*(3), 659–685.

Campbell, P. (2001). Gendered human rights: The international community's failed response to the persecution of women. *Politics and Policy, 29*(1), 121.

Charlesworth, H. (2005). Not waving but drowning: Gender mainstreaming and human rights in the United Nations. *Harvard Human Rights Journal, 18*(1), 18.

Charlesworth, H., Chinkin, C., & Wright, S. (2001). Feminist approaches to international law. *American Journal of International Law, 85*(4), 613–645.

Christensen, A., & Siim, B. (2010). Citizenship and politics of belonging: Inclusionary and exclusionary framings of gender and ethnicity. *Kvinder, Kon & Forskning, 2–3,* 8–17.

Clark, S. (2007). Female subjects of international human rights law: The hejab debate and the exotic other female. *Global Change, Peace and Security, 19*(1), 35–48.

Crawley, H. (1999). Women and refugee status; beyond the public/private dichotomy in UK asylum policy. In D. Indra (Ed.), *Engendering forced migration: Theory and practice*. Berghahn Books.

Crawley, H. (2000). Gender, persecution and the concept of politics in the asylum determination process. *Forced Migration Review,* (9), 17–20.

Crawley, H. (2001). *Refugees and gender: Law and process.* Jordan Publishing.

Crawley, H., & Lester, T. (2004). *Comparative analysis of gender-related persecution in national asylum legislation and practice in Europe.* Evaluation and Policy Analysis Unit Evaluation Report EPAU/2004/05. Geneva: UNHCR.

Crawley, H., & Setrana, M (forthcoming). 'The evolution of the 'global refugee regime': From rights to 'burden sharing''. In E. Carmel, K. Lenner and R. Paul (Eds.), *Handbook on the governance and politics of migration.* Edward Elgar.

Crawley, H., & Skleparis, D. (2017). Refugee, migrant, neither, both: Categorical fetishism and the politics of bounding in Europe's 'migration crisis'. *Journal of Ethnic and Migration Studies, 44*(1), 48–64.

Crenshaw, K. (1989). Demarginalizing the intersection of race and sex: A Black feminist critique of antidiscrimination doctrine, feminist theory and antiracist politics, *University of Chicago Legal Forum*, https://chicagounbound.uchicago.edu/uclf/vol 1989/iss1/8.

Davis, A. N. (2015). Intersectionality and international law: Recognizing complex identities on the global stage. *Harvard Human Rights Journal, 28*(1), 205–242.

Edwards, A. (2010). Transitioning gender: Feminist engagement with international refugee law and policy 1950–2010. *Refugee Survey Quarterly, 29*(2), 21–45.

European Parliament (2012). *Gender related asylum claims in Europe: A comparative analysis of law, policies and practice focusing on women in nine eu member states,* brussels https://www.endfgm.eu/resources/asylum/european-parliament-gender-related-asylum-claims-in-europe-a-comparative-analysis-of-law-policies-and-practice-focusing-on-women-in-nine-eu-member-states-2012/.

Firth, G., & Mauthe, B. (2013). Refugee law, gender and the concept of personhood. *International Journal of Refugee Law, 25*(3), 470–501.

Foster, M. (2007). *International refugee law and socio-economic rights: Refuge from deprivation.* Cambridge: Cambridge University Press.

Freedman, J. (2015). *Gendering the international asylum and refugee debate* (2nd ed.). Basingstoke: Palgrave Macmillan.

_____. (2008). 'Women seeking asylum'. *International Feminist Journal of Politics, 10*(2), 154–172.

______. (2009). 'Protecting women asylum seekers and refugees: from international norms to national protection'. *International Migration, 48*(1), 175–198.

Geddes, A. (2003). *The politics of migration and immigration in Europe.* London: Sage.

Gibney, M. J. (2004). *The ethics and politics of asylum.* Cambridge: CUP.

Greatbach, J. (1989). The gender difference: Feminist critiques of refugee discourse. *International Journal of Refugee Law, 1*(4), 519–527.

Hancock, A-M. (2015). When is fear for one's life race-gendered? An intersectional analysis of the Bureau of Immigration Appeal's in re AR-C-G- Decision. *Fordham Law Review, 83*(6), 2977–2998. http://ir.lawnet.fordham.edu/flr/vol83/iss6/5.

Hennebry, J. L., & Petrozziello, A. J. (2019). Closing the gap? Gender and the global compacts for migration and refugees. *International Migration, 57*(6), 115–138.

Hyndman, J. (2010). Introduction: The feminist politics of refugee migration. *Gender, Place and Culture, 17*(4), 453–459.

Indra, D. (Ed.) (1999). *Engendering forced migration: Theory and practice.* New York: Berghahn Books.

Johnson, H. L. (2011). Click to donate: Visual images, constructing victims and imagining the female refugee *Third World Quarterly, 32*(6), 1015–1037.

Kneebone, S. (2005). Women within the refugee construct: 'Exclusionary inclusion' in policy and practice the Australian experience. *International Journal of Refugee Law, 17*(1), 7.

Macklin, A. (1995). Refugee women and the imperative of categories. *Human Rights Quarterly, 17*(2), 213–277.

McKinnon. (2010). Excavating gender in women's early claims to political asylum in the United States. *Women's Studies in Communication, 33*(2), 79–95.

Mullally, S. (2011). Domestic violence asylum claims and recent development in international human rights law: A progress narrative? *International and Comparative Law Quarterly, 60*(2), 459–484.

Musalo, K. (2010). A short history of gender asylum in the United States: Resistance and ambivalence may very slowly be inching towards recognition of women's asylum claim. *Refugee Survey Quarterly, 29*(2), 46–63.

Oswin, N. (2010). Rights spaces: An exploration of feminist approaches to refugee law. *International Feminist Journal of Politics, 3*(3), 347–364.

Otto, D. (2013). 'International human rights law: towards rethinking sex / gender dualism and asymmetry' (Melbourne Legal Studies Research Paper No. 620).

Oxford, C. (2014). Where are the women? In E. Arbel, C. Dauvergne, & J. Millbank (Eds.), *Gender in refugee law: From the margins to the centre* (pp. 157–175). Abington, Oxon: Routledge.

——. (2005). Protectors and victims in the gender regime of asylum. *NWSA Journal, 17*(3), 18.

Spijkerboer, T. (2000). *Gender and refugee status*. Aldershot: Ashgate.

Squire, V. (2009). *The exclusionary politics of asylum*. Basingstoke: Palgrave Macmillan.

UNHCR. (2002). *Guidelines on international protection: Gender-related persecution within the context of Article 1A (2) of the 1951 Convention and/or its 1967 Protocol relating to the status of refugees* Geneva: UNHCR https://www.unhcr.org/publications/legal/3d58ddef4/guidelines-international-protection-1-gender-related-persecution-context.html.

UNHCR. (2020). *Global trends: Forced displacement in 2019*. Geneva: UNHCR https://www.unhcr.org/globaltrends2019/.

Yuval-Davis, N. (2011). *The politics of belonging: Intersectional contestations*. London: Sage.

Heaven Crawley joined the Centre for Trust, Peace and Social Relations (CTPSR) at Coventry University in September 2014 from where she leads MIDEQ, the world's largest migration research project focusing on relationships between migration, inequality and development in the context of the Global South. Educated at the Universities of Sussex (1989-1994) and Oxford (1995-1999), Heaven has nearly 30 years' experience of undertaking research on international migration in a wide range of institutional settings. She has published extensively on a wide range of asylum and migration issues including the drivers of migration and migrant decision-making, gender issues in forced migration, refugee and migrant rights, the experiences of children and young people on the move, attitudes towards migration and migrants and politics of migration policy-making.

CHAPTER 23

"Aberrant" Masculinity: Men, Culture and Forced Migration

Rose Jaji

Abstract Masculinization of violence essentially feminizes and infantilizes victimhood in ways that are exclusionary to male victimhood. Images of women and children fleeing violence have become ingrained in the collective imagination leading to incredulity towards male victimhood and quest for asylum. Male refugees are generally depicted in anti-refugee rhetoric as an aberration at two levels. The first level relates to the perceived incompatibility between flight and masculinity while the second depicts male refugees' cultural backgrounds as fostering a backward masculinity that is incongruous with life in "civilized" host societies. Drawing specifically from Africa, this chapter argues that men's experience of victimhood and their quest for asylum can be understood within the logic of patriarchal governance of male conduct and the division of labour among masculinities which is observable in violent conflict situations. Flight is consistent with rather than contrary to the patriarchal logic that underpins the social positioning of men in families and performance of masculinity in specific socio-cultural and political contexts.

Introduction

Femininity and childhood, conflated into *womenandchildren* (Enloe, 1992), have become the quintessential representation of the vulnerability and suffering that violent conflict and forced migration entail. Focus on men in

R. Jaji (✉)
University of Zimbabwe, Harare, Zimbabwe
e-mail: rjaji@sociol.uz.ac.zw

C. Mora and N. Piper (eds.), *The Palgrave Handbook of Gender and Migration*, https://doi.org/10.1007/978-3-030-63347-9_23

conflict situations predominantly concentrates on their "violences" (Myrttinen, Khattab, & Naujoks, 2017) the impression created being that manhood and masculinity are immune to victimhood. Both women and men suffer because of violence, persecution, gross human rights violations and the ensuing flight but in gender specific ways. Considering that contemporary violent conflicts increasingly situate their battlegrounds in civilian spaces, unarmed or non-combatant men inhabiting these spaces are as vulnerable to violence as women and children. In this regard, continued feminization and infantilization of flight obscure male victimhood which leads to flight by men being treated as an aberration observable in the paradoxical situation of men who flee being simultaneously not "man enough" and dangerous. Contrary to the ridicule and scepticism anti-refugee/migrant discourses in Western Europe direct especially to recent waves of asylum seekers from the Middle East and parts of Africa, flight by men is as legitimate and culturally coherent as flight by women, children and other "inherently" vulnerable populations such as the aged and infirm. This legitimacy and coherence can be understood by situating male flight within heterogeneous exercise of masculinity which creates a division of labour among masculinities by which some masculinities perpetrate violence and benefit from it while others shun it because its incidence renders their socio-economic position tenuous. This chapter uses African cultures and their intersection with colonization to problematize and critique the current anti-refugee rhetoric's caricatured depiction of "other" cultures with regard to masculinity, violence and victimhood.

Violence and Masculinity in Cultural Context

Since Connell's (1995) seminal work, there has been growing consensus that masculinity is a heterogenous concept encompassing multiple versions (see, for example, Beynon, 2002; Haywood & Ghaill, 2003). Various masculinities co-exist in a continuum of relations or non-relations ranging from latent and occasionally overt contestation for hegemony to mutual indifference that can potentially morph into mutual exploitation and dependence in various situations. For example, political elites' normative or socially approved masculinity built around power, wealth and influence in countries that experience violent political contestation tend to dissociate itself from violent and deviant masculinity by consigning perpetration of election violence on the elites' behalf to young economically disempowered men (David & Manu, 2015; Umar, 2003). The young men in turn manipulate and exploit their relationship with political elites in pursuit of their own agendas in their communities.

Masculinities' contestation for hegemony can have physical manifestations through episodes of violence that become ubiquitous and gruesome in war situations. In various cultures, violence is generally associated with masculinity as the antithesis of femininity and boyhood, whose culturally determined meekness relegates them to a state of perpetual vulnerability. However, many cultures institute norms intended to curb wanton violence by drawing the parameters within which violence is legitimate or is for "good reasons" as construed in the specific cultural framework. In many traditional African societies, male violence was culturally sanctioned in line with men's role as warriors and providers of security for their communities and families (Gilmore, 1990). The anti-refugee rhetoric attributes the violent conflicts ejecting Africans out of their countries to "their culture" and disregards the disruptive cultural and political impact of colonialism on the continent's contemporary political configuration.

In many African cultures, men's capacity for violence was counterbalanced by self-restraint which was also an integral attribute of what was regarded as normative masculinity. This balance between traditional masculinization of violence and self-restraint is still essential for contemporary normative masculinity in many parts of Africa and elsewhere around the world. The logic embedded in this balance derived from the contradictory but internally coherent patriarchal norm that men's capacity for violence on its own should suffice to control subordinates and adversaries. "Real" men were culturally expected to be able to stamp their authority without having to actually unleash violence except in cases where refraining from violence was attributed to cowardice which brought their manhood and masculinity into question or disrepute. The exercise of normative masculinity particularly in the public sphere was governed by the expectation on men to conform to cultural norms that drew the parameters within which violence was legitimate. While men were generally taught to be fearless, violence for its own sake subverted normative masculinity and, in many African cultures, dignified men were those who were capable of violence but shunned it in pursuit of dialogue and peace. Even with the training to become fighters as witnessed in "warrior" cultures, to be a man was not to be anti-peace and, indeed, normative masculinity was not inevitably about militarism and violence.

Civilian men in conflict situations still use cultural attributes of hegemonic masculinity to promote non-violent ways of building consensus and peace (Kunz, Myrttinen & Udasmoro, 2018). In traditional "warrior" societies such as the Maasai and Samburu of Kenya, men were taught to fight not for violence's own sake but out of necessity, for example, in self-defence and for honour, prestige and rectitude which were the *sine qua non* of normative masculinity (Gilmore, 1990). This norm can be observed in other cultures that condemn or scorn wanton violence but extol and regard as heroic violence perpetrated in defence of the weak or one's country. "Real" men are those who use violence to defend rather than fight the weak. Many cultures contain within their linguistic repertoire terminology which chides masculinity for being either effeminate or unrestrained as opposed to being heroic and dignified (see Suerbaum, in this volume).

In contemporary armed conflicts, normative masculinity which is predominantly exercised by civilian men often finds itself subordinated by gun-toting men whose deviant masculinity thrives on normlessness and anarchy. Perversion of masculinity's capacity for violence is an essential trait of deviant masculinity which becomes pervasive among armed groups especially where rules of engagement are disregarded leading to war crimes that target civilians. The just war-making principle provides immunity to non-combatants but many civilians are not protected especially from sexual and gender-based violence (Banwell, 2014; see also Carlsen, 2009; Gottschall, 2004) and indiscriminate killings. In so far as cultural norms of masculinity have unravelled and rules of engagement in combat go unheeded, wars are fought in civilian spaces and civilian men, like women, children, the elderly and the infirm, are caught between armed sides. Fighting becomes an archetypal attribute of deviant masculinity when it encroaches into civilian space where it maims and kills unarmed citizens and forces survivors to flee their homes, communities and the country altogether. In this context, non-violent masculinities' choice is limited to either staying and getting killed or fleeing.

"Other" Cultures and Stereotyped Masculinity

Africa underwent a significant political and socio-cultural change in the nineteenth and twentieth centuries when much of the continent came under colonial rule and was transformed through political centralization, conversion to Christianity, Western education and migrant labour (McKittrick, 2003). The nascent monetized economies, formal education and wage employment provided alternative routes to manhood and masculinity beyond traditional rites of passage. The political, social and economic changes that colonialism, Christianization, Western education and wage employment engendered transformed the role of men in marriage and households (Miescher & Lindsay, 2003) and provided new avenues of becoming senior men (Lindsay, 2003). Colonialism, Christianization, Western education and wage employment led to disintegration of traditional masculinities and created new pathways to masculinity that resuscitated traditionally ridiculed and marginalized masculinities, for example, those exercised by physically weak men. These external forces eroded, in Durkheimian terms, the mechanical solidarity and mutual identification in homogenized masculinities that bound together traditional African societies. New heterogeneous and non-conformist versions of masculinity emerged and sought legitimacy and hegemony outside the traditional cultural norm. The violent masculinities currently wreaking havoc in parts of the continent are not as much sanctioned by African cultures as they illustrate colonization's dilution, perversion and disintegration of the traditional ethos regulating masculinity and violence coupled with contemporary global and regional geo-politics.

Although local patriarchies and gender norms persevered, they were now mediated by European ideologies that created "two male genders" that consisted of traditional and modern masculinities (White, 2003). The former perpetuated precolonial cultural practices built around gerontocracy or rule by the elders while the latter adopted colonial European ideologies that emphasized education and training for the colonial bureaucracy (see Hodgson, 2003). The colonial experience transformed traditional societies built upon gerontocracy to modern societies that thrive on individual autonomy and professional relevance to the novel political economy. This diversified routes to manhood and masculinity in which masculinity could now be achieved outside traditional structures of social organization. Colonial transformation engendered a considerable degree of homogeneity across culture in the colonized parts of the world. New and modern masculinities emerged and they owed their allegiance to the colonial political economy from which they drew sustenance. Modern masculinities were initially scorned and described in derogatory terms such as *ormeek* among the Maasai of Kenya which means modern or rather effeminate (Gilmore, 1990; Hodgson, 2003). However, attitudes became positive with recognition that modern masculinities led to wealth and a higher social status through opportunities for employment in the colonial system (Miescher & Lindsay, 2003).

An aftermath of the colonial experience is that some of the societies that are currently generating refugees are not necessarily traditional in the classical anthropological sense but a hybrid of precolonial and colonizing cultures. Nonetheless, some sections of both mainstream and social media stereotypically homogenize, exoticize and demonize refugee masculinities and depict them as a violent and misogynistic polar opposite of law-abiding masculinities in the host country. This exoticization of the "Other" is not confined to countries in the global North because it is also prevalent in host countries in the global South where it generates xenophobic attacks such as those witnessed in South Africa.[1] The media plays a role in dehumanizing refugees by obscuring their individuality and substituting it with an overall image which casts refugees as a security threat and the embodiment of violence, criminality and misogyny (Bleiker, Campbell, Hutchison, & Nicholson, 2013; Scheibelhofer, 2017). The violent conflicts currently generating asylum seekers and refugees are in the global South which remains the home of "other" cultures that continue to be the subject of fear consistent with their depiction in classical anthropology's terminology and in the media. Notwithstanding colonialism's dilution of "other" cultures, masculinities from these cultures continue to be perceived and depicted in various media as alien, inherently wild, uncivilized and violent. In the context of classical anthropology and contemporary global geo-politics, the term culture as it refers to the global South has a pejorative undertone which suggests backwardness and anachronism as opposed to civilization and modernity. Cultures in this region are "othered" in a homogenized way and assigned a compulsive force that suggests lack of individual choice and moral

[1] In the numerous xenophobic attacks that have occurred in South Africa especially since 2008, perpetrators of xenophobic attacks explain their actions by stating that migrant men "steal" their jobs and women, spread diseases and commit crimes such as drug trafficking and promoting child prostitution.

reasoning for the people who live them. To come from "other" cultures is to ostensibly embody a regressive and homogenized masculinity that is incompatible with life in cultured society.

In Europe, anti-refugee discourses curiously portray African and Muslim cultures as trapped in a time warp as implied in terminology such as "Africanization" or "Islamicization" of Europe and the trepidation that this language generates in host countries. To be an African or Muslim man is presumably to be the antithesis of "the Western man". Notwithstanding diversity and differences among masculinities from different societies and cultures, masculinities from "other" places especially those that are characterized by violence and lawlessness are homogenized as having the overall characteristic of being backward, violent and lawless. One of the concerns expressed in Kenya regarding Somali asylum seekers and refugees is that they brought weapons into the country and fuelled armed crimes in addition to engaging in violent conflict themselves (see Jaji, 2013). The irony is that the very deviant masculinity that forced refugee men to flee their countries is superimposed on them thus blurring the line between victim and perpetrator. At a global level, the regionalized and dialectical conceptualization of masculinity translates into South to North migration being synonymous with relocation of dangerous masculinity from "primitive" to "civilized" cultures and places where it becomes an aberration because it is deemed both cowardly and uncultivated and therefore out of place. This leads to the institution of mechanisms to keep refugee men out through various physical, political and legal barriers or, if allowed in, monitor and keep them in check under an overall migration apparatus modelled around what Fassin (2011: 221) refers to as "policiarization of immigration".

Recent waves of forced migration from war zones in the Middle East and parts of Africa to Western Europe were initially met with hospitality. However, concern has been growing about credibility of their asylum claims especially in the wake of sexual assaults on women typified by those that occurred on New Year's Eve in 2015 in Cologne, Germany and a rise in crime blamed on migrants or refugees. This led to sexualization, politicization and securitization of refugees which generated strident calls for closure of the borders, for example, in Austria (see Greussing & Boomgaarden, 2017; Scheibelhofer, 2017). Some sections of both mainstream and social media had already expressed hostility towards refugees even prior to these assaults. The latter fed into incredulity towards male flight from conflict zones and this scepticism was expressed in the exhortation that men fight to take back their country instead of fleeing. For example, these sentiments were expressed on the Twitter hashtag #refugees NOTwelcome (see Kreis, 2017; Rettberg & Gajjala, 2016). The widespread view that refugees did not want to work but wanted to enjoy welfare benefits in asylum countries conjured up the image of a lazy, freeloading and anomalous masculinity sharply contrasted to a presumably self-reliant and independent normative masculinity in the global North. Young men in particular were labelled cowards taking advantage of and infiltrating "genuine" refugees for welfare benefits and nefarious reasons such as terrorism and crime in receiving countries in Western Europe (Hughes, 2017; Narkowicz, 2018; Rettberg & Gajjala, 2016; Scheibelhofer 2017). Whereas refugees in the global North are accused of sponging off the welfare system, their industriousness in the global South often leads to accusations of taking away jobs from locals or displacing them out of businesses. These stereotypical images of refugee men obscure diversity of socio-economic or class backgrounds. Indeed, the incredulity directed to refugees whose socio-economic circumstances do not conform to the quintessential image of refugees as poor emanates from this oblivion to diversity in cultural and class backgrounds characteristic of refugees.

In conformity to feminization and infantilization of vulnerability and victimhood, men's claims to refugee status are delegitimized by the absence of women and children. Without the latter, male asylum seekers become suspicious as implied by labels such as terrorists, rapists and criminals. They are also labelled cowards because the situation supposedly cannot be that bad if women can remain there. Although the anti-refugee rhetoric is agitated by most of the men arriving being young, it, without a hint of self-contradiction, expects these young men most of whom may not be married to arrive with women and children. It interprets being male and single as anomalous to seeking asylum because male victimhood supposedly renders itself visible and credible by attaching itself to women and children. Many anti-refugee arguments also racialize victimhood and dismiss claims to refugee status by men who do not conform to the essentialist view of "refugeeness" predominantly embodied by poor Africans (Flahaux & De Haas, 2016). Over the years, the visual representation of refugees has revolved around racialization and feminization of insecurity (Johnson, 2011) coupled with poverty. For example, poor African women with usually malnourished babies and toddlers in dusty refugee camps made up of makeshift shelters have become the symbol of persecution and the image of "genuine" refugees. In this scheme of things, a combination of masculinity and material comfort is antithetical to this essentialist view of refugees. As more refugees arrived in Europe, many anti-refugee comments expressed on various online platforms filtered insecurity out of material comfort and expressed the sentiment that young men arriving with backpacks and smartphones were not refugees because refugees were "those poor African women and children".[2] Where refugees are poor and African

[2]A comment on a Unite the Right video entitled "There's a lot of smartphones on this boat of 'refugees' on its way to Europe!" reads, "Who gave them the smartphones? Those things are not cheap." https://www.youtube.com/watch?v=0RqsZk61zIo, accessed 28 Aug 2019. Most of the comments on a CNN video in 2011 on African refugees entitled "Life inside the world's largest refugee camp" stated that the solution to the plight of the refugees was having fewer children. https://www.youtube.com/watch?v=cZ80D lch2E, accessed 20 Mar 2019.

but male, the latter transforms them into an "invasion force from Wakanda".[3] Discourses that portray masculinities from war-torn countries and different cultural backgrounds as violent place refugee men in "a situation of questioning" (Herz, 2019: 443), which compels them to deflect suspicion, normalize their masculinity and allay their interlocutors' fears by redirecting attention to their aspirations.

Rethinking "Aberrant" Masculinity and Flight

The "new" and diversified masculinities that emerged with creation of the modern nation-state, centralization of governance and relinquishing of military duties to centralized authority have arguably become the norm notwithstanding their earlier denigration. With modernity, killing a lion, raiding neighbouring communities for booty and hunting as ways to prove that one was a "real" as was the case among the Samburu and Maasai (see Gilmore, 1990) have been supplanted by professional careers and other non-violent expressions of masculinity by which men still amass wealth and acquire power, honour, social status and prestige as expected of them in traditional societies. Masculinities are not rigid and in contemporary times, they have become even more diverse, complex and detached from place as they embrace cosmopolitan, transnational and globalized identities. Masculinities from "other" cultures in the classical anthropological sense of the term are no longer as "exotic" and self-contained as they presumably used to be. They have become fluid and adaptive in order to remain functional in varying socio-political and economic contexts such that performing traditional masculinity does not curtail the capacity to exercise a modern masculinity depending on the context. When men migrate, they can choose to relinquish or adapt masculinities that are detrimental to their successful integration in the receiving country and adopt those that enable them to blend with the host population (see Herz, 2019).

In addition to availability of alternative pathways to manhood and masculinity, not all men are capable of violence or have the inclination towards it. Men are not born but made fighters through culture-specific notions of masculinity. The idea that men have a natural proclivity to fight can be regarded as a myth because perpetration of violence does not emanate from nature but from culture which teaches men to overcome fear and lose inhibition for killing (Gilmore, 1990). Men's socialization into violence through its cultural association with masculinity does not automatically mean that all men perpetrate violence. Without the compulsion of cultural rites of passage, it is not unusual for men to choose not to fight or kill. Instead, they choose masculine pursuits such as providing for the family, non-military professions, political

[3]This is a comment on a France 24 English story entitled "Europe refugee crisis: Number of migration arriving in Spain triples year on year", https://www.youtube.com/watch?v=UGqwE72NPl0, accessed 28 Aug 2019. It is a quip on the movie Black Panther set in an imaginary technologically advanced African kingdom called Wakanda.

careers, excelling in sports and accumulating wealth or economic success for which peace is a prerequisite. Cultures also regulate the fighting spirit they instil in men and many men do not embrace the deviant masculinity that terrorizes and brutalizes civilians in armed conflict. They choose to continue with their lives in non-violent occupations and flee when the situation becomes untenable. In this light, men from countries experiencing violent conflict are not naturally and automatically violent as the stereotype suggests.

The violence unleashed by soldiers, militias, rebels and even criminals and social misfits who take advantage of the lawlessness that accompanies war affects civilian men's capacity to play the culturally defined role of providing for and protecting family members and communities who look up to them for security. In many conflict situations, men fail to protect their wives, daughters and other female relatives from sexual and gender-based violence perpetrated by armed men (Christian, Safari, Ramazani, Burnham, & Glass, 2011; Kelly et al., 2012). Rape of women by armed men is in itself an assertion of dominance over other masculinities by gun-toting deviant masculinity. There are also cases of men, together with their wives and daughters, who have been raped by soldiers and rebels alike (Christian et al., 2011). Even so, the silence surrounding male victimhood regarding sexual and gender-based violence contributes to the invisibility of men's suffering during war. While men's experiences of sexual violence are increasingly recognized, they are still not treated in depth which leads to neglect, isolation and abandonment of male survivors of sexual violence (Chynoweth, Freccero, & Touquet, 2017; Schulz, 2018; Solangon & Patel, 2012; Touquet & Gorris, 2016). In the battle for dominance during war; the masculinity that rejects wanton violence is forced to cede physical and socio-political space.

In contrast to the dominant narrative justifying anti-refugee (men) sentiments, male flight is not necessarily an aberration or a sign of cowardice. Violent conflict does not only mean loss of life; people also lose their homes or witness atrocities that make it difficult for them to stay. Many see flight as the only way they can leave behind traumatic experiences and seek healing and restoration. This is also consistent with normative masculinity and the expectation in many cultures that it engages in self-help and shun wallowing in self-pity. Flight is not essentially effeminate because "real" men are culturally expected to know when to advance or retreat as well as when to stand their ground or escape. Indeed, escape from a strong opponent as opposed to getting killed shows good judgement which constitutes the essence of masculinity. For men with families, flight is consistent with rather than contrary to the culturally sanctioned male responsibility to keep the family safe, secure and well-provided for. Men's quest for refuge in this case is not only about individual security but also about family obligations and quest for a new life in an environment that is conducive for non-violent pursuits such as going to work and raising children in a safe home. Men who are caught up in war find themselves under pressure to fulfil their roles as economic providers and heads of the family (Lwambo, 2013). War destroys livelihoods and material possessions yet to be a man is to have possessions and status. Flight is an effort by men to recover lost possessions and recuperate and restore masculinity in an environment where it has the opportunity and potential to thrive and fulfil its cultural obligations.

While anti-refugee public discourse has raised alarm at mostly young men fleeing violent conflicts in the Middle East and parts of Africa and cast them as invaders or a security threat, flight by young men has a cultural and historical rationale. In many non-nomadic cultures, mobility was historically a masculine venture and this was reinforced by colonial ideologies that targeted men for labour migration and wage employment. More women are now involved in migration but migrant routes that involve danger are largely masculinized. Forced migration involves initiative, risk-taking and fearlessness as people embark on journeys to usually unknown destinations with small backpacks or only the clothes on their backs. Dealing with people smugglers, hiding under haulage truck cargo, being packed in car trunks and crossing the high seas on unsafe boats constitute risk-taking behaviour associated with masculinity although women also embark on these perilous journeys. Risk-taking is observable in the aquatic crossings from Asia to Australia, from Turkey to Greece or from North Africa to Southern Europe. While some of the boats and ferries capsize along the way and significant numbers have perished on these perilous voyages, a "real" man dies trying. A man cannot wait to die or get deterred by fear when providing for the family is "a measure of his manhood" (Gilmore, 1990: 117). Even when it is faced by the atrocities of war, masculinity has the cultural obligation to be resourceful. Men's responsibility to confront the unknown and assess the migration route and destination in order to ensure that they do not put their families in more danger affirms rather than detract from masculinity. Conversely, masculinity is also about sending the family to safety first when the route is deemed less dangerous than staying.

Readers and viewers' comments on online news coverage of refugee men often wonder why the refugees are men and not women.[4] Anti-refugee (men) views expressed in public discourse on various online platforms overlook the cultural rationale or patriarchy's inherently self-perpetuating tendencies that result in more young men than women fleeing as witnessed in some of the arrivals in Europe in recent years. Many cultures still consider sons as more important than daughters because they are the "right" people to carry the family name, keep it alive and perpetuate the family tree or lineage. This means that for many young refugee men, the decision to flee may have been made by the whole family even against their will. In the event of war, many families invest in sons' flight in order to give them the chance to survive, build their

[4]Below a Unite the Right video entitled "There's a lot of smartphones on this boat of 'refugees' on its way to Europe!" one viewer wondered, "Where are the women and children? Oh that's right, the cowardly 'men' left them behind!!!!" https://www.youtube.com/watch?v=0RqsZk61zIo, accessed 28 Aug 2019.

lives and become the family's future. Sons are more useful alive than dead in the logic of patriarchal ideologies that value men over women. It is also sons whose education is prioritized over that of daughters and in the event of war, the sons are sent away before the other members of the family so that they can continue with their education away from the disruption caused by war. Women still do not travel alone in some of the cultures that the refugee men come from mainly because men are deemed more adapted to the hazards of flight. Against these gender norms, male asylum seekers and refugees are not incongruous with how masculinity is constructed, performed, interpreted and adapted to war situations.

It is also not only the poor who flee. Refugees are made up of people from a continuum of socio-economic statuses or classes. Indeed, many poor people remain in the country of origin as internally displaced persons or flee to neighbouring countries because they lack adequate resources to undertake long and costly journeys that traverse continents or are intra-continental but expensive and dangerous. Traversing continents requires considerable financial investment and strong will which explains why some of the people who arrived in Europe do not conform to the stereotypical images of "genuine" refugees. It is easier for people from stable economic backgrounds than for the extremely poor to move. Some of the young refugee men may have had jobs or come from economically stable backgrounds that enabled them to possess the middle-class trappings that render their claims to asylum anomalous and suspect. Possession of smartphones which, among others, has emerged as a red flag regarding credibility of asylum claims ensures digital connectivity which plays an important role in the planning, navigation and recording of journeys (Gillespie, Osseiran, & Cheesman, 2018). They also ensure that the refugees maintain contact with left-behind families.

Conclusion

The anti-refugee rhetoric depicts refugee men as an aberration not only because flight "feminizes" men but also because the cultures they come from are presumably archaic, transgressive of "civilized" gender norms and criminal. "Other" places supposedly produce "other" masculinities that need to be policed and re-socialized so that they do not transplant to the host country patriarchal tendencies that are backward if not illegal in "civilized" societies. Refugee masculinities thus provide the arena in which the politics surrounding the "self" and "others" play out. Yet, depiction of male flight as an aberration not only overlooks how colonization mediated "other" cultures in terms of notions of masculinity such that the "Other" is not necessarily as alien as suggested by the anti-refugee rhetoric. Apart from regulating masculinity and violence, traditional cultures have been superseded by diversified, globalized and cosmopolitan versions of masculinity capable of adapting to host country cultures. Men who exercised non-military masculinities and have the resources to undertake long journeys are among the people who flee violent conflicts in search of destinations that are compatible with their exercise of masculinity. Flight by men is not necessarily an aberration when situated within the patriarchal logic in which men's responsibility to face danger ahead of their families is counterbalanced by the privilege to enjoy patriarchal benefits such as young men's lives, welfare and future being prioritized over the needs of young women. While cultures that extol the virtues of a self-sacrificing masculinity scorn cowardice, flight by civilian men, instead of being an aberration, is a logical course of action where unarmed masculinities are confronted by an armed opponent.

References

Banwell, S. (2014). Rape and sexual violence in the Democratic Republic of Congo: A case study of gender-based violence. *Journal of Gender Studies, 23*(1), 45–58.

Beynon, J. (2002). *Masculinities and culture*. Buckingham, England: Open University Press.

Bleiker, R., Campbell, D., Hutchison, E., & Nicholson, X. (2013). The visual dehumanisation of refugees. *Australian Journal of Political Science, 48*(4), 398–416.

Carlsen, E. (2009). Ra/pe and war in the Democratic Republic of the Congo. *Peace Review, 21*(4), 474–483.

Christian, M., Safari, O., Ramazani, P., Burnham, G., & Glass, N. (2011). Sexual and gender based violence against men in the Democratic Republic of Congo: Effects on survivors, their families and the community. *Medicine, Conflict and Survival, 27*(4), 227–246.

Chynoweth, S., Freccero, J., & Touquet, H. (2017). Sexual violence against men and boys in conflict and forced displacement: Implications for the health sector. *Reproductive Health Matters, 25*(51), 90–94.

Connell, R. (1995). *Masculinities*. Sydney, Australia: Polity Press.

David, N., & Manu, Y. (2015) Democracy, youth and violent conflicts in Nigeria's fourth republic: A critical analysis. *Research on Humanities and Social Sciences, 5*(2), 159–171.

Enloe, C. (1992). The gendered gulf. In C. Peters (Ed.), *Collateral damage: The "New World Order" at home and abroad* (pp. 93–110). Boston, MA: South End Press.

Fassin, D. (2011). Policing borders, producing boundaries: The governmentality of immigration in dark times. *Annual Review of Anthropology, 40*, 213–226.

Flahaux, M.-L., & De Haas, H. (2016). African migration: Trends, patterns, drivers. *Comparative Migration Studies, 4*(1), 1–25.

Gillespie, M., Osseiran, S., & Cheesman, M. (2018). Syrian refugees and the digital passage to Europe: Smartphone infrastructures and affordances. *Social Media + Society, 1*, 1–12.

Gilmore, D. (1990). *Manhood in the making: Cultural concepts of masculinity*. New Haven, CT: Yale University Press.

Gottschall, J. (2004). Explaining wartime rape. *The Journal of Sex Research, 41*(2), 129–136.

Greussing, E., & Boomgaarden, H. G. (2017). Shifting the refugee narrative? An automated frame analysis of Europe's 2015 refugee crisis. *Journal of Ethnic and Migration Studies, 43*(11), 1749–1774.

Haywood, C., & Ghaill, M. (2003). *Men and masculinities: Theory, research and social practice*. Buckingham, England: Open University Press.

Herz, M. (2019). "Becoming" a possible threat: Masculinity, culture and questioning among unaccompanied young men in Sweden. *Identities, 26*(4), 431–449.

Hodgson, D. (2003). Being Maasai men: Modernity and the production of Maasai masculinities. In L. Lindsay & S. Miescher (Eds.), *Men and masculinities in modern Africa* (pp. 211–229). Portsmouth, NH: Heinemann.

Hughes, G. (2017). The chastity society: Disciplining Muslim men. *Journal of the Royal Anthropological Institute, 3*(2), 1–18.

Jaji, R. (2013). Somali asylum seekers and refoulement at the Kenya–Somalia border. *Journal of Borderlands Studies, 28*(3), 355–368.

Johnson, H. (2011). Click to donate: Visual images, constructing victims and imagining the female refugee. *Third World Quarterly, 32*(6), 1015–1037.

Kelly, J., Kabanga, J., Cragin, W., Alcayna-Stevens, L., Haider, S., & Vanrooyen, M. (2012). "If your husband doesn't humiliate you, other people won't": Gendered attitudes towards sexual violence in eastern Democratic Republic of Congo. *Global Public Health, 7*(3), 285–298.

Kreis, R. (2017). #refugeesnotwelcome: Anti-refugee discourse on Twitter. *Discourse & Communication, 11*(5), 498–514.

Kunz, R., Myrttinen, H., & Udasmoro, W. (2018). Preachers, pirates and peace-building: Examining non-violent hegemonic masculinities in Aceh. *Asian Journal of Women's Studies, 24*(3), 299–320.

Lindsay, L. (2003). Money, marriage, and masculinity on the colonial Nigerian railway. In L. Lindsay & S. Miescher (Eds.), *Men and masculinities in modern Africa* (pp. 138–155). Portsmouth, NH: Heinemann.

Lwambo, D. (2013). "Before the war, I was a man": Men and masculinities in the eastern Democratic Republic of Congo. *Gender & Development, 21*(1), 47–66.

McKittrick, M. (2003). Forsaking their fathers? Colonialism, Christianity, and coming of age in Ovamboland, northern Namibia. In L. Lindsay & S. Miescher (Eds.), *Men and masculinities in modern Africa* (pp. 33–51). Portsmouth, NH: Heinemann.

Miescher, S., & Lindsay, L. (2003). Introduction: Men and masculinities in modern African history. In L. Lindsay & S. Miescher (Eds.), *Men and masculinities in modern Africa* (pp. 1–29). Portsmouth, NH: Heinemann.

Myrttinen, H., Khattab, L., & Naujoks, J. (2017). Re-thinking hegemonic masculinities in conflict-affected contexts. *Critical Military Studies, 3*(2), 103–119.

Narkowicz, K. (2018). "Refugees not welcome here": State, church and civil society responses to the refugee crisis in Poland. *International Journal of Politics, Culture, and Society, 31,* 357–373.

Rettberg, J., & Gajjala, R. (2016). Terrorists or cowards: negative portrayals of male Syrian refugees in social media. *Feminist Media Studies, 16*(1), 178–181.

Scheibelhofer, P. (2017). "It won't work without ugly pictures": Images of othered masculinities and the legitimisation of restrictive refugee-politics in Austria. *NORMA, 12*(2), 96–111.

Schulz, P. (2018). The "ethical loneliness" of male sexual violence survivors in northern Uganda: Gendered reflections on silencing. *International Feminist Journal of Politics, 20*(4), 583–601.

Solangon, S., & Patel, P. (2012). Sexual violence against men in countries affected by armed conflict. *Conflict, Security & Development, 12*(4), 417–442.
Touquet, H., & Gorris, E. (2016). Out of the shadows? The inclusion of men and boys in conceptualisations of wartime sexual violence. *Reproductive Health Matters, 24*(47), 36–46.
Umar, M. (2003). The role of youth in politicians in political violence in north-east Nigeria. In A. Olasupo (Ed.), *Electoral violence in Nigeria: Issues & perspectives* (pp. 207–216). Lagos, Nigeria: Friedrich Ebert Stiftung.
White, L. (2003). Matrimony and rebellion: Masculinity in Mau Mau. In L. Lindsay & S. Miescher (Eds.), *Men and masculinities in modern Africa* (pp. 177–191). Portsmouth, NH: Heinemann.

Rose Jaji is senior lecturer in the Department of Sociology at the University of Zimbabwe. Her research areas of interest are migration/refugees, gender, conflict, and peace-building. She has published on migrant/refugee masculinities and femininities, refugees and social technology, identity, refugee hosting and border crossing as well as gender and peace-building. Her latest publication is a book entitled *Deviant Destinations: Zimbabwe and North to South Migration* (Lexington Books, 2020).

CHAPTER 24

Constructions of Masculinities, Class and Refugee Status Among Syrian Refugee Men in Egypt

Magdalena Suerbaum

Abstract This chapter explores constructions and negotiations of masculinities in contexts of (forced) migration. Geographically, I focus primarily on the Middle East, especially on Egypt as a destination country, and the situation of Syrian refugees who settled there. This chapter engages with three dimensions that affected Syrian refugee men's lives: their claim of belonging to the middle class, their difficulty in coming to terms with the refugee label, and the challenges related to unemployment and social exclusion and how they defined notions of manhood during forced migration. In this chapter, I illustrate the significance of an intersectional analysis and argue that masculinities in times of forced migration need to be understood as inherently relational, actively constituted and responsive to changes in the everyday life.

Introduction

This chapter is dedicated to the study of masculinities in relation to forced migration. It traces how academic attention to men and masculinities in (forced) migration contexts has developed and changed over the last decades. Drawing on existing literature from the fields of (forced) migration studies, Middle Eastern studies and gender studies, as well as my own work on constructions of masculinities among Syrian refugee men in Egypt, this chapter explores how masculinities are challenged and subsequently subject to change

M. Suerbaum (✉)
Max-Planck-Institute for the Study of Religious and Ethnic Diversity, Göttingen, Germany
e-mail: Suerbaum@mmg.mpg.de

C. Mora and N. Piper (eds.), *The Palgrave Handbook of Gender and Migration*, https://doi.org/10.1007/978-3-030-63347-9_24

as a result of forced migration.[1] In the process, the importance of an intersectional analysis will be highlighted. This chapter will focus on three dimensions of masculinities and the way they play out in male refugees' lives: firstly, class belonging; secondly, the refugee label and its incompatibility with male ideal role models; and thirdly, the challenges of finding work and remaining in the position of the provider in relation to male refugees' sense of manhood.

In my own research on Syrian refugees in Egypt, applying an intersectional lens foregrounds the significance of social class belonging. For interviewed Syrian men, proving one's success as a man is intimately related to being accepted as someone from the middle class.[2] Furthermore, this chapter stresses that most men during forced migration have an uneasy relationship with the label 'refugee'; in fact, they struggle with it and, as a result, attempt to distance themselves from it. Syrian men in Egypt, for instance, make use of various strategies to clarify why they do not self-define as refugees. Lastly, this chapter discusses the importance men attribute to paid labour and social inclusion. Circumventing unemployment and thereby managing to be the provider for one's family is proof of success for Syrian refugee men interviewed in Egypt. Ultimately, this chapter demonstrates that constructions of masculinities in relation to forced migration should be understood as intersectional, relational and context-specific.

My arguments are based on ethnographic research conducted in Cairo in 2014/2015. I collected sixty-one semi-structured interviews (conducted mainly in Arabic) and engaged in countless informal conversations and daily participant observation when teaching English classes, visiting families and attending events organised by local NGOs. Most of the interviews were recorded. Discussions of gender were mostly implicit, even though I sometimes asked during interviews what interlocutors perceived as the specific challenges of forced migration for men and women. Research participants came from various cities and villages in Syria. Most of them had arrived in Egypt in 2012 or early 2013 and described themselves as middle class, as educated and from an affluent background.[3] They were between 18 and 70 years old. Among the research participants were many students who were

[1]An extended discussion of this chapter can be found in Suerbaum (2020).

[2]Class is inherently about differentiation and gender has always been a way to articulate differences (Scott, 1999: 60). Moreover, one's sense of belonging to the social space is classed and gendered. Hence, these two axes influence the making of masculine (and feminine) subjectivities and define access to opportunities and resources (Ghannam, 2013: 165).

[3]It is important to stress that the group of Syrians that arrived in Egypt after the uprising is not homogenous. The majority of research participants seemed to have a similar socio-economic background, however, I also got in touch with Syrians who I locate, based on their narratives, at the margins of middle-classness, among them men whose families possessed big companies, or others who came from the poor Syrian countryside. Furthermore, several Syrian families had lost their savings since their arrival in Egypt and were severely impoverished.

unmarried. Older interlocutors were mostly married and active in the Egyptian labour market.

As far as the use of the term 'refugee' in this chapter is concerned, it is understood as a definition and marker of identity that has legal, material and symbolic implications. Following Malkki (1995: 496), the classification as refugee has analytical usefulness only as a descriptive legal rubric with a plethora of socio-economic statuses, biographies and psychological conditions hiding behind its façade. Through the dynamic and not always consensual process of labelling as a refugee, one 'becomes' rather than one 'is' a refugee (Sauerbaum 2020). Thus, the meaning of 'refugeeness' (Malkki, 1995)—the state of being a refugee—in a specific social context is determined not only by external factors, but also by individual agency and experiences. While refugees are subjected to a series of determining codifications and administrative rulings that govern how 'they should be assessed and treated, and how they should think of themselves and their actions', they are simultaneously engaged in self-making by struggling against imposed knowledges and practices (Ong, 2003: 16).

Because of its focus on men in times of forced migration, this chapter needs to engage briefly with the prominent, ongoing debate on the question of categories. Following Crawley and Skleparis (2018), the differentiation between 'refugees' and 'migrants' is for many reasons problematic. These categories do not capture the messy reality of most people who are on the move. Moreover, they imply a linearity and clarity regarding people's choices and motivations. The differentiation between those terms infers that there is a simple, neutral ordering process that manages to tie people's belonging to one of the two categories. While Crawley and Skleparis (2018: 52) consequently argue that forced migration should not be perceived and studied as an exceptional form of migration, there are also good reasons to foreground forced displacement and highlight how it differs from other migration processes. Long (2013: 21), for instance, stresses that refugees 'cannot return "home"' and do not possess the rights that an intact state-citizen relationship can guarantee to them. Jansen (2008: 186) argues that a violent and sudden form of forced displacement may cause a harsh experience of loss in terms of contact to loved ones, wealth, status and recognition. Following Lubkemann (2008: 193), forced displacement relates to "the transformation of lifescapes in ways that render essential life projects harder to achieve". Thus, the conditions of displacement include loss of access to social, material and symbolic resources that were once available. In my research with Syrian refugee men, it was the extreme violence of the Syrian civil war, the speed with which Syrians had to flee after the outbreak of the uprising, the accumulation of various losses that they experienced as well as the war's uncertain outcome that justify the classification of this phenomenon as 'forced migration'.

In this chapter, I argue that the study of refugee men from a gender perspective is necessary because an analysis of how the reformulation of life projects and coping with various losses affects men has often been non-prioritised, Yet, such an analysis adds to our overall understanding of constructions of masculine ideals and trajectories as well as men's perception of and relations with women. By asking how men try to meet expectations and attempt to materialise the social norms that define them as men in a context of forced displacement, we learn a lot about changing gender regimes, strategies of (re-)constructing masculinities and coping mechanisms in times of stress, uncertainty and fear. This does not mean that there is a homogeneous 'experience' of being a male refugee, rather men respond to forced displacement differently depending on their social class background, nationality, age, education, etc.

This chapter begins with an overview on the developments of the study of masculinities over the past decades, moves on to the study of men and masculinities in the Middle East, and then engages with some literature that applies a gendered lens to the study of (forced) migration. Subsequently, I introduce intersectionality as a useful approach in order to analyse contexts in which men become refugees. This chapter then discusses men's ambivalent reactions to the refugee label and the importance of work in times of (forced) displacement.

The Development of Masculinity Studies

The study of masculinities has flourished over the last four decades. The introduction of the concept of 'hegemonic masculinity' (Carrigan, Connell, & Lee, 1985; Connell, 1995) was a significant moment in the developing discipline. This concept, which defines the existence of various masculinities competing for power based on the subordination of women, received wide recognition. However, it was also severely criticised with critiques referring to: the need to account for men's various strategies to construct themselves as men (e.g. Wetherell & Edley, 1999), the concept's ignorance of the potential existence of several hegemonic masculinities (see Cornwall & Lindisfarne, 1994: 19) and the constructed dualism of hegemonic versus non-hegemonic masculinities (Demitrou, 2001). In the 1990s, a wave of masculinity studies came into being, influenced by the advent of post-structural theory that has a common focus on the social and cultural construction of masculinity (Edwards, 2006). Another aspect that has concerned the study of manhood was the question of whether there was an ongoing crisis of masculinity (2006: 22). Recently, the study of masculinities is more often framed in the plural showing that there are multiple forms of manhood at play rather than a single form. It focuses on both individuals and institutions, such as the state, security services and capital markets, which impact on the social order (Connell, 2014).

Inspired by various theories and concepts (Butler, 1999; Connell, 1995; Haywood & Mac an Ghaill, 2003; Kimmel, 1994; Wentzell, 2015), I suggest understanding masculinities as plural, complex and actively constituted with men combining various images of ideal manhood. Conceptualising masculinities as diverse, changing and intersectional, it is additionally useful to perceive of constructions of masculinities as a constant fight for recognition and approval by other men, while women are frequently relegated to a 'currency' which men use to improve their ranking on the masculine social scale (Kimmel, 1994: 128/129). It should be noted, however, that this currency is not always women and that the audiences in front of which men try to prove themselves are not always homosocial. Furthermore, masculinities need to be conceived as subjectively and actively constituted (Haywood & Mac an Ghaill, 2003). They are composites of various acts, attitudes and relationships that are available to men at a given time in a specific context. These composites are then weaved by the individual into a "coherent masculine selfhood" (Wentzell, 2015: 179)

Studying Masculinities in the Middle East

The study of masculinities in the Middle East has developed significantly over the past decades (see for example Ghoussoub & Sinclair-Webb, 2000; Inhorn, 2012; Monterescu, 2006; Naguib, 2015). Recent studies have focussed on men's multiple responses to changes in their lives, such as marriage and fatherhood, ageing and illness (Inhorn, 2012); how men who belong to a minority reframe their subject positions in heterogeneous places (Monterescu, 2006); and the importance men ascribe to their role as nurturers and providers for their families (Naguib, 2015). Of particular relevance is Ghannam's (2013: 7) notion of 'masculine trajectories' which she introduces to analyse constructions of masculinities in urban Egypt. She highlights that men, over the course of their lives, identify with different ideals, norms and values. Hence masculine trajectories are neither neat nor fixed in the order of their stages and they are often ambiguous and contradictory. Another recurring focus in the discipline is the engagement with aspects of masculinities in contexts of war and militarization. Haugbolle (2012), for instance, analyses how men who fought in the Lebanese civil war relate to the public discourse on masculinity in the post-war period, while Peteet (1994) and Jean-Klein (2000) discuss the violence Palestinian men experience in their fight against the Israeli occupation. Peteet (1994) observes that young men turn the violence they have experienced at the hand of the Israeli military into signs of brave heroic manhood. Jean-Klein (2000) found in her study that mothers on behalf of their sons narrate the violence they endured, while the young men remain silent and thus adhere to the expectation in men to show 'humility and modesty' (2000: 115). All of these studies turn the spotlight on the effects of socio-economic, cultural and political processes on constructions of masculinities. Authors stress that being a man follows the logic and specific experiences of everyday life. In her critique of contemporary scholarship dealing with masculinities in the Middle East, Hasso (2018) condemns ahistorical naturalisations of Arab or Muslim masculinities and thus seeks to avoid romanticising and producing false binaries, such as East versus West and traditional versus modern. She argues that masculinities need to be conceptualised as plural and changing, impacted by 'capitalism, colonialism and imperialism, but also 'local' systems and ideas'

(2018). Furthermore, she calls for a close analysis of men's embodiments and emotions rather than an over-emphasis on 'the abstract workings of ideology, law, institutions and systems'. These points, especially the focus on embodiments and emotions, resonate with my own study and analysis of Syrian male refugees who fled to Egypt.

Masculinities in (Forced) Migration Scholarship

When reviewing the literature conducted on migration contexts in the past decades, one recognises a notable scholarly interest in gendered approaches to the study of migration beginning in the 1980s. After decades of scholarly interest in migrants who happened to be mostly male, female (im)migrants became the focus of scholarly attention because of their increasing number, and their new position as wage earners or heads of households rather than 'dependants' (Nolin, 2006: 4). Scholarship on gender and migration went through several stages, with the first stage often being attempts to complement the picture by bringing female migrants' experience to the surface (Donato, Gabaccia, Holdaway, Manalansan, & Pessar, 2006: 13). This stage was preceded by a focus on the whole migration process as a gendered phenomenon and a search for ways to re-think predominant theories in migration studies from a gender perspective (2006). Over the last years, several scholars have pointed to the lack of gendered analysis of male migration. Mahler and Pessar (2006: 51) recognise an over-emphasis on female migrants and call for a correction of this imbalance by taking into consideration men's experiences and gender relations during migration (see also Jaji in this volume). Donaldson and Howson (2009: 210) support this view claiming that migration needs to be studied as providing with potential for change among male migrants and renegotiation of 'hegemonic masculine identifications'. Charsley and Wray (2015: 404) stress that especially scholarship on the regulation of migration lacks a focus on men and masculinities and that there is a need for the deconstruction of the essentialised, negative image of the 'migrant man'. Among the existing studies of the last decade that applied a gendered lens to the study of male migration, there is a tendency to present men's migration as linear transformations into adulthood. Monsutti (2007: 169), for instance, portrays Afghan men's migration to Iran as 'a way of life' to broaden their social network. The dangerous journey and the social separation of young migrants from their families are conceptualised as 'a rite of passage' to manhood. In a similar vein, Osella and Osella (2000: 118) suggest that migration from South India to the Gulf may move men closer to what is widely accepted and appreciated as 'mature manhood'.

Focusing explicitly on the experience of forced migration, Indra (1999) called, two decades ago, for a gendered, context-dependent, intersectional, in-depth analysis of the lives of male and female forced migrants. Early literature suggests gendered challenges for men and a potential for redefinition of gendered roles in contexts of refugee migration. Matsuoka and Sorensen

(1999: 226), for instance, argue that neither women nor men live in gender isolation and therefore analyse how forced migration change gendered spaces for men and women. McSpadden (1999: 246) found a dialectic relationship between the 'situational and relational construction of gender' and the 'shifting nature of the refugee experience'. Despite the problematic use of 'refugee experience' in the singular,[4] the value of her analysis lies in the importance she ascribed to the study of refugee men, their social places, pursuits and ideals during forced migration. Analysing the context of the British asylum system and its management of men whose asylum application were turned down, Griffiths (2015: 469) argues that the study of refugees is not only predominantly depoliticised and dehistoricised but also often 'degendered'. She finds that the immigration struggle is frequently intimately interlinked with asylum-seeking men's sense of self and with the perceptions of masculinities among decision makers in the asylum system of the country of destination (2015). Ultimately, Griffiths shows the stigma and shame that are often involved in the process of seeking asylum for men.

In the so-called refugee crisis in Europe, there has been a focus on the intersection of masculinity and age with a specific interest in unaccompanied male minors, and their representation as a 'threat', especially in terms of their sexuality and views on gender defined by their culture of origin (Allsopp, 2017; Herz, 2019). Unaccompanied male minors are often interpreted through a racialised, gendered lens that differs significantly based on the male refugee's age: while children are predominantly seen as mere victims, boys and young men are perceived as inherently dangerous by virtue of their gender (Pruitt, Berents, & Munro, 2018). Directly and indirectly, engagement with young unaccompanied male refugees brings to the fore the focus on gendered notions of vulnerability and how these are static, misleading and penalising to those refugees and asylum seekers who do not conform to the dominant image (Palillo, 2018).

An Intersectional Analysis

Most of the reviewed literature on gendered approaches to (forced) migration urges researchers to apply an intersectional lens (Charsley & Wray, 2015; Hyndman, 2010; Indra, 1999; Mahler & Pessar, 2006). Intersectionality is a concept and theory that was introduced by black feminist and legal scholar Kimberley Crenshaw (1991: 1245), who argues that a focus on intersectionality is needed 'to account for the multiple grounds of identity when considering how the social world is constructed'. Stressing that oppression can never be reduced to a single cause, Crenshaw shows that oppression

[4]It is problematic to speak of a 'refugee experience' because it suggests that there is an essentialised, coherent, unified situation that all people who are labelled refugees simultaneously encounter and endure. Several scholars critiqued refugee studies for producing such a single, essentialized, transhistorical refugee condition (e.g. Chimni, 2009; Hayden, 2006; Malkki, 1995).

and discrimination have their roots in overlapping social grounds and highlights that different social identities influence each other. Importantly, an intersectional approach does not ask for the creation of a 'list of social divisions' that is assumed to describe and define the individual. Instead, it is 'an important corrective to essentialising identity constructs that homogenise social categories' (Anthias, 2012: 3). Hence, researchers need to pay tribute to differences among individuals and the influence of specific social contexts which define access to resources and cause diverse forms of oppression (Yuval-Davis, 2006: 199/200). The wide use of intersectionality has been subject to critique: Hasso (2018), for instance, emphasises that intersectionality ignores the impacts of colonial and imperial relations as well as modern state laws on gender. Henry (2017) wonders, when looking at the application of intersectionality to the study of marginalised men in the military, whether 'the concept of intersectionality has become detached from identity politics and those racialized subjects for whom it was written'. She urges researchers to avoid using the concept to merely capture how multiple differences affect individuals. Instead, the focus should be set on the differences that cause multiple and intersecting oppressions among those who are already marginalised (Henry 2017).

Applying an intersectional lens to the study of masculinities means analysing how notions of manhood are shaped and affected by forms of social differentiation, such as class, race or age (see also Lutz & Amelina in this volume). From an intersectional perspective, ideas, images and practices of masculinities are perceived to vary for different groups. The specific intersections of masculinities with other social divisions can be potentially damaging or strengthening (Christensen & Jensen, 2014: 69). With the help of an intersectional approach, the spotlight can be turned on the strategic and performative elements of formations of masculine identities. Masculinities are context dependent, respond to the surrounding sensual realms and draw on the specific resources available (Hopkins & Noble, 2009: 813–816). Engaging in an intersectional analysis of how male Syrian refugees in Egypt construct themselves as men show the significance of social class, especially the association with middle-class life. Furthermore, an intersectional analysis brings to the fore Syrian men's ambivalent relationship with the Egyptian and Syrian states. Their plight as men who recently fled from a war-torn country is thus defined by their struggle to reclaim middle-class belonging and their insecure position vis-à-vis the state (Suerbaum 2020).

Masculinities and Middle-Class Identity in Forced Migration

Applying intersectionality as a lens to the analysis of constructions of masculinities among Syrian men in Egypt helps to identify the relevance of social class, that is a phenomenon 'associated with group experiences of socioeconomic difference' (Heiman, Liechty, & Freeman, 2012: 8) in forced migration contexts. An analysis that pays attention to middle-class belonging needs to account for middle-classness as a category combining aspiration, longing and anxiety (2012: 7).

In my research, Syrian men displayed clear identifications as middle class. In their narratives about the past and their aspirations for present and future, Syrian men clarified that they wanted to be seen as middle-class men. Their wish to belong to and reclaim middle-classness in Egypt affected their thinking, ambitions, judgements and behaviour. Their identity as men was intimately related to middle-classness and they predominantly valorised conservative gendered rules, norms and complementary gender roles (Suerbaum, 2020).[5] While being middle class was such a central claim among Syrian refugee men interviewed in Egypt, attention paid to social class in the fields of refugee studies has been limited. An early ethnographic study by Finnan (1981) shows the importance of refugees' occupation for self-identification, sense of class belonging and adaptation in the host country. Refugees engaged in creating a middle-class image of their occupation and drew classed boundaries within the industry they worked in. Moreover, several studies delineate that a refugee remains a 'classed' person (McSpadden, 1999: 251; Van Hear, 2006), that the start in the receiving society is dependent on one's socio-economic background and can cause frustration and misrecognition if one cannot hold on to one's middle- or upper-class background (Kleist, 2010: 198), and that it is challenging to transfer a celebrated social status from one's country of origin to the country of resettlement (Jansen, 2008: 182). McSpadden's (1999: 251) analysis of the impact of class on constructions of masculinities in forced migration contexts shows the relevance of class and status for men's judgement of 'good' and 'bad' lives. She highlights that upper-class men 'favoured class-congruent life trajectories and at least verbally rejected what they judged to be class inappropriate'. Focusing on transnational contexts, Levitt and Glick Schiller (2004: 1015) stress that refugees and migrants often 'occupy different gender, racial and class positions within different states at the same time'. Among Zimbabwean migrants in the UK, McGregor (2008: 467) found that class belonging was established in relation to various aspects appertaining to their lives in both host and home countries, such as migrants' current workplaces, neighbourhoods and the host state's legal categories, as well as family networks and social judgements in the homeland. Following Bonjour and Chauvin (2018: 9), classed conceptions of masculinities are at the core of the men's migration trajectories, their strategies and motivations.

Coming back to the case of Syrian refugee men in Egypt, belonging to the middle class was an important and recurring theme in conversations as part of my research. Most of them defined themselves as middle class on

[5]The idea of 'correct' femininity plays a crucial role in gendered constructions of middle-class belonging (Jones, 2012: 146). It is often times through static, complementary, conservative gender roles that class divisions are marked and naturalised (Donner, 2015; Vera-Sanso, 2016).

the basis of properties, such as a chalet at the beach or a beautiful house their family owned in Syria. Syrian men and women alike referred to their family's former status and influence and highlighted that middle-classness was dependent on the region one hailed from as well as on one's upbringing and education. Moreover, middle-classness was based on work or employment: a man's status was defined based on his income, type of profession, productivity, inventiveness and success (Suerbaum, 2020). Conversely, a woman could claim middle-classness if she did not have to work. This shows that middle-classness was expressed through certain morals and values and manifested itself in forms of accepted (gendered) behaviour, certain expectations of how one should be seen and treated by others and in people's dreams and aspirations. Syrians men's use of middle-classness was thus a marker of identity, an approach to life, as well as a major aspect of their understanding and construction of masculinities and their dealing with forced displacement. Nevertheless, it is important to keep in mind that belonging to the middle class is a widespread aspiration combining a continuum of different socio-economic and symbolic positions (Schielke, 2012: 36).

Being Challenged by the Label 'Refugee'

In addition to refugee men's sense of class belonging, their positioning as objects of humanitarian assistance requires attention. Szczepanikova (2010: 463) stresses that there is a lack of knowledge about the gendered character of being dependent on assistance, for example in relation to NGOs' role as interlocutor between the asylum seeker or refugee and the state imposing performances of apolitical, victimised and infantilised refugeeness seen as required in order to lead to successful outcomes (see also Jaji, in this volume). Her research of refugees in the Czech Republic demonstrates that the gendered performances of refugeeness are part of women's realm, while both refugee men and women believed that it was undignified for men to admit to one's dependence and need for assistance in an emotional way (2010: 465). According to prevailing humanitarian understandings, women are considered to be the primary recipients of care (Hyndman & Giles, 2011). As per Turner's research (2019a), when humanitarian actors were asked about refugee men's needs as objects of humanitarian aid, they signalled interest but gave no answer. His analysis of the Syria refugee response by the humanitarian sector in Jordan shows that a perception of men as objects of humanitarian care would disrupt dominant, binary ideas of women as needy and men as agential. Moreover, the difficulties in grasping Syrian men's positions as recipients of humanitarian aid were intertwined with particular conceptions of 'male Arabness', for instance, that Syrian men were independent and always able to find work (2019a: 14). The ambiguity over Syrian men's positioning as objects of humanitarian aid can however not be approached, if they are simply labelled as 'vulnerable' (Turner, 2019b). Instead, Turner suggests that questioning the centrality of the notion of vulnerability in humanitarian aid and the domination and disempowerment that defines humanitarian work in refugee contexts would be the way forward.

Since the refugee label is so clearly associated with victimisation, infantilising and denial of agency (Hyndman, 2010; Szczepanikova, 2010; Turner 2019a, 2019b), most people who are categorised as 'refugees' struggle with the imposed label. In my study, the Syrian men interviewed in Egypt refused the label for as long as possible and consequently avoided approaching the UNHCR for registration. In a study conducted by Ayoub and Khallaf (2014: 22), Syrian refugees expressed that the main reason for their refusal to approach the UNHCR was that the refugee title was connoted with weakness, desperation and misery. It was because of their rising social and economic insecurity in Egypt that the majority eventually registered with the UNHCR and accepted to hold the title 'refugee'. Nevertheless, upon their categorisation as refugees, they engaged in various acts to 'unbecome' refugees (Suerbaum, 2018b). For instance, they described their economic success, stamina and inventiveness to prove their incompatibility with the stigmatising refugee label. Moreover, they engaged in 'othering' processes, defining those who travelled to Europe to seek asylum there as the 'real refugees'. In addition to erecting boundaries to prove themselves as proper and respectable middle-class men, they applied reverse stigmatisation to the Egyptian host population after having experienced public denunciation and discrimination (2018a).

Unemployment, Social Exclusion and Marginalisation

Apart from struggling with the refugee label, refugee men are likely to suffer from the inability to secure employment and counter marginalisation. Jaji (2009: 182) argues that refugee men in Nairobi from the Great Lake region are deeply challenged by unemployment, which not only impacts on their economic well-being, but also on their ability to conform to dominant cultural definitions of masculinity. Furthermore, the lack of economic capital translates into social exclusion, since these young refugee men need financial means to be accepted by the peer group and by women who come into their lives. Additionally, they are unable to provide from the distance for their families at home (2009: 183/189). Thus, these men cannot meet pre-flight goals and ideas of masculinity. Instead, they develop alternative masculinities, discarding former ideals because of the dissonance they perceive between these values and the reality of forced migration (2009: 190). In a similar vein, Kleist (2010) shows in her study of Somali men in the diaspora how they had to readjust aspirations and status perceptions. They predominantly perceived their life in exile as humiliating and experienced the ambivalent reception, as well as their unemployment and exclusion 'as a situation of social misrecognition' (Kleist, 2010: 191). In order to deal with this misrecognition, Somali men engaged in distancing themselves from 'other' failed and marginalised Somali men, naturalised the male breadwinner as the ideal, and tried 'to establish a "traditional" gender baseline' (2010: 200). Moreover, some Somali men got involved in associations or community activities that helped them to restore 'individuality, status and respectable masculinity'. Jansen (2008: 183) highlights how Bosnian men who fled the civil war experience a form of status loss in exile that is not recoverable, since it relates to their past position in a setting that no longer exists. These 'misplaced' Bosnian men held forms of masculinity dearly that were considered irrelevant in their current context, in which they had lost their self-image as heads of their households due to their dependence on benefits, their wives' employment or their own low paid jobs (2008: 187).

In my study among Syrian refugee men in Egypt, I found too that the value of employment and income generating opportunities was immense. Unemployment and the inability to provide one's female kin with the luxury to stay at home were perceived as men's ultimate failure. Consequently, Syrian men invested a lot of energy to circumvent and discursively condemn unemployment and come across as successful, innovative and hard-working. Many affluent Syrian men used 'the currency'[6] of managing one's life upon displacement to create a hierarchy between themselves and others, labelling only those who, in their perception, were the least successful—namely those who fled to Europe—as refugees (Suerbaum, 2018a: 672). As with the refugee label, work and unemployment were used to create hierarchies: between Syrians who participated in the Egyptian labour market and those who had not found work, as well as between Syrians as a group and the Egyptian host population.

Conclusion

This chapter has traced the development of the study of men and masculinities in the context of (forced) migration. It has highlighted the importance of an intersectional lens to properly understand how men (distinguished by class, age, origin) are affected by forced migration. An intersectional analysis provides the tool for the identification of aspects and markers that explain why men perceive their position as challenged in times of forced migration.

The chapter has discussed three dimensions of masculinities and the ways they play out in male refugees' lives: I identified the relevance of social class belonging and the significance of association with middle-classness in order to counter stereotypes associated with refugeeness. Furthermore, I presented how ideals, such as being the sole provider for the family, were valorised and required negotiation in contexts of forced migration, where such ideals clash with the 'victim' image of the prototype of a refugee. As a result, being categorised as a refugee is often comparable with an affront and consequently evokes strong reactions on the part of Syrian men. Finally, I showed the importance of engaging in paid work in order to counter marginalisation, exclusion and non-recognition. Refugee men tend to suffer when being unable to prove

[6] I am inspired to use this term by Kimmel (1994) who speaks about women as 'currency' in men's homosocial enactment.

their inherent value through paid work in a respected profession. Lack of access to economic capital also frequently means social exclusion, marginalisation and loss of previously held privileges based on being the breadwinner for one's family.

By providing a glimpse into the themes of class belonging, labelling and employment, I have shown how men find ways to adapt and re-create acceptable versions of masculinities for themselves. Thus, constructions of masculinities in forced migration can only be understood as fluid, plural and responsive to changes in everyday life. Furthermore, those constructed gender identities should be understood as intersectional, relational and context-specific.

References

Allsopp, J. (2017). Agent, victim, soldier, son: Intersecting masculinities in the European "refugee crisis". In J. Freedman, Z. Kivilcim, & N. Özgür Baklacioglu (Eds.), *A gendered approach to the Syrian refugee crisis* (pp. 155–174). London and New York: Routledge.

Anthias, F. (2012). Intersectional what? Social divisions, intersectionality and levels of analysis. *Ethnicities, 13*(1), 3–19.

Ayoub, M., & Khallaf, S. (2014). *Syrian refugees in Egypt: Challenges of a politically changing environment*. Paper No. 7, Cairo Studies on Migration and Refugees, American University in Cairo. http://schools.aucegypt.edu/GAPP/cmrs/Documents/Final_Syrian%20refugees.pdf.

Bonjour, S., & Chauvin, S. (2018). Social class, migration policy and migration strategies: An introduction. *International Migration, 56*(4), 5–18.

Butler, J. (1999). *Gender trouble: Feminism and the subversion of identity*. New York and London: Routledge.

Carrigan, T., Connell, B., & Lee, J. (1985). Towards a new sociology of masculinity. *Theory and Society, 14*(5), 551–604.

Charsley, K., & Wray, H. (2015). Introduction: The invisible (migrant) man. *Men and Masculinities, 18*(4), 403–423.

Chimni, B. S. (2009). The birth of a 'discipline': From refugee to forced migration studies. *Journal of Refugee Studies, 22*(1), 11–29.

Christensen, A., & Jensen, S. Q. (2014). Combining hegemonic masculinity and intersectionality. *NORMA: International Journal for Masculinity Studies, 9*, 60–75.

Connell, R. (1995). *Masculinities*. Cambridge: Polity.

Connell, R. (2014). Margin becoming centre: For a world-centred rethinking of masculinities. *NORMA: International Journal for Masculinity Studies, 9*(4), 217–231.

Cornwall, A., & Lindisfarne, N. (1994). Introduction. In A. Cornwall & N. Lindisfarne (Eds.), *Dislocating masculinity: Comparative ethnographies* (pp. 1–10). London and New York: Routledge.

Crawley, H., & Skleparis, D. (2018). Refugees, migrants, neither, both: Categorical fetishism and the politics of bounding in Europe's 'migration crisis'. *Journal of Ethnic and Migration Studies, 44*(1), 48–64.

Crenshaw, K. (1991). Mapping the margins: Intersectionality, identity politics, and violence against women of color. *Standford Law Review, 43*(6), 1241–1299.

Demitrou, D. (2001). Connell's concept of hegemonic masculinity: A critique. *Theory and Society, 30*(3), 337–361.

Donaldson, M., & Howson, R. (2009). Men, migration and hegemonic masculinity. In M. Donaldson, R. Hibbins, R. Howson, & B. Pease (Eds.), *Migrant men: Critical studies of masculinities and the migration experience* (pp. 210–217). New York and London: Routledge.

Donato, K., Gabaccia, M. D., Holdaway, J., Manalansan, M., & Pessar, P. R. (2006). A glass half full? Gender in migration studies. *International Migration Review: Gender and Migration Revisited, 40*(1), 3–26.

Donner, H. (2015). Making middle-class families in Calcutta. In J. G. Carrier & D. Kalb (Eds.), *Anthropologies of class: Power, practice and inequality* (pp. 131–148). Cambridge: Cambridge University Press.

Edwards, T. (2006). *Cultures of masculinity*. London: Routledge.

Finnan, R. (1981). Occupational assimilation of refugees. *The International Migration Review, 15*(1/2), 292–309.

Ghannam, F. (2013). *Live and die like a man: Gender dynamics in urban Egypt*. Stanford, CA: Stanford University Press.

Ghoussoub, M., & Sinclair-Webb, E. (2000). *Imagined masculinities: Male identity and culture in the modern Middle East*. London: Saqi.

Griffiths, M. (2015). "Here, man is nothing!": Gender and policy in an asylum context. *Men and Masculinities, 18*(4), 468–488.

Hasso, F. (2018, October 15). Decolonizing Middle East men and masculinities scholarship: An axiomatic approach. *Arab Studies Journal Online*. Available at: http://www.arabstudiesjournal.org/asj-online/decolonizing-middle-east-men-and-masculinities-scholarship-an-axiomatic-approach.

Haugbolle, S. (2012). The (little) militia man: Memory and militarized masculinity in Lebanon. *Journal of Middle East Women's Studies, 8*(1), 115–139.

Hayden, B. (2006). What's in a name? The nature of the individual in refugee studies. *Journal of Refugee Studies, 19*(4), 471–487.

Haywood, C., & Mac an Ghaill, M. (2003). *Men and masculinities: Theory, research and social practice*. Buckingham and Philadelphia: Open University Press.

Heiman, R., Liechty, M., & Freeman, C. (2012). Introduction: Charting an anthropology of the middle classes. In R. Heiman, M. Liechty, C. Freeman (Eds.), *The global middle classes: Theorizing through ethnography* (pp. 3–33). Santa Fe: School for Advanced Research Press.

Henry, M. (2017). Problematizing military masculinity, intersectionality, and male vulnerablity in feminist critical military studies. *Critical Military Studies, 3*(2), 182–199.

Herz, M. (2019). 'Becoming' a possible threat: Masculinity, culture and questioning among unaccompanied young men in Sweden. *Identities, 26*(4), 431–449.

Hopkins, P., & Noble, G. (2009). Editorial: Masculinities in place, situated identities, relations and intersectionality. *Social and Cultural Geography, 10*(8), 810–819.

Hyndman, J. (2010). Introduction: The feminist politics of refugee migration. *Gender, Place & Culture: A Journal of Feminist Geography, 17*(4), 453–459.

Hyndman, J., & Giles, W. (2011). Waiting for what? The feminization of asylum in protracted situations. *Gender, Place & Culture, 18*(3), 361–379.

Indra, D. (1999). *Engendering forced migration: Theory and practice*. New York: Berghahn.

Inhorn, M. (2012). *The New Arab man: Emergent masculinities, technologies, and Islam in the Middle East*. Princeton, NJ: Princeton University Press.

Jaji, R. (2009). Masculinity on unstable ground: Young refugee men in Nairobi, Kenya. *Journal of Refugee Studies, 22*(2), 177–194.

Jansen, S. (2008). Misplaced masculinities: Status loss and the location of gendered subjectivities amongst 'non-transnational Bosnian Refugees'. *Anthropological Theory, 8*(2), 181–200.

Jean-Klein, I. (2000). Mothering, statecraft and subjectivity in the Palestinian Intifada. *American Ethnologist, 27*(1), 100–127.

Jones, C. (2012). Women in the middle: Femininity, virtue, and excess in Indonesian discourses of middle classness. In R. Heiman, C. Freeman, & M. Liechty (Eds.), *The global middle classes: Theorizing through ethnography* (pp. 145–168). Santa Fe: School for Advanced Research Press.

Kimmel, M. (1994). Masculinity as homophobia: Fear, shame, and silence in the construction of gender identity. In H. Bord & M. Kaufman (Eds.), *Theorizing masculinities* (pp. 119–142). Thousand Oaks, CA: Sage.

Kleist, N. (2010). Negotiating respectable masculinity: Gender and recognition in the somali diaspora. *African Diaspora, 3*(2), 185–206.

Levitt, P., & Glick Schiller, N. (2004). Conceptualizing simultaneity: A transnational social field perspective on society. *International Migraiton Review, 28*(3), 1002–1039.

Long, K. (2013). When refugees stopped being migrants: Movement, labour and humanitarian protection. *Migration Studies, 1*(1), 4–26.

Lubkemann, S. (2008). *Culture in Chaos: An Anthropology of the Social Condition in War*. Chicago: University of Chicago Press.

Mahler, S., & Pessar, P. (2006). Gender matters: Ethnographers bring gender from the periphery toward the core of migration studies. *International Migration Review, 40*(1), 127–163.

Malkki, L. (1995). Refugees and exile: From 'refugee studies' to the national order of things. *Annual Review of Anthropology, 24*, 495–523.

Matsuoka, A., & Sorensen, J. (1999). Eritrean Canadian refugee households as sites of gender renegotiation. In D. Indra (Ed.), *Engendering forced migration: Theory and practice* (pp. 218–241). New York: Berghahn.

McGregor, J. (2008). Abject spaces, transnational calculations: Zimbabweans in Britain navigating work, class and the law. *Transactions of the Institute of British Geographers, New Series, 33*(4), 466–482.

McSpadden, L. A. (1999). Negotiating masculinity in the reconstruction of social place: Eritrean and Ethiopian refugees in the United States and Sweden. In D. Indra (Ed.), *Engendering forced migration: Theory and practice* (pp. 242–260). New York: Berghahn.

Monsutti, A. (2007). Migration as a rite of passage: Young Afghans building masculinity and adulthood in Iran. *Iranian Studies, 40*(2), 167–185.

Monterescu, D. (2006). Stranger masculinities: Gender and politics in a Palestinian-Israel 'third-space'. In L. Ouzgane (Ed.), *Islamic masculinities* (pp. 123–142). New York and London: Zed Books.

Naguib, N. (2015). *Nurturing masculinities: Men, food and family in contemporary Egypt*. Austin: University of Texas Press.

Nolin, C. (2006). *Transnational ruptures: Gender and forced migration*. Hampshire and Burlington: Asghate.

Ong, A. (2003). *Buddha is hiding: Refugees, citizenship, the New America*. Berkeley: University of California Press.

Osella, C., & Osella, F. (2000). Migration, money and masculinity in Kerala. *Journal of the Royal Anthropological Institute, 6*(1), 117–133.

Palillo, M. (2018). 'If I Die here, I'm a Hero!' On masculinity and vulnerability among male asylum seekers. *Journal of Extreme Anthropology, 2*(1), 28–45.

Peteet, J. (1994). Male gender and rituals of resistance in the Palestinian "Intifada": A cultural politics of violence. *American Ethnologist, 21*(1), 31–49.

Pruitt, L., Berents, H., & Munro, G. (2018). Gender and age in the construction of male youth in the European migration "crisis". *Signs: Journal of Women in Culture and Society, 43*(3), 687–709.

Schielke, S. (2012). Capitalist ethics and the spirit of Islamization in Egypt. In S. Schielke & L. Debevec (Eds.), *Ordinary lives and grand schemes: An anthropology of everyday religion* (pp. 131–145). New York and Oxford: Berghahn Books.

Scott, J. W. (1999). *Gender and the politics of history* (rev ed.). New York: Columbia University Press.

Suerbaum, M. (2018a). Defining the other to masculinize oneself: Syrian men's negotiations of masculinity during displacement in Egypt. *Signs: Journal of Women in Culture and Society, 43*(3), 665–686.

Suerbaum, M. (2018b). Becoming and 'unbecoming' refugees: Syrian men's negotiations of masculinity and refugeeness in Egypt. *Men and Masculinities, 21*(3), 363–382.

Suerbaum, M. (2020). *Masculinities and displacement in the Middle East: Syrian refugee men in Egypt*. London: I.B.Tauris.

Szczepanikova, A. (2010). Performing refugeeness in the Czech Republic: Gendered depoliticisation through NGO assistance. *Gender, Place & Culture: A Journal of Feminist Geography, 17*(4), 461–477.

Turner, L. (2019a). Syrian refugee men as objects of humanitarian care. *International Feminist Journal of Politics, 21*(4), 595–616. https://doi.org/10.1080/14616742.2019.1641127.

Turner, L. (2019b). *The politics of labelling refugee men as "vulnerable"*. Social Politics: International Studies in Gender, State and Society. https://doi.org/10.1093/sp/jxz033.

Van Hear, N. (2006). 'I went as far as my money would take me': Conflict, forced migration and class. In F. Crepeau, D. Nakache, M. Collyer, N. Goetz, A. Hansen, R. Modi, et al. (Eds.), *Forced migration and global processes: A view from forced migration studies* (pp. 125–158). Oxford: Lexington Books.

Vera-Sanso, P. (2016). Taking the long view: Attaining and sustaining masculinity across the life course in South India. In A. Cornwall, F. Karioris, & N. Lindisfarne (Eds.), *Masculinities under neoliberalism* (pp. 80–98). London: Zed Books.

Wentzell, E. (2015). Enhancing fathering through medical research participation in Mexico. In M. Inhorn, W. Chavkin, & J.-A. Navarro (Eds.), *Globalized fatherhood* (pp. 177–196). New York and Oxford: Berghahn Books.

Wetherell, M., & Edley, N. (1999). Negotiating hegemonic masculinity: Imaginary positions and psycho-discursive practices. *Feminism & Psychology, 9*(3), 335–356.

Yuval-Davis, N. (2006). Intersectionality and feminist politics. *European Journal of Women's Studies, 13*(3), 193–209.

Magdalena Suerbaum is a post-doctoral researcher at the Max-Planck-Institute for the Study of Religious and Ethnic Diversity in Göttingen, Germany. Her current research focuses on the social implications of legal statuses among recent asylum-seekers in Germany. Specifically, she is interested in mothering practices in times of forced migration and legal precarity. She is the author of 'Masculinities and Displacement in the Middle East: Syrian refugees in Egypt' (I.B. Tauris 2020). Magdalena holds a Ph.D. in Gender Studies from the School of Oriental and African Studies in London. She completed an M.A. at the University of Exeter in Middle East and Islamic Studies and a B.A. at the Philipps University Marburg in Islamic Studies.

CHAPTER 25

Gender and Refugee Resettlement: The Role of Proximal and Distal Stressors in the Experiences of Survivors of Sexual and Gender-Based Violence

Jenny Phillimore, Sandra Pertek, and Lailah Alidu

Abstract Migration and indeed forced migration are gendered as are resettlement processes. This chapter examines the gendered nature of refugee integration focusing upon refugee survivors of sexual and gender-based violence. The chapter focuses on the impact of violence, from distal events of physical and sexual violence, which happen during conflict and flight, to proximal everyday violence including interpersonal violence in detention, camps and resettlement and finally the structural and symbolic violence associated with gendered asylum processes. We argue that interventions are required to support the integration of refugee women which account for, and address, the multiple forms of violence they face. These include recognising the gendered nature of integration processes and specifically taking account of the state's role as a perpetrator of violence against refugee women.

J. Phillimore (✉) · S. Pertek
School of Social Policy, Sociology and Criminology, University of Birmingham, Birmingham, UK
e-mail: j.a.phillimore@bham.ac.uk

S. Pertek
e-mail: s.m.pertek@pgr.bham.ac.uk

L. Alidu
School of Law, Social and Behavioural Sciences, Kingston University, London, UK

C. Mora and N. Piper (eds.), *The Palgrave Handbook of Gender and Migration*, https://doi.org/10.1007/978-3-030-63347-9_25

Introduction

The number of refugees in the world in 2016 reached the highest point since records began (United Nations High Commission for Refugees, UNHCR, 2016) as conflicts across the globe forced millions of people to flee their homes. Some 70.8 million people were said to be displaced in 2018 (UNHCR, 2018). Data indicates that more men have fled the crisis than women but it is evident that increasing numbers of women and children are fleeing conflict using routes which involve multiple risks (UNHCR, 2016). While refugees seek to escape violence including sexual and gender-based violence (SGBV), in their countries of origin, women and children, continue to be vulnerable to SGBV across the refugee journey. Such threats can come from many places: police and security services, smugglers, other refugees and even family members and partners.

SGBV is defined by UNHCR as 'any harmful act that is perpetrated against one person's will and that is based on socially ascribed (gender) differences between males and females. It includes acts that inflict physical, mental, or sexual harm or suffering, threats of such acts, coercion and other deprivations of liberty, whether occurring in public or in private life' (2011: 6). There is no data available reporting the full extent of SGBV against refugees but it is known to be gendered, with women and children forming the majority of survivors. It is well acknowledged that many people do not disclose their experiences, but also there is no common recording mechanism. Freedman (2016) showed that the closing of EU borders to refugees has coincided with increased levels of sexual violence being recorded, for example, as some refugees were forced to cover smuggling costs with transactional sex. Norredam, Mygind, and Krasnik (2005) highlighted how some asylum seekers were harassed and abused by strangers, people in authority and those allocated to their protection in camps, detention centres and host countries.

Given the estimated prevalence of SGBV across the refugee journey,[1] experiences need to be conceptualised as an ongoing and multi-faceted experience of trauma, with both immediate and long-term consequences for refugee resettlement. The aim of this chapter is to identify firstly the state of knowledge about the impacts of refugees' experiences of SGBV on their resettlement in countries of refuge and to argue for the need to include consideration of the effects of every day acts of symbolic violence occasioned by the state on the integration of survivors of SGBV. This approach moves us beyond the focus of SGBV in conflict or flight, to think critically about what happens in resettlement and how resettlement may be associated with a continued vulnerability to SGBV, as well as, or even instead of refuge.

This chapter draws upon a literature review undertaken as part of the Sexual and gender-based violence in the refugee crisis: from displacement to arrival

[1]The estimates of SGBV prevalence in the refuge journey differ. For example, one study found that up to 28.6% of male and 69.3% of female migrants have experienced sexual and gender-based violence (SGBV) since their arrival to Europe (Keygnaert & Guieu, 2015).

(SEREDA) project[2] which focused on the experiences of forced migrants who can include asylum seekers, refugees, the family or spouse of refugees, undocumented migrants with asylum claim refused, as well as internally displaced people.[3]

The literature search crossed multiple disciplines and highlighted a lack of studies focusing specifically on the effect of SGBV on integration in countries of refuge. Rather than looking across integration processes studies, our review tended to examine the psychological and physical effects of SGBV on individuals perhaps referring to a specific dimension of integration such as employment. Lack of comprehensive studies about refugee survivors' integration experience meant that we had to pull together piecemeal evidence from across disciplines. Also, most studies focused upon women refugees, asylum seekers and migrants. No studies on migrant men and SGBV in countries of resettlement were identified highlighting a key gap in existing knowledge.

SGBV and Refugee Integration

The UK Government consider that integration has been achieved when 'people, whatever their background, live, work, learn and socialise together, based on shared rights, responsibilities and opportunities' (HM Government, 2018). There is no widely accepted scholarly definition of integration although there is some agreement that integration has multiple meanings in different contexts (Grzymala-Kazlowska & Phillimore, 2017). Fundamentally integration is a context specific set of processes that are multi-dimensional and depend on the interaction of multiple factors (Ndofor-Tah et al., 2019). Integration is multi-directional involving adjustments by everyone and it is a responsibility shared between newcomers, receiving communities and governments. SGBV experience, combined with other personal, cultural and experiential factors and local opportunity structures, are some of the factors influencing the pace and direction of integration processes and refugee experiences. To underpin our analysis of the evolving relationship between SGBV and refugee integration in countries of refuge, we deploy Ager and Strang's (2008), conceptual framework of integration developed as part of the *Indicators of*

[2] https://www.birmingham.ac.uk/research/activity/superdiversity-institute/research/projects/sereda.aspx.

[3] A review was conducted using the; PsycARTICLES Full Text, Embase (Classic), HMIC Health Management Information Consortium, Journals@Ovid Full Text, Ovid MEDLINE(R) In-Process & Other Non-Indexed Citations and Ovid MEDLINE(R), PsycINFO, Social Policy and Practice databases. Further searches were made on Web of Science and Scopus. Key terms used included: experience, violence, sexual violence, gender-based violence, internally displaced person*, refugee*, immigrant*, or asylum seeker*, impact of SGBV, SGBV experience, trauma, weapon of war, well-being and integration. Broad and narrow search methods were deployed to assure inclusion of a wide spectrum of research enquiry. Backward (reference) and forward (citation) searches were made with the articles to identify papers that might have been omitted. Some 93 articles were included from between 2000 and 2018.

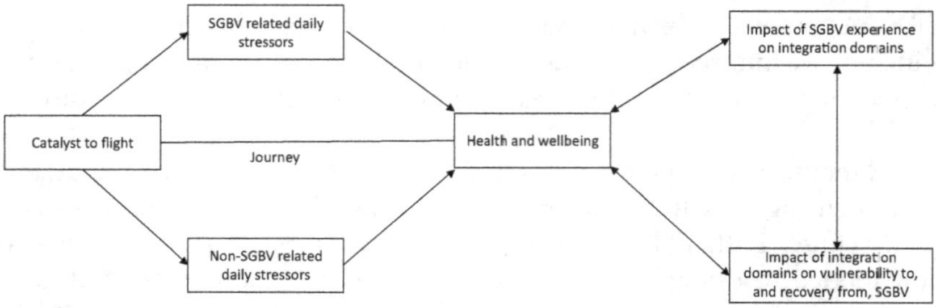

Fig. 25.1 SGBV and refugee integration model (Authors, inspired by Miller and Rasmussen, 2010)

Integration study commissioned by the UK Home Office in 2002. This framework provides an analytical structure in which we can examine the relationship between SGBV and integration incorporating integration indicators from across multiple disciplines including social policy, sociology, political science and migration studies.[4]

The framework consists of four integrative domains; (1) Means and markers, (2) Social connections, (3) Facilitators and (4) Foundations (Ager & Strang, 2008). Each of these domains includes indicators, in the respective order; (1) Employment, health, housing, education; (2) Social bonds, bridges and links; (3) Language and cultural knowledge, safety and stability; (4) Rights and citizenship.

We argue that experiences of SGBV are likely to shape individuals' ability to integrate and use the findings from our literature review to examine the impacts of such experiences on integration across the domains. We might describe the impact of SGBV as distal impacts—referring to traumatic but largely sporadic events that happened in the past. However, our review revealed that refugees' exposure to, and ability to recover from, SGBV is shaped by experience after arrival across integration domains. We might describe these effects as the proximal impacts of integration on SGBV survivors' lives. Proximal impacts are often ongoing and occur within the country of refuge. The interaction between integration and proximal and distal factors is illustrated in Fig. 25.1.

We begin by highlighting the importance of health in influencing all other integration domains before discussing the impacts of SGBV on refugees' ability to integrate and the impact of the domains on their susceptibility to, or prevention of, continued SGBV and upon their ability to recover from SGBV. We finish the paper with a short conclusion outlining the importance of taking account of proximal effects.

[4]The updated Indicators of Integration framework (Ndofor-Tah et al., 2019) that has recently been published was not available at the time of writing this chapter.

Means and Markers: Health, Employment, Housing and Education

Health

Health and well-being are perhaps the fundamental factors shaping SGBV survivors' ability to integrate, while factors undermining ability to recover health, or exacerbating physical and/or psychological well-being will also shape integration opportunities, hence the highlighting of the central role of health and well-being in Fig. 25.1. Health interacts with many of the other indicators as we will see later in the chapter. As such being healthy, as Ager and Strang (2008) argue, is a key *means* to integration.

A) Distal impacts

Refugees from poor communities, living in shelters or detention and camps, younger/teenage children from a low socio-economic background or single parent families are said to be most vulnerable to SGBV (Bonewit & Shreeves, 2016; UNHCR, 2016). People who have been survivors of SGBV or have witnessed SGBV are at an increased risk of further trauma (Keygnaert, Vettenburg, & Temmerman, 2012).

Trauma may result in physical, psychological, social and spiritual consequences either in the long or short term (Tavara, 2006) and can have distal impacts on survivors' ability to integrate. The physical consequences of sexual violence range from bodily injuries, infections to disability (Keygnaert et al., 2012). Reproductive health consequences included sexually transmitted infections (STIs), HIV, symptoms of sexual disorders, unwanted pregnancy, miscarriage, forced abortion, fistulas and maternal and neonatal deaths (Barkho, Fakhouri, & Arnetz, 2011). Psychological problems associated with surviving SGBV can include anxiety, depression, sleep difficulties, which can develop into long-term psychological problems such as Post-Traumatic Stress Disorder (PSTD). Women refugees with trauma have been found to report poorer physical health and higher medication usage than those without trauma

Factors shaping access to treatment and recovery.

A) Proximal impacts

While evidence about the importance of disclosing SGBV experiences for recovery is inconclusive, there are a number of factors known to prevent women from seeking help including patriarchal gender norms, poverty and cultural expectations. Discussing anything of a sexual nature is taboo in some societies: survivors may feel guilt and shame (Tankink & Richters, 2007). If these norms continue within refuge societies, they can exacerbate mental and physical health problems (Keygnaert et al., 2012).

The circumstances surrounding resettlement may adversely affect the health of refugee SGBV survivors. Being unable to access treatment for psychological health problems increases the risk of further SGBV and, because having 'no self-confidence', can affect ongoing vulnerability to further violence (Keygnaert et al., 2012). Miller and Rasmussen (2010), argue that daily stressors have the most overall influence on refugee health and recovery from trauma. High rates of psychological distress among refugee survivors are exacerbated by everyday stressors many of which are imposed by the state (ibid). Studies highlight that the cumulative effect of unemployment, inadequate housing and uncertain legal status, have been reported to be more harmful in the long-term than the original trauma because they tend to be interminable with no end in sight (see Miller & Rasmussen, 2010). Stressful conditions imposed by states actively seeking to create a hostile environment impede recovery processes, perpetuating the cycle of exclusion and disadvantage. Poverty and debt exacerbate SGBV survivors' mental health conditions (Tankink & Richters, 2007).

B) Distal impacts

SGBV survivors score significantly highly on PTSD scales and face major obstacles to treatment and care (Amone-P'Olak, Elklit, & Dokkedahl, 2018). Untreated mental health conditions generate impacts on survivors' capacity to integrate resulting in memory problems, poor well-being, and socio-economic exclusion. Such conditions also result in poor levels of concentration that hinder learning and the development of language skills which are generally associated with good integration outcomes (Gušić, Cardeña, Bengtsson, & Søndergaard, 2017).

The continued mental and physical aftermath of SGBV can hamper individuals' ability to work, implying long-term socio-economic consequences. Numerous studies have demonstrated that survivors take more sick days from work and have decreased productivity. Experiences of SGBV affects women's capacities to engage in productive activities. The cultural context of SGBV survivors often influences economic participation. Patriarchal cultures can impose additional social and cultural norms as some SGBV survivors are not permitted to work and are expected to rely on a male breadwinner which can contribute to women's marginalisation (Akua-Sakyiwah, 2016). Thus, both the physical and psychological impacts and the social impacts of SGBV can reduce access, or ability, to work undermining social status and compounding poverty—therein distal impacts are connected to proximate impacts perhaps perpetuating a vicious cycle of poverty and hopelessness.

Employment

A) Proximal impacts

Some categories of forced migrants are not permitted to work, while others struggle to make a living (Phillimore & Goodson, 2006). Among those permitted to work refugees often cannot find work or are under-employed with their qualifications not recognised in countries of refuge. Poverty is a key risk factor and proximal impact which increased exposure to further SGBV. Without permission to work and for undocumented migrants, with no other means of support, some have little choice but to work in the informal economy risking labour and sexual exploitation (Keygnaert et al., 2012). There is evidence that destitute refugee women are forced to engage in sex work for survival or transactional is sex expected by employers or landlords (Samari, 2017).

Unemployment among male refugees undermines their self-esteem, impacts upon their sense of belonging and reduces their well-being (see also Suerbaum and Jaji in this volume), all of which have been associated with increased perpetration of IPV (Christian, Safari, Ramazani, Burnham, & Glass, 2011). Economic independence can help women escape abuse, but access to work is difficult for mothers as access to childcare can be poor, there are few opportunities to engage in vocational training and spousal migrants may not have legal permission to work.

Housing

A) Proximal impacts

We found no evidence outlining the impact of SGBV on refugees' access to safe and secure housing in countries of resettlement. Certainly, there was no evidence that any special provision is made for SGBV survivors. However, there is ample evidence showing that providing such housing improves the functioning and well-being of SGBV survivors (Cesario, Nava, Bianchi, McFarlane, & Maddoux, 2014). Findings from our review indicate that survivors mainly rely on public housing and shelters often segregated from mainstream communities making it difficult for them to make connections with local people which can be important in aiding survivors to access services.

The nature of housing can influence refugees' ability to recover from, and further exposure to, SGBV. For example, European asylum reception facilities have been found to be unsafe with high numbers of cases of SGBV reported (Oliveira, Keygnaert, Oliveira Martins, & Dias, 2018). Housing insecurity proximally effects the lives of refugees and is a major barrier to escaping domestic violence (Thurston et al., 2013). Good quality shelters can address SGBV survivors' immediate housing needs but are often useful only on a short-term basis (Cesario et al., 2014).

Overcrowded housing and lack of privacy increases SGBV survivors' vulnerability and exposure to violence because they often involve sharing with unknown men (Oliveira et al., 2018). Sexual harassment in European reception centres has been reported as occurring on a daily basis (Bonewit & Shreeves, 2016). Single women felt more vulnerable to sexual assault than married women because they lacked male protection. Rooms can lack locks and toilets are sometimes located some distance from rooms increasing refugees' vulnerability (ibid). Women SGBV survivors living in inadequate conditions have been shown to experience a deterioration in their mental health (Bonewit and Shreeves, 2016).

Education

A) Distal impacts

Several studies have emphasised the importance of education in assisting integration (Gesemann, 2006; Ouald Chaib, 2011 in Akua-Sakyiwah, 2016). There are few studies examining the interaction between SGBV, refugees and education. Work by Phillimore (2011a) showed that SGBV associated trauma was associated with poor levels of concentration which undermined individuals' ability to study.

B) Proximal impacts

Keygnaert et al. (2012: 513) found one third of respondents with refugee, asylum seeking or undocumented migrant background identified lack of knowledge and information as a risk factor. This included 'not knowing the language and culture of the host country' and lack of 'sexual knowledge' and 'self-defence skills'. Acquiring education is challenging for SGBV survivors economically dependent on their spouses, and not able to cover their own costs (Shabbar, 2012) including childcare costs. Cultural expectations can further determine whether refugee women are permitted to attend classes (Akua-Sakyiwah, 2016).

Social Connections: Social Bonds, Social Bridges and Social Links

Social Bonds

A) Proximal impacts

Social and family support have been shown to be protective factors for many refugees (Gorman, Brough, & Ramirez, 2003), increasing their psychological well-being (Sherwood and Liebling-Kalifani, 2012). Refugees with social networks are better able to access social and material support and less likely to report psychological distress (McMichael & Manderson, 2004; Cheung & Phillimore, 2017). The risk of depression and other psychological problems appears to increase with length of time away from close relatives (Keygnaert et al., 2012). Seeking advice and interventions from extended family, and from community elders or from religious leaders were the first steps taken to resolve IPV in the West African community in the USA (Kalunta-Crumpton, 2017). Only when support from social networks failed, did survivors turn to public services.

Family and friends in some contexts act as bridges and encourage survivors to seek support, while in others try to prevent them from seeking external support (Tankink & Richters, 2007; Byrskog, Essén, Olsson, & Klingberg-Allvin, 2016). Survivors of IPV sometimes seek support from religious leaders, especially among faith communities where they are the first point of contact. On the one hand, religious leaders can hold patriarchal views encouraging women to endure abuse, and on the other, can also play a facilitating role in supporting survivors (Kulwicki, Aswad, Carmona, & Ballout, 2010). Faith networks can also be crucial in supporting feelings of continuity for SGBV survivors after flight (Byrskog et al., 2016; Parsitau, 2011).

B) Distal impacts

As noted above many SGBV survivors suffer from trauma and PTSD which have social implications (Gušić et al., 2017). In many, socio-cultural settings SGBV survivors reported wanting to keep their experiences secret to avoid shame and stigma. Yet, disclosure has been noted by some to be a prerequisite for recovery, meaning those who keep quiet may struggle unaided (Tankink & Richters, 2007). The emotional and psychological pain of SGBV have direct implications for social isolation and can be described as a distal impact on the ability to integrate. Keygnaert et al. (2012), found survivors described themselves as being 'depressed', 'a psychological wreck', 'dispirited' or 'very insecure'. They often isolated themselves from others, not trusting anybody. The social consequences of SGBV can be particularly severe for children born of sexual violence who can face stigmatisation from their communities (Josse, 2010). Isolation can exacerbate SGBV survivors' feelings of anxiety, shame, anger, frustration and hatred (Josse, 2010). Feelings of guilt prevented survivors from disclosing and seeking support from service providers (Tankink & Richters, 2007; Keygnaert et al., 2012).

Men and boys who experience sexual abuse also struggle socially (Christian et al., 2011). They experienced fear, ridicule, disrespect and marginalisation, forcing them to separate from their communities and sometimes their families (Christian et al., 2011). Sturm, Baubet, and Moro (2007) show how male asylum seekers' traumatic memories can overwhelm and destabilise them diminishing their capacity 'to represent his experiences and to maintain or create social bonds' (p. 216).

Such disruption has been shown to sometimes result in tensions escalating to domestic violence and forced isolation (Shabbar, 2012). SGBV has a distal impact upon individuals' ability and willingness to form social networks with peers which is problematic for integration given that social networks contribute to a sense of well-being and help refugees to develop understanding of a new society (Byrskog et al., 2016).

Men and women survivors of SGBV are sometimes also subjected to torture which compounds the effects of SGBV related trauma. Memories of torture are associated with re-traumatisation (Cleveland, Kronick, Gros, & Rousseau, 2018), severe depression, loneliness, anger and low self-esteem and a tendency to withdraw from networks (Liebling, 2018). In some countries asylum seekers are housed in detention centres or refugee camps indefinitely generating feelings of helplessness that can result in or exacerbate depression, anxiety and isolation (Kirmayer et al., 2011).

Social Bridges

A) Distal impacts

Relationships outside of refugees' own communities have been little studied in relation to recovery from SGBV. 'Outsiders' (of ethnic communities) were perceived by some refugees as easier and safer to disclose to than survivors within their own communities in Australia and in the Netherlands (Ogunsiji, Wilkes, Jackson, & Peters, 2012; Tankink & Richters, 2007).

B) Proximal impacts

SGBV survivors have been shown to build friendships with local women from host communities and eventually breaking their 'silence'. Trusted local women acted as social bridges guiding survivors to support services. Ogunsiji et al. (2012) describe the case of a woman experiencing continued domestic abuse receiving encouragement to leave her husband from an Australian friend, who told her that she would receive government support breaking her dependency. Practical assistance from the wider community contributes to feelings of stability for survivors and aids their recovery (Byrskog et al., 2016).

Many SGBV survivors have weak social networks and suffer in silence because they lack knowledge about social and health resources (Hyman et al., 2006; Ogunsiji et al., 2012). This lack of awareness prolonged victimisation and reduced access to protection or encouragement to escape from abuse (Thurston et al., 2013). Lack of local language proficiency reduces women's chance of sharing experiences, increasing the risk of low self-esteem, loneliness and depression (Shishehgar, Gholizadeh, DiGiacomo, Green, & Davidson,

2017). Often refugees were not aware of, or informed about, their rights or the availability of resources available to survivors of domestic abuse such as counselling and hospital treatment needing to access this information from people more familiar with local welfare systems (Kalunta-Crumpton, 2017).

Social Links

A) Distal impacts

Although there is evidence that access to treatments can be beneficial, many SGBV survivors were reluctant to link with organisations offering support as they associated seeking mental health care with stigma. Some rape survivors were reluctant to disclose sexual violence to medical practitioners for fear of having their family's name in the medical records (Tankink & Richters, 2007). In addition, some countries charge for services if individuals do not have the appropriate immigration status, making them reluctant or unable to access treatment.

B) Proximal impacts

Some host countries offer a variety of support services for SGBV survivors. For example, in Spain social workers help survivors to access employment, housing, education and tackle social exclusion through referrals to health centres, hospitals, local social services, emergency care services, subsidised housing and the courts (Martinez-Roman, Vives-Cases, & Pérez-Belda, 2017). A wide range of NGOs assist refugees who are at increased risk of violence and social exclusion, additionally, migrant organisations can play a significant role in providing context specific counter-SGBV initiatives (ibid). Despite efforts to prevent and respond to SGBV, there remain insufficient support services for survivors. Reluctance to seek help often related to local services being inadequately equipped or culturally insensitive (Rossiter et al., 2018).

Facilitators: Language and Cultural Knowledge, Safety and Stability

Language and Cultural Knowledge

Competency in the language of the country of resettlement is fundamental for integration (Shishehgar et al., 2017). Language skills enable survivors of SGBV to understand and access information about their rights, to access services, employment and become self-reliant.

A) Proximal impacts

Language barriers can prevent survivors from seeking support and have proximal impacts on refugees' lives. Thurston et al. (2013) found that refugee IPV survivors were vulnerable to homelessness because information about domestic violence shelters was not available in their language. Language and cultural knowledge to some extent influence SGBV survivors' degree of engagement with host communities which we show above are important facilitators. Communication difficulties between officials and refugees generate proximal impacts and have serious consequences for both survivors and perpetrators of SGBV, for example inappropriate arrest or release of a perpetrator (Kulwicki et al., 2010). Female survivors often find it difficult to work with male interpreters. Particularly, interpreters can increase vulnerability if interpretation is poor or personal opinions are interjected (Martinez-Roman et al., 2017). Survivors with communication disabilities whose needs, such as sign language, are not adequately met, face further safety issues (Hyman et al., 2006). There is evidence that poverty, gender and disability intersect to create vulnerabilities to SGBV for refugees with disabilities especially if they are financially, emotionally or physically dependent on the perpetrator (ibid).

B) Distal impacts

Some abusive partners do not permit their spouses to attend language classes wanting them to remain isolated and unable to seek help (Kalunta-Crumpton, 2018). Low local language competency also reduces access to training and employment prospects and thus reduced the prospect of independence (Thurston et al., 2013; Kapai, 2012).

Safety and Stability

Insecurity and unsafe environments where refugees reside often pose concerns regarding refugee safety, especially in countries where anti-immigration discourse is common and/or physical violence against migrants has been reported. Hostile host communities and public policy create a difficult environment for refugee survivors to recover from trauma. Increased securitisation associated with the so-called 'war on terror' restricts migrants' freedoms and choices.

A) Proximal impacts

Feeling unsafe can undermine refugee recovery from trauma while failure to ensure refugee safety can expose them to further abuse. Safety and stability can be compromised by poor living conditions. Survivors struggle to develop social networks locally if subject to racism or discrimination. Anti-migrant discourse can make them feel unwelcome and fearful of further attack (Phillimore, 2011b). Insecurity can be exacerbated by stereotyping communities as perpetuating gender-based violence, stigmatising them and perhaps using those stereotypes to justify exclusionary immigration policies (Jesuthasan et al., 2018). Service providers sometimes lack cultural knowledge and do not know how to effectively and sensitively challenge stigma, taboo and shame. Support workers may lack competence in engaging with the cultural norms and values that abusers use to justify violence (Kapur et al., 2017) and consequently may be unable to help prevent further abuse.

Foundation: Rights and Citizenship

Access to rights and citizenship frequently depend on immigration status. But even where individuals have full rights and entitlements many have insufficient knowledge about rights, suggesting that much more could be done to inform them about their entitlements. Such knowledge often depends on refugees being able to access services in appropriate languages.

A) Distal impacts

Experiences of SGBV can have a distal impact on survivors' ability to deal with their asylum claim. SGBV related mental health conditions, in particular untreated trauma-related complications, can result in memory problems or present as engaging in tactics to avoid discussing traumatic events. Yet, a positive outcome from an asylum claim is frequently predicated on being able to offer a consistent and coherent self-narrative (Gušić et al., 2017). Inconsistency in the asylum 'story' is viewed as a sign that it is falsified and thus a reason to deny refugee status and to detain and/or deport.

B) Proximal impacts

Proximal impacts of limitations in access, to rights, entitlements and citizenship can hamper refugee recovery. A number of studies have found that survivors not aware of their legal rights were less likely to use social services (Hyman et al., 2006; Martinez-Roman et al., 2017), partly because of language problems (Oliveira et al., 2018). Such lack of awareness contributes to low levels of SGBV reporting. Often those in precarious immigration positions lack any rights and are subject to increased residence-related risk factors and vulnerability (Oliveira et al., 2018; Keygnaert et al., 2012). In many countries, only those with the appropriate documentation can access to healthcare services, depriving undocumented immigrants of basic services.

Asylum determination processes decrease levels of resilience and coping in SGBV survivors; '*Fear of being sent back to their country of origin and uncertainties about their future in the UK were mentioned by all women as well as being unable to work whilst waiting for asylum applications to process'* (Sherwood & Liebling-Kalifani, 2012). Frustration and stress related to immigration procedures have been found to increase rates of IPV (Oliveira et al., 2018). Conditions tied to spousal visas can leave women vulnerable to abuse. Undocumented or illegal immigrants are often trapped in abusive relationships fearing deportation if they leave the relationship or contact the police (Kulwicki et al., 2010; Martinez-Roman et al., 2017). Some legal exemptions have been introduced to mitigate the harmful effect of laws on women subjected to SGBV but these require women to meet strict requirements regarding for example proof of abuse. In general, states in the Global North are responsible for the protection of refugee women under the principle of universality in international human rights, when refugee women are aware of such provisions and those around women's rights, they express appreciation for mandated freedom and rights (Byrskog et al., 2016; Martinez-Roman et al., 2017).

Conclusions

Using Ager and Strang's (2008) integration indicators as an analytical framework has enabled us to outline how the experience of SGBV influences refugees' mental and physical health and their ability to integrate in countries of refuge. The framework is useful in that it provides a structure within which to consider multiple integration domains. Evidence about the relationship between SGBV and integration is patchy because as yet there are no studies focused upon multi-faceted indicators of integration and their interaction with experiences of SGBV. However, looking across the literature on SGBV we have been able to draw some tentative conclusions about the relationship between SGBV and integration. The multi-dimensionality and multi-directionality of integration processes are evident in that there are clear interactions between the different indicator areas and the roles and responsibilities of different stakeholders. It is important to note that the vast majority of evidence identified focuses upon the experiences of women. We could find no papers that offered a gendered analysis comparing the experiences of men and women, none that adopted an intersectional perspective and almost no mention of the experiences of LGBTQi refugees. While these groups are attracting more attention in terms of describing trauma, as yet little is known about the relationship between their traumatic experiences and integration.

Our analysis shows that the experience of SGBV clearly can and does hamper survivors' attempts to resettle and integrate into a new life through a number of distal impacts. In addition, integration domains and embedded within these structural inequalities, can expose women in particular to further abuse or undermine their recovery. These interactions can generate numerous proximal impacts on the lives of refugee SGBV survivors. Survivors can experience disadvantage across integration domains largely because of effects of multi-faceted experience of trauma, a distal stressor, with its immediate and long-term health and socio-economic implications. Lack of appropriate support to address trauma and continued traumas, such as IPV, once in countries of refuge, can be exacerbated by the combined proximal non-SGBV and SGBV related stressors such as uncertain immigration status or structural violence associated with hostile environments designed by the state to discourage asylum seeking and undocumented immigration. Immigration regimes appear to have significant influence over survivors' ability to recover *from* SGBV but also their ongoing vulnerability *to* SGBV. It is evident that structural factors, such as insecure status, poor asylum accommodation, denial of the right to work and/or regulations that make women's presence in a country dependent on another person, actually increase vulnerability to SGBV in the country of refuge. Preventing survivors' access to resources, that can help recovery or failing to make those resources suitably accessible, offer further structural barriers which mitigate against recovery.

The SGBV experiences of refugees should be recognised as spatially and temporally complex. Trauma can and does occur pre-conflict, during conflict, in flight and in refuge and it is exacerbated by the state. Daily stressors related to life in a new country with uncertain status and lack of socio-economic resources can have a compounding effect on vulnerable refugees' psychological well-being. There is growing body of evidence that structural factors, for example poverty and destitution are greater predictors of refugees' mental health condition, than 'big' life changing traumatic events. There is also evidence that daily stressors decrease people's resilience to cope with potentially traumatic life events (Kubiak, 2005). Therefore, addressing the impacts of SGBV experiences on refugees requires consideration and mitigation of all potential proximal and distal stressors in addition to those associated with trauma, while strengthening refugees' inherent capacities and resources to recover (Betancourt & Williams, 2008; Boothby et al., 2006 in Miller & Rasmussen, 2010). In particular there is a need for Governments to implement more humane asylum systems which support vulnerable forced migrants to recover rather than exposing them to further risk.

There is a lack of systematic analysis of the impacts of SGBV experiences on refugees' ability to integrate. We have pieced together a picture from many sources, most of which just cover one or two aspects of integration. It is unsurprising that most attention focuses upon refugees' mental and physical health. What is surprising is the lack of knowledge about how these health problems shape short- and long-term life chances. While there are numerous possible effects without in-depth and longitudinal research it is not possible to know which effects are experienced in what circumstances, and how those effects might be reduced. We have also identified some contradictions that require further consideration including the contrasting positive and negative role of faith leaders in addressing SGBV, the appropriateness of existing solutions for refugee women experiencing domestic violence in the context of limited familiar support in new countries of residence, the appropriateness of Western-based psychotherapies and diagnoses for individuals from non-Western backgrounds. Also, the extent of commitment to a survivor-centred approach and respecting survivor's own choices throughout interventions deserve further scrutiny. We were unable to identify any evidence about the types of policy and practice that can best support the integration of SGBV survivors so further research is needed in this area.

More knowledge is needed about the ways in which experiences of the asylum process, with its associated expectation of retelling trauma, can re-traumatise refugees and impact upon their ability to resettle. Longitudinal studies would be particularly helpful in developing an understanding of short- and long-term impacts as well as the effectiveness of different interventions. Finally, studies should integrate gender sensitivity and adopt a gender analysis to understand the differences in needs, outcomes and experiences of all refugees regardless of their background.

References

Ager, A., & Strang, A. (2008). Understanding integration: A conceptual framework. *Journal of Refugee Studies, 21*(2), 166–191.

Akua-Sakyiwah, B. (2016). Education as cultural capital and its effect on the transitional issues faced by migrant women in the diaspora. *Journal of International Migration and Integration, 17*(4), 1125–1142. https://doi.org/10.1007/s12134-015-0455-8.

Amone-P'Olak, K., Elklit, A., & Dokkedahl, S. B. (2018). PTSD, mental illness, and care among survivors of sexual violence in Northern Uganda: Findings from the WAYS study. *Psychological Trauma: Theory, Research, Practice, and Policy, 10*(3), 282–289. https://doi.org/10.1037/tra0000295.

Barkho, E., Fakhouri, M., & Arnetz, J. E. (2011). Intimate partner violence among Iraqi immigrant women in metro detroit: A pilot study. *Journal of Immigrant and Minority Health, 13*(4), 725–731. https://doi.org/10.1007/s10903-010-9399-4.

Betancourt, T. S., & Williams, T. (2008). Building an evidence base on mental health interventions for children affected by armed conflict. *Intervention (Amstelveen, Netherlands), 6*(1), 39.

Bonewit, A., & Shreeves, R. (2016). *Reception of female refugees and asylum seekers in the EU case study Germany*. Brussels: Directorate General for International Policies, Policy Department C: Citizens' Rights and Constitutional Affairs. Available at: http://www.europarl.europa.eu/RegData/etudes/STUD/2016/536497/IPOL_STU(2016)536497_EN.pdf. Accessed 26 May 2018.

Byrskog, U., Essén, B., Olsson, P., & Klingberg-Allvin, M. (2016). 'Moving on' violence, wellbeing and questions about violence in antenatal care encounters: A qualitative study with Somali-born refugees in Sweden. *Midwifery, 40,* 10–17. https://doi.org/10.1016/j.midw.2016.05.009.

Cesario, S. K., Nava, A., Bianchi, A., McFarlane, J., & Maddoux, J. (2014). Functioning outcomes for abused immigrant women and their children 4 months after initiating intervention. *Revista Panamericana de Salud Publica, 35*(1), 8–14.

Cheung, S. Y., & Phillimore, J. (2017). Gender and Refugee Integration: A quantitative analysis of integration and social policy outcomes. *Journal of Social Policy, 46*(2), 211–230. https://doi.org/10.1017/S0047279416000775.

Christian, M., Safari, O., Ramazani, P., Burnham, G., & Glass, N. (2011). Sexual and gender-based violence against men in the Democratic Republic of Congo: Effects on survivors, their families and the community. *Medicine, Conflict and Survival, 27*(4), 227–246. https://doi.org/10.1080/13623699.2011.645144.

Cleveland, J., Kronick, R., Gros, H., & Rousseau, C. (2018). Symbolic violence and disempowerment as factors in the adverse impact of immigration detention on adult asylum seekers' mental health. *International Journal of Public Health, 63*(8), 1001–1008. https://doi.org/10.1007/s00038-018-1121-7.

Freedman, J. (2016). Sexual and gender-based violence against refugee women: A hidden aspect of the refugee "crisis". *Reproductive Health Matters, 24*(47), 18–26. https://doi.org/10.1016/j.rhm.2016.05.003.

Gesemann, F. (2006). Indicators of integration in education. *Migration citizenship education, network migration in Europe eV*. Retrieved from http://www.migrationeducation.org/22.1.html.

Gorman, D., Brough, M., & Ramirez, E. (2003). How young people from culturally and linguistically diverse backgrounds experience mental health: Some insights for mental health nurses. *International Journal of Mental Health Nursing, 12*(3), 194–202. https://doi.org/10.1046/j.1440-0979.2003.00289.x.

Grzymala-Kazlowska, A., & Phillimore, J. (2017). Rethinking integration: New perspectives on adaptation and settlement in the era of super-diversity. *Journal of Ethnic and Migration Studies.*

Gušić, S., Cardeña, E., Bengtsson, H., & Søndergaard, H. P. (2017). Dissociative experiences and trauma exposure among newly arrived and settled young war refugees. *Journal of Aggression, Maltreatment & Trauma, 26*(10), 1132–1149. https://doi.org/10.1080/10926771.2017.1365792.

HM Government. (2018). *Integrated communities strategy green paper*. Available at: https://assets.publishing.service.gov.uk/government/uploads/system/uploads/attachment_data/file/696993/Integrated_Communities_Strategy.pdf. Accessed 3 Apr 2019.

Hyman, I., Forte, T., Du Mont, J., Romans, S., & Cohen, M. M. (2006). The association between length of stay in Canada and intimate partner violence among immigrant women. *American Journal of Public Health, 96*(4), 654–659. https://doi.org/10.2105/AJPH.2004.046409.

Jesuthasan, J., Sönmez, E., Abels, I., Kurmeyer, C., Gutermann, J., Kimbel, R., ... & Wollny, A. (2018). Near-death experiences, attacks by family members, and absence of health care in their home countries affect the quality of life of refugee women in Germany: A multi-region, cross-sectional, gender-sensitive study. *BMC Medicine, 16*(1). https://doi.org/10.1186/s12916-017-1003-5.

Josse, E. (2010). 'They came with two guns': The consequences of sexual violence for the mental health of women in armed conflicts. *International Review of the Red Cross, 92*(877), 177–195.

Kalunta-Crumpton, A. (2017). Attitudes and solutions toward intimate partner violence: Immigrant Nigerian women speak. *Criminology & Criminal Justice, 17*(1), 3–21. https://doi.org/10.1177/1748895816655842.

Kalunta-Crumpton, A. (2018). *Race and drug trials: The social construction of guilt and innocence*. New York: Routledge.

Kapai, P. (2012). Minority women. In B. Baines, D. Barak-Erez, & T. Kahana (Eds.), *Feminist constitutionalism* (pp. 336–356). Cambridge: Cambridge University Press. https://doi.org/10.1017/cbo9780511980442.025.

Kapur, S., Zajicek, A., & Hunt, V. (2017). Immigration provisions in the violence against women act: Implications for Asian Indian marriage migrants. *Journal of Women, Politics & Policy, 38*(4), 456–480. https://doi.org/10.1080/1554477X.2016.1268872.

Keygnaert, I., & Guieu, A. (2015). What the eye does not see: A critical interpretive synthesis of European Union policies addressing sexual violence in vulnerable migrants. *Reproductive Health Matters, 23*(46), 45–55.

Keygnaert, I., Vettenburg, N., & Temmerman, M. (2012). Hidden violence is silent rape: Sexual and gender-based violence in refugees, asylum seekers and undocumented migrants in Belgium and the Netherlands. *Culture, Health & Sexuality, 14*(5), 505–520. https://doi.org/10.1080/13691058.2012.671961.

Kirmayer, L. J., Narasiah, L., Munoz, M., et al. (2011). Common mental health problems in immigrants and refugees: General approach in primary care. *CMAJ, 183*(12), E959–E967. https://doi.org/10.1503/cmaj.090292.

Kubiak, S. P. (2005). Trauma and cumulative adversity in women of a disadvantaged social location. *American Journal of Orthopsychiatry, 75*(4), 451–465. https://doi.org/10.1037/0002-9432.75.4.451.

Kulwicki, A., Aswad, B., Carmona, T., & Ballout, S. (2010). Barriers in the utilization of domestic violence services among Arab immigrant women: Perceptions of professionals, service providers & community leaders. *Journal of Family Violence, 25*(8), 727–735. https://doi.org/10.1007/s10896-010-9330-8.

Liebling, H. (2018). Survivors of conflict and post-conflict sexual and gender-based violence and torture in the Great Lakes region of Africa: A holistic model of care. In C. O'Manique, & P. Fourie (Eds.), *Global Health and Security: Critical Feminist Perspectives*. Routledge. https://doi.org/10.4324/9781315559568-3.

Martinez-Roman, M.-A., Vives-Cases, C., & Pérez-Belda, C. (2017). Immigrant women suffering from IPV in Spain: The perspectives of experienced social workers. *Affilia, 32*(2), 202–216. https://doi.org/10.1177/0886109916667158.

McMichael, M., & Manderson, L. (2004). Somali women and well-being: Social networks and social capital among immigrant women in Australia. *Human Organization, 63*(1), 88–99.

Miller, K. E., & Rasmussen, A. (2010). War exposure, daily stressors, and mental health in conflict and post-conflict settings: Bridging the divide between trauma-focused and psychosocial frameworks. *Social Science & Medicine (1982), 70*(1), 7–16. https://doi.org/10.1016/j.socscimed.2009.09.029.

Ndofor-Tah, C., Strang, A., Phillimore, J., Morrice, L., Michael, L., Wood, P., & Simmons, J. (2019). *Home Office indicators of integration framework 2019*. London: Home Office.

Norredam, M., Mygind, A., & Krasnik, A. (2005). Access to health care for asylum seekers in the European Union—A comparative study of country policies. *The European Journal of Public Health, 16*(3), 285–289. https://doi.org/10.1093/eurpub/cki191.

Ogunsiji, O., Wilkes, L., Jackson, D., & Peters, K. (2012). Suffering and smiling: West African immigrant women's experience of intimate partner violence. *Journal of Clinical Nursing, 21*(11–12), 1659–1665. https://doi.org/10.1111/j.1365-2702.2011.03947.x.

Oliveira, C., Keygnaert, I., Oliveira Martins, M. R. O., & Dias, S. (2018). Assessing reported cases of sexual and gender-based violence, causes and preventive strategies in European asylum reception facilities. *Globalization and Health, 14*(1). https://doi.org/10.1186/s12992-018-0365-6.

Parsitau, D. S. (2011). The role of faith and faith-based organizations among internally displaced persons in Kenya. *Journal of Refugee Studies, 24*(3), 493–512.

Phillimore, J. (2011a). Monitoring for equality? Asylum seekers and refugees' retention and achievement in ESOL. *International Journal of Inclusive Education, 15*(3), 317–330. https://doi.org/10.1080/13603110902866145.

Phillimore, J. (2011b). Refugees, acculturation strategies, stress and integration. *Journal of Social Policy, 40*(3), 575–593. https://doi.org/10.1017/S0047279410000929.

Phillimore, J., & Goodson, L. (2006). Problem or opportunity? Asylum seekers, refugees, employment and social exclusion in deprived urban areas. *Urban Studies, 43*(10), 1715–1736. https://doi.org/10.1080/00420980600838606.

Rossiter, K. R., Yercich, S., Baobaid, M., Al Jamal, A., David, R., Fairbairn, J., … Jaffe, P. (2018). *Domestic violence in immigrant and refugee populations: Culturally-Informed Risk and Safety Strategies* (4). London, ON: Canadian Domestic Homicide Prevention Initiative.

Samari, G. (2017). Syrian refugee women's health in Lebanon, Turkey, and Jordan and recommendations for improved practice. *World Medical & Health Policy, 9*(2), 255–274. https://doi.org/10.1002/wmh3.231.

Shabbar, F. (2012). Protecting our non-citizens: Iraqi women on Australian temporary spouse visas. *The Sociological Review, 60*(1), 149–168. https://doi.org/10.1111/j.1467-954X.2011.02046.x.

Sherwood, K., & Liebling-Kalifani, H. (2012). A grounded theory investigation into the experiences of African women refugees: Effects on resilience and identity and implications for service provision. *Journal of International Women's Studies, 13*(1), 86–108. Available at: http://vc.bridgew.edu/jiws/vol13/iss1/6. Accessed 5 May 2018.

Shishehgar, S., Gholizadeh, L., DiGiacomo, M., Green, A., & Davidson, P. M. (2017). Health and socio-cultural experiences of refugee women: An integrative review. *Journal of Immigrant and Minority Health, 19*(4), 959–973. https://doi.org/10.1007/s10903-016-0379-1.

Sturm, G., Baubet, T., & Moro, M. R. (2007). Mobilising social and symbolic resources in transcultural therapies with refugees and asylum seekers: The story of Mister Diallo. In *Voices of Trauma* (pp. 211–231). New York: Springer.

Tankink, M., & Richters, A. (2007). Silence as a coping strategy: The case of refugee women in the Netherlands from South-Sudan who experienced sexual violence in the context of war. In *Voices of Trauma* (pp. 191–210). New York: Springer.

Tavara, L. (2006). Sexual violence. *Best Practices Research in Clinical Obstetrics & Gynaecology, 20*(3), 395–408. https://doi.org/10.1016/j.bpobgyn.2006.01.011.

Thurston, W. E., Roy, A., Clow, B., Este, D., Gordey, T., Haworth-Brockman, M., … Carruthers, L. (2013). Pathways into and out of homelessness: Domestic violence and housing security for immigrant women. *Journal of Immigrant & Refugee Studies, 11*(3), 278–298. https://doi.org/10.1080/15562948.2013.801734.

UNHCR. (2011). *Action against sexual and gender-based violence: An updated strategy*. Available online at: https://cms.emergency.unhcr.org/documents/11982/51689/UNHCR%2C+Action+Against+Sexual+and+Gender-based+Violence.+An+

Updated+Strategy%2C+2011/4f9d2a1c-280e-4ac8-a832-1a789de63d46. Accessed 7 June 2018.

UNHCR. (2016). *Initial assessment report: Protection risks for women and girls in the European refugee and migrant crisis*. Available at: http://www.unhcr.org/protection/operations/569f8f419/initial-assessment-report-protection-risks-women-girls-europeanrefugee.html. Accessed 21 May 2018.

UNHCR. (2018). *Trends at a glance*. Available at: https://www.unhcr.org/globaltrends2018/.

Jenny Phillimore is Professor of Migration and Superdiversity at the University of Birmingham where she was founding director of the Institute for Research into Superdiversity. Jenny's research focuses on refugee integration with a particular emphasis on gender. She was co-author of the UK's Integration Indicators and advises European and national governments on refugee integration. Jenny is currently the Project Lead of the multi-country SEREDA project which examines forced migration, sexual and gender-based violence, resilience and vulnerability across the refugee journey. She also leads an independent evaluation of the UK's Community Sponsorship programme. Jenny has published widely in journals such as *Journal of Refugee Studies, Journal of Ethnic and Migration Studies, Social Science and Medicine, Journal of Social Policy* and *Policy and Politics*.

Sandra Pertek is a Doctoral Researcher at the University of Birmingham and part of the SEREDA project. Her research explores the intersection of sexual and gender-based violence (SGBV) and religion in forced migration. As a gender specialist, she consulted organisations at the international and national level in Africa, the Middle East and in the UK. Previously, she served as Senior Policy Adviser on Gender at Islamic Relief Worldwide spearheading gender mainstreaming strategy into policy and programmes. She serves in the UN Women Roster of Experts—Ending Violence Against Women and Girls. She holds an M.Sc. in Social Development Practice from University College London and a B.A. in European Studies from the University of Warsaw.

Lailah Alidu is a Research Fellow with the School of Law, Social and Behavioural Sciences, Kingston University. Her research focuses on health, wellbeing and migration, including the SEREDA project (sexual and gender-based violence in the refugee crisis). Lailah is currently working on a UKRI/NHIR funded project on Ethnicity and Covid-19, which aims to understand and reduce health inequalities and COVID-19 risk by developing a range of culturally appropriate health interventions to Black and South Asian communities in the UK. She holds a Ph.D. in Health Psychology from the University of Birmingham and a M.Phil. in Population and Health from the University of Cape Coast, Ghana. She is an Associate Fellow of the Higher Education Academy, UK.

CHAPTER 26

Slavery Versus *Marronage* as an Analytic Lens on "Trafficking"

Julia O'Connell Davidson

Abstract Dominant discourse on "trafficking" and "modern slavery" has been heavily criticised by scholars who argue that its framing of the problem as a criminal justice issue both overlooks the political and structural roots of vulnerability to violence, exploitation and abuse (such as criminalisation and stigmatisation of sex workers, immigration regimes, austerity, neoliberal economic reform) and encourages the criminalisation and/or immobilisation of marginalised groups, including sex workers, irregular migrants, child labourers, child migrants and runaway youth. This chapter explores how histories of fugitivity and *marronage*—the process of extricating oneself from slavery—might provide a more helpful starting point from which to theorise and research the contemporary experience (both positive and negative) of migrants who appear as vulnerable to "trafficking" and "modern slavery" in mainstream discourse.

INTRODUCTION

At the end of the twentieth century, global north state actors were increasingly anxious about the unauthorised cross-border movement of goods (especially drugs and guns) and people, which they packaged together as a problem of "transnational criminality" and regarded as a threat to national sovereignty and security. The unauthorised movement of people was subdivided, however, into

J. O'Connell Davidson (✉)
School of Sociology Politics & International Studies (SPAIS),
University of Bristol, Bristol, UK
e-mail: julia.oconnelldavidson@bristol.ac.uk

C. Mora and N. Piper (eds.), *The Palgrave Handbook of Gender and Migration*, https://doi.org/10.1007/978-3-030-63347-9_26

those migrants assumed to be complicit with the criminals (smuggled persons) and those assumed to be their victims (trafficked persons). The United Nations Convention on Transnational Organised Crime (2000) was accompanied by supplementary protocols on both smuggling and trafficking (also one on arms). Where smuggling was said to be "a crime against the state", human trafficking was "a crime against the person" and so a human rights violation. Thus, although "trafficking" entered into international law as a problem of crime control not human rights, it was also discursively constructed as a humanitarian issue (Aradau, 2004; Mai, 2018). The latter framing was supported by an emphasis on "trafficking" as a form of violence that especially affected women and children, and by repeated reference to "trafficking" as a form of "modern slavery".

By the 2000s, antislavery NGOs were insisting that "trafficking" *is* "the modern-day slave trade—the process of enslaving a person" (FTS, 2007–2014). Although this claim is incompatible with the definition of "trafficking" offered by the UN Protocol on Trafficking (UN OHCHR, 2000), which lists slavery as just one of several possible outcomes of trafficking, North American, Australian and EU politicians ran eagerly with the idea that "trafficking" is "nothing less than modern day slavery" (IOM, n.d.). In 2015, Theresa May (then Britain's Home Secretary) introduced a Modern Slavery Act, which includes more punitive sentences for "trafficking" offences, and similar legislation has subsequently been introduced in Australia, the Netherlands and France, and is under consideration in Canada (de Haas, 2019).

Legal definitions of both "trafficking" and "modern slavery" are extremely vague. The UN Protocol describes "trafficking" as a process, rather than a specific act, and one that can be arranged in multiple ways and lead to a variety of different outcomes. Furthermore, for a process to count as "trafficking", a number of elements, such as "exploitation", "force", "deception", and so on, must be present, yet none of these sub-elements are defined in the Protocol (Anderson & O'Connell Davidson, 2003). The term "modern slavery" is even more slippery. It is used as an umbrella term for various disparate phenomena (trafficking, forced labour, forced marriage, worst forms of child labour, etc.), each of which itself presents serious definitional challenges (O'Connell Davidson, 2015). Hence, there can be legal disputes as to whether an individual who has experienced a given constellation of harms is or is not a victim of "trafficking" or "modern slavery", and even as to whether the same individual is a "modern slave" or a "slave master" (Koch, 2019). More generally, the elasticity of the term "trafficking" means it is used in different ways by different actors. It operates, Grupo Davida (2015) observe, as a "floating signifier", and gender and childhood have been and remain nodal points in on-going struggles between different actors attempting to fix its meaning.

Critical scholarship argues that "trafficking" and "modern slavery" discourse is an interpretative grid which makes (some) members of (some) marginalised groups—including sex workers, irregular migrants, child labourers, child migrants and runaway youth—visible as "victims of trafficking" while rendering invisible the political and structural roots of group marginalisation (criminalisation and stigmatisation of sex workers, immigration regimes, austerity, neoliberal economic reform, and so on) (Bernstein, 2018; Cruz, O'Connell Davidson, & Sanchez Taylor, 2019; Day, 2010; Kempadoo, 2015; Kempadoo et al., 2005; Shih, 2016). This chapter continues that critique and reflects on how histories of fugitivity and *marronage*—the process of extricating oneself from slavery—can provide a more helpful starting point from which to theorise and research the contemporary experience (both positive and negative) of migrants who appear as vulnerable to "trafficking" and "modern slavery" in mainstream discourse.

From Slavery to *Marronage*—Switching the Frame, Seeing the State

Comparing contemporary forms of migration to the transatlantic slave trade makes for powerful rhetoric. It also deflects attention from the contradiction between states' efforts to control and limit migration, and their humanitarian obligations. Indeed, the concept of "modern slavery" functions as a bridge between anti-immigrant and humanitarian sentiment. If the humanitarian problem is caused by a slave trade that moves immigrants illegally across borders, then cracking down on the criminal "slavers" simultaneously prevents illegal immigration and relieves human suffering. Yet the comparison does not stand up to even basic scrutiny. The transatlantic slave trade was a large scale, profitable and legally sanctioned business that flourished for some four centuries, and it relied on overwhelming physical force at every stage. Its African victims had no pre-existing desire to move to the Americas, and the outcome of transportation was always catastrophic for them. By contrast, the vast majority of the people who are today described as "trafficked" actively, often urgently, *wanted* to move. Indeed, whether migrating to earn enough to pursue a life project, or to escape war or persecution or domestic violence, or simply to travel and see more of the world, they want to move to realise greater *freedom*. Moreover, the processes today described as "trafficking" do not inevitably lead to a situation that is subjectively perceived as worse than remaining at the point of departure, since on the one hand, the conditions people flee may be horrific, and even life-threatening (see Kook, 2016), while on the other, unfreedoms at the point of destination may be temporary (see Bastia & McGrath, 2011).

Rather than taking the transatlantic slave trade that transported Africans *into* slavery as a point of reference, a more helpful historical comparison is between migrants and refugees today and enslaved people who attempted to escape *from* slavery (O'Connell Davidson, 2015). Among them, we see a common, strong desire for mobility and a common quest to secure certain rights and freedoms. There is also commonality in the sense that these are groups of people whose independent mobility is criminalised. Importantly, through this lens, other historical parallels come into focus, especially between the states that supported slavery in the past and contemporary states. What these states have in common is a desire to thwart certain groups' pursuit of freedom and to restrict human mobility, and the techniques they employ are remarkably similar.

Most of the mobility control strategies currently being used by global north states were anticipated and deployed by American states to control the mobility of the slave population, including: passports, visas, fences, walls, border patrols and checkpoints, carrier sanctions, detention, the manipulation of affective ties between children and parents and husbands and wives. Hungary has gone so far as to licence the use of sniffer dogs at its borders, and there are reports of migrants being hunted down and attacked by those dogs (Byrne, 2016; Pasha-Robinson, 2017). Last but not least, there are a growing number of examples of humanitarian actors being prosecuted and fined under people smuggling and trafficking laws for assisting refugees and migrants making the journey to Europe and to the USA, which strongly echo prosecutions under American fugitive slave law used to criminalise those who harboured or assisted runaway slaves (O'Connell Davidson, 2015; Sinha, 2019).

The criminalisation of people who offered humanitarian assistance to fugitive slaves alerts us to the fact that states that supported slavery historically had to do more than just "build that wall". Measures to prevent human mobility, no matter how brutal, are rarely 100% effective. Even slave states could not ensure that no enslaved person ever escaped. Knowing this, they doubled down by trying to make certain that if "fugitive slaves" ran, they would find nowhere to hide. Again, there are striking parallels between the ways in which environments were made "hostile" to escapees from slavery (in "free" northern states of the US as well as southern slave states) and the ways in which contemporary states create "hostile environments" for irregular migrants (Bowling & Westenra, 2018).

To focus only on the treatment of the enslaved as things or commodities is to miss what Saidiya Hartman (1997) identifies as the central ambiguity of New World slavery, namely, the double character of the slave as both person *and* thing. That double character was especially clear in Fugitive Slave Law in the US, a body of law that made the runaway slave criminally liable as a *person* for stealing herself as a *thing* or piece of property (Best, 2004). Slavery was not merely the treatment of a human being as a piece of property. It also designated "a relation to law, state, and sovereign power; a condition of disfigured personhood, civil incapacitation, and bare life" (Best & Hartman, 2005: 10). This had to be so, because without an edifice of law and law enforcement to restrain and punish the enslaved as *persons*, one that controlled slave mobility in particular, slaveholders' *property* rights in slaves would have been empty and unenforceable.

Southern states of the US in early nineteenth century, with their thoroughly "modern" legal systems, sustained slavery by constructing the enslaved in law as ephemeral beings: "the slave was always a kind of afterlife, a form of legal being that was neither birthed by law, nor extinguished by law, but nonetheless present in the law" (Han, 2015). Irregular migrants and asylum seekers in the contemporary world share something of this ghostly character. As Lyndsey Stonebridge (2018) has recently argued, rather than becoming subjects of human rights law in the post-World War II era as European refugees did, the displaced and dispossessed peoples of the global south became objects of humanitarian attention, separate and unequal from the "international community" that claims to act on their behalf. To this we can add that when "illegally" present in the territory of global north states, migrants, like slaves, are criminally culpable persons but denied other aspects of what constitutes socially recognised personhood in those countries. Indeed, they are frequently criminalised for undertaking any and all acts necessary to support life itself, from taking employment to renting housing to accessing banking services. Like fugitive slaves who managed to escape to urban cities in the slave south or to "free" northern states, their standing in relation to law is zombie-like. They are flesh and blood humans who can commit crimes but who cannot lawfully love, sleep, work, or dream.

To live suspended in this hostile environment is necessarily to be vulnerable to abuse and exploitation since it creates dependency on others to mediate access to the means of life. Sometimes those who offer assistance are good, honest and decent; sometimes they are quite the opposite. This was true for escapees from slavery in the past, and it is true for irregular migrants and asylum seekers today. In both cases, vulnerability to abuse and exploitation is created by the laws and systems that deny the full panoply of rights and protections afforded to those who enjoy socially recognised personhood (see also Crawley, in this volume). Switching the point of comparison from the slave trade to *marronage* or fugitivity allows for an analysis that reveals, rather than obscures, immigration regimes as systems of domination. It also allows for more complex engagement with the problem of agency in relation to phenomena dubbed "trafficking" and "modern slavery".

Slavery, *Marronage* and the Problem of Agency: The Case of "Sex-Trafficking"

For contemporary antislavery thinkers, the treatment of human beings as "things" that can be traded, owned and possessed literally objectifies them—it strips them of free will and agency; they cannot walk away from the situation of exploitation and violence (or its threat) in which they find themselves. They are "slaves" (Bales, 2004; Bales & Soodalter, 2009). The idea of enslavement as entailing the literal eradication of individual agency, the transformation of one human being into the extension of another's will, is crucial to contemporary policy on "trafficking and modern slavery". It means that victims of trafficking or "modern slaves" are not regarded as knowing, agential participants in unlawful activities, such as unauthorised border crossings, prostitution (where selling sex is criminalised), illegal working and so on. They are "innocent victims", deserving of certain (limited) forms of protection and assistance; they are disidentified from categories of person more usually feared, despised or debased in political rhetoric—"illegal immigrants", "whores" and "criminals" (Aradau, 2004).

Because "trafficking" is read as "slavery", and "slavery" is understood to entail the erasure of will and agency, gender and age play an important role in its identification. Women and children, who have long been imagined as naturally dependent, innocent, gullible and tractable, are much more readily dressed in the garb of victimhood than adult men. This is evident in policy and political discourse on "trafficking". The UN OHCHR (n.d.), for example, links women and girls' greater vulnerability to "trafficking" to discrimination and poverty, which "results in fewer and poorer life choices, and may lead certain individuals to take risks and make decisions that they would never have done if their basic needs were being met". A 2016 European Parliament Briefing paper explains women and girls' vulnerability as follows:

> One common method [of recruitment] consists of exploiting the lack of economic opportunities in the local community and promising employment. Women and girls can be 'offered' jobs as models or dancers, as well as to participate in beauty contests or 'study abroad' programmes. Matrimonial agencies, also called 'mail-order bride agencies', offering to arrange a marriage abroad, sometimes serve only as fronts for sex trafficking. Yet another method, mainly used by male perpetrators to abuse female victims, especially young girls, involves feigned romantic interest (the 'lover-boy' method), where the trafficker enters into a romantic relationship with the victim, just to gain her trust, and then manipulates or coerces her into sexual exploitation. (EPRS, 2016: 6)

Once duped into taking a foolish risk, women and girls are reduced to mere objects of trade by their "traffickers", "used, controlled and exploited for commercial or personal gain" (Freedom Fund, 2018). Anti-trafficking campaign materials abound with images of objectified and commoditized women and girls. They feature young women bottled, or barcoded, or wrapped in cellophane and placed on plastic trays like meat in a supermarket, or transformed into dolls and marionettes (Andrijasevic, 2007). More recently, it has become vogue to visually represent "trafficked" women and girls as stripped of agency by picturing them as voiceless—gagged, or with their mouths covered by a hand, or taped shut. The message is that "labour migration is always risky or reckless for women, and that their inviolability is always threatened by moving abroad" (Brace, 2013: 881). The solution is not more and safer opportunities for migration, rather:

> Strengthening women and girls' "agency" – their ability to influence and make decisions that affect their lives – is key. Impactful programs teach women and girls life skills that allow them to make informed decisions that affect their lives. (Freedom Fund, 2018)

Agency, or the lack thereof, is thus central to the dominant discourse of "trafficking and modern slavery" promulgated by both state and civil society actors. Its presumed absence simultaneously makes the "victim of trafficking" into an object of humanitarian concern and prevents her from being perceived as a criminal or security threat. Like victims of the transatlantic slave trade, she was taken from her home, thingified and exploited for profit. State and civil society can therefore enter an alliance to protect her.

To focus on slavery only as the reduction of human beings to objects of market exchange is to disregard the bifurcated character of the enslaved in law as both persons and things, and so to miss the central role of the state and law in producing and reproducing slavery as a system of domination. Does switching the frame to *marronage* help with the problem of agency? Neil Roberts' (2015) *Freedom as Marronage* provides one answer. Roberts is critical of Western philosophical and political traditions that imagine freedom in fixed and static terms—freedom as a "thing" that you either do or do not possess; and he draws on Frederick Douglass' idea of movement as "a cornerstone of the human condition and essential to reform and progress... Movement is inevitable. Flight from slavery a continual process of release from bondage" (41). For Roberts, *marronage* represents "a flight from the negative, subhuman realm of necessity, bondage and unfreedom toward the sphere of positive activity and human freedom". A focus on the experience of flight, he says, opens up possibilities for a new, dialectical and more relational understanding of freedom, an understanding of "freedom as *marronage*".

Unlike the concept of "modern slavery", *marronage* focuses our attention on people's efforts to negotiate the spaces between domination and freedom, and in this respect, it can helpfully be applied to many of the forms of mobility currently discussed under the rubric of "trafficking and modern slavery". We could, for instance, apply it to research on Nigerian women's sex work migration, which so often features as an exemplar of the evil of "trafficking" and "modern slavery" in research and media reports (e.g., Aronowitz, 2009; Baarda, 2016; Taliani, 2012). In these accounts, Nigerian women's movement to Europe is orchestrated by "traffickers" by means of coercion and deception, and once in Europe, they suffer extensive and routine violence at the hands of the Madams to whom they are bound by a combination of debt and "juju". Through this lens, prosecuting "traffickers" and shutting down possibilities for them to move their victims out of Nigeria and into Europe are measures that will combat slavery, protect women's human rights and even save lives.

And yet there is a great deal of research that reveals a more complicated and nuanced picture of Nigerian women's experience of sex work in Europe, one that not only draws attention to the time-limited nature of their unfreedoms, but also shows that migrants often simultaneously acknowledge the serious restrictions on their freedom imposed by the third parties who arranged their movement, *and* express gratitude towards them for having facilitated their migration (Mai, 2016; Plambech, 2017a; Testai, 2008). Such research draws attention to women's desire to move freely, not merely to access better earning opportunities but also simply to be able to see for themselves. One of Irene Peano's (2013: 127) interviewees eloquently expressed the universality of this desire:

> [It] is like you, maybe since you've been in Europe you've been hearing of "Africa, Africa", you've not really seen Africa [and] have been wishing to see Africa. "Ah, this is Africa! I really want to know what is special about this Africa, you know, I want to know what is really going on there". Same thing with Africans, they want to go there. "Ah, is this Europe, Europe, Europe…I really want to go there and see". Even if somebody is telling you that "is not what you think-o!", you don't even want to listen. You want to see for yourself.

If we foreground this desire for mobility, it is very clear why the measures currently being pursued by the EU in the name of combatting "trafficking", namely intensifying and externalising border controls, do not work to protect human rights but rather the reverse. Making it harder for women to travel safely from Nigeria to Europe by plane has not stopped them from attempting to migrate, merely pushed them into heavier dependence on third parties and into taking riskier journeys, such as the Mediterranean crossing (Plambech, 2016), or seeking alternative points from which to fly. Nigeria is Africa's most populous country and the largest economy on the continent, yet the Nigerian passport is extremely weak, currently in the bottom quarter of the annual Henley Passport Index. Nigerian passport holders enjoy visa free access to fewer countries today than they did in 2010, and their applications for visas to study or visit the UK, Canada and the US are increasingly being rejected (Kazeem, 2020).

It is in this context, and having been frustrated by EU countries' draconian (and very expensive) visa application processes and decisions that the Nigerian women participants in Okyere's (2021, forthcoming) research decided to use the services of smugglers. Without access to safe or reliable means to move directly from Nigeria to Europe, they had taken the new route through Benin, Togo and Ghana, from where they hoped to fly to Europe. As Okyere (2021, forthcoming) notes, contrary to the notion that "traffickers" or smugglers recruit "victims"/travellers, the women in his study had all "approached the smugglers to help facilitate their travel", and did so in the knowledge and with the intention of paying off the extremely high costs of the smuggling services through participation in sex work.

There are many other studies that show women's migration between global south or from global south to north to earn through sex work can be subjectively understood as a kind of freedom project, even when it implies temporary restrictions on freedom of a type that in dominant discourse appear as "trafficking" (e.g., Lainez, 2019; Levy & Lieber, 2009; Piscitelli, 2012; Plambech, 2017b; Sobieszczyk, 2002). In this respect, thinking with *marronage*, as defined by Roberts, may offer a way in which to recognise and grasp the agency of those whose personhood is disfigured in law as described above, and yet who remain active, agential subjects rather than being literally "thingified" by the system of domination that seeks to control them. Nonetheless, I will argue below that *marronage* is an analytical concept that requires handling with care if we are to avoid inverting, as opposed to transcending the binaries that underpin discourse on "trafficking as modern slavery".

Beware of the Binaries

For the past 40 odd years in the US and Europe, most academic research on sex work has been framed by or contributed to a highly polarised and very divisive and emotive feminist debate between those who regard prostitution as an expression of patriarchal power and a form of male violence against women akin to rape; and those who regard women sex workers as rebels against the patriarchal strictures which demand sexual "purity" and "innocence" of women. There is an interesting parallel as regards approaches to the history of slavery, since studies of transatlantic slavery "often divide between works that emphasize the overwhelming power of the institution and scholarship that focuses on the resistant efforts of the enslaved" (Brown, 2009: 1235). The latter approach frequently rests upon what Walter Johnson (2003: 115) describes as a "misleading entanglement of the categories of 'humanity' and (liberal) 'agency'", through which "the bare fact (as opposed to the self-conscious assertion) of enslaved "humanity" has come to be seen as "resistance" to slavery". And yet the former can reproduce an equally problematic vision in which the bare fact of slaves' objectification as legal property comes to be seen as erasing their volition and free will, and so *literally* de-humanising them (O'Connell Davidson, 2015; Rinehart, 2016). Slavery, as much as any other social phenomenon, presents us with a problem at the heart of all sociological enquiry, namely the relationship between structure and agency.

So far as slavery is concerned, the problem is made all the more politically troubling by the racialisation of the categories of freedom and agency as white in dominant liberal thought, for these are the categories through which the normatively human was, and is still, imagined and socially recognised. In this context, it is difficult to stress the restraining violence of slavery without suggesting that its victims were reduced to mere bodies, the passive objects of structural forces, and therefore also hard to avoid reproducing the racist association between blackness, dependency and dishonour. And yet to attempt to counter this by emphasising the agency, courage and ingenuity of those who led or participated in revolts and revolution, or who made daring escapes from slavery, also carries its own risks. It leaves, for instance, little space to acknowledge the quiet valour of those women, men and children who somehow managed to make their lives within the confines of the institution (Brown, 2009; Wong, 2009). And a story in which the enslaved were resistant heroes makes it extremely difficult to admit the evidence that shows enslaved people often capitulated to, and were sometimes complicit with, the power of slaveholders (Midlo Hall, 1998) and/or even exercised de facto powers of ownership over other enslaved people (Beckles, 1987; Schafer, 2018).

Maroon and fugitive slaves are associated with a kind of glamour, embodying the kind of bravery and dignity that most of us wish, but somehow doubt, we would have displayed in the face of slavery's tyranny. And for contemporary theorists, concepts of fugitivity and *marronage* apparently provide the basis for hope. They fit with a politics of change and freedom as opposed to the politics of despair. This is also a feature of writings on migration that adopt what Peano (2013: 128) terms the "method of hope", an orientation that acknowledges structural constraints, but does "not reduce subjects to structures' epiphenomena, nor… subscribe to overly functionalist or individualist accounts of intentionality and action". Yet in relation to *marronage*, as Wilder (2017) observes, the historical reality overlooked in hopeful accounts is that most Maroons existed "under perpetual threat of discovery, social destruction, and re-enslavement", living "a vulnerable and conditional autonomy that bound them to the system they fled". *Marronage* did not necessarily lead to a condition markedly different from what Roberts (2015) terms the "negative, subhuman realm of necessity". Likewise, Peano (2013) notes that while those who apply the method of hope to migration typically acknowledge "suffering as also being part of migrants' experiences, they nonetheless proceed to elide any reference to it in their account". And yet, she continues, "hope itself carries defeat as a potential condition within it, and is born of ill-being and uncertainty" (2013: 128).

The concept of *marronage* is preferable to that of "modern slavery" as a starting point for analysing contemporary forms of migration currently are discussed under the rubric of "trafficking", but still needs to be used with caution. If "the modern maroon" is substituted for "the modern slave" and read as an ideal, heroic figure, we see agency but miss the fact that as agents, those at the sharp end of systems of domination can still only choose between options that are not of their choosing. This, after all, is part of what is wrong with the Freedom Fund's (2018) assertion that the "agency" of women and girls who are "modern slaves" or vulnerable to "modern slavery" needs to be strengthened. The problem for women and girls who undertake risky debt-financed forms of sex work migration is not their weak decision-making capacities, but rather the absence of a panoply of safe and attractive opportunities to decide between.

Conclusion

In 1869, Frederick Douglass made a speech in Boston advocating the acceptance of Chinese migration to US, arguing:

> There are such things in the world as human rights. They rest upon no conventional foundation, but are external, universal, and indestructible. Among these, is the right of locomotion; the right of migration; the right which belongs to no

> particular race, but belongs alike to all and to all alike… I know of no rights of race superior to the rights of humanity, and when there is a supposed conflict between human and national rights, it is safe to go to the side of humanity. (Blackpast, 2007)

The forms of abuse, violence and exploitation that are discussed under the umbrella of "trafficking" do not only affect cross-border migrants. However, immigration regimes that deny the right of locomotion to vast numbers of people, especially those from the global south, are one of a number of intersecting systems of domination that foster what is dubbed "trafficking" and "modern slavery". Instead of addressing this, anti-trafficking policy assumes a line between those who have actively chosen to move and consented to the type of work they will undertake and the conditions under which it will be undertaken (cast as agents), and those who have been tricked, cheated, bullied and trapped (cast as non-agential victims). It conceives of migrants as *either* willing selves, acting on the basis of their own freely made choices, *or* as forced to submit to the will of another.

This transports us onto territory that is profoundly gendered and aged, since women and children's grip on their own wills is understood to be fragile and tenuous. It also reflects a preoccupation with the form of compulsion produced by the exercise of direct, personalistic power, and so a very narrow and distinct understanding of "force". It reproduces a liberal vision of a world in which people are *either* abject, passive objects and slaves *or* freely contracting subjects, thereby missing the unseen, structural factors that force fates on men as well as women and children under the social relations of capitalism. It privileges a very particular kind of "freedom"—what G. A. Cohen described as the "the bare bourgeois freedom which distinguishes the most abject proletarian from the slave" (1995: 101).

Marx's basic insight about the relationship between structure and agency—people make history, but not in circumstances of their own choosing—needs to be connected to a different vision of the relation between agency and force. We need to transcend the gendered and aged division between willing subjects and will-less objects, retaining a concern with the systems of domination that constrain choice but approaching agency "as a socio-culturally situated capacity for action allowing people to differently inhabit and perform norms, or a capacity for people to act that is always created and enabled by specific relations of subordination rather than by abstract canons of freedom" (Mai, 2016; Shah, 2014). Reframing "trafficking" through the lens of histories of *marronage* and fugitivity, rather than the slave trade, might help to move research in that direction.

Acknowledgements The support of the European Research Council ('Modern Marronage: The Pursuit and Practice of Freedom in the Contemporary World', ERC ADG 788563), and the insights of my colleagues Angelo Martins Junior and Sam Okyere, are gratefully acknowledged.

References

Anderson, B., & O'Connell Davidson, J. (2003). Is trafficking in human beings demand driven? A multi-country pilot study. *IOM Migration Research Series, No. 15.*

Andrijasevic, R. (2007). Beautiful dead bodies: Gender, migration and representation in anti-trafficking campaigns. *Feminist Review, 86*(1), 24–44.

Aradau, C. (2004). The perverse politics of four-letter words: Risk and pity in the securitisation of human trafficking. *Millennium: Journal of International Studies, 33*(2), 251–277.

Aronowitz, A. (2009). *Human trafficking, human misery: The global trade in human beings*. London: Praeger.

Baarda, C. (2016). Human trafficking for sexual exploitation from Nigeria into Western Europe: The role of voodoo rituals in the functioning of a criminal network. *European Journal of Criminology, 13*(2), 257–273.

Bales, K. (2004). *New slavery: A reference handbook.* Santa Barbara, CA: ABC-CLIO.

Bales, K., & Soodalter, R. (2009). *The slave next door.* Berkeley: University of California Press.

Bastia, T., & McGrath, S. (2011). Temporality, migration and unfree labour: Migrant Garment Workers, Working Paper No. 6. *Manchester Papers in Political Economy.* Manchester: University of Manchester.

Beckles, H. (1987). *Black rebellion in Barbados.* Bridgetown, Barbados: Carib Research & Publications.

Bernstein, E. (2018). *Brokered subjects.* Chicago: University of Chicago Press.

Best, S. (2004). *The fugitive's properties: Law and the poetics of possession.* Chicago: University of Chicago Press.

Best, S., & Hartman, S. (2005). Fugitive justice. *Representations, 92*(1), 1–15.

Blackpast. (2007). *(1869) Frederick Douglass describes the "Composite Nation".* https://www.blackpast.org/african-american-history/speeches-african-american-history/1869-frederick-douglass-describes-composite-nation/.

Bowling, B., & Westenra, S. (2018). 'A really hostile environment': Adiaphorization, global policing and the crimmigration control system. *Theoretical Criminology, 24*(2), 163–183.

Brace, L. (2013). Inhuman commerce: Anti-slavery and the ownership of freedom. *European Journal of Political Theory, 12*(4), 466–482.

Brown, V. (2009, December). Social death and political life in the study of slavery. *American Historical Review, 144*(5), 1231–1249.

Byrne, B. (2016, August 30). Syrian refugees share aftermath of dog attack at Hungary's Border. *Vocativ.* https://www.vocativ.com/354794/syrian-refugees-dog-attack-hungary-border/index.html.

Cohen, G. A. (1995). *Self-ownership, freedom and equality.* Cambridge: Cambridge University Press.

Cruz, K., O'Connell Davidson, J., & Sanchez Taylor, J. (2019). Tourism and sexual violence and exploitation in Jamaica: Contesting the "trafficking and modern slavery" frame. *Journal of the British Academy, 7*(1), 189–214.

Day, S. (2010). The re-emergence of 'trafficking': Sex work between slavery and freedom. *Journal of the Royal Anthropological Institute, 16*(4), 816–834.

De Haas, A. (2019). Keeping up with the times: Why Canada should enact a modern slavery act. *McGill Centre for Human Rights and Legal Pluralism.* https://www.

mcgill.ca/humanrights/article/70th-anniversary-universal-declaration-human-rights/keeping-times-why-canada-should-enact-modern-slavery-act.

EPRS. (2016, February). *The gender dimension of human trafficking* (European Parliamentary Research Briefing). https://www.europarl.europa.eu/RegData/etudes/BRIE/2016/577950/EPRS_BRI(2016)577950_EN.pdf. Accessed 23 Mar 2020.

Freedom Fund. (2018, March 8). We must empower women and girls to fight modern slavery. *Press Release*. https://freedomfund.org/press-release/we-must-empower-women-and-girls-to-fight-modern-slavery/. Accessed 23 Mar 2020.

FTS. (2007–2014). *Glossary*. https://www.freetheslaves.net/page.aspx?pid=305. Free the Slaves. Accessed 11Sept 2014.

Grupo Davida. (2015). Trafficking as a floating signifier: The view from Brazil. *Anti-Trafficking Review, 4*, 161–166. www.antitraffickingreview.org.

Han, S. (2015). Slavery as contract: *Betty's Case* and the question of freedom. *Law & Literature, 27*(3), 395–416.

Hartman, S. (1997). *Scenes of subjection: Terror, slavery and self-making in nineteenth century America*. Oxford: Oxford University Press.

IOM. (n.d). *Human trafficking is a modern day form of slavery*. Dublin: International Organization for Migration. https://ec.europa.eu/anti-trafficking/sites/antitrafficking/files/train_the_trainers_manual_1.pdf. Accessed 23 Mar 2020.

Johnson, W. (2003). On agency. *Journal of Social History, 37*(1), 113–124.

Kazeem, Y. (2020, January 13). Nigeria has dropped the most in the ranking of powerful passports over the last decade. *Quartz Africa*. https://qz.com/africa/1784107/nigerias-passport-drops-most-in-global-travel-rankings/. Accessed 22 June 2020.

Kempadoo, K. (2015). The modern-day White (wo)man's burden: Trends in anti-trafficking and anti-slavery campaigns. *Journal of Human Trafficking, 1*(1), 8–20.

Kempadoo, K., Sanghera, J., & Pattanaik, B. (Eds.). (2005). *Trafficking and prostitution reconsidered*. London: Paradigm.

Koch, I. (2019, December 5). *Who are the slave masters of today? County lines, drug trafficking, and modern slavery policies*. Centre for Crime and Justice Studies. https://www.crimeandjustice.org.uk/resources/who-are-slave-masters-today-county-lines-drug-trafficking-and-modern-slavery-policies.

Kook, K. (2016). North Korean migrants in China: neither trafficked nor smuggled. *Migration and Mobility: Beyond Trafficking and Slavery Short Course, 5*, 89–92.

Lainez, N. (2019). Social structure, relationships and reproduction in quasi-family networks: Brokering circular migration of Vietnamese sex workers to Singapore. *Journal of Ethnic and Migration Studies, 45*(9), 1631–1649.

Levy, F., & Lieber, M. (2009). La sexualité comme ressource migratoire: Les Chinoises du Nord à Paris. *Revue française de sociologie, 50*(4), 719–746.

Mai, N. (2016). 'Too much suffering': Understanding the interplay between migration, bounded exploitation and trafficking through Nigerian sex workers' experiences. *Sociological Research Online*.

Mai, N. (2018). *Mobile orientations: An intimate autoethnography of migration, sex work, and humanitarian borders*. Chicago: University of Chicago Press.

Midlo Hall, G. (1998). *Africans in Colonial Louisiana*. Baton Rouge: Louisiana State University Press.

O'Connell Davidson, J. (2015). *Modern slavery: The margins of freedom*. London: Palgrave.

Okyere, S. (2021, forthcoming). The gendered dynamics of migration deterrence and anti-trafficking interventions: The case of Nigerian sex workers in Kumasi, Ghana. *Crime & Delinquency*.

Pasha-Robinson, L. (2017, January 25). Refugee children suffering dog bites and 'violent' beatings by border police in Balkans. *Independent*. https://www.independent.co.uk/news/world/europe/children-attacked-dogs-violently-beaten-border-police-brutality-balkans-serbia-belgrade-bulgaria-a7546366.html.

Peano, I. (2013). Escaping the exception: Migrant sex workers between subjectification and excess. *The Greek Review of Social Research, 140*, 119–132.

Piscitelli, A. (2012). Revisiting notions of sex trafficking and victims. *Vibrant: Virtual Brazilian Anthropology, 9*(1), 274–310.

Plambech, S. (2016, October 10). Becky is dead. *Beyond Trafficking and Slavery*. https://www.opendemocracy.net/en/beyond-trafficking-and-slavery/becky-is-dead/.

Plambech, S. (2017a). Sex, deportation and rescue: Economies of migration among Nigerian sex workers. *Feminist Economics, 23*(3), 134–159.

Plambech, S. (2017b, September 29). My body is my piece of land. *Beyond Trafficking and Slavery*. https://www.opendemocracy.net/en/beyond-trafficking-and-slavery/my-body-is-my-piece-of-land/.

Rinehart, N. (2016). The man that was a thing: Reconsidering human commodification in slavery. *Journal of Social History, 50*(1), 1–23.

Roberts, N. (2015). *Freedom as marronage*. Chicago: The University of Chicago Press.

Schafer, D. (2018). *Anna Madgigine Jai Kingsley: African Princess, Florid Slave, plantation slaveowner*. Gainsville: University of Florida Press.

Shah, S. (2014). *Street corner secrets: Sex, work, and migration in the city of Mumbai*. London: Duke University Press.

Shih, E. (2016). Not in my "backyard abolitionism" vigilante rescue against American sex trafficking. *Sociological Perspectives, 59*(1), 66–90.

Sinha, M. (2019, July 17). The new fugitive slave laws. *The New York Review of Books*. https://www.nybooks.com/daily/2019/07/17/the-new-fugitive-slave-laws/.

Sobieszczyk, T. (2002). Risky business: Debt bondage international labour migration from Northern Thailand. In *Communication présentée à la IUSSP Regional Population Conference on Southeast Asia's Population in a Changing Asian Context*.

Stonebridge, L. (2018). *Placeless people*. Oxford: Oxford University Press.

Taliani, S. (2012). Coercion, fetishes and suffering in the daily lives of young Nigerian women in Italy. *Journal of the International African Institute, 82*(4), 579–608.

Testai, P. (2008). Victim protection policy in Italy: between 'emancipation' and 'redemption'. *St Antony's International Review, 4*(1), 37–57.

UN OHCHR. (2000). *Protocol to prevent, suppress and punish trafficking in persons especially women and children, supplementing the United Nations convention against transnational organized crime*. https://www.ohchr.org/en/professionalinterest/pages/protocoltraffickinginpersons.aspx. Accessed 22 June 2020.

UN OHCHR. (n.d.). *Human rights and trafficking in persons*. https://www.ohchr.org/EN/Issues/Trafficking/TiP/Pages/Index.aspx. Accessed 23 Mar 2020.

Wilder, G. (2017, February). The promise of freedom and the predicament of marronage: On Neil Roberts' freedom as marronage. *SX Salon*, 24.

Wong, E. (2009). *Neither fugitive nor free*. New York, NY: New York University Press.

Julia O'Connell Davidson is Professor of Social Research at the University of Bristol. Her research questions the use of the slave trade as a metaphor in relation to contemporary migration, and instead examines similarities between the history of state controls over the mobility of enslaved people and the techniques used by contemporary states to restrict the mobility of migrants and refugees. She currently holds a five-year ERC Advanced Grant that is, among other things, gathering data on subSaharan Africans' experience of irregular migration to Europe and to Brazil. These data will be analysed in relation to historical evidence on enslaved people's flight from slave states, and the maroon or *quilombo* communities formed by escapees from slavery. Julia has also recently completed research in Jamaica that examined the unintended and very negative impact of United States Trafficking in Persons Reports and policy on vulnerable populations, including sex workers and LGBTQ communities. She is author of *Modern Slavery: The Margins of Freedom* (Palgrave, 2015) and *Children in the Global Sex Trade* (Polity, 2005), among others.

CHAPTER 27

Refugees, Gender and Disability: Examining Intersections Through Refugee Journeys

Lucy Fiske and Chrisanthi Giotis

Abstract 'Disability' is inherently intersectional. A displaced person will always have a socially constructed gender, the implications of which will be different in each geographic and cultural location and never fixed anywhere. The displaced person may be internally displaced and still retain formal citizenship, or may have crossed an international border to a refugee camp or an urban location. He may be in the hands of a smuggler attempting a clandestine border crossing. She may be incarcerated in a detention centre. They may be displaced also from their family and social networks, and from the attendant support and enablement. In this chapter we highlight factors necessary to grapple with as a *starting* point for mapping the ways in which gender, disability and forced migration interact. We set out some key facts and definitions, discuss tensions engaged in these intersections and, illustrate these by drawing on others' and our own empirical work. The task here is to make the intersectional identities, experiences and lives of refugees with impairments visible in a more holistic manner through considering their social embodiment in particular places and at specific times. In the conclusion we discuss the necessity for process-focused, rather than programmatic, thinking and practice.

L. Fiske (✉)
University of Technology Sydney (UTS), Broadway, NSW, Australia
e-mail: lucy.fiske@uts.edu.au

C. Giotis
University of Technology Sydney (UTS), Ultimo, NSW, Australia
e-mail: chrisanthi.giotis@uts.edu.au

C. Mora and N. Piper (eds.), *The Palgrave Handbook of Gender and Migration*, https://doi.org/10.1007/978-3-030-63347-9_27

Introduction

We have been charged with writing about the intersection of gender, disability and forced migration, a Sisyphean task as the intersections of these three major structures produce an infinite range of possible social locations and lived experiences. Goodley (2013: 632) notes that while disability may well be the starting point for theorising, it is never the end point. 'Disability' is inherently intersectional. A displaced person will always have a socially constructed gender, the implications of which will be different in each geographic and cultural location and never fixed anywhere. The displaced person may be internally displaced and still retain formal citizenship, or may have crossed an international border to a refugee camp or an urban location. He may be in the hands of a smuggler attempting a clandestine land or sea crossing. She may be incarcerated in a detention centre within or outside the border of a *Refugees Convention* signatory country. They may be displaced also from their family and social networks, and from the support and enablement that comes with these relationships. He may have a newly acquired impairment or she may have developed effective adaptive strategies over many years with their impairment. The possible permutations of social, geographic, legal and economic contexts in which a displaced person's impairment may become disabling are too varied to possibly design effective theory or practice procedures, and this complexity is precisely why scholars and practitioners must turn our attention to this difficult task.

In this chapter we highlight factors necessary to grapple with as a *starting* point for mapping the ways in which gender, disability and forced migration interact. We set out some key facts and definitions, discuss tensions engaged in these intersections and, illustrate these by drawing on others' and our own empirical work. The task here is to make the identities, experiences and lives of refugees with impairments visible in a more holistic manner; and in conclusion we discuss the necessity for process-focused, rather than programmatic, thinking and practice.

Prevalence and Lack of Statistics

There is a paucity of information about disability and impairment within refugee populations (Crock, Smith-Khan, McCallum, & Saul, 2017), impacting on the ability to monitor 'the implementation of the UN Conventions' and 'ensure the human rights of people with disability' (Eide & Loeb, 2016, 52). The UNHCR first identified refugees with disabilities as a group to whom it owed obligations (arising from the *Convention on the Rights of Persons with Disabilities* [CRPD]) in its 2010 ExCom 'Conclusion on refugees with disabilities' and, in 2011 issued a field guidance note on *Working with Persons with Disabilities in Forced Displacement* (UNHCR, 2011; UNHCR Executive Committee, 2010). Despite this recognition and some improvements in identifying disability within reception and registration processes, 'refugees with disabilities remain under-identified in most circumstances' (Crock et al., 2017: 20). Refugees with visible impairments, such as missing limbs or an inability to walk are more likely to be recognised than refugees with less visible communication, intellectual or psychological impairments yet these impairments can be just as disabling and may significantly increase a person's vulnerability to additional harms (Crock et al., 2017; Marshall & Barrett, 2018). Nine years after the Executive Committee Conclusion, the UNHCR's annual statistical publication, *Global Trends*, does not report on disability, much less disaggregated by sex, impairment type, age or other relevant categories (cf. UNHCR, 2019) (see also Boyd, in this volume).

The World Health Organisation (WHO) estimates that approximately 15% of the world's population has a disability (Dowling, 2016: 3). Extrapolating from UNHCR global displacement figures indicates more than 10 million displaced people will have a disability of some kind (UNHCR, 2019). This figure however, likely underestimates the incidence. WHO reported that 80% of people with disabilities (PWD) live in the Global South (Grech, 2016: 3). Conflict, dangerous refugee journeys and, conditions of displacement are all risk factors in causing disability. The few empirical studies that do exist point to a much higher incidence of disability among refugee populations. A 2014 study found 30% of Syrian refugees in Jordan and Lebanon had a disability, rising to 77% among refugees aged over 60 (HelpAge & HI, 2014: 4). Disability among Afghan refugees in Pakistan is reported at 15% (in line with WHO estimates), but rises to 46% among refugees over 60, and 'disability prevalence is substantially higher amongst adult women than adult men' (Smith-Khan & Crock, 2015: 6).

Even with the current potted picture it is clear that there is a sizeable population of refugees and displaced people with disabilities; this is numerically, as well as ethically, a significant issue. That so little is empirically known about this population is perhaps a reflection of the power position of People With Disabilities (PWD) more broadly, power relations that withstand the many disruptions of displacement (see Phillimore, Pertek, and Alidu, in this volume).

Approaches to 'Disability'

The biomedical model has traditionally dominated approaches to disability (Connell, 2011; Smart & Smart, 2006). This approach sees disability as a medical 'problem' and any limitations a person with a disability may face, as a direct consequence of that medical problem. Interventions are then decided upon by professionals and seek to fix or treat the problem within the person. Disability activists have fought hard against this approach, arguing that it reduces people to their impairments, fails to capture lived experiences of disability, over-looks people's agency and, locates control over people's lives and bodies with 'experts' rather than people with disabilities (Smart & Smart, 2006). Disability rights activists pushed for a social model approach, arguing that a person may have an impairment (such as impaired mobility, sight, cognition etc.), but how disabling this impairment becomes is determined by the interaction of the impairment with the social, legal and physical environment in which the person lives. It is ableist social attitudes and environmental and procedural designs that determine how disabling an impairment will be. It is the social model that is used in the CRPD, which states:

> Persons with disabilities include those who have long-term physical, mental, intellectual or sensory impairments which in interaction with various barriers may hinder their full and effective participation in society on an equal basis with others. (UNGA, 2006: Art 1)

This approach recognises the legal and ontological equality of people with disabilities and articulates the obligations of states to make positive provisions necessary for equal participation in society. The social model paves a way for people with disabilities to escape the role of passive recipient of expertise and charity and, instead to make claims as citizens and rights-bearing individuals upon states.

Critical Disability Studies (CDS) develop the social model further, to recognise that disability entails more than materiality and draws upon a range of disciplines to interrogate disability within identity politics, queer theory, post-colonial studies or security studies. Disability is, as Goodley (2013: 632) argues, the 'space from which to think through a host of political, theoretical and practical issues that are relevant to all'. Critical Disability Studies recognises disability as just one aspect of identity and social location; that a person's sexuality, gender, class, nationality and other traits and experiences all work to produce particular social locations and that theory from all these fields is needed to adequately understand disability. For example, CDS enables the impact of Global North states' securitisation of borders on PWD (further distancing PWD from effective protection, causing impairments through violence routinely meted out at the border, dangerous journeys made necessary by the fortification of borders) to be brought within the analytical frame (Pisani & Grech, 2015).

Intersectionalities and Social Embodiment

The wider lens that CDS brings enables more complex power dynamics to be analysed. Mansha Mirza (2011) observes that the push to de-medicalise approaches to disability is largely driven by movements in the Global North and reflect the desires of PWD in those settings. Such desires rest upon a presumption that certain needs are met as a matter of course, needs that one cannot presume are met in the Global South, much less among refugee populations. Reporting from the Global South, Mirza notes that refugees with disabilities often make very different claims and their voices are not readily heard in international disability rights discussions (2011: 1533). That global inequalities of power work *within* the disability rights movement, ought not be surprising. Crenshaw (1989) called attention to how difference operates within both feminist (to privilege the concerns of white middle class women over those of women of colour) and anti-racism movements (to centre on the concerns of men of colour at the expense of women's concerns). Pisani and Grech (2015: 421) trace precisely this trend within forced migration and critical disability studies:

> forced migration studies, as well as humanitarian practice continue to be premised on and adopting an ableist approach focused on heteronormative productive bodies, while disability studies, with a corpus of work premised on an assumption of citizenship, has failed to critically engage with issues of sovereignty, borders and bodies that lie beyond the protection of the Nation State.

There is a nascent body of work beginning to recognise the complex nexus that refugees with disabilities exist within (Crock et al., 2017; Mirza, 2011, Pisani & Grech, 2015) but much remains to be done, and it is important that how gender works (upon men and women) is included in these analyses.

In outlining intersectionality, Crenshaw (1989: 140) cautions against an 'additive' approach, emphasising the mutually constitutive dynamism of intersecting identities and social relations; these intersections create new social and political locations that require theorising from the ground up. Working within pre-existing models and simply adding them together risks ignoring or erasing elements of a person's identity or experience that do not 'fit' in models of disability, gender or displacement. For example, in gender programmes the role of the family unit is often subsumed to discourses around individual rights however, in many parts of the world women are effectively able to access rights only through male relatives; how benevolent or otherwise significant male relatives are determines freedom from violence, access to land and other economic resources, access to education and so on (Fiske & Shackel 2015: 111). Enhancing women's independence (and reducing her reliance on male relatives) is a priority in many international aid and development programmes. People with disabilities (in both Global South and North countries) similarly often rely on social relationships for meeting basic needs and rights, but in this discourse 'a strong family unit is the most effective support and safeguard for a person with a disability' (APH, 2010: sec 5.23). Forced migration very often ruptures social relationships as people become separated during refugee flight. Social relationships are a vital source of protection for refugees—for assistance accessing food, information, transport or safety, and the rupture of these relationships poses significant risks for any refugee. Conversely, these relationships may also be sources of stress and constriction as ruptured familial responsibilities and gender norms emerge. An additive approach cannot capture the dynamism of these relationships or resolve tensions when one paradigm comes into conflict with another.

Including relationships within the analytical frame is, for Raewyn Connell, necessary if we are to understand 'social embodiment'. Connell (2011: 1371) argues that to 'understand social embodiment we need to recognise the agency of bodies, [and] their productive power in social relationships' and, that different bodies (sex, gender, ability, age) generate different possibilities and limitations. To acknowledge the specificity of each person's situation is not to drown in the particular but to cultivate an awareness necessary in both practical and theoretical contexts. The insights from intersectionality and social embodiment theories suggest critical social and biological factors include:

- legal status as non-citizens (refugees) or quasi-citizens (internally displaced people) (including law, policy and attitudes pertaining to forced migrants in that location)
- type of physical, intellectual, communication, psychological impairment
- physical environment
- gender status (in culture of origin, current and (potential) future location)
- social attitudes (social construction of disability in culture of origin, current location)
- time (as refugee, living with impairment)
- social networks.

Tracing the power relations inherent in any social location requires recognition of the influence of dominant discourses. Particular discourses define prescribed ways of knowing a subject, moreover, they tend to ascribe causes and solutions. For example, the dominant way of 'knowing' the refugee experiencing impairment is through a discourse of need and vulnerability, thus overlooking agency and ability. That even well-intentioned organisations can fall prey to this way of knowing is perhaps, ironically, seen through initiatives driven by refugees with disabilities in sites where NGOs have been absent or latecomers. Mirza (2011: 1531–1533) reports several initiatives including Care Villa, a facility established and run by and for refugee landmine survivors in on the Thai-Burma border; *Umoja*, a collaboration between refugees with disabilities from Rwanda, Somalia, DRC and other countries that advocates for disability rights and access within Dzaleka camp in Malawi; and, the Gulu Disabled Person's Union in Uganda began in 1979, 'a time when institutional actors such as UNHCR barely recognised the presence of people with disabilities among refugee populations'. She further describes disability acting as a basis for solidarity action between displaced and host communities in Jordan, Tanzania and Nepal: 'By connecting across disability, gender, ethnicity and nationality, such initiatives not only disrupt divisive tendencies and authoritative power relations but also defy institutional practices of treating categories of difference, such as gender and disability, in an insular, fixed and isolated manner' (Mirza, 2011: 1533). Mirza concludes that 'institutional humanitarian response to disability has been both slow in development and inconsistent in application between principle and practice' yet at the same time there have 'emerged some promising examples of disability advocacy and organising within refugee settlements' (Mirza, 2011: 1532–1533). These examples disrupt established views of disability as a vulnerability and demonstrate how disability can in fact be a strength, assisting transgression of otherwise difficult divides of nationality, religion, gender and citizenship status.

Gender and Disability in Refugee Law and Policy

Refugee law, like other areas of national and international law, is a patriarchal institution designed with 'masculine experience as the norm' (Edwards, 2010: 23). This can be seen by the exclusion of sex or gender as a protected category in the *Refugees Convention*, however, the patriarchal assumptions extend beyond this (see Crawley, in this volume). Refugee law incorporates androcentric assumptions that feminists have been struggling against for decades (Bunch, 1990; Charlesworth, Chinkin, & Wright, 1991). Perhaps chief among these is the importation of the divide between public and private spheres and an acceptance that law must carefully justify any intrusion upon the private domain. This has meant that women have more difficulty having their actions recognised as political (one ground of *Convention* protection) and much of the harm that women more commonly face (such as violence committed by non-state actors such as family members, economic harm including exclusion from work or property ownership, or sexual violence) is too easily seen as not constituting persecution (Edwards, 2010: 30) or, is seen as a private matter not engaging *Convention* protection (Kneebone, 2005: 8, 26).[1] Women 'are largely represented in refugee law ... as part of a family unit, as mothers and wives and sisters in need of male protection' and, although entitled to make individual claims, are most commonly included as dependents on male relatives' applications (Kneebone, 2005: 10–11). Women, especially if accompanied by an adult male relative, are often not asked about any claims she might have independent of the family unit.[2]

[1]Alice Edwards (2010) cites a 2004 study of asylum claims in 41 European jurisdictions which found that sexual violence was recognised as persecution in only 41.5% of jurisdictions, 33% of jurisdictions did not accept persecution by non-state actors as falling within the scope of the Convention, and only 36.5% of jurisdictions recognised women as members of a particular social group.

[2]While Kneebone was writing in 2005, this practice continues. During fieldwork in Indonesia this year (2019) I assisted a woman write her statement of claims just days before her Embassy resettlement interview. No advocate, NGO workers or UNHCR officers had asked anyone other than her husband about their claims during the six years of the family's refugee status determination process. Although present at the Embassy interview, she was not asked any questions, despite having claims of her own. The absence of an

Refugee policy areas, which largely govern offshore resettlement processes, also reflect ableist and patriarchal structures. Most countries have 'capacity to settle' tests in their selection of refugees for prospective resettlement. Such tests typically prefer refugees with 'likely employment prospects, taking into account their work history, qualifications and English language ability' (Karlsen, 2016: Note 10) thus compounding the effects of inequalities in countries of origin and transit. Many studies have shown that both women and PWD have lower levels of education, foreign language learning and paid work opportunities in most countries around the world (Price & Goyal, 2016: 306). Within disability rights organisations, representative positions are 'limited mostly to educated men with physical impairments', meaning that the impacts of such policies on people with communication or cognitive impairments and on women with disabilities are seldom put forward (Berghs & Kabbara, 2016: 273). Inequalities pre-displacement are then replicated rather than remedied in refugee transit and in resettlement processes.

Since at least the mid-1990s, refugee women have become a priority for the UNHCR. Much of this concern however, has mirrored developments in International Criminal Law in which women enter legal and policy concern primarily as victims of sexual and gender-based violence (SGBV) (Grewal, 2015). While high level recognition of SGBV is laudable, it has had a number of unintended side-effects. This approach reinforces the 'hierarchy of harm' which prioritises civil and political harms over economic, social and cultural harms, disadvantaging both PWD and 'women claimants who are more likely to be adversely affected by poverty and social and cultural marginalization' (Edwards, 2010: 26; Price & Goyal, 2016, 306).[3]

The international focus on SGBV has also meant that the term 'gender' has come to mean 'women' which has been further narrowed to largely mean 'victim of sexual violence', a view that obscures other dimensions of women's lives and, women's agency and capacity. This approach (focusing on vulnerability) echoes the deficiency approach to disability, and the combining of 'gender' and 'disability' paradigms has found common ground in which each logic reinforces the other, leaving little room for alternate views in policy formation or practice procedures.

The reduction of 'gender' to 'women' additionally ignores masculinity as a gender and masks the gender-based harms that men face. Men are more likely to be targeted in conflicts for forced military recruitment, killing or deliberate disablement (Edwards, 2010: 41). Likewise, men and boys can also be targeted for sexual violence with the aim of 'emasculating' them (Kastner & Roy-Trudel 2019: 150–151). The same gendered beliefs also act as a barrier to men disclosing any sexual violence they have experienced. Once

appeal mechanism in resettlement processes only heightens the importance of recognising women as individuals with rights extending beyond the family unit.

[3]A recent study showed that only 20% of women with disabilities globally are in paid employment, 'compared with 53 per cent of disabled men' (Price & Goyal, 2016: 306).

displaced, gendered expectations of men as providers and protectors for the family exposes refugee men to heightened risks from efforts to fulfil these roles in a range of ways such as by attempting dangerous border crossings or engaging in work with little or no protection against exploitation, injury or death (Crock, Saul, McCallum, & Smith-Khan, 2015; Pisani & Grech, 2015).

Experiences Through the Journey

Both disability and gender are key dynamics in all stages of displacement journeys. As Crock et al. (2017: 77) note, PWD *do* travel, and their impairments will have different enabling and disabling effects in different contexts. More needs to be known about the ways in which gender, disability, law and policy work in displacement situations globally. Legal and policy frameworks, particularly those set by governments hostile to refugees, can play causal roles in producing both physical and psychological impairments. A comprehensive global overview is beyond the scope of this chapter, so we focus here on examples in different transit settings and resettlement processes to illustrate issues for practice and policy.

A foundational challenge for policy and practice is the lack of data about refugees with disabilities; if people's existence and needs are not identified, they cannot be addressed. Data collected by UNHCR Malaysia showed a disability prevalence rate of less than 0.2%, a rate at odds with global estimates of disability and justified by UNHCR staff speculating that refugees with disabilities do not travel (Crock et al., 2015: 8). This view, however, ignores the importance of social networks that can make travel not only possible but necessary, as maintaining family networks is crucial. In some situations, impairments may be a driver in forced migration. Handicap International found a motivating factor for Syrians with disabilities to become refugees was the inability to access medical needs in Syria (HelpAge International & Handicap International, 2014). Under-reporting of disability likely stems from limited understanding and awareness of disability, particularly less visible disabilities.

Both Pearce and Lee (2018) and Crock et al. (2015) argue for more detailed assessment. For Pearce and Lee it is important for practitioners to better understand 'intersecting vulnerabilities' (2018: 32). They give the example of a gender- and disability- blind cash programme for Syrian refugees in Lebanon in which female-headed households with caring responsibilities for a disabled family member were disadvantaged. A lack of visibility and other factors, meant that many such households did not receive cash payments despite having fewer income-generating opportunities and higher medical expenses (p. 32). Standard reception and data gathering methods failed to capture the existence of such households, despite well-established evidence in both disability and gender fields indicating the likelihood of their existence and that 'standard operating procedures' would result in their exclusion from programmes.

For Crock et al. (2015) detailed data collection reveals important structural issues impacting refugees in transit. Researchers found that of 209 refugees with disabilities interviewed in Malaysia and Indonesia, 60% had acquired their disability as an adult and about one-third had acquired their disability during or after refugee flight. Further investigation showed that of impairments acquired in Malaysia, accidents were the most common cause, and most of these accidents occurred at work (Crock et al., 2015: 39). Such statistics indicate that the conditions of refuge (including no legal right to work combined with no financial support) contribute to producing disability. Disease was the second most common source of impairment (p. 39) which could indicate refugees' lack of access to health services plays a causal role. In Malaysia refugees are able to access health services but must pay the significantly higher cost prescribed for internationals.

Syrian refugees in Turkey formally have rights to health services, however these rights are unattainable for most due to 'notable practical obstacles' including language, transport and inadequate information regarding registration processes (Cantekin 2019: 202). Such obstacles are likely compounded for women refugees and refugees with disabilities who, in general, face greater hurdles in obtaining information and travelling independently. Cantekin's female interviewees, particularly those unaccompanied by male relatives, highlighted their reluctance to leave their accommodation for fear of harassment and that this limited their ability to learn the local language. They reported not feeling confident to speak to officers and so missed out on goods, services and information otherwise available (2019: 211). Again, this is a structural issue and recognising it as such allows us to envisage solutions such as employing more female officers. This example also demonstrates the importance of social networks for women who access society through familial relationships, which may be ruptured in displacement. Given the importance of the family network for people with impairments, in a gendered society, the loss of males in the household can create added stress in cases of disability. In Cantekin's research one disabled woman decided she would no longer ask her sister's help to leave the tent due to gendered harassment adding anxiety to an already stressful situation. Cantekin (p. 214) states this confinement likely exacerbated the psychological problems of the tent-bound woman.

Psychological impairments are seldom recognised as disabilities in refugee transit situations, despite the prevalence in comparable populations in resettlement countries (Silove, Ventevogel, & Rees, 2017). Conflict, torture and displacement are all associated with a range of mental health problems including Post-Traumatic Stress Disorder (PTSD), anxiety and depression. McNevin and Missbach (2018: 17) highlight the importance of time in refugee situations and as a strategy of migration management, noting that 'chronic waiting' is a structural feature of contemporary displacement. This chronic waiting can be seen as both a key cause of mental impairments *and* a barrier to recovery (Procter, Kenny, Eaton, & Grech, 2018). The brother of a young Afghan refugee in Indonesia who developed schizophrenia during transit told us that he attributes his sister's condition entirely to the prolonged and stressful uncertainty. He explained that

> I know for certain because her symptoms all revolved around her life as a refugee and resettlement. She was coming up with different scenarios on how to get out of Indonesia or thinking she has to do certain hidden things to be resettled, like posting very specific things online, sending signals by wearing a particular colour of dress, donating all her things to specific people, so many things to get resettled. She thought there were immigration spies following her, watching her every move, controlling her phone and her mind. (Interviewed 15 August 2019)

The Australian government justifies funding the containment of refugees in Indonesia, Nauru and Papua New Guinea as discouraging dangerous asylum journeys to Australia. However, the journey to Indonesia can be a cause of disability; as was the case for 10% of interviewees in the research by Crock et al. (2015: 38). Moreover, the issue is complicated as the long-term limbo inflicted on asylum seekers and refugees in the name of 'safety' is in fact causing mental impairments and suicides (McNevin & Missbach 2018: 18).

For Syrians living on the border in Turkey the more time passes the more debilitating the restrictions and social isolation of the refugee camp [including from other family members in other camps due strict sign-in/out procedures] become. The situation is compounded by the proximity to fighting as audible bomb attacks trigger episodes of PTSD. Most refugees reported they had not expected the war to last this long, hence taking refuge close to the border in anticipation of their return (Cantekin, 2019). As time passes the proximity of the camp has gone from being an advantage to a disadvantage. Moreover, as time passes the uncertain legal status of Syrians in Turkey has become a major stressor. As well as being caught by Turkey's geographic exception which provides only limited refugee recognition[4] several Turkish provinces have suspended registrations for Syrian refugees, making them ineligible for health care and other rights and, leaving them vulnerable to deportation (HRW, 2018). As Europe closes its borders to refugees (justified in part as preventing dangerous boat journeys) one of the few refugee routes left puts refugees at risk of impairment through landmines in Croatia from the former war there (Gözübenli, 2018).

For the majority of refugees, the 'chronic waiting' is unlikely to result in effective protection unless a refugee's country of origin becomes safe for return, an unlikely prospect in the foreseeable future for the major refugee producing countries. And while gender and disability status may heighten

[4]As discussed by both Cantekin (2019) and Human Rights Watch, Syrians in Turkey are not able to access full asylum 'Although Turkey ratified the 1951 Refugee Convention and its 1967 Protocol, the country maintains a geographical limitation that excludes anyone not originally from a European country from full refugee recognition. That means it does not fully grant asylum to people fleeing violence or persecution in Syria and any other non-European country' (HRW, 2018).

one's need for protection through resettlement, both gender and disability can make resettlement more difficult to obtain. Resettlement is available to fewer than 1% of refugees each year. The UNHCR must therefore select which refugees to refer for prospective resettlement. While the UNHCR may focus on vulnerability and protection needs, and have determined to submit refugees with disabilities for resettlement on an equal basis to non-disabled refugees, they are also bound by resettlement states' policies (Crock et al., 2015: 60). As discussed above, 'capacity to settle tests' negatively impact both women and PWD. Each resettlement country additionally has health criteria that can disadvantage refugees with disabilities. Australia, USA and Canada all have exclusion provisions based on public health grounds. Australia further restricts resettlement eligibility for people with conditions that would incur 'undue costs to the government in health care or community services' (UNHCR, 2018: 8). This provision has been used to refuse entry to PWD, including children with an Australian permanent resident parent (APH, 2010). This has sometimes resulted in tragic outcomes. Shahraz Kiane was granted refugee status in Australia, but multiple applications for his family to join him were refused due to his then eight-year-old daughter, who had cerebral palsy, failing the health test (APH, 2010, Case study 5.2). The 'applications were in part rejected after the department estimated it would cost taxpayers up to $750,000 to care' for his daughter (AAP, 2005). Mr. Kiane fatally self-immolated on the steps of Parliament House in 2001.

Conclusion

Accounting for such encompassing, and mutable, structures in people's lives as gender, displacement and disability is difficult. These structures intersect with individuals' lives in time and place to produce unique social locations. Therefore, theorising from the ground up is crucial, and process is central to any such attempt. As discussed by Connell (2011) social embodiment is a productive relationship capable of generating new possibilities and vulnerabilities, not just for the person experiencing social disablement but for their social network; for example, a caregiving relationship may provide a sense of purpose in a difficult waiting situation while simultaneously imposing limitations on travel or income-generating opportunities. Accurate and detailed data is crucial for identification of PWD and for the analysis of structural issues such as the impact of unsafe work conditions, poor health conditions or medical access and how gender norms might limit access to formal rights.

In focusing on social locations, we are advocating a 'generative' rather than 'additive' approach. Sketching potentialities, such as those plotted in the matrix below, reveal dynamic intersections and allow key concerns to come to the fore. For example, asking how one's legal status impacts on displacement, gender and disability may assist in revealing multi-layered risks and opportunities arising within a person's social location. At the same time it becomes apparent that each square interacts dynamically with different points in the matrix and, like other such matrixes focused on plotting lived experience, the 'only strategy that really works is to keep the tension moving dialectically across all positions in the matrix' (Harvey, 2006: 147). In both practice and theory *welcoming* this dialectic tension is crucial (Table 27.1).

Table 27.1 Mapping intersections of displacement, gender and disability

	Time	*Legal Status*	*Attitudes towards*	*Physical environment*	*Social network*
Displacement	How long has the person and population been displaced? For how long will they likely be displaced? Impacts of this?	Access to work? Health services? Asylum seeker, refugee, IDP? Deportation risk? Access to resettlement?	Political and social attitudes of hosts? Support, ambivalence, hostility? NGO presence? Charity or rights-based approach?	Camp, urban, semi-urban? Detention? Safety, shelter, water, population density?	In-tact or ruptured family? Ethnic or faith networks? Networks with host community?
Gender	Rupturing or changes in gender norms over time? Conservatisation of gender norms?	RSD process—does the woman have claims, has she been asked?) Resettlement process? Freedom of movement in host society?	Real access to rights? Gender roles and responsibilities (work, caring)? Host community gender norms? Dominant discourses around men and women in NGOs?	Safety concerns (men and women)? Access to opportunities?	In-tact or ruptured family? Gender-based networks? Is agency recognised?
Disability	New impairment or old? Structure of displacement causing disablement?	Access to income, health, legal services? Accommodations needed in RSD? Resettlement policies?	Host community attitudes to impairment? Local disability groups? NGO awareness? Agency recognised?	Accessible infrastructure (public and residential buildings, roads, transport, ablutions)? What support would be most enabling?	In-tact or ruptured family/care networks? Host society disability groups?

An intersectional and social location approach helps one move between the particular and the structural, making visible that which is often hidden. Major changes are necessary to address policies designed around able-bodied men and tackle disabling structures such as chronic waiting. Even if structural changes are geopolitically unlikely in the short term, people working within key organisations can cultivate a critical attitude which enables multi-layered, complex experiences of vulnerability and capacity to become visible. Crock et al.'s (2017: 21) study noted that staff training is a relatively low-cost intervention that can deliver tangible benefits to refugees with disabilities through improved practice processes. The many structural intersections potentially impacting men and women refugees with disabilities requires that policy officers, practitioners and academics proactively work to ensure that policies and procedures reflect the complexity and dynamism of peoples' lived experiences.

References

AAP. (2005). Coroner clears officials over refugee's death. *The Age*. Retrieved from https://www.theage.com.au/national/coroner-clears-officials-over-refugees-death-20050202-gdzhcx.html.

APH. (2010). *Enabling Australia: Inquiry into the migration treatment of disability*. Canberra: Parliament of Australia. https://www.aph.gov.au/Parliamentary_Business/Committees/Joint/Completed_Inquiries/mig/disability/report.

Berghs, M., & Kabbara, N. (2016). Disabled people in conflicts and wars. In S. Grech & K. Soldatic (Eds.), *Disability in the global south* (pp. 269–283). Cham: Springer.

Bunch, C. (1990). Women's rights as human rights: Toward a re-vision of human rights. *Human Rights Quarterly, 12*(4), 486–498.

Cantekin, D. (2019). Syrian refugees living on the edge: Policy and practice implications for mental health and psychosocial wellbeing. *International Migration, 57*(2), 200–220.

Charlesworth, H., Chinkin, C., & Wright, S. J. (1991). Feminist approaches to international law. *The American Journal of International Law, 85*(4), 613–645.

Connell, R. (2011). Southern bodies and disability: Re-thinking concepts. *Third World Quarterly, 32*(8), 1369–1381.

Crenshaw, K. (1989). Demarginalizing the intersection of race and sex: A black feminist critique of antidiscrimination doctrine, feminist theory and antiracist politics. *The University of Chicago Legal Forum, 1*, 139–167.

Crock, M., Saul, B., McCallum, R., & Smith-Khan, L. (2015). *Refugees and asylum seekers with disabilities: A preliminary report from Malaysia and Indonesia*. https://www.researchgate.net/publication/281853950_Refugees_and_Asylum_Seekers_with_Disabilities_A_Preliminary_Report_from_Malaysia_and_Indonesia.

Crock, M., Smith-Khan, L., McCallum, R. C., & Saul, B. (2017). *The legal protection of refugees with disabilities: Forgotten and invisible?*. Cheltenham, UK: Edward Elgar.

Dowling, S. (2016). Disability and unsafe migration: Data and policy, understanding the evidence. *IOM Global Migration Data Analysis Centre. Data Briefing Series No. 7*. https://publications.iom.int/system/files/pdf/gmdac_data_briefing_series_issue_7.pdf.

Edwards, A. (2010). Transitioning gender: Feminist engagement with international refugee law and policy 1950–2010. *Refugee Survey Quarterly, 29*(2), 21–45.

Eide, A. H., & Loeb, M. (2016). Counting disabled people: Historical perspectives and the challenges of disability statistics. In S. Grech & K. Soldatic (Eds.), *Disability in the global south* (pp. 51–68). Cham: Springer.

Fiske, L., & Shackel, R. (2015). Gender, poverty and violence: transitional justice responses to converging processes of domination of women in eastern DRC, northern Uganda and Kenya. *Women's Studies International Forum, 51,* 110–117.

Goodley, D. (2013). Dis/entangling critical disability studies. *Disability & Society, 28*(5), 631–644.

Gözübenli, A. S. (2018). Landmines and refugees: Out of the frying pan, into the fire. *Annual of Institute for Sociological, Political and Juridical Research, 42,* 9–19.

Grech, S. (2016). Disability and development: Critical connections, gaps and contradictions. In S. Grech & K. Soldatic (Eds.), *Disability in the global south* (pp. 3–19). Cham: Springer.

Grewal, K. (2015). International criminal law as a site for enhancing women's rights? Challenges, possibilities, strategies. *Feminist Legal Studies, 23*(2), 149–165.

Harvey, D. (2006). *Spaces of global capitalism.* New York: Verso.

HelpAge International, & Handicap International. (2014). *Hidden victims of the Syrian crisis: Disabled, injured and older refugees.* Retrieved from HelpAge International and Handicap International, London and Lyon. www.helpage.org.

HRW. (2018, July 16). *Turkey stops registering Syrian asylum seekers.* Retrieved from https://www.hrw.org/news/2018/07/16/turkey-stops-registering-syrian-asylum-seekers.

Karlsen, E. (2016). *Refugee resettlement to Australia: What are the facts?* Canberra: Parliament of Australia. https://www.aph.gov.au/About_Parliament/Parliamentary_Departments/Parliamentary_Library/pubs/rp/rp1617/RefugeeResettlement#_Toc461022106.

Kastner, P., & Roy-Trudel, E. (2019). Addressing masculinities in peace negotiations: An opportunity for gender justice. In R. Shackel & L. Fiske (Eds.), *Rethinking transitional gender justice: Transformative approaches in post-conflict settings* (pp. 154–164). Abingdon: Palgrave.

Kneebone, S. (2005). Women within the refugee construct: Exclusionary inclusion in policy and practice—The Australian experience. *International Journal of Refugee Law, 17*(1), 7–42.

Marshall, J., & Barrett, H. (2018). Human rights of refugee-survivors of sexual and gender-based violence with communication disability. *International Journal of Speech-Language Pathology, 20*(1), 44–49.

McNevin, A., & Missbach, A. (2018). Luxury limbo: Temporal techniques of border control and the humanitarianisation of waiting. *International Journal of Migration and Border Studies, 4*(1–2), 12–34.

Mirza, M. (2011). Disability and humanitarianism in refugee camps: The case for a travelling supranational disability praxis. *Third World Quarterly, 32*(8), 1527–1536.

Pearce, E., & Lee, B. (2018). From vulnerability to resilience: Improving humanitarian response. *Forced Migration Review, 57,* 31–33.

Pisani, M., & Grech, S. (2015). Disability and forced migration: Critical Intersectionalities. *Disability and the Global South, 2*(1), 421–441.

Price, J., & Goyal, N. (2016). The fluid connections and uncertain spaces of women with disabilities: Making links across and beyond the global south. In S. Grech & K. Soldatic (Eds.), *Disability in the global south* (pp. 303–321). Cham: Springer.

Procter, N., Kenny, M. A., Eaton, H., & Grech, C. (2018). Lethal hopelessness: Understanding and responding to asylum seeker distress and mental deterioration. *International Journal of Mental Health Nursing, 27*(1), 448–454.

Silove, D., Ventevogel, P., & Rees, S. (2017). The contemporary refugee crisis: An overview of mental health challenges. *World Psychiatry, 16*(2), 130–139.

Smart, J. F., & Smart, D. (2006). Models of disability: Implications for the counselling profession. *Journal of Counseling & Development, 84*(1), 29–40.

Smith-Khan, L., & Crock, M. (2015). *Refugees with disabilities in Pakistan: An introductory report*. Retrieved from https://www.researchgate.net/publication/281854024_Refugees_with_Disabilities_in_Pakistan_An_Introductory_Report, https://doi.org/10.13140/rg.2.1.2009.3922.

UNGA. (2006). *Convention on the rights of persons with a disability (CRPD)*. Retrieved from United Nations General Assembly, New York. https://www.un.org/disabilities/documents/convention/convoptprot-e.pdf.

UNHCR. (2011). *Working with persons with disabilities in forced displacement*. Retrieved from UNHCR, Geneva. https://www.refworld.org/docid/4e6072b22.html.

UNHCR. (2018). *UNHCR resettlement handbook—Country chapters: Australia*. Retrieved from UNHCR, Geneva. https://www.unhcr.org/en-au/protection/resettlement/4a2ccf4c6/unhcr-resettlement-handbook-country-chapters.html.

UNHCR. (2019). *Global trends 2018*. Retrieved from UNHCR, Geneva. https://www.unhcr.org/search?comid=56b079c44&&cid=49aea93aba&tags=globaltrends.

UNHCR Executive Committee. (2010). Conclusion on refugees with disabilities and other persons with disabilities protected and assisted by UNHCR. No. 110 (LXI)—2010. In *Executive Committee 61st session*. Contained in United Nations General Assembly document A/AC.96/1095.

Lucy Fiske is a Senior Lecturer in Social and Political Sciences at the University of Technology Sydney (UTS). She has worked in both academia and practice with refugees and asylum seekers for over 20 years. She has published widely on human rights, refugees, transitional justice and gender including *Human Rights, Refugee Protest and Immigration Detention* (Palgrave, 2016) and *Rethinking Transitional Gender Justice: Transformative Approaches in Postconflict Settings* (co-edited with Rita Shackel, Palgrave, 2019). Her current research examines social relations and informal protection strategies among refugees in South East Asia.

Chrisanthi Giotis is a postdoctoral fellow at the Centre for Media Transition, University of Technology Sydney. She teaches and publishes in both journalism and social and political sciences. Her research focuses on bringing marginalised voices into public discourse. This includes examination of the structured interactions between refugees and interveners, the types of narratives these produce and the gendered impacts of these narratives.

CHAPTER 28

"I'm a Refugee in My Own Country"! Gendering Internal Displacement & Trauma

Malathi de Alwis

Abstract This chapter explores a variety of debates and scholarly discussions that have arisen due to the large-scale displacement/forced migration of people *within* national borders, as a result of civil war and conflict. Drawing on material from Sri Lanka, a country which experienced three decades of civil war, the chapter analyses various linguistic artefacts of displacement, the concepts of "refugee camp" and "home" that have not only been gendered but also inflected with violence, and several gendered and ethnicised effects of displacement, predominant among them being "collective trauma" and "suffering" It concludes by considering the "doubleness" of displacement; while it brings about despair, dispossession and diminution, it simultaneously offers new possibilities and thus enables innovation and change.

> Mammoty in hand we flee
> With a few other essentials
> To a life of displacement
>
> Although the war chases and chases
> And displaces
> We retain our pride and hope
> We shall farm new lands
> And survive

M. de Alwis (✉)
University of Colombo, Colombo, Sri Lanka

C. Mora and N. Piper (eds.), *The Palgrave Handbook of Gender and Migration*, https://doi.org/10.1007/978-3-030-63347-9_28

The latest news shattered our resolve
Multi-barrel rockets have replaced single ones

Now, death to many is assured
Who will be there to bury them?
Mammoty in hand we flee
They are handy for burials

Mammoty by Jeyanthy Thalayasingham. Tamil translation by Udhayani Navaratnam. Adapted by Malathi de Alwis.

Introduction

In August 2019, the United Nations High Commission for Refugees announced that we are now witnessing "the highest levels of displacement on record" with 37,000 people *per day* being forced to flee their homes due to conflict or persecution (UNHCR, 2019, emphasis author's). Of the currently 70.8 million forcibly displaced persons worldwide, 41.3 million persons are internally displaced along with 25.9 million refugees and 3.5 million asylum-seekers (UNHCR, 2019). The latter set of statistics highlight a fact that has been consistent throughout the twentieth century, and now also the twenty-first, that there are more IDPs (Internally Displaced Persons) than refugees. However, it is refugees who are more visible globally as they cross national borders and thus become an "international problem" worthy of media attention. The tragic image of three-year-old Syrian refugee Alan Kurdi face down on the sand on a Turkish beach, after he drowned on 2nd September 2015 during his family's failed attempt to smuggle themselves to Greece and then onto Canada, is not only irrevocably etched in our minds but caused a dramatic upturn in international concern for Syrian refugees and a surge in donations to charities that support refugees (Henley, Grant, Elgot, McVeigh, & O'Carroll, 2015).

IDPs, on the other hand, are less visible as their violation and displacement takes place within national borders and often has fewer witnesses as local and international media access as well as an international humanitarian presence is at the discretion of the nation-state which is frequently the perpetrator of the violence resulting in displacement. Unlike refugees, IDPs do not have a separate legal status as they remain under the jurisdiction of their nation-state and thus may not claim any additional rights (Vincent, 2001). Since most IDPs are also dependent on an oppressive and discriminatory state for the provision of refuge, return or resettlement, their positioning is doubly precarious and makes them doubly in need of legal safeguards. As a result, scholars and practitioners have debated the pros and cons of differentiating IDPs from refugees as well as IDPs and other civilians affected by conflict (for a discussion of this debate, see Brun, 2005: 3–4). The UN officially recognized "internal displacement" as a distinct category only in the 1990s. The Guiding Principles on Internal Displacement defines IDPs as "persons or groups of persons who have been forced or obliged to flee or leave their homes or places of habitual residence, in particular as a result of or in order to avoid the effects of armed conflict, situations of generalized violence, violations of human rights or natural or human-made disasters, and who have not crossed an internationally recognized State border" (Deng, 1998: 5). It is important to keep in mind that displacement not only affects those who are forced to flee but also those who are left behind/choose to stay back and those who receive/host the displaced. Those who are "emplaced" also undergo changes and experience feelings of displacement (Brun, 2008; de Alwis, 1993, 2004).

Globally, 80% of refugees and IDPs are children and women (Qayum, Mohmand, & Arooj, 2012). International recognition of the gendered dimensions of displacement has also been gradually increasing, since the 1990s, largely due to the indefatigable efforts of feminist scholars, activists and practitioners (for a discussion of some of these efforts, see Baines, 2004; Giles, 2013). Early discourses focused on "women" and/or *women and children* (Enloe, 1990), with the latter category having serious consequences for policy making and service delivery as women and children have distinct identities and needs and should not be grouped together. This was also coupled with a concern for "protecting" women rather than addressing structural inequalities and patriarchal power relationships that result in their abuse and violation (Yazid & Natania, 2017). A shift towards "gender" and "gender mainstreaming" sought to rectify some of these imbalances and lacunae but has not been particularly successful (Giles, 2013).

The intensification of armed conflict, especially in the south, has also increased the brutalization and violation of women and girls who have been repeatedly raped, enslaved and impregnated. Odwe, Undie, and Obare (2018: 1), refer to this as a "silent epidemic" with approximately 1 in 5 displaced women experiencing sexual and gender-based violence (SGBV) which significantly impacts their social, mental and economic well-being. SGBV in war settings is frequently an assertion of power and a strategy to degrade and humiliate entire communities by those who seek to wrest control of territory and peoples. As Pittaway and Bartolomei (2018) have pointed out, the consequences of SGBV are itself often gendered. Women are not only vulnerable during conflicts and while attempting to escape them but also when they seek refuge in camps and even return or resettle post-conflict. Feminists have developed new terminology to capture these realities such as women having to resort to "survival sex" to pay for passes and travel documents, to bribe the police and aid workers or to feed themselves and their children; women and girls having to enter into "protection marriages" or "forced marriages" to avoid sexual assaults or conscription within camps and resettled areas, and lesbian and transgendered women being subjected to "corrective rape" by camp guards and fellow IDPs (Simon-Butler & McSherry, 2019). Sexual violence perpetrated on men and boys during flight, at make-shift detention centres and informal work contexts, an under-reported and under-studied phenomenon, has also gained more visibility this past decade (Chynoweth, 2017; Jensen, 2019) (see also Jaji, in this volume).

This chapter, which draws on over fifteen years of ethnographic research in different regions of Sri Lanka, will explore several gendered effects of "internal displacement" due to war and ethnic conflict by focusing on how it unfolded in Sri Lanka over the longue durée. Studies of "people out of place" across time and space, I argue, enables us to interrogate their homogenization, dehumanization and dehistoricisation. In Sri Lanka, different ethnic communities were displaced at different points in time in different regions of the country for different reasons. The disparate and unequal socio-political positioning of each ethnic community along with gender/sexuality, religion, caste, class, age and occupation influenced not only how and to what extent individuals, families and communities were affected by displacement but also how they responded to displacement and were treated by host communities, humanitarian agencies, the media and the state. While it is beyond the scope of this chapter to address all these complexities, it will attempt to highlight a few intersections.

As Robert Zetter (1991) has aptly noted, labels have consequences in real life. Thus, this chapter also considers a variety of labels, categories and concepts associated with IDPs in Sri Lanka, and how our understanding of this terminology and ideology has changed over the years as internal displacement has increased and lengthened, and along with it ethnographic, feminist and other forms of social scientific research. The consequences of labels and acronyms that delineate particular spaces, how some of these spaces have been specifically ethnicised and gendered, and how concepts such as "trauma" and "suffering" have been problematized, in recent years, will also be discussed. The chapter also hopes to reflect on how new research questions and concerns have further complicated our understanding of internal displacement and its consequences for IDP women in particular who have mobilized, disputed and re-negotiated their precarious spatial and social positioning, across the decades.

Internal Displacement in Sri Lanka

Much of the displacement that has taken place in Sri Lanka has been due to intermittent ethnic riots, from the 1950s onwards, an armed insurrection in the southern regions of the country in 1971 and again from the late 1980s to the early 1990s, and a protracted civil war that cast an ominous shadow across the island for almost three decades, from 1982 to 2009.[1] The latter resulted in the displacement of all three major ethnic groups in Sri Lanka—the Sinhalese constitute approximately 78% of the population, the Tamils 12% and the Muslims (Moors and Malays) 10%. However, the Tamil population

[1]A significant population has also been displaced due to development projects and natural disasters. Of the latter, those displaced by the 2004 tsunami amounted to several hundred thousand.

has been the most affected by riots and the civil war with one in two Tamils having been displaced at some point in their life (Sriskandarajah, 2004). Most Tamils have experienced multiple displacements and have not been able to return home for over twenty or twenty-five years (the latter has also been the case for Northern Muslims and some Sinhala IDPs). While the majority of this displacement took place within Sri Lanka, over one and a half million Tamil refugees are also now scattered across the globe with several thousand still languishing in refugee camps in southern India.

Internal displacement was a continuous process throughout the war years but there have also been several occasions when mass displacement took place on a very large scale. The first such event took place in October 1990 with the ethnic cleansing of all Muslims from the Northern Province, by the LTTE (Tamil militants). Approximately 100,000 Muslims were "transformed into homeless, poverty-stricken refugees within a matter of hours" during what came to be known as "The Eviction" (de Alwis, 2004: 217). Around 65,000 of the displaced were housed in around 200 refugee camps in the Puttalam District while thousands of others sought refuge with family and friends in other parts of the country. The majority of these IDPs were only able to return to their homes twenty years later, once the war was ended. Many chose to resettle where they had been displaced as they had "put down roots" due to the protracted nature of their displacement and/or were fearful of living among Tamils again.

The second major displacement of civilians occurred when the LTTE forced the entire population of the Jaffna Peninsula to move to the Vanni, a region under their control, as the Sri Lankan army approached Jaffna town, in October 1995. Approximately 450,000 civilians left Jaffna on foot, along the Chemmani-Kandy road, during the height of the northern monsoon. The road soon became littered with the bodies of those who were too weary to carry on. "It took the LTTE two weeks to clear a recalcitrant, unwilling population out of Jaffna town" notes anthropologist Sharika Thiranagama (2011: 66), one of the few scholars to have collected and analysed narratives of what is now described as "The Exodus". Most returned to Jaffna after a year, but there were some families that chose to resettle in the Vanni or move to Colombo or further afield, as the war continued to rage unabated.

The third major displacement of civilians occurred at the end of the war, in May 2009. Approximately 350,000 Tamils either fled on foot or were evacuated by the military from the "No Fire Zone". This displacement resulted in the creation of "the world's largest camp for internally displaced people" at the 700-hectacre Menik Farm (Vavuniya District) that held over 225,000 people (UNHCR, 2012). Heavy security restrictions were imposed on this camp complex on the grounds that the military needed to "weed out" LTTE combatants from among the civilians through a long drawn out screening process that resulted in a variety of human-rights violations (University Teachers for Human Rights [Jaffna], 2009). Scholars were refused access with the exception of a team of psychiatrists (Somasunderam, 2010).[2] The Menik Farm complex was finally shut down in 2012 (UNHCR, 2012).

During the 1980s and 1990s, the term "refugee" and its Tamil equivalent *ahathi* and Sinhala equivalent *anatha*, the latter two terms connoting orphaning, destitution and homelessness, continued to be used to describe the internally displaced in Sri Lanka, with many displaced persons also using the English term "refugee" to describe themselves (de Alwis, 2004: 228). Anthropologist Joke Schrijvers (1997: 64) coined the term "internal refugee" to describe internally displaced peoples as she wished to retain the association of agentive individuals in search of refuge and security while arguing that the term "displaced" falsely suggested a "natural connection between people and places" that she wished to avoid. However, by the end of the war, the English acronym "IDP" and its Tamil equivalent *idam peyrntha*, had become firmly entrenched (Hewamanne, 2015; Thanges, 2008; Thiranagama, 2011). Many of those who have returned or resettled continue to bear the stigma of "IDP-hood", as it is stamped on their National Identity Card (NIC). As Vasuki (pseudonym), a resettled Tamil IDP woman lamented: "I no longer think of myself as an IDP. I have lots of different identities, as a single mother, a teacher, a Hindu woman, but because the word 'IDP' is stamped on my NIC, I continue to be seen as one and face discrimination at military checkpoints and government offices".

Clearly, internal displacement has different histories and socio-political connotations for diverse groups of Tamil, Sinhala and Muslim displaced populations. The opening title of this chapter, "I am a refugee in my own country"! a plaint cum plea often articulated by Muslim women during the early years of their displacement from the north, captures the bewilderment and anguish felt by these women about a war they felt they had no part in but were nonetheless suffering its brutal consequences. Sorensen (2001) notes with approval how Northern Muslims pooled their resources to help their less fortunate compatriots, unlike displaced Tamils. However, my own field research showed that this kind of supportive behaviour was fractured when Muslims from different parts of the Northern Province were inter-mixed in refugee camps (de Alwis, 1993). I also observed (de Alwis, 1993, 2004) that the out-of-placeness and liminality of internal refugees has frequently resulted in them being perceived as "polluting" and disruptive and thus "other" by their host communities despite commonalities of ethnicity and religion. From among Tamil IDPs, those who were most oppressed and tended to languish the longest in IDP camps, be it in Vavuniya or Jaffna, were those from the lower castes, many of whom had been originally displaced from the hill country regions during racial riots in the 1970s and 1980s (Schrijvers, 1997, 1999; Thanges, 2008).

[2]Some journalists and aid workers who did gain access posted "eye witness" accounts on social media which engendered heated debates due to their divergent reportage (see for example the comments and counter-postings at: https://groundviews.org/2009/07/02/an-eye-witness-account-of-idp-camp-conditions-in-sri-lanka/).

They were not allowed to draw water, buy land or enter high caste temples in their places of origin as well as displacement.

The civil war also disrupted our understanding of the physical topography of the island by overlaying it with "a topography o violence" (de Alwis, 2004: 213). "Front lines" and "welfare camps" constantly expanded and contracted in correspondence with the vagaries of warfare between the Sri Lankan state and the Tamil militants. Such terms not only became part of everyday discourse but signalled the displacement of people, goods and livestock (de Alwis, 2004: 213). They also recall Henri Lefebvre's (1991) articulation of "social space" that highlights the co-constitutive nature of people, commodities and spaces and the central role played by spatial practices and representations. (de Alwis, 2004: 213). A "front line" or "border village" suggests imminent displacement whenever battle lines shift or generates "night IDPs"—those who spend the day in their homes and fields and the night in the jungle or a neighbouring village to avoid shelling by the military or attacks by Tamil militants. Similarly, "cleared" (under Sri Lankan government rule) and "uncleared" (under LTTE rule) areas could change overnight depending on which side was victorious in a battle and was able to extend their control (Hyndman & de Alwis, 2005). The above poem, for example, is voiced by a young Tamil woman fleeing her home in an "uncleared area" as the Sri Lankan Army advances. Her resolve to start life anew by farming new lands with her mammoty, wherever she is displaced, is rudely shaken when she discovers that multi-barrel guns, which can fire about 40 rounds at high speed, are being used. Her mammoty now becomes handy for burying those whose bodies litter the land.

A "High Security Zone" (HSZ) means that it has become the sole preserve of security forces entailing the displacement of all civilians living within it. A "No-Man's Land", on the other hand, means the complete absence of any people—be they civilians, military or militants. Such a de-militarized zone was created in the Omanthai region, on the A-9 Road, between the last checkpoint of the Sri Lankan government and the first checkpoint of the LTTE, during the war. A different kind of de-militarized zone—the "No-Fire Zone"—to ensure the safety of civilians was created during the last stages of the war, in 2008–2009, but neither party adhered to it with the LTTE moving into this zone along with Tamil civilians, and the LTTE and Sri Lankan government exchanging fire resulting in a tremendous loss of civilian life. The number of those who died in this zone continues to be disputed and thousands remain missing (see UTHR(J), 2009). The "refugee camp" or its euphemistically named equivalent in Sri Lanka, the "welfare camp", delineates a space where those who are displaced are congregated. While it was not a new phenomenon in Sri Lanka, having been mobilized previously in the aftermath of riots and natural disasters, it nonetheless became a permanent fixture upon the Sri Lankan landscape with the onset of the civil war and thus became an exemplification of internal displacement.

"Welfare Camps" and Social Control

Anthropologist Liisa Malkki (1996) has argued that the refugee camp which has its antecedents in post-World War II Europe functions as a "technology of care and control" through the management of peoples and spaces (1996: 444). Similarly, geographer Jennifer Hyndman (2000) has noted parallels between the strategies adopted by colonial administrators to police and control their subject populations and the ways in which the spatial layout of refugee camps as well as humanitarian efforts facilitate the surveillance and regulation of refugees. Indeed, the production of maps, statistics and assessments as well as the control of mobility through government-issued documents such as special identity cards and passes were quintessential technologies of management and surveillance that were mobilized in Sri Lankan "welfare camps". Tamil women, long-term inmates of camps in Vavuniya, in the 1980s and 1990s, complained that the most oppressive part of camp life was the "pass system" that "forced them to get permission to leave the camp, to identify themselves at each and every checkpoint, and to report back in the camp on the same day after a varying number of hours" (Schrijvers, 1999: 190). In the Menik Farm camp complex, even these forms of restricted mobility were denied to Tamils. They were incarcerated behind barbed wire fences under military surveillance with families often being separated from each other. Additionally, the possession and use of mobile phones was banned, visiting other camps within the complex was forbidden and access to visiting relatives was either restricted or denied (Somasunderam, 2010; UTHR(J), 2009). Muslim and Sinhala IDPs, on the other hand, faced no such restrictions with the latter only having armed guards patrolling their camp if the government perceived them to be under threat from the LTTE.

Though Muslim women IDPs movements were not restricted by a pass system, they faced other forms of policing and surveillance by male family members as well as the all-male Camp Committees and Mosque Committees that administered the camps. Most of the women supplemented meagre camp rations by working as day labourers in the onion and chilli fields owned by their host community, while their men idled inside the camps. This inversion in the pattern of income generation was a result of the men refusing to work for such low wages, but it made women's positioning "within pre-existing patriarchal power structures" where the majority of them worked within the home while the men worked in the fields or fished, "a fraught one" (de Alwis, 2004: 223). As a result, women were subjected to new forms of surveillance and discipline inside the camps: "My husband insists on following me even when I go to the toilet", complained one young woman while many others felt constrained to adopt the *hijab* when previously they had only loosely covered their heads with their sarees (de Alwis, 2004: 223). As I've noted previously, the Muslim IDP woman had become "a cypher for all that was lost as well as what must be preserved for the future; her community's purity of displacement [was] imbricated in her moral purity" (de Alwis 2004: 227).

While the policing of Muslim women's bodies (unheard of in their own homes) was framed by "a discourse of security and safety", it also provided "an opportunity for their male kin to reassert their authority and masculinity" within the unfamiliar space of the camp (de Alwis, 2004: 223). Shanmugaratnam and Zackariya (2002: 8), report how the Mosque Committee in one of the camps had begun to complain that "women were roaming about, didn't obey the men, were going about their work and that men felt insulted and inferior as they were not in a position to control their women". When the women convened a meeting among themselves to discuss these accusations, the Mosque Committee expressed its displeasure at this show of initiative and independence. However, the women countered by noting that it is time "for us to assert our rights and to tell our men that we did not want to be treated like slaves but as human beings" (Shanmugaratnam & Zackariya, 2002: 8). A few days later, the Mosque Committee announced that "women could hold meetings and that women also had some rights"! (Shanmugaratnam & Zackariya, 2002: 9).

Women IDPs from all three ethnic groups also frequently complained that their husbands had become listless and depressed due to loss of social status and lack of economic independence and would often take out their frustration on their wives and children. As Elizabeth Colson (2003: 11) has aptly noted, "[i]n the humiliation of the devalued lies the roots of much of the violence, especially against women and children." Feminists have also expanded sociologist Michael Johnson's (1995) term "patriarchal terrorism", used to analyse domestic violence, to also refer to broader social structures (that are historically asserted and culturally reproduced) that promote the idea that men have ownership of women and can control and discipline them as they wish (Jasinski, 2001; Jensen, 2019; Mahoney, Williams, & West, 2001; Ward, 2002). Indeed, as one Muslim man who had lost all his wealth which had been tied up in immovable property informed me: "The only thing that belongs to me now is my wife" (de Alwis, 2004: 223).

However, as the anecdote discussed above shows, "welfare camps" did also provide unforeseen opportunities. The Muslim women's close proximity to each other within the confines of the camp provided them with an opportunity to work collectively to struggle against at least some patriarchal strictures. Faathima (pseudonym), now a dedicated women's rights activist who returned to her home after twenty years of displacement, reminisced recently of how she not only had been pushed to work outside her home, for the first time in her life, but had also learned, along with other women in her camp, how to negotiate for better wages and for more social recognition within her community. "I hate to admit it", she said with a rueful laugh, "but it's only after I was displaced that I found my voice and learned I was strong enough to face any adversity"! Similarly, two Sinhala women who had honed their leadership skills by continuously lobbying local politicians and administrators for the provision of better living conditions, rations and voting rights, during their displacement, decided to contest in the first post-war local government elections, and won!

Gendering "Suffering" and "Trauma"

During these past two decades, the "privileged object of attention" of anthropologists, notes Joel Robbins (2013: 450–453), has gradually shifted from the "savage subject" to the "suffering subject," from "analytic distance and critical comparison" to one of "empathic connection and moral witnessing based on human unity." Such a shift, Robbins argues, is the result of a crisis in Anthropology that was precipitated by decolonizing movements and post-colonial critiques. The "suffering slot", in the 1990s, provided anthropology with a new, politically and ethically acceptable object of study. Concomitantly, "trauma" emerged as the bridge between cultures because the "human being suffering from trauma…became the very embodiment of our common humanity" (Fassin & Rechtman, 2009: 23). Trauma "represents a suffering without borders, a suffering that knows no cultural barriers" (Fassin & Rechtman, 2009: 239). Because of the universal qualities of trauma, anthropologists as well as reporters, politicians, aid workers, TV viewers etc., are confident in our ability to "know it when we see it" and to feel empathy with those who suffer in "a sort of communion in trauma" (Fassin & Rechtman, 2009: 18; Robbins, 2013: 453). Clearly, it was such a form of "communion" that resulted in the outpouring of concern as well as donations, after images of Alan Kurdi was flashed across the globe, as discussed above.

It is now common to describe entire populations in war-torn regions as "traumatized" (Agger, Vuk, & Mimica, 1995; de Jong, Mulhern, Ford, van der Kam, & Kleber, 2000; de Vries, 1998). Daya Somasunderam (2010: 24), one of the few psychiatrists who was allowed access to the Menik Farm complex observed that "a striking theme" that emerged from many of the displacement narratives he collected in the camps and hospitals was "the collective nature of the trauma" that had been undergone by this populace. Somasunderam's work in previous Tamil IDP camps also noted "psychological disturbances particularly depressive symptoms were much more common in displaced families than in those living in their own homes" (2007: 9). Displaced children in particular suffered from "separation anxiety… cognitive impairment, conduct disorders and sleep disturbances" while family dynamics were severely disrupted with an increase in disputes between husband and wife "due to economic stress, lack of privacy and interference of others in over-crowded camps" (2007: 9).

Somasunderam's observations have been corroborated in displacement camps across the globe with scholars noting how entire communities have suffered acute degrees of disintegration as community structures, social networks and even kin groups have been dispersed across the country and even abroad. Recent scholarship has also begun to describe protracted displacement as a form of "slow violence" that particularly affects domestic/familial relations especially when children are separated from parents (de Leeuw, 2016; Hyndman, 2019; Nixon, 2011). Additionally, practices of refugee screening and resettlement can also be "imbued with traumatic stressors" (Loyd, Ehrkamp, & Secor, 2018: 1). The destruction of affective ties between individuals/communities and their material environments is also severely debilitating but is an aspect that is less discussed in the literature (however, see de Alwis, 2004; Sanaathanan, 2011). Most war-displaced Tamils in Sri Lanka had also witnessed the death/disappearance/dismemberment of loved ones, been betrayed/violated/tortured, had to leave severely injured family members behind in order to safeguard the rest of the family etc. Thus, "survivor guilt", mistrust and suspicion, anger and bitterness, listlessness as well as high levels of alcoholism and sexual laxity among displaced individuals were common (Somasunderam, 2010: 23–24).

Somasunderam's recourse to describing these pervasive symptoms as manifestations of "collective trauma" due to "loss of communality" and the destruction of the "social fabric" is understandable and makes an important political point in the face of the Sri Lankan state's refusal to acknowledge the physical and psychic wounding of the Tamil populace. However, as I have noted elsewhere (de Alwis, 2017), I agree with Derek Summerfield (1998: 1581), that to label an entire population as "traumatized" is to lose sight of the fact that "suffering per se is not pathology" and that for most survivors, "PTSD features simply represent the epiphenomena of normal distress". This is not to discount the fact that wars and conflicts can and do trigger "clear cut psychiatric disorders" especially among those "with underlying vulnerability" but this is usually only a very small percentage of the population (Summerfield 1997: 1568; 1998: 1581). The "pathologization of distress" *en masse* (Pupavac, 2001: 362) thus inappropriately imputes "a sick role" to war-affected peoples and "transforms the social into the biopsychomedical" (Summerfield, 1999: 771). This only leads to the distortion of survivors' perceptions of themselves and hinders their ability to rebuild meaningful lives.

Undoubtedly, IDP camps are particularly detrimental to the rebuilding of meaningful lives as they epitomize the out-of-placeness of people. The insidious relationship between trauma and dependency is further reinforced as space and mobility is constrained and livelihood options are extremely limited or non-existent. This often resulted in individuals being pushed "to behave as if they were dependent and helpless, which assists the perpetuation of an institution largely composed of workers whose role is to respond to people who have problems and are in need" (Hitchcox, 1990: 173). In the Muslim IDP camps where many women had become the breadwinners of their families, aid workers nonetheless frequently spoke paternalistically about their "ignorance and 'backwardness', their docility and helplessness, and their 'annoying habit' of constantly asking for things" (de Alwis, 2004: 224). My analysis of why these women played into "a universalized image of the victimized refugee woman" led me to argue that the women's constant demand for kitchen utensils such as buckets and pails, pots and pans reflected their desire to not only recreate a semblance of a familiar domestic world that had been disrupted but also "to reterritorialize their longing and identity in the space of displacement" (de Alwis, 2004: 226).

I would like to end this chapter with the story of an ordinary woman whose extraordinary response to the many violations and losses she had suffered not only enabled her to extricate herself from her "traumatic state" but offered succour to hundreds of others. When I met Saraswathy (pseudonym), she had been displaced 58 times during her sixty-two years of life. During the last stages of the war, she was displaced 16 times in 12 months. Saraswathy had endured rape, the death of her older son who had joined the LTTE voluntarily, and the forcible conscription of her younger son by the LTTE. The latter is currently listed as missing. Saraswathy's greatest anguish was that she had had to leave her husband, severely wounded but still alive, and flee the No Fire Zone with her daughter and grandson who were then tragically killed "before my very eyes", by a military missile. She was diagnosed as being "severely traumatized" during a medical screening at the Menik Farm camp complex. Bereft of family, she barely ate or slept and was unable to recognize former neighbours and relatives for almost six months. Though Saraswathy kept to herself and "sat in a daze" during most of the day and night, she gradually acceded to requests to perform cowrie shell divinations—throwing cowrie shells and making predictions based on how they fall—for those who were desperately seeking news of their missing or forcibly disappeared loved ones. Though destitute, she refused to be remunerated for her services informing her "clients" that they could reward her later if her predictions were proved correct. Saraswathy became quite renowned as a diviner during her time at Menik Farm and now has a large following in her resettled village.

Concluding Reflections

"It is striking how often the abundant literature claiming refugees as its object of study", notes Lisa Malkki (1996: 8), "located 'the problem' not first in the political oppression or violence that produces massive territorial displacements of people but within the bodies and minds of people classified as refugees". This chapter has sought to rectify this imbalance by not only highlighting the continuum between violence and displacement over the longue durée but also how labels, categories and concepts can have consequences. It has also sought to problematize the usual assumption that IDPs are only knowable through their need or trauma or that they are merely "victims" or "beneficiaries" by attempting to offer you a glimpse of how women IDPs, in this case, are actors in their own right and have similar desires and dreams as the rest of us (see also Malkki, 1997). The women IDPs determination to reconstruct their lives and livelihoods against great odds pushes us to be open to the "doubleness" of displacement: while displacement brings about despair, dispossession and diminution it simultaneously offers new possibilities and thus enables innovation and change (Brun, 2008: 1–2).

References

Agger, I., Vuk, S., & Mimica, J. (1995). *Theory and practice of psychosocial projects under war conditions in Bosnia-Herzegovina and Croatia*. Brussels: ECHO/ECTF.

Baines, E. (2004). *Vulnerable bodies: Gender, the UN, and the global refugee crisis*. Aldershot: Ashgate.

Brun, C. (2005, January). *FMO research guide: Internal displacement*. Oxford: Forced Migration Online.

Brun, C. (2008). *Finding a place: Local integration and protracted displacement in Sri Lanka*. Colombo: Social Scientists' Association.

Chynoweth, S. (2017). *"We keep it in our heart": Sexual violence against men and boys in the Syria crisis*. Geneva: UNHCR.

Colson, E. (2003). Forced migration and the anthropological response. *Journal of Refugee Studies, 16*(1), 1–18.

de Alwis, M. (1993). *Socio-economic and psycho-cultural survey of the muslim displaced populations at Alankudah*. Unpublished Research Report submitted to Italian Cooperation, Health Assistance Programme to Displaced Populations. Colombo: December 1993.

de Alwis, M. (2004). The 'purity' of displacement and the re-territorialization of longing: Muslim women refugees in North-Western Sri Lanka. In W. Giles & J. Hyndman (Eds.), *Sites of violence: Feminist politics in conflict zones*. Berkeley: University of California Press.

de Alwis, M. (2017). Trauma, memory, forgetting. In D. Bass & A. Amarasingham (Eds.), *Sri Lanka: The struggle for peace in the aftermath of war*. London: Christopher Hurst.

De Jong, K., Mulhern, M., Ford, N., van der Kam, S., & Kleber, R. (2000). The trauma of war in Sierra Leone. *Lancet, 355*(9220), 2067–2070.

de Leeuw, S. (2016). Tender grounds: Intimate visceral violence and British Columbia's colonial geographies. *Political Geography, 52*, 14–23.

Deng, F. (1998). *Guiding principles on internal displacement*. E/CN.4/1998/53/Add.2. New York: United Nations. https://www.ohchr.org/EN/Issues/IDPersons/Pages/Standards.aspx. Accessed 1 Sept 2019.

de Vries, F. (1998). To make a drama out of trauma is fully justified. *Lancet, 351*(9115), 1579–1580.

Enloe, C. (1990, September 25). Women and children: Making feminist sense of the Persian Gulf crisis. *The Village Voice*.

Fassin, D., & Rechtman, R. (2009). *The empire of trauma: An inquiry into the condition of victimhood*. Princeton: University Press.

Giles, W. (2013). Women forced to flee: Refugees and internally displaced persons. In C. Cohn (Ed.), *Women and wars*. Cambridge: Polity Press.

Henley, J, Grant, H., Elgot, J., McVeigh, K., & O'Carroll, L. (2015, September 3). Britons rally to help people fleeing war and terror in Middle East. *The Guardian*. https://www.theguardian.com/uk-news/2015/sep/03/britons-rally-to-help-people-fleeing-war-and-terror-in-middle-east. Accessed 1 Sept 2019.

Hewamanne, S. (2015). Complicated belonging: Gendered empowerment and anxieties about 'returning' among internal displaced Muslim women in Puttalam, Sri Lanka. In H. Ahmed-Ghosh (Ed.), *Asian Muslim women: Globalization and local realities*. Albany: SUNY Press.

Hitchcox, L. (1990). *Vietnamese refugees in South East Asian camps*. London: Macmillan.

Hyndman, J. (2000). *Managing displacement: Refugees and the politics of humanitarianism*. Minneapolis: University of Minnesota Press.

Hyndman, J. (2019). Unsettling feminist geopolitics: Forging feminist political geographies of violence and displacement. *Gender, Place and Culture, 26*(1), 3–29.

Hyndman, J., & de Alwis, M. (2005). Performing the pass: Conflict, mobility and displacement in sri Lanka. In B. P. Lorente, N. Piper, S. Sui-Hua, & B. Yeoh (Eds.), *Asian migrations: Sojourning, displacement, homecoming and other travels*. Singapore: Asian Research Institute.

Jasinski, J. (2001). Theoretical explanations for violence against women. In C. Renzetti, J. Edleson, & R. Bergen (Eds.), *Sourcebook on violence against women*. London: Sage.

Jensen, M. (2019). Gender-based violence in refugee camps: Understanding and addressing the role of gender in the experiences of refugees. *Inquiries, 11*(2), 1–13.

Johnson, M. (1995). Patriarchal terrorism and common couple violence: Two forms of violence against women. *Journal of Marriage and Family, 57*(2), 283–294.

Lefebvre, H. (1991). *The production of space*. Oxford: Blackwell.

Loyd, J., Ehrkamp, P., & Secor, A. (2018). A geopolitics of trauma: Refugee administration and protracted uncertainty in Turkey. *Transactions of the IBG, 43*(3), 377–389.

Mahoney, P., Williams, L., & West, C. (2001). Violence against women by intimate relationship partners. In C. Renzetti, J. Edleson, & R. Bergen (Eds.), *Sourcebook on violence against women*. London: Sage.

Malkki, L. (1996). National geographic: The rooting of peoples and the territorialization of national identity among scholars and refugees. In G. Eley & R. Suny (Eds.), *Becoming national*. Oxford: Oxford University Press.

Malkki, L. (1997). Speechless emissaries: Refugees, humanitarianism, and dehistoricization. In K. Olwig, & K. Hastrup (Eds.), *Siting culture: The shifting anthropological perspective*. London: Routledge.

Nixon, R. (2011). *Slow violence and the environmentalism of the poor*. Cambridge, MA: Harvard University Press.

Odwe, G., Undie, C.-C., & Obare, F. (2018). Attitudes towards help-seeking for sexual and gender-based violence in humanitarian settings: The case of Rwamwanja refugees settled in Uganda. *BMC International Health and Human Rights, 18*(15), 1–12.

Pittaway, E., & Bartolomei, L. (2018). Enhancing the protection of women and girls through the global compact on refugees. *Forced Migration Review, 57*, 77–79.

Pupavac, V. (2001). Therapeutic governance: Psycho-social intervention and trauma risk management. *Disasters, 25*(4), 358–372.

Qayum, M., Mohmand, S., & Arooj, H. (2012). Frequency and physical factors associated with gender-based violence in the internally displaced people of Pakistan. *Journal of the College of Physicians and Surgeons Pakistan, 22*(1), 63–65.

Robbins, J. (2013). Beyond the suffering subject: Toward an anthropology of the good. *Journal of the Royal Anthropological Institute (N.S.), 19*(3), 447–462.

Sanaathanan, T. (2011). *The incomplete Thombu*. New York: Raking Leaves.
Shanmugaratnam, N., & Zackariya, F. (2002). *Stepping out: Women surviving amidst displacement and deprivation*. Colombo: Muslim Women's Research and Action Forum.
Schrijvers, J. (1997). Internal refugees in Sri Lanka: The interplay of ethnicity and gender. *European Journal of Development Research, 9*(2), 62–82.
Schrijvers, J. (1999). Constructing 'womanhood', 'tamilness', and 'the refugee': Internal refugees in Sri Lanka. In S. Thiruchandran (Ed.), *Women, narration and nation: Collective images and multiple identities*. WERC: Colombo.
Simon-Butler, A., & McSherry, B. (2019). *Defining sexual and gender-based violence in the refugee context* (IRiS Working Paper Series No. 28). Birmingham.
Somasunderam, D. (2007). Collective trauma in northern Sri Lanka: A qualitative psycho-social ecological study. *International Journal of Mental Health Systems, 1*(5), 1–27.
Somasunderam, D. (2010). Collective trauma in the Vanni: A qualitative inquiry into the mental health of the internally displaced due to the civil war in Sri Lanka. *International Journal of Mental Health Systems, 4*(22), 1–31.
Sorensen, B. (2001). Developing new livelihoods in the shadow of war: Displaced, relocated and resettled Muslims. In M. Vincent & B. Sorensen (Eds.), *Caught between borders: Response strategies of the internally displaced*. London: Pluto Press.
Sriskandarajah, D. (2004). *Forced migration online: Sri Lanka*. http://www.forcedmigration.org/research-resources/expert-guides/sri-lanka. Accessed 1 Sept 2019.
Summerfield, D. (1997). Legacy of war: Beyond 'trauma' to the social fabric. *Lancet, 349*(9065), 1568.
Summerfield, D. (1998). 'Trauma' and the experience of war: A reply. *Lancet, 351*(9115), 1580–1581.
Summerfield, D. (1999). Bosnia and Herzogovina and Croatia: The medicalisation of the experience of war. *Lancet, 354*(9180), 771–772.
Thanges, P. (2008). Caste and social exclusion of IDPs in Jaffna society. *Colombo Review, 1*(2), 1–27.
Thiranagama, S. (2011). *In my mother's house: Civil war in Sri Lanka*. Philadelphia: University of Pennsylvania Press.
UNHCR. (2012, September 27). *Sri Lanka's displacement chapter nears end with closure of Menik Farm*. https://www.unhcr.org/506443d89.html. Accessed 19 Aug 2019.
UNHCR. (2019). *Figures at a glance*. https://www.unhcr.org/figures-at-a-glance.html.Accessed 8 Aug 2019.
University Teachers for Human Rights (Jaffna). (2009). *Special report 34: Let them speak*. Available at http://www.uthr.org/SpecialReports/Special%20rep34/Uthr-sp.rp34part6.htm. Accessed 14 Aug 2019.
Vincent, M. (2001). Introduction and background. In M. Vincent & B. Refslund Sørensen (Eds.), *Caught between borders: Response strategies of the internally displaced*. London: Pluto Press.
Ward, J. (2002). *If not now, when? Addressing gender-based violence in refugee, internally displaced, and post-conflict settings: A global overview*. New York: The Reproductive Health for Refugees Consortium.
Yazid, S., & Natania, A. (2017). Women refugees: An imbalance of protecting and being protected. *Journal of Human Security, 13*(1), 34–42.

Zetter, R. (1991). Labeling refugees: Forming and transforming an identity. *Journal of Refugee Studies, 4*(1), 39–62.

Malathi de Alwis received her Ph.D. in Socio-Cultural Anthropology from the University of Chicago and is currently affiliated with the Faculty of Graduate Studies, University of Colombo. She has written extensively on displacement, militarisation, humanitarianism, maternalism and memorialization. She is the co-editor of *Tsunami in a Time of War: Aid, Activism and Reconstruction in Sri Lanka and Aceh* (2009), *Feminists Under Fire: Exchanges Across War Zones* (2003) and *Embodied Violence: Communalising Women's Sexuality in South Asia* (1996).

Part VI

Gendering Migrant Rights, Social Welfare and the Politics of Inclusion

CHAPTER 29

Social Protection, Gender and International Migrations: From National Worlds to Transnational Quests

Sònia Parella and Thales Speroni

Abstract What are the interfaces between migration, social protection and gender? In this chapter, we approach this issue from critical standpoints on gender relations and transnational dynamics. First, we introduce the feminist critique that calls into question the nationalist and masculinist assumptions that characterise the agenda that conceived welfare regimes as "national worlds." Esping-Andersen's framework highlights the relationships between states and markets, relegating the family (and therefore women) to a subsidiary level in the welfare provision and neglecting the importance of the internationalisation of productive and reproductive work. Then, we present the research agenda on transnational social protection (TSP) that emerged from the necessity to understand the new forms of production and (re)distribution of social protection. Finally, from the perspectives on migration and the gendered nature of the labour markets, the circulation of care and the intersectionality of inequalities, we argue that TSP takes the form of a gendered quest, insofar it is only made possible through the continuous, reflexive and risky (often sacrificial) action of women involved in transnational strategies of social protection.

S. Parella (✉) · T. Speroni
Universitat Autònoma de Barcelona, Barcelona, Spain
e-mail: sonia.parella@uab.cat

T. Speroni
Pontifical Catholic University of Rio Grande do Sul, Porto Alegre, Brazil

C. Mora and N. Piper (eds.), *The Palgrave Handbook of Gender and Migration*, https://doi.org/10.1007/978-3-030-63347-9_29

Introduction

We live in a world characterised by an unequal distribution of risk factors and social protection resources. As a result, the risks to which people are most significantly exposed, as well as their capacities to face them, are not random, the result of individual misfortune or the unpredictable forces of nature. The place of birth, residence and citizenship, as well as the dynamics of gender, generation, class, race and ethnicity, are fundamental to explain these inequalities. Undoubtedly, when it comes to social protection, we are not all "in the same boat," and for many women—who tend to accumulate multiple forms of vulnerability and care responsibilities—international migration is a risky but necessary way to seek protection. Within this context, how can we understand the interfaces between social protection, gender and international migration? In this chapter, we approach this issue from critical standpoints on gender relations and transnational dynamics, which allows us to problematise the nationalist and masculinist assumptions that characterise much of the mainstream frameworks on social protection.

This chapter is organised into three sections. First, we approach the feminist critique of the "worlds of welfare" perspective. Esping-Andersen's approach revealed the crucial importance of socio-historical elements for understanding the different ways in which the state, the market and families are articulated for the provision of welfare on a national scale. Nevertheless, feminist critiques questioned the ideals of citizenship underlying Esping-Andersen's framework, which ends up neglecting the role of women (especially immigrant women) in the provision of welfare. The recognition of the importance of care practices and unpaid/low-paid work performed by women led to the affirmation of social welfare as not reducible to the rights of citizens or the income of the breadwinner man (Lister, 1997). The feminist critique meant that welfare systems came to be seen more as complex constellations of social protection, with formal and informal components not limited to the national scale (Franzoni, 2008).

The second section introduces the research agenda on transnational social protection (TSP). The TSP refers to the set of capacities and strategies to respond to social risks in the spheres of care, work, health, education and social security, which can be developed through the combination of resources from multiple sources (national states, markets, families and civil society) even if they are located in territories of different national states (Bilecen, Barglowski, Faist, & Kofman, 2019; Gómez, 2019; Levitt, Viterna, Mueller, & Lloyd, 2017; Parella & Speroni, 2018). While mainstream frameworks on welfare regimes tended to approach exclusively institutional arrangements, the TSP agenda focuses on the availability and possible combinations of formal and informal resources in strategies of (im)mobility, portability and simultaneity of actors and resources of social protection. In this sense, the studies on TSP assume that social protection is not only a result of institutional configurations, but is also a product of the continuous, reflective and potentially risky action performed by social actors dispersed across borders.

As the globalisation of care, the asymmetric distributions of reproductive responsibilities and the intersectionality of multiple strands of inequality make TSP more dependent on the risky efforts of protection made by women, we argue, in the final section, that the TSP should be understood as a "gendered quest." Therefore, we present the frameworks of care circulation, gendered labour markets and scalar intersectionalities, highlighting the heuristic potential of these approaches for understanding the TSP from a gender standpoint.

Feminist Critiques of the National Worlds of Welfare

Esping-Andersen's seminal work "The Three Worlds of Welfare Capitalism" (1990) had an immediate and substantial impact. The principal contribution made by the author was the development of a sophisticated sociological framework on the interaction between states, markets and families in shaping welfare regimes. According to Esping-Andersen, the welfare state assumed multiple contents, which correspond to different political histories that have produced heterogeneous institutional configurations. The analysis of 18 OECD countries allowed him to propose the existence of three national worlds of welfare: the liberal, the conservative, and the social-democratic. The study has become not only one of the most influential on the international comparison of social policies but also a contemporary classic in the social sciences. In the last three decades, the relevance of this work has encouraged both numerous studies and critical debates.

Regarding the analysis of the interfaces between gender, migration and social protection, Esping-Andersen's framework has two major interconnected inadequacies. On the one hand, it is based on the epistemological nationalism that considers society as strictly national (Wimmer & Glick Schiller, 2002). On the other hand, it neglects the relevance of the dynamics of gender and social reproduction to the different welfare regimes. Concerning this last aspect, the feminist critique highlighted three problematic elements.

First, Esping-Andersen's approach tends to focus on the relationship between states and markets, relegating the family to a subsidiary level in the welfare provision (Lewis, 1992, 1997; O'Connor, 1993; Orloff, 1993). This analytical position implies severe consequences for the study of welfare regimes in which the family (a euphemism for women) is responsible for almost all social protection. Therefore, the analyses restricted to the interfaces between states and markets have promoted distorted interpretations, as these articulations are shaped by the ties with the domestic units (Lewis, 1997: 161). The negligence of the family is also expressed when it is presumed to be formed by symmetric and harmonious gender relations. As gender and generation variables produce asymmetries in the distribution of obligations and rights to welfare within family relationships, families cannot be considered as homogeneous actors (Sainsbury, 1999).

Second, the analysis of welfare regimes proposed by Esping-Andersen presupposes a conception of citizenship based on the model of salaried man. To the extent that women tend to have more responsibilities than men in unpaid areas of social life, particularly those linked to care, formal equality between men and women as citizens does not mean de facto equality (Kilkey & Bradshaw, 1999: 148; O'Connor, 1993). Therefore, authors like Lister (1997) state that women's economic independence to men and the market forces are a necessary condition for the integral status of their citizenship (see also Bassel, in this volume). In this sense, a gender analysis must be able to recognise the socially differentiated capacities of access to well-being among wives, mothers, grandmothers and female workers (Trifiletti, 1999). In this way, specific but recurrent situations must be considered, such as the case of single mothers. They are especially exposed to the risk of falling into a "welfare gap" caused by the combination of multiple care responsibilities and the absence of employment and income (Gornick, Meyers, & Ross, 1997). This situation often leads women to emigrate as a way to make a living, a strategy that could generate subjective suffering and social stigma.

Third, feminist criticism emphasises that Esping-Andersen's framework is limited when considering the effects that the state has on gender relations. From a feminist perspective, it is necessary to recognise the quality of social rights and their ability to reproduce or transform gender relations (Sainsbury, 1999). Consequently, it is vital to scrutinise the potential impact of social policies considering not only the number of resources invested but also the different components and normative foundations that shape those policies (Franzoni, 2008).

The feminist critique of the worlds of welfare has encouraged the creation of two main research programmes (Sainsbury, 1997: 2). The first assumes that feminist research should contribute to the theoretical advancement of hegemonic approaches to welfare regimes through the incorporation of gender relations into existing interpretive frameworks (O'Connor, 1993; Orloff, 1993). The second denies the strategy of adding women into normative frameworks that consider men as the norm. Therefore, for this second group, it is necessary to formulate new models and theories (Lewis, 1992; Sainsbury, 1997). Despite the different strategies and goals, these authors recognise the gendered nature of the nation-states redistributive action, which demands to consider gender as a relevant variable to investigate welfare regimes, social policies and their effects (Lewis, 1997: 161). In this sense, researchers as O'Connor (1993), Orloff (1993) and Franzoni (2008) have proposed alternative welfare regimes typologies, incorporating different variables permeated by the gender dynamics such as the sexual division of paid and unpaid work, the positions of married and single women in labour markets, and the availability of resources and the provision of services that allow the defamiliarisation of women.

One of the main consequences of the feminist critique to the worlds of welfare framework was the recognition of the significance of non-state welfare providers (Gough & Wood, 2004). It is at this point that feminist criticism meets the transnational perspective of migration. International mobility can be understood as a strategy of social protection, a way to access not only formal resources but also to generate informal resources through transnational exchanges of care and remittances. To this extent, there is no doubt that the feminist critique served as the basis for diagnosing the transnationalisation of specific components of welfare regimes, primarily through the migration of women and the globalisation of care.

It is in this sense that Franzoni proposed to understand the welfare regimes as "constellations of state, markets and families" (Franzoni, 2008). The author intended to broaden the concept of welfare regimes to include components that might not be strictly national due to international migration and transnational networks. This conceptual nuance is especially relevant in addressing both the importance of women and international migrations in the provision of welfare. That is vital in regions, such as Latin America, where the states' role tends to be contingent.

In order to interpret the constellations of welfare in Latin America, Franzoni (2008) elaborates a typology based on the variation of three principles: the national and transnational commodification of the labour force, and the income levels generated; the decommodification of resources and welfare services; and the defamiliarisation of care, in other words, the ability to respond to care responsibilities independently of unpaid female work (Franzoni, 2008).

In this way, the author identifies three types of welfare regimes in Latin America. The state-productivist, that encompasses countries such as Chile and Argentina, and the state-protectionist, that alludes to countries like Brazil, Costa Rica, Mexico and Uruguay, have in common a high national commodification of the labour force with different remunerating capacities (higher in the first regime mentioned). In contrast, the informal-familiarist regime—the most frequent in Latin America and identified in countries such as Bolivia, El Salvador, Ecuador, Nicaragua and Paraguay—is characterised by the low efficiency of the national labour market when it comes to absorbing and remunerating the workforce, which generates a tendency towards its transnationalisation through broad-spectrum emigration flows (Franzoni, 2008: 122). Within the scope of this third regime, families develop a simultaneously productive, reproductive and redistributive role, something that is only possible through migration (internal and international), self-employment and the direct provision of care by family members (Franzoni & Voorend, 2009: 144).

The concept of "constellations of welfare" and the typology proposed by Franzoni (2008) contribute to interpreting the complexity of welfare configuration in Latin America. Since it is not a homogeneous region, the constellations between states, markets and families are plural, both in their formats and in their intensities. The analysis proposed by Franzoni demonstrates how the feminist critique of welfare regimes leads to the consideration of the possibility of its transnationalisation, particularly in contexts that combine relevant migratory flows and the weakness of states in the provision of welfare.

The analytical frameworks developed from the critique of Esping-Andersen's work have made a valuable contribution to the analysis of the interfaces between migration and social protection since they allowed us to understand both the national and international scenarios in which social protection resources are distributed and encourage family strategies of mobility (Sabates-Wheeler & Feldman, 2011; Sainsbury, 2006). Despite this crucial contribution, recognising the transnationalisation of certain components of welfare regimes through the migration of women is a significant but insufficient interpretive shift. It is also essential to articulate this "top-down" approach on institutions and social policies with a "bottom-up" perspective focused on the individual and collective strategies of social protection through international mobility (Gubaray, 2019). It was with this purpose that, in recent years, the research agenda on TSP has been developed.

Transnational Interstices of Social Protection

In the last decade, researchers such as Peggy Levitt, Jean-Michel Lafleur and Thomas Faist have established a research agenda on the TSP. These initiatives emerged from the necessity to understand the new forms of production and (re)distribution of social protection that are not satisfactorily captured by frameworks—as the "worlds of welfare" approach—that conceive welfare resources as always territorialised and grounded in the relationship between nation-states and their citizens (De Guchteneire & Pécoud, 2009: 6). Therefore, the TSP agenda proposes an articulation between the studies on welfare that emphasises the relevance of the dynamics of gender and mobility and the investigations that mobilise a transnational gaze to address the cross-border circulation of rights, care, remittances and sociocultural conceptions.

Levitt et al. (2017) define TSP as institutional actions, policies and programmes, as well as individual and collective strategies, which provide protection resources in *transnational ways* (Levitt et al., 2017: 6). The TSP focuses on the absence and weakness of protection systems, which impels migrants (their family and personal networks) to negotiate and combine transnationally dispersed social protection resources, whether they are formal or informal (Levitt et al., 2017: 3). The availability of resources of protection depends on the characteristics of the sources of protection (capacity of states, the nature of the markets and the presence of the third sector), as well as the attributes of the individuals and their families (place of origin, place of residence, gender, race, ethnicity, religion, income, education and the breadth of social networks) (Levitt et al., 2017: 6–7). The TSP has the potential to identify the possible limitations of institutional arrangements, as well as to propose public policies capable of promoting more efficient sets of protection resources and services (Levitt et al., 2017: 3). In order to make this type of analysis possible, Levitt et al. (2017) propose the concept of "resource environment," which refers to the combination of all social protection resources (formal and informal) to which individuals have access, both in origin and destination.

The formal social protection is generated by state regulations promoted by public funds, which are institutionalised through social policies and legal frameworks that operate at various scales (national, sub-national and supranational) and apply different eligibility criteria (Bilecen & Barglowski, 2015). Nevertheless, formal social protection is incapable of satisfying the transnational demands of social protection (Righard & Boccagni, 2015). The disjunction between the transnationalisation of social life and the forms of access to formal protection entails growing processes of social exclusion (Speroni, 2019). The widespread conception of social protection as a citizenship right serves as a legal and political justification to prevent non-citizens (immigrants) and non-residents (emigrants) from accessing formal social protection. Despite this panorama, there is currently an expansion of alternative ways to access formal social protection, which includes the portability of rights, and a higher tolerance towards dual nationality (Faist & Gerdes, 2008). Moreover, it is also noticeable the proliferation of multilateral and bilateral social security agreements between nation-states (Avato, Koettl, & Sabates-Wheeler, 2010; Lafleur & Lizin, 2015; Sabates-Wheeler, Koettl, & Avato, 2011).

There are different studies focused on the bases upon which states establish and legitimise TSP (Lafleur & Vivas-Romero, 2018; Paul, 2017) and on the portability of rights as a part of the supranational framework of formal social protection (Fingarova, 2019: 48) In respect of access and portability of social insurance benefits, Hujo and Piper (2010) highlight the differences between South-North and South-South migrants. The latter face more disadvantages due to the poorly developed social security systems in both origin and receiving countries and the high rates of unauthorised or undocumented migration within these contexts (Avato et al., 2010). Besides, countries with insufficient social protection systems combined with low and volatile levels of income and employment, give shape to the major push factors for emigration (Hujo & Piper, 2010: 3).

The study of TSP cannot be restricted to analysing the institutional gaps, but it is also necessary to understand how informal resources compensate, complement or replace formal resources. Certainly, a lack of access to formal welfare in both origin and receiving countries lead migrants to rely on their social networks for support (Serra Mingot & Mazzucato, 2018). Thus, informal social protection refers to a set of resources and strategies "from below," which play a crucial role in the provision of social protection. Informal social protection emerges from social and care networks based on gendered and asymmetrical ties of solidarity, reciprocity and obligations (Bilecen & Barglowski, 2015) and constitutes, according to Boccagni (2016: 7), "(...) the less visible but more substantive side of migrant social protection". The circulation of different forms of care and monetary remittances in the social and family networks allows—with the combination of formal social protection resources—the development of strategies in the spheres of education, health, labour and social security. Thereby, informal TSP is consistent with the idea of migration as a family livelihood strategy, but it also highlights the interaction among demands, risks and (de)territorialised provisions (Boccagni, 2016).

TSP not only refers to different types of resources but also to the ways they are combined by mobile and immobile actors through cross-border relationships and strategies. It is in this sense that Bilecen and Barglowski (2015) describe the TSP as a multi-sited assemblage of formal and informal resources that allows access to tangible and intangible elements to face social risks. The notion of assemblage denotes the union of disparate elements that become coherent at a particular moment and in a certain circumstance (Lacroix, 2014: 653). Speroni emphasises that the "artisanal" process of assembly is crossed by deep reflexivity since it is not a random or opportunistic union of dissimilar components, but rather it is a process where subjective meanings play a decisive role (Speroni, 2017). In the same way, to Fingarova (2019), the assemblage is an expression of the migrant agency since it requires a knowledge of multiple formal systems and informal dynamics, and the understanding of what is adequate in each situation.

Through the reconstruction of resource environments (Levitt et al., 2017) and the interpretation of the efforts of assemblage (Bilecen & Barglowski, 2015), the TSP agenda promotes a fruitful synergy between the analysis of transnational care and the comparative studies of social policies. However, it should be noted that both resource environments and the assemblages of social protection are processes that are deeply shaped by asymmetric gender dynamics (Gubaray, 2019). It is for this reason that we propose to recognise TSP as a gendered quest.

Transnational Social Protection as a Gendered Quest

The on-going debates on the interfaces between social protection and international migration require to consider gender as a transversal process that shapes the resource environments as well as the assemblages of TSP. We argue that TSP takes the form of a gendered quest, insofar it is only made possible through the continuous, reflexive and risky (often sacrificial) action of women involved in complex (im)mobility strategies to access and combine multi-sited resources of formal and informal social protection. Hence, if we examine the different periods of gender analysis from a transnational standpoint (Mahler & Pessar, 2006), we can distinguish some vital aspects for the analysis of social protection in migratory contexts.

In this section, we will address three perspectives, emphasising their contributions to the topic at hand: (i) Transnational families and "care circulation," (ii) Globalised labour markets and feminised migrations, and (iii) Scalar Intersectionalities. From a theoretical and methodological point of view, they share the ambition to integrate macro-, meso- and micro-levels, conceiving social protection arrangements from a multi-actor and relational approach. Let us comment briefly on their differences, but also their potential complementarities.

First, although the transnational perspective has the significant potential for analyses of gendered processes in multiple socio-geographical scales (global, supranational, national, subnational and local), institutions (social movements, national-states and families) and dimensions (cultural, economic, political and psychosocial) (Erel & Lutz, 2012), most of the research has focused exclusively on how gender is present in migrants' family life. This tendency has generated an extensive literature on "transnational families" (Alicea, 1997; Bryceson & Vuorela, 2002; Hondagneu-Sotelo & Avila, 1997). This set of diverse approaches seeks to analyse care relationships within its location in institutional settings (Kilkey & Merla, 2014) and different scales of the international social reproduction system (Parreñas, 2001).

The circulation of care approach is one of the most recent developments in studies in transnational families. Baldassar and Merla (2014: 25) define care circulation as "(…) the reciprocal, multidirectional and asymmetrical exchange of care that fluctuates over the life course within transnational family networks subject to the political, economic, cultural and social contexts of both sending and receiving societies." The authors sought to go beyond the analysis of care in transnational families provided by the theory of global care chains. Therefore, this perspective does not focus exclusively on women migrants who provide care but instead considers all actors (people, third sector organisations and state institutions) involved in different types of caring relationships across borders.

The analysis of the circulation of care is fundamental for the understanding of TSP, as it allows us to comprehend the dynamics of simultaneity and non-linearity that characterise transnational protective strategies. Undoubtedly, the care circulation framework provides a holistic interpretation that can be mobilised to understand the entanglements between formal and informal protection operating at different levels inside multi-local and unequal family relationships (Merla & Baldassar, 2016).

For instance, Dankyi, Mazzucato and Manuh (2017) analyse the case of Ghana and the complexities of transnational caregiving arrangements when remittances are not enough to guarantee adequate living standards for migrants' children and to provide social protection within transnational families. Female caregivers may go back home through the mobilisation of interpersonal networks of support, as a way to make the mobility-protection nexus possible. Thus, these women constitute an essential pillar of TSP systems based on reciprocity ties between migrants and their families back home and elsewhere.

Serra Mingot (2019) goes beyond the provision of care between parents and children and addresses the intricacies of care dynamics within extended families. Her research focuses on how female care-receivers mobilise the care provided by different male family members abroad while protecting themselves and their children. Through a multi-sited ethnography with Sudanese transnational families across the Netherlands, the UK and Sudan, the results show the different strategies of women to manage the reception of unwanted care by avoiding conflict and affirming personal autonomy (including over one's own body).

Secondly, another relevant research front on international migration had been the study of labour markets and the feminisation of international migration. The increasing demand for women's labour in reproductive and productive sectors has consolidated female niches of precarious work. The precariousness of these activities—such as domestic service, sex work and labour in sweatshops—and the obstacles to pursue alternative ways of integration in the labour markets mean a contingent and fragile access to formal social protection (Bilecen et al., 2019: 5). These situations take on even more alarming levels if we consider all the possible intersections with generational, ethnic and religious factors.

The study of the demand for women's labour is a powerful way for understanding not only gendered employment patterns but also to explain some of the main demographic and geographic tendencies of contemporary international migrations and the impacts of such on life chances and the access to protective resources. For instance, Castellani and Martín-Díaz (2019) describe the Ecuadorian migration to Spain as an emblematic case of the feminisation of international migration. This migration flow has been shaped by transnational female social networks in which different types of capital circulate, which provide social protection for people involved in them, both migrants and non-migrants. Nevertheless, the role of these women in social protection provision fluctuates over time. First, the migration of these women who become principal economic providers puts at risk their primary role of caregivers assigned by the "traditional" gender division of roles: informal social protection to the household in Ecuador. However, on the contrary, these women, in their roles as employees (most of them domestic workers), transform formal resources of social protection (such as unemployment insurances, family reunification permissions and portability of rights) into a different kind of capital to be put into circulation within the transnational household.

Lastly, a third group of female researchers proposed "scaling intersectionalities" as a way to examine gender dynamics at the transnational level, considering the complex conjunctions with other factors of inequality. The intersectional analysis was primarily developed by black women in the United States (Crenshaw, 1991; McCall, 2005), and it has been widely adopted within gender research. To this perspective, it is critical to understand the multiple intersectionalities of class, race, ethnicity, generation, religion, sexuality and others in order to explain gender relations (Erel & Lutz, 2012; Thimm & Chaudhuri, 2019). According to an intersectional standpoint, migrants are socially positioned regarding the interfaces of various axes of differentiation that shape power relationships. The ways that these axes interact with each other are largely plural due to different and complex social contexts (Thimm & Chaudhuri, 2019).

Mahler, Chaudhuri and Patil (2015) ponder that the intersectional approach represents a crucial heuristic development, that tends to be mobilised for studies confined to a specific country, generating interpretations limited to a "domestic intersectionality." Alternatively, the authors recommended a multi-scalar and intersectional approach. The objective of this framework is to examine the intersections of gender, class, ethnicity, race and national origin/citizenship in multiple social scales, including the intimate (whether as individual or families), local, national and transnational. While Mahler et al. (2015) proposed "scaling intersectionality" to the study of transnational families, it could be employed to understand any individual or organisation embedded in transnational networks and spaces. Concerning social protection debates, the intersectional approach enables to rethink the dynamics through which formal and informal social protection intersect with inequalities explained by gender divisions together with other social hierarchies (Amelina, 2017), which operates in social and intra-family relationships.

For instance, Lafleur and Vivas-Romero (2018) examined the transnational healthcare arrangements constituted by Andean family networks. The authors demonstrate how these arrangements are the product of negotiations between family members living in different countries, who have different resources, conceptions and strategic repertoires (2018: 2). The main contribution of Lafleur and Vivas-Romero's research is the incorporation of an intersectional approach to capture immigrants' multiple inequalities in access to social protection. They conclude that beyond the classic "gender, race and class trilogy" adopted in intersectional studies, "a thorough understanding of transnational health care arrangements by immigrant families also need to take in consideration additional heterogeneity markers such as religiosity, generation and their level of transnational engagement" (Lafleur & Vivas-Romero, 2018: 2).

Comparative methodological strategies are an excellent way to approach intersectionality. This quality is demonstrated by the study conducted by Landolt and Da (2005), in which they compared Salvadoran working low-income families and migrants from the Republic of China. The authors identified how the organisation of social protection within spatially ruptured practices of migrant families was marked by significant transformations in gender roles and gender ideologies. Although Landolt and Da (2005) do not explicitly refer to an intersectional lens, the two case studies bear ample evidence that beyond gender arrangements, a combination of factors (such as family's personal resources, social capital, origin/destination nation-state and cultural concepts) can determine family's ability to combine diverse social protection resources.

Final Considerations

This chapter has demonstrated the need to look at new forms of production and (re)distribution of social protection from a transnational standpoint. Migrants, and their geographically distant family members, are particularly affected by the challenge of access and provide social protection (Bilecen et al., 2019). In order to understand how migrants produce TSP assemblages of formal and informal nature, it is essential to take not only a transnational gaze but also a gender perspective. In the same way that feminist research has rectified the bias concerning welfare regimes analysis in comparative social policy (the absence of gender relations in theoretical frameworks and empirical studies), it is necessary to bring gender lines and TSP closer together.

The theoretical perspectives and empirical studies on the dynamics of transnational care, the gendered nature of globalised labour markets and the intersectionality of gender with other vectors of inequality, emphasise, in multiple but complementary ways, how gender, as a relational category, shape both the provision of social protection (not only care) and its implications. In this sense, we argue that social protection becomes a gendered quest, in the scope of which, there is a tendency for women with migration background to be, simultaneously, those who have to provide the most protection and those who are most exposed to social risks. Finally, it should be highlighted that in order to capture the complexity of the experience of women navigating the transnational interstices as a risky way of generating and accessing social protection, the imprint of intersectional analysis offers a fruitful path for a necessary critical reflection.

References

Alicea, M. (1997). "A Chambered Nautilus": The contradictory nature of Puerto Rican women's role in the social construction of a transnational com-munity. *Gender and Society, II*(5), 597–626.

Amelina, A. (2017). *Transnationalizing inequalities in Europe: Sociocultural boundaries, assemblages and regimes of intersection*. London: Routledge.

Avato, J., Koettl, J., & Sabates-Wheeler, R. (2010). Social security regimes, global estimates, and good practices: The status of social protection for international migrants. *World Development, 38*(4), 455–466. https://doi.org/10.1016/j.worlddev.2009.10.003.

Baldassar, L., & Merla, L. (Eds.). (2014). *Transnational families, migration and the circulation of care: Understanding mobility and absence in family life*. New York and London: Routledge.

Bilecen, B., & Barglowski, K. (2015). On the assemblages of informal and formal transnational social protection. *Population, Space and Place, 21*(3), 203–214.

Bilecen, B., Barglowski, K., Faist, T., & Kofman, E. (2019). Gendered dynamics of transnational social protection. *Comparative Migration Studies, 7*(47), 1–14.

Boccagni, P. (2016). Searching for well-being in care work migration: Constructions, practices and displacements among immigrant women in Italy. *Social Politics, 23*(2), 284–306. https://doi.org/10.1093/sp/jxv031.

Bryceson, D., & Vuorela, U. (Eds.). (2002). *The transnational family: New European frontiers and global networks*. Oxford: Berg.

Castellani, S., & Martín-Díaz, E. (2019). Re-writing the domestic role: Transnational migrants' households between informal and formal social protection in Ecuador and in Spain. *Comparative Migration Studies, 7*(1), 1–18. https://doi.org/10.1186/s40878-018-0108-0.

Crenshaw, K. (1991). Mapping the margins: Intersectionality, identity politics, and violence against women of color. *Stanford Law Review, 43*, 1241–1299.

Dankyi, E., Mazzucato, V., & Manuh, T. (2017). Reciprocity in global social protection: Providing care for migrants' children. *Oxford Development Studies, 45*(1), 80–95. https://doi.org/10.1080/13600818.2015.1124078.

De Guchteneire, P., & Pécoud, A. (2009). Introduction: The UN convention on migrant workers' rights. In R. Cholewinski, P. De Guchteneire, & A. Pecoud (Eds.), *Migration and human rights: The United Nations convention on migrant workers' rights* (pp. 1–44). Cambridge: Cambridge University Press. https://doi.org/10.1017/cbo9780511811272.001.

Erel, U., & Lutz, H. (2012). Gender and transnationalism. *European Journal of Women's Studies, 19*(4), 409–412; 409–412. https://doi.org/10.1177/1350506812461466.

Esping-Andersen, G. (1990). *The three worlds of welfare capitalism*. https://doi.org/10.2307/2073705.

Faist, T., & Gerdes, J. (2008). Dual citizenship in an age of mobility. *Transatlantic Council on Migration, 63*(1), 1–21.

Fingarova, J. (2019). *Agency in transnational social protection: Practices of migrant families between Bulgaria and Germany*. Berlin: Frank & Timme.

Franzoni, J. (2008). *Domesticar la incertidumbre en América Latina: mercado laboral, política social y familias*. San José: Editorial UCR.

Franzoni, J. M., & Voorend, K. (2009). ¿Es posible (des) encadenar la desigualdad de género en América latina? *Documentos de Trabajo (Fundación Carolina), 32*, 143–152.

Gómez, C. G. G. (2019). *Social protection on the move: A transnational exploration of Nicaraguan migrant women's engagement with social protection in Spain and Nicaragua* (Working Paper, No. 648). International Institute of Social Studies.

Gornick, J. C., Meyers, M. K., & Ross, K. E. (1997). Supporting the employment of mothers: Policy variation across fourteen welfare states. *Journal of European Social Policy, 7*(1), 45–70.

Gough, I., & Wood, G. (2004). *Insecurity and welfare regimes in Asia Africa and Latin America: Social policy in development contexts*. Cambridge: Cambridge University Press.

Gubaray, C. G. (2019). *Social protection on the move: A transnational exploration of Nicaraguan migrant women's engagement with social protection in Spain and Nicaragua* (Working Paper, No. 48). The Haye: Institute of Social Studies.

Hondagneu-Sotelo, P., & Avila, E. (1997). I'm here, but I'm there' the meanings of Latina transnational motherhood. *Gender and Society, 11*(5), 548–571.

Hujo, K., & Piper, N. (2010). *South-South migration implications for social policy and development*. New York: Plagrave Macmillan.

Kilkey, M., & Bradshaw, J. (1999). Lone mothers, economic well-being, and policies. In D. Sainsbury (Ed.), *Gender and welfare state regimes* (pp. 147–184). Oxford: Oxford University Press.

Kilkey, M., & Merla, L. (2014). Situating transnational families' caregiving arrangements: The role of institutional contexts. *Global Networks, 14*(2), 210–229. https://doi.org/10.1111/glob.12034.

Lacroix, T. (2014). Conceptualizing transnational engagements: A structure and agency perspective on (hometown) transnationalism. *International Migration Review, 48*(3), 643–679.

Lafleur, J.-M., & Lizin, O. (2015). Transnational health insurance schemes: A new avenue for Congolese immigrants in Belgium to care for their relatives' health from abroad? *Following the Flows: Transnational Approaches to Intangible Remittances, 3/2015*(3). Retrieved from http://seminars.wcfia.harvard.edu/files/tsi/files/3-lafleur2015_tsiworkingpaper.pdf.

Lafleur, J., & Vivas-Romero, M. (2018). Combining transnational and intersectional approaches to immigrants' social protection: The case of Andean families' access to health. *Comparative Migration Studies, 6*(14), 1–18. https://doi.org/10.1186/s40878-018-0073-7.

Landolt, P., & Da, W. (2005). The spatially ruptured practices of migrant families: A comparison of immigrants from El Salvador and the People's Republic of China. *Current Sociology, 53*(4), 625–653. https://doi.org/10.1177/0011392105052719.

Levitt, P., Viterna, J., Mueller, A., & Lloyd, C. (2017). Transnational social protection: Setting the agenda. *Oxford Development Studies, 45*(1), 2–19. https://doi.org/10.1080/13600818.2016.1239702.

Lewis, J. (1992). Gender and the development of welfare regimes. *Journal of European Social Policy, 2*(3), 159–173. https://doi.org/10.1177/095892879200200301.

Lewis, J. (1997). Gender and welfare regimes: Further thoughts. *Social Politics: International Studies in Gender, State & Society, 4*(2), 160–177. https://doi.org/10.1093/sp/4.2.160.

Lister, R. (1997). Citizenship: Towards a feminist synthesis. *Feminist Review, 57*(1), 28–48.

Mahler, S. J., & Pessar, P. (2006). Gender matters: Ethnographers bring gender from the periphery toward the core. *International Migration Review, 40*(1), 27–63.

Mahler, S. J., Chaudhuri, M., & Patil, V. (2015). Scaling intersectionality: Advancing feminist analysis of transnational families. *Sex Roles, 73*(3), 100–112. https://doi.org/10.1007/s11199-015-0506-9.

Mccall, L. (2005). The complexity of intersectionality. *Signs: Journal of Women in Culture, 30*(3), 1771–1800.

Merla, L., & Baldassar, L. (2016). Concluding reflections: 'Care circulation' in an increasingly mobile world: Further thoughts. *Papers: Revista de Sociologia, 101*(2), 275–284. https://doi.org/10.5565/rev/papers.2268.

O'Connor, J. S. (1993). Gender, class and citizenship in the comparative analysis of welfare state regimes: Theoretical and methodological issues. *British Journal of Sociology, 44*(3), 501–518.

Orloff, A. S. (1993). Gender and the social rights of citizenship: The comparative analysis of gender relations and welfare states. *American Sociological Review, 58*(3), 303–328.

Parella, S., & Speroni, T. (2018). Las perspectivas transnacionales para el análisis de la protección social en contextos migratorios. *Autoctonía Revista de Ciencias Sociales e Historia, 2*(1), 37–46.

Parreñas, R. S. (2001). *Servants of globalization: Women, migration and domestic work*. Redwood City: Standford University Press.

Paul, R. (2017). Welfare without borders: Unpacking the bases of transnational social protection for international migrants. *Oxford Development Studies, 45*(1), 1–14. https://doi.org/10.1080/13600818.2016.1271868.

Righard, E., & Boccagni, P. (2015). Mapping the theoretical foundations of the social work–migration nexus. *Journal of Immigrant & Refugee Studies, 13*(3), 229–244.

Sabates-Wheeler, R., & Feldman, R. (2011). *Migration and social protection: Claiming social rights beyond borders*. London: Springer.

Sabates-Wheeler, R., Koettl, J., & Avato, J. (2011). Social security for migrants: A global overview of portability arrangements. In R. Sabates-Wheeler & R. Feldman (Eds.), *Migration and social protection: Claiming social rights beyond borders* (pp. 91–116). London: Springer.

Sainsbury, D. (Ed.). (1997). *Gendering welfare states*. London: Sage.

Sainsbury, D. (1999). *Gender and welfare state regimes*. Oxford: Oxford University Press.

Sainsbury, D. (2006). Immigrants' social rights in comparative perspective: Welfare regimes, forms in immigration and immigration policy regimes. *Journal of European Social Policy, 16*(3), 229–244.

Serra Mingot, E. (2019). The gendered burden of transnational care-receiving: Sudanese families across The Netherlands, the UK and Sudan. *Gender, Place & Culture, 27*(4), 546–567. https://doi.org/10.1080/0966369x.2019.1611546.

Serra Mingot, E., & Mazzucato, V. (2018). Moving for a 'better welfare'? The case of transnational Sudanese families. *Global Networks, 19*(2), 139–157. https://doi.org/10.1111/glob.12224.

Speroni, T. (2017). It is not only about access: Transnational Bolivian families in Barcelona and their meanings of social protection. *RIEM, 7*(3), 74–103.

Speroni, T. (2019). *Desencajes y Bricolajes de la Protección Social: Las familias Transnacionales Bolivianas en Barcelona y São Paulo*. Doctoral thesis. Universitat Autònoma de Barcelona—UAB; Universidade Federal do Rio Grande do Sul—UFRGS.

Thimm, V., & Chaudhuri, M. (2019, February 24). Migration as mobility? An intersectional approach. *Applied Mobilities*, 1–16. https://doi.org/10.1080/23800127.2019.1573780.

Trifiletti, R. (1999). Southern European welfare regimes and the worsening position of women. *Journal of European Social Policy, 9*(1), 49–64. https://doi.org/10.1177/095892879900900103.

Wimmer, A., & Glick Schiller, N. (2002). Methodological nationalism and beyond: Nation-state building, migration and the social sciences. *Global Networks, 2*(4), 301–334. http://dx.doi.org/10.1111/1471-0374.000.

Sònia Parella Ph.D., is a professor in the Sociology Department of the Autonomous University of Barcelona and coordinator of the GEDIME / CER-M research group. Throughout her career, she has conducted researches on gender and migration, migration policies, and transnational practices in the contexts of Southern Europe, North Africa, and Latin America.

Thales Speroni Ph.D., is a researcher at the GEDIME/CER-M research group (Autonomous University of Barcelona) and at the Medical School of the Pontifical Catholic University of Rio Grande do Sul (PUCRS). His recent doctoral thesis approaches, from a transnational perspective, the conjunctions between international mobilities and social protection in the case of Bolivian migrations to Spain and Brazil.

CHAPTER 30

Gender, Naturalisation and Deserving Citizenship

Leah Bassel

Abstract This chapter focuses on citizenship policies in Europe that require migrants to prove that they 'deserve' citizenship, e.g. through naturalisation processes which require proof of 'good character' and language proficiency. In this chapter, drawing on work largely informed by intersectional scholarship—specifically Black feminist traditions—I consider what happens when the study of naturalisation in Europe is analysed at the intersection of gender, sexuality, race and class. Analysing naturalisation, and specifically deserving citizenship, at these intersections exposes the deeply entrenched colonial power relations underlying naturalisation processes and provides insights into what it would mean to fundamentally challenge 'deservingness' and citizenship.

'Deserving' Citizenship: Introduction

The call to demonstrate that one 'deserves' citizenship shapes public debates over citizenship tests and ceremonies. To access citizenship, migrants have to prove their ability and willingness to be 'good citizens' (Fassin & Mazouz, 2007; Schinkel & Van Houdt, 2010) who are model neo-liberal subjects, self-governing, responsible, productive and trustworthy (Rose, 1999; Turner, 2014). I had the opportunity to study the naturalisation process in the United Kingdom, exploring the experiences of people from a range of nationalities at different stages in the process of becoming British citizens in a four year

L. Bassel (✉)
University of Roehampton, London, UK
e-mail: Leah.bassel@roehampton.ac.uk

C. Mora and N. Piper (eds.), *The Palgrave Handbook of Gender and Migration*, https://doi.org/10.1007/978-3-030-63347-9_30

mixed-methods study (Bassel, Monforte, Bartram, Khan, & Misztal, 2017; Bassel, Monforte, & Khan, 2018).

Our work on deservingness (Monforte, Bassel, & Khan, 2019) joins a host of studies that have analysed the significance for migrants of the injunction of deservingness in naturalisation or legalisation processes. Through in-depth interviews with migrants in two cities (Leicester and London), we show that many of the participants in our study use narratives in which they distinguish between 'deserving citizens' and the 'undeserving Others' when they reflect upon their experience of becoming citizens. In so doing, they negotiate new hierarchies of inclusion and exclusion which, we argue, reflect broader neo-liberal and ethos-based conceptions of citizenship.

In a similar vein, as we have previously discussed, Fassin and Mazouz (2007) and Mazouz (2012) have shown that the nature of the naturalisation process in France leads migrants to show their attachment to what they perceive as the dominant cultural norms. This process leads them to distinguish themselves symbolically from those 'who don't want to integrate' and 'who speak bad French' (Mazouz, 2012: 148). Byrne (2014) and Fortier (2008, 2013) show that, although they can be sometimes critical of dominant discourses around integration, migrants in the UK strategically present themselves as 'deserving' and emotionally engaged during their naturalisation process. Berger (2009) and Villalon (2010) show how women who apply for legalisation through the Violence Against Women Act in the United States present themselves through narratives that fit with what state representatives expect (Menjívar & Lakhani, 2016; see also Chauvin & Garcés-Mascarenas, 2012; Coutin, 2000; Nicholls, 2013. As shown by Menjívar and Lakhani (2016) in the American case, the reproduction of dominant discourses on deservingness has transformative effects: in the prospect of their legalisation, migrants undertake personal and social changes to fulfil their position of deserving citizens, and also broader life projects. Looking at the language requirements in the British case, Khan (2019) shows how migrants reproduce dominant monolingual discourses during the naturalisation process, but also—countering the assimilatory aims of these policies—use bilingual resources to navigate through. Morrice (2017) analyses women's experiences of 'British values' in the naturalisation process, showing that those who are comfortable with secular principles of liberal individualism are more likely to succeed in demonstrating the sort of 'shared values' that underpin dominant ideas about 'life in the UK'.

This work documents some of the ways in which, during naturalisation, people strategise, and then distinguish themselves as deserving relative to others, as well as shedding light on the potentially transformative, unforeseen effects of these processes. Yet drawing on work largely informed by intersectional scholarship, specifically Black feminist traditions, it becomes possible to show the ongoing colonial legacies that underpin policy instruments.

My own social location—as a white, cisgender, heterosexual, socio-economically privileged woman—requires me to engage respectfully with these traditions to avoid whitening and otherwise appropriating and distorting them

(Bilge, 2013). The erasure of Black women from popularised versions of narratives of activist movements—as discussed by activists leading Rhodes Must Fall Oxford—and from academic knowledge production through pervading forces of patriarchy and white supremacy (Collins, 1990; Rhodes Must Fall Oxford, 2018) requires this explicit reminder of the centrality of Black feminism in intersectionality and decolonial scholarship.

Beginning with and carrying this awareness through the study of citizenship and naturalisation recasts the requirement that migrants prove they 'deserve' citizenship. At the intersection of gender, sexuality, race and class, it is possible to identify the longstanding power relations at play in becoming a citizen and to challenge the concept of 'deservingness'. This opens the path to studying the ongoing colonial legacies that underpin policy instruments and shape experiences of the citizenship process as well as the kind of citizen one becomes (see also Palmary, in this volume).

Intersectionality and Deservingness

This approach calls into question the foundations of the study of citizenship and integration more generally. It draws into relief how the power relations of citizenship acquisition might shift beyond the deservingness of neoliberalism into the frame of nation-states created out of empires. The study of naturalisation could start instead with the recognition of citizenships, sovereignties and nations that have been displaced, through the violence of settler colonialism or of occupation and extraction from colonial metropoles.

Naturalisation and citizenship in Europe and North America have always been bound up and part of colonial histories though, as I will discuss below, current policies are often presented as something new. While the focus of this review is on Europe, I note that the white settler colonies of the United States and Canada are not exempt from this scrutiny. For instance, in her book *Mohawk Interruptus: Political Life Across the Borders of Settler States* Audra Simpson examines refusal by the Mohawks of Kahnawà:ke of 'the "gifts" of American and Canadian citizenship; they insist on the integrity of Haudenosaunee governance[1]' (2014: 7). This work forcefully demonstrates the politics of refusal of colonialism that survives in a settler form (2014: 7). 'There are alternative citizenships to the state that are structured in the present space of intracommunity recognition, affection, and care, outside of the logics of colonial and imperial rule' (2014: 109).

Postcolonial and decolonial scholarship challenge 'the insularity of historical narratives and historiographical traditions emanating from Europe' and the

[1]'The Mohawks of Kahnawà:ke are nationals of a precontact Indigenous polity that simply refuse to stop being themselves. In other words, they insist on being and acting as peoples who belong to a nation other than the United States or Canada…Their genealogical and political connectedness is part of a covenant – the decision-making Iroquois Confederacy called Haudenosaunee – which is made up of clans that spread across territory' (Simpson, 2014: 2).

'parochial character of arguments about the endogenous European origins of modernity' to instead consider 'the emergence of the modern world in the broader histories of colonialism, empire, and enslavement' (Bhambra, 2014: 115). Citizenship and naturalisation policies—and the constitutive hierarchies and formations of race, class, gender sexuality—are shaped directly by these broader histories.

The colonial entanglements of gender and sexuality have been powerfully explored through Jasbir Puar's concept of homonationalism (Puar, 2007). Sirma Bilge notes that the concepts of homonationalism (Puar, 2007), and relatedly gay imperialism (Haritaworn, Tauqir, & Erdem, 2008) and homopatriotism (Kuntsman, 2008) 'converge in their problematising of power relations undergirding current inclusions of some gay-and-lesbian sexualities to the national body – a homonormative addendum to heteronormative nationalism rather than its substitution' (Bilge, 2012: 305).

In the study of migration policy, these questions have been applied to critique the asylum process and who is deemed to deserve protection. For example, Francesca Ammaturo analyses the 'Pink Agenda' as a set of judicial, social and political instruments employed by both nation-states and international human rights institutions, such as the Council of Europe to reinforce a concept of European Sexual Citizenship that is strongly homonationalist in nature. The homonationalist paradigm of sexual citizenship is applied to strengthen the divide between tolerant and intolerant member states and to suggest the existence of a difference between a *queer-friendly* 'West' and homophobic and transphobic non-western countries (Ammaturo, 2015; Puar, 2007). 'Deserving' protection in asylum processes means meeting the demands for particular gender and sexual legibilities as articulated in a system which demands—and continually produces—identities 'in ways that are deemed legible to adjudicating states, (trans)national publics, and supranational bodies…even though there are no universal or transhistorical sexualities and genders into which diverse migrants could be expected to fit' (Luibhéid, 2014).

Challenging the gender binary and its origins in European 'modernity' shows the arbitrary and particular understandings of genders and sexualities that one needs to 'fit' in order to be deemed deserving of protection and asylum, which intersect with the race/class dynamics of border control. However, as Alyosxa Tudor argues 'Racial and religious categorisation systems, misogyny, homophobia and a compulsory gender binary have come into existence through modernist ideas of Europeanness and can therefore not be displaced to an elsewhere' (Tudor, 2018: 1069).

When bringing this questioning of gender and sexuality to citizenship as a legal status and a formal institution, Manuela Boatcă and Julia Roth argue that gender, revealed as a colonial construction, has informed 'the modern/colonial institution of citizenship from its emergence. Gender and citizenship have been embedded in highly complex ways with other dimensions of stratification and inequality such as racialisation and enslavement, placing men, women and transgender persons at very different positions in racialized colonial hierarchies....To this unequal and gendered mode of organization corresponded a structural distribution of unequal and gendered citizenship rights that shaped the modern/colonial institution of citizenship from its inception and that gets continuously rearticulated in new forms and understandings of citizenship today' (Boatcă & Roth, 2016).

Boatcă and Roth build on María Lugones' highly influential work on the coloniality of gender (2008), which problematises gender and points to its origins in European 'modernity'. 'Biological dimorphism, heterosexual patriarchy are all characteristic of what I call the "light" side of the colonial/modern organisation of gender. Hegemonically these are written large over the meaning of gender' (Lugones, 2008). However, this approach raises its own questions in relation to intersectional Black feminist scholarship and activism, including intersectional decolonial movements such as Rhodes Must Fall Oxford, discussed above. These points cannot be resolved in this chapter, but I note the risks of using the term 'decolonial' without robust interrogation of diverse underlying traditions. In elaborating her account of decolonial feminism, María Lugones (2010) argues that 'The intersection of "woman" and "black" reveals the absence of black women rather their presence...Intersectionality is important when showing the failures of institutions to include discrimination or oppression against women of color. But here I want to be able to think of their presence as being both oppressed and resisting. So, I have shifted to the coloniality of gender at and from the colonial difference' (2010: 757). This rejection of intersectionality as a site of Black women's resistance risks the erasure that Rhodes Must Fall Oxford activists fight against.

Instead, in this chapter, it not only the legacy of colonialism that is brought to the fore at the intersections of gender, sexuality, race and class in citizenship and naturalisation policies. Resistance and knowledge production can be centred. This chapter, therefore, explores an intersectional approach to the study of deserving citizenship. By this I mean an approach in which intersectionality not only names oppression and erasure, but is a space of knowledge production and resistance. I take an expansive view of 'women', to refer to cis and trans women, non-binary, gender non-conforming folks who identify with and relate to this term in diverse ways (see also Emejulu & Bassel, 2020). As in previous work, I define 'women of colour' as women who 'experience the effects of processes of racialisation, class and gender dominations as well as other sources of inequality, particularly hierarchies of legal status' (Bassel & Emejulu, 2017: 6).

Becoming a Deserving Citizen: The Naturalisation Process

The section focuses on the ways in which naturalisation processes require the 'right' kind of integration and of citizen. It shows how gender and sexuality intersect with race and class when migrants confront these requirements, and how a longer view of underlying power relations, forged under colonialism, recasts both the concept of deservingness and indeed, citizenship itself. The UK is used as an example.

I begin by rethinking the conventional chronology through which the 'story' of naturalisation in the UK is told, and how it is often presented as something new. This account often begins by observing the recent turn in Europe toward language and citizenship tests as new requirements for access to long-term residence or naturalization since the early 2000s. In the UK in 2005, in the context of intense debates on diversity and integration, the Labour government changed the process through which migrants become British citizens. In order to 'foster the "right kind" of integration' (MacGregor & Bailey, 2012: 366), it required migrants to demonstrate their ability to speak and read English, to take a citizenship test (the Life in the UK Test), and to attend a ceremony.[2] Migrants who want to become British citizens were to undertake a 'journey' (Home Office, 2011) that begins with the application for citizenship and ends with receiving the British passport. At the core of the 'journey' were the language test and the citizenship test in which candidates have to answer a set of questions in order to prove their knowledge of British 'history, culture and traditions' (Kostakopoulou, 2010; Schinkel & Van Houdt, 2010). These requirements have been analysed as a new paradigm in British immigration and integration policies (Joppke, 2007; Ryan, 2008; Vink & de Groot, 2010).

However, this focus on the period from 2005 onwards can be quite misleading. This becomes evident through a longer historical perspective that challenges the framing of history that erases colonialism. As Nadine El-Enany (2020) points out, it is in fact the 1981 Immigration Act and previous post-war immigration controls that 'crafted as white supremacist a Britain as possible, short of introducing an explicit "White Britain policy"' (2020: 130). Through the 1981 Immigration Act, 'racialised colony and Commonwealth citizens were told that their present or historical connections to the British Empire neither entitled them to Britishness as an identity nor access to Britain as a place' (2020: 130). This prevented entry to Britain but also deprived colonised populations 'of access to stolen colonial wealth as it manifests in Britain, in the form of infrastructure, employment, healthcare, welfare, safety and opportunity. Colonialism entailed the extraction and accumulation of material and temporal resources. Colonial theft of intangibles such as economic growth prospects, opportunities, life chances and futures is difficult to discern in colonial relationships which are traditionally understood to have come to an end…Yet Britain, built on wealth accumulated via colonial dispossession, *is* the spoils of colonialism' (2020: 130–131).

[2]These requirements were then extended to those who applied to Indefinite Leave to Remain.

The Life in the UK test, by requiring people (from former colonies and more widely) to prove their knowledge of 'British values', extends longstanding Othering processes. It does not create them anew.

Scholars in the UK have focused recently on the Hostile/Compliant Environment policy implemented through the 2014 and 2016 Immigration Acts, through which landlords, GPs, universities and other social actors have been turned into border guards through 'everyday bordering' (Jones et al., 2017; Yuval-Davis, Wemyss, & Cassidy, 2018). With a longer timeframe, law and policies that govern borders are seen to always have been 'hostile environments' and internal bordering is recast as 'a new mode of colonialism, producing and sustaining the post-colonial project of white Britain' (El-Enany, 2020: 132). These 'hostile environment' measures have included citizenship policies, which challenges the understanding of naturalisation policies as tools to 'foster the "right kind" of integration' (MacGregor & Bailey, 2012: 366). Instead of a tool for determining the mobility and integration of new arrivals, 'Immigration law… is the modality through which Britain transitioned from being a colonial power in the traditional overseas extractive sense, to a space of domestic colonialism masquerading as a post-colonial nation' (El-Enany, 2020: 131).

Gender and race are part of the colonial logic that has shaped citizenship processes through which one must now seek naturalisation. Following on from the 'war on terror', Muslim men are targeted through specific measures—citizenship deprivation, deportation and extradition—which Kapoor characterizes as '21st century state extremism' (Kapoor, 2018; Kapoor & Narkowicz, 2019). The persistence of colonial logic is demonstrated, for example, through the 'ease with which the two systems, colonial counterinsurgency and domestic racialised policing, ordinarily conceived of as distinct, operate together in tandem' (Kapoor, 2018: 56) indicating the long-standing processes through which citizenship has been securitised and deprived. 'The first incorporates counterterrorism policing techniques and infrastructures established in the British state system in response to Northern Ireland as well as to anti-colonial resistance in other parts of the Empire' (2018: 56). However, these origins are camouflaged and instead 'war stories about the need to secure identity are used to implement a host of citizenship policies but in ways that camouflage the very racialized, gendered, and classed nature of these policies' (Rygiel, 2006: 146).

The internal bordering logics El-Enany identifies underpin processes of naturalisation. Kapoor and Narkowicz note the rise of the 'character' of a subject across immigration and citizenship border policing. The requirement to prove good character 'was institutionalised not only in immigration legislation but in nationality legislation, effectively extending the border from the point of entry and admittance into the nation-state to a more fluid point of inclusion/exclusion encroaching into everyday life of racially marginalised communities' (2019: 654). This notion 'has been overlaid onto the already institutionalised, complex interconnections between race and criminality, supplementing the check against criminal conduct with a broader assessment of civil and community behaviour and standing' (2019: 653). They argue that this measure 'operates as a racialised exclusionary mechanism that works to supplement and legitimate race/class dynamics of immigration/citizenship exclusion...governing by "character" furthers the project of biopolitical disciplinary paradigms in order to sustain and enhance systems of racial exclusion' (2019: 653).

Within naturalisation processes, this colonial scrutiny to produce the right kind of citizens manifests in citizenship tests that will correct or exclude incompatible values, through tests and requirements. Even the language requirement is, itself, arguably a tool of empire. Recent studies of the UK note that citizenship testing, and the language requirement itself, contain traces of imperialism in their formation through the colonial spread of English (Byrne, 2014; Fortier, 2018; Khan, 2019). For example, in her study of the UK naturalization process, Anne-Marie Fortier reveals the role of colonial legacies 'lining the language requirements in Britain' (2018: 1266). Testing language skills 'is a willing of the nation in the face of fears about cohesion and national security, much like the standardisation of English as the national language took place in response to 19th-century imperial expansion' (2018: 1265). The effects of this testing are unequal. Heller and McElhinney (2017) identify how countries give citizenship preference to nationals from countries that share those languages; thus, those who fall outside of this linguistic classification face more onerous routes to becoming citizens (for further discussion, see Khan, 2019).

In addition to knowledge of English, knowledge of 'British values' need to be displayed through the Life in the UK test, and performing deservingness in interactions in social, political and economic spheres of everyday life. The gender and sexuality of racialized Others become explicit areas for testing and screening. For instance, the third edition of the Life in the UK handbook issued by the Home Office (2013) for applicants for naturalisation or permanent residency to review in order to pass the multiple choice 'Life in the UK' test includes a section on 'Fundamental Principles'. This section notes: 'UK laws ensure that people are not treated unfairly in any area of life or work because of their age, disability, sex, pregnancy and maternity, race, religion or belief, sexuality or marital status'. The following section discusses domestic violence, then Female Genital Mutilation [sic] and forced marriage (2013: 149–150), presumably as key illustrations as violations of these 'Fundamental Principles' (see also Morrice, 2017). Alyosxa Tudor notes the ways in which this 'constant externalisation prevents seeing sexism, misogyny, homo/trans/queerphobia and anti-Semitism as problems that everyone must address as inherent to the nation states, institutions, families and communities they live in. Solely throwing this struggle on migrant, Muslim, Black and brown bodies and the Global South is a migratist and/or racist instrumentalisation' (2018: 1069).

Muslim women, in particular, have been repeatedly targeted in the UK in high-profile calls to enable integration and counter-extremism, with a specific focus on their English language proficiency. The political response to 'riots' in Northern English cities in 2001 linked understanding English with social cohesion and ultimately led to the naturalisation process in the United Kingdom which involves tests, a ceremony, and administrative procedures. English language was to act as the 'panacea' (Greenwood & Robins, 2002: 507; Sasse, 2005: 678) to the ills caused by a lack of integration and serve as an antidote to radicalisation (Khan, 2019). The frame is of 'new migrants' for whom the 'right kind of integration' must be facilitated/enforced.

This narrative continues on repeat. For example, in 2016, in the context of young British people travelling to Syria, Prime Minister David Cameron called for an end to the 'passive tolerance' of separate communities which left many Muslim women facing discrimination and social isolation, and proposed the launch of a £20 million language fund (The Guardian, 2016). Thus, migrant women would need to demonstrate sufficient English if in the United Kingdom on a five-year spousal visa, with the possibility that they might be forced to leave if they were unable to provide evidence of improvement in the language. The underlying rationale was that lack of English at home was leading to young people's rejection of British values. The shift this time, in 2016, was to focus on young people's travel to Syria to join Da'esh/ISIS. In this way, naturalisation policy converges with counter-extremism: the assumption is that learning English will equip people with the critical thinking that enables them to resist radicalisation.

From this perspective, in the UK, Muslim migrant women (and particularly Pakistani and Bangladeshi Muslim migrant women) are victims of social isolation who need English to do their job of socialising young people better (Bassel, 2016; more generally see Luibhéid, 2006; Tyler, 2013). They are the vehicle for this panacea of English language as biological and social 'reproducers' of the nation (Anthias & Yuval-Davis, 1989; Yuval-Davis, 1997), who will, it is argued, raise young people with the correct values (see also Yuval-Davis, Anthias, & Kofman, 2005). The role of a decade of austerity measures and cuts to language provision is invisible—this is a pathological, colonially rooted understanding of deficiency that intersects with neoliberal welfare conditionalities. More generally 'class as a visible marker of differentiation in immigration policy is increasingly prominent, while intersecting with differentiation on the basis of ethnicity' (Sirriyeh, 2015: 236). 'Deservingness' is thereby anchored within domestic welfare conditionalities—which cut more deeply under a decade of austerity following the 2008 financial crisis—and 'a normative and moralising discourse of the family' (2015: 243).

Deservingness requires activating colonial tropes: as victims to be 'saved' from communities (Bassel & Emejulu, 2017), or 'deserving' to be granted citizenship when reproducing European nations correctly (Farris, 2016). The UK is but one example, with its own new and old austerities and hostile environments under different names and guises. The 'new' European turn in civic integration policies is 'animated by an older colonial anxiety' (2016, n.p.), with, as noted above, 'a homonormative addendum to heteronormative nationalism rather than its substitution' (Bilge, 2012: 305).[3] Welfare conditionalities shape the movements and lives of individuals and families who are migrant and citizens.[4] Challenging deservingness exposes these fissures in a move toward the rejection of colonial tropes and ongoing exclusion that are shaped within and across national borders.

Conclusion: Challenging Deservingness at the Intersections

This chapter has sought to challenge notions of 'deservingness' at the intersections by drawing on scholarship that locates current naturalisation processes in colonial starting points. Categories of colonial knowledge production (Simpson, 2017)—gender and sexuality, race, class, and indeed citizenship itself as a legal status—remain intact if deservingness is only analysed within the horizon of neoliberal subjectivities and policies, such as the austerity measures following 2008 in the UK or its official Hostile Environment policies of 2014 onwards. Instead, this chapter underscores work at the intersections which centres colonial legacies rather than erasing them. Thus, deservingness becomes another entry point into challenging ongoing relations of exclusion and domination forged under empire.

Different horizons can open as a result of such a challenge. Scholars have recently amplified the call to refashion legal frameworks governing both citizenship and migration based on the recognition of colonialism and neocolonialism. E. Tendayi Achiume (2019) argues for an understanding of 'migration as decolonisation'. This thesis 'posits imperial interconnection and the extraterritorial co-sovereignty relations it entails as the baseline from which Third World migration to the First World must be assessed and negotiated. What is vital here is that First and Third World peoples, as co-sovereigns, bear an equal right and stake in First World national borders' (2019: 1567). Because they are not 'political strangers…First World nation-states have no right to exclude Third World peoples, and creating a world that reflects this fact requires a complete reimagining of national borders and the institutions of political inclusion' (2019: 1573). Citizenship plays a modest, and remedial rather than fully reparatory role in this reimagining, granted to recognise 'cosovereign status and correct an unethical set of power relations by redistributing the power to achieve greater prospects of equality. There is a

[3] See also Puar (2015) on trans(homo)nationalism, linking trans hailing and normativisation by the US state and homonationalism (2015: 46).

[4] See also Sara Farris' (2017) concept of 'femonationalism' which addresses the rearticulation of male Other depicted as sexual threats and female Other as sexual victims and the property of western 'saviours' in the changed context of neoliberalism and rising Islamophobia (2017: 11–12).

corrective justice dimension to this, in the sense that redistribution is made necessary by a historical wrong and its continuing, present-day legacy' (2019: 1553). This reimagining is a far cry from the violence of border control and its extension into naturalisation in Europe today.

References

Achiume, E. (2019). Migration as decolonization. *Stanford Law Review, 71*, 1509–1574.

Ammaturo, F. (2015). The 'pink agenda': Questioning and challenging European homonationalist sexual citizenship. *Sociology, 49*(6), 1151–1166. https://doi.org/10.1177/0038038514559324

Anthias, F., & Yuval-Davis, N. (1989). *Women, nation, state*. Basingstoke: Macmillan.

Bassel, L. (2016). *The Casey review on opportunity and integration: Reinventing the wheel: Discover society*. Available at: https://discoversociety.org/2016/12/09/the-casey-review-on-opportunity-and-integration-re-inventing-the-wheel/.

Bassel, L., & Emejulu, A. (2017). *Minority women and austerity: Survival and resistance in France and Britain*. Bristol: Policy Press.

Bassel, L., Monforte, P., Bartram, D., Khan, K., & Misztal, B. (2017). *The UK citizenship test process: Exploring migrants' experiences*. Final Project Report. Leicester: University of Leicester. https://www2.le.ac.uk/departments/sociology/research/uk-citizenship-process/final-report.

Bassel, L., Monforte, P., & Khan, K. (2018). Making political citizens? Migrants' narratives of naturalization in the United Kingdom. *Citizenship Studies, 22*(3), 225–242. https://doi.org/10.1080/13621025.2018.1449808.

Berger, S. (2009). Un-worthy: Latina battered immigrants under VAWA and the construction of neoliberal subjects. *Citizenship Studies, 13*(3), 201–217.

Bhambra, G. (2014). Postcolonial and decolonial dialogues. *Postcolonial Studies, 17*(2), 115–121. https://doi.org/10.1080/13688790.2014.966414.

Bilge, S. (2012). Mapping Québécois sexual nationalism in times of 'crisis of reasonable accommodations'. *Journal of Intercultural Studies, 33*(3), 303–318. https://doi.org/10.1080/07256868.2012.673473.

Bilge, S. (2013). Intersectionality undone: Saving intersectionality from feminist intersectionality studies. *Du Bois Review, 10*(2), 405–424.

Boatcă, M., & Roth, J. (2016). Unequal and gendered: Notes on the coloniality of citizenship. *Current Sociology, 64*(2), 191–212.

Byrne, B. (2014). *Making citizens*. Basingstoke: Palgrave Macmillan.

Chauvin, S., & Garcés-Mascareñas, B. (2012). Beyond informal citizenship: The new moral economy of migrant illegality. *International Political Sociology, 6*(3), 241–259.

Collins, P. H. (1990). *Black feminist thought: Knowledge, consciousness and the politics of empowerment*. New York: Routledge.

Coutin, S. (2000). *Legalizing moves: Salvadoran immigrants' struggle for U.S residency*. Ann Arbor: University of Michigan Press.

El-Enany, N. (2020). *(B)ordering Britain: Law, race and empire*. Manchester Manchester University Press.

Emejulu, A., & Bassel, L. (2020). The politics of exhaustion. *City, 24*(1–2), 400–406 https://doi.org/10.1080/13604813.2020.1739439.

Farris, S. (2016). Dispossessing the private sphere? Civic integration policies and colonial legacies. *Darkmatter: In the ruins of imperial culture, 14*(4). http://www.darkmatter101.org/site/2016/05/16/dispossessing-the-private-sphere-civic-integration-policies-and-colonial-legacies/.

Farris, S. (2017). *In the name of women's rights: The rise of femonationalism*. Durham: Duke University Press.

Fassin, D., & Mazouz, S. (2007). Qu'est-Ce Que Devenir Français? La Naturalisation Comme Rite D'institution Républicain [What is it to become French? Naturalization as a rite of Republican Institution]. *Revue Française De Sociologie, 48*(4), 723–750.

Fortier, A.-M. (2008). *Multicultural horizons: Diversity and the limits of the civil nation*. London: Routledge.

Fortier, A.-M. (2013). What's the big deal? Naturalization and the politics of desire. *Citizenship Studies, 17*(6–7), 697–711.

Fortier, A.-M. (2018). On (not) speaking English: Colonial legacies in language requirements for British citizenship. *Sociology, 52*(6), 1254–1269.

Greenwood, J., & Robins, L. (2002). Citizenship tests and education: Embedding a concept. *Parliamentary Affairs, 55*(3), 505–522.

Haritaworn, J., Tauqir, T., & Erdem, E. (2008). Gay imperialism: The role of gender and sexuality discourses in the 'war on terror'. In E. Miyake & A. Kuntsman (Eds.), *Out of place: Queerness and raciality* (pp. 71–95). York: Raw Nerve Books.

Heller, M., & McElhinny, B. (2017). *Language, capitalism, colonialism: Toward a critical history*. New York: University of Toronto Press.

Jones, H., Gunaratnam, Y., Bhattacharyya, G., Davies, W., Dhaliwal, S., Forkert, K., et al. (2017). *Go home? The politics of immigration controversies*. Manchester: Manchester University Press.

Joppke, C. (2007). Beyond national models: Civic integration policies for immigrants in Western Europe. *West European Politics, 30*(1), 1–22.

Kapoor, N. (2018). *Deport, deprive, extradite: 21st Century state extremism*. London: Verso Books.

Kapoor, N., & Narkowicz, K. (2019). Characterising citizenship: Race, criminalisation and the extension of internal borders. *Sociology, 53*(4), 652–670. https://doi.org/10.1177/0038038518820250.

Khan, K. (2019). *Becoming a citizen: Linguistic trials and negotiations*. London: Bloomsbury.

Kostakopoulou, D. (2010). Matters of control: Integration tests, naturalization reform and probationary citizenship in the United Kingdom. *Journal of Ethnic and Migration Studies, 36*(5), 829–846.

Kuntsman, A. (2008). Genealogies of hate, metonymies of violence: Immigration, homophobia, homopatriotism. In E. Miyake & A. Kuntsman (Eds.), *Out of place: Queerness and raciality* (pp. 97–118). York: Raw Nerve Books.

Lugones, M. (2008). The coloniality of gender. *Worlds & Knowledges Otherwise, 2*(Spring), 1–17.

Lugones, M. (2010). Toward a decolonial feminism. *Hypatia, 25*(4), 742–759.

Luibhéid, E. (2006). Sexual regimes and migration controls: Reproducing the Irish nation-state in transnational contexts. *Feminist Review, 83*(1), 60–78.

Luibhéid, E. (2014). Afterword: Troubling identities and identifications. *Sexualities, 17*(8), 1035–1040.

MacGregor, S., & Bailey, G. (2012). British citizen or other? Reflections on new labour's reforms to the UK citizenship process. *British Politics, 7*(4), 365–388.

Mazouz, S. (2012). "Une faveur que vous a accordée la République" Comment la naturalisation se mérite. In D. Fassin & J. Eideliman (Eds.), *Economies morales contemporaines* (pp. 137–154). Paris: La Découverte.

Menjívar, C., & Lakhani, S. (2016). Transformative effects of immigration law: Immigrants' personal and social metamorphoses through regularization. *American Journal of Sociology, 121*(6), 1818–1855.

Monforte, P., Bassel, L., & Khan, K. (2019). Deserving citizenship? Exploring migrants' experiences of the 'citizenship test' process in the United Kingdom. *British Journal of Sociology, 70*(1), 24–43.

Morrice, L. (2017). British citizenship, gender and migration: The containment of cultural differences and the stratification of belonging. *British Journal of Sociology of Education, 38*(5), 597–609.

Nicholls, W. (2013). Making undocumented migrants into legitimate political subjects: A comparison of France and the United States. *Theory, Culture and Society, 30*(3), 82–107.

Office, Home. (2011). *Life in the UK test: A journey to citizenship*. London: TSO.

Office, Home. (2013). *Life in the United Kingdom: A guide for new residents* (3rd ed.). London: TSO.

Puar, J. (2007). *Terrorist assemblages: Homonationalism in queer times*. Durham, NC: Duke University Press.

Puar, J. (2015). Bodies with new organs: Becoming trans, becoming disabled. *Social Text, 33*(3 [124]), 45–73.

Rhodes Must Fall Oxford. (2018). *Rhodes must fall: The struggle to decolonise the racist heart of empire*. London: Zed Books.

Rose, N. (1999). Inventiveness in politics. *Economy and Society, 28*(3), 467–493.

Ryan, B. (2008). Integration requirements: A new model in migration law. *Journal of Immigration Asylum and Nationality Law, 22*(4), 303–316.

Rygiel, K. (2006). Protecting and proving identity: The biopolitics of waging war through citizenship in the ost-9/11 era. In K. Hunt & K. Rygiel (Eds.), *(En)gendering the war on terror: War stories and camouflaged politics* (pp. 145–168). Aldershot, Hampshire: Ashgate.

Sasse, G. (2005). Securitization or securing rights? Exploring the conceptual foundations of policies towards minorities and migrants in Europe. *Journal of Common Market Studies, 43*(4), 673–693.

Schinkel, W., & Van Houdt, F. (2010). The double helix of cultural assimilationism and neoliberalism: Citizenship in contemporary governmentality. *The British Journal of Sociology, 61*(4), 696–715.

Simpson, A. (2014). *Mohawk interruptus: Political life across the border of settler states*. Durham, NC: Duke University Press.

Simpson, L. B. (2017). *As we have always done: Indigenous freedom through radical resistance*. City: University of Minnesota Press.

Sirriyeh, A. (2015). 'All you need is love and £18,600': Class and the new UK family migration rules. *Critical Social Policy, 35*(2), 228–247.

The Guardian. (2016, January 18). Muslim women to be taught English in £20m plan to beat "backward attitudes". *The Guardian*. Available at: https://www.theguardian.com/uk-news/2016/jan/18/muslim-women-to-be-taught-english-in-20m-plan-to-beat-backward-attitudes.

Tudor, A. (2018). Cross-fadings of racialisation and migratisation: The postcolonial turn in Western European gender and migration studies. *Gender, Place & Culture, 25*(7), 1057–1072. https://doi.org/10.1080/0966369x.2018.1441141.

Turner, J. (2014). Testing the liberal subject: (In) security, responsibility and "self-improvement" in the UK citizenship test. *Citizenship Studies, 18*(3–4), 332–348.

Tyler, I. (2013). *Revolting subjects: Social abjection and resistance in neoliberal Britain*. London: Zed Books.

Villalon, R. (2010). *Violence against Latina immigrants: Citizenship, inequality, and community*. New York: New York University Press.

Vink, M., & de Groot, G. (2010). Citizenship attribution in Western Europe: International framework and domestic trends. *Journal of Ethnic and Migration Studies, 36*(5), 713–734.

Yuval-Davis, N. (1997). *Gender and nation*. London: Sage.

Yuval-Davis, N., Anthias, F., & Kofman, E. (2005). Secure borders, safe haven and the gendered politics of belonging: Beyond social cohesion. *Ethnic and Racial Studies, 28*(3), 513–535.

Yuval-Davis, N., Wemyss, G., & Cassidy, K. (2018). Everyday bordering, belonging and the reorientation of British immigration legislation. *Sociology, 52*(2), 228–244.

Leah Bassel (pronouns: she/her) is Professor of Sociology at the University of Roehampton, UK. Leah's research interests include the political sociology of migration, intersectionality and citizenship. Leah's books include *Refugee Women: Beyond Gender versus Culture* (Routledge, 2012), *The Politics of Listening: Possibilities and Challenges for Democratic Life* (Palgrave, 2017), and *Minority Women and Austerity: Survival and Resistance in France and Britain* co-authored with Akwugo Emejulu (Policy Press 2017). Leah has led projects funded by the Open Society Foundation, Leverhulme Trust, Economic and Social Research Council and the British Academy. She is Associate Editor of the journal *Citizenship Studies*. Before pursuing an academic career, Leah was an emergency outreach worker in Paris, where she provided humanitarian assistance to asylum seekers and created a circus camp project for refugee youth.

CHAPTER 31

The Impact of Immigration Regulations and Visa Policies on the *Gendered* Nature of International Migration

Luisa Feline Freier and Nieves Fernández Rodríguez

Abstract Visa requirements are a strong indicator of the objectives and priorities of a country's immigration policy. This chapter explores the gendered dimensions of work, student, and family reunification visas. The requirements commonly in place for different residence categories tend to appear neutral in gender terms. However, there is a consensus in the literature that they impact immigration flows and integration processes, which include the masculinisation and feminisation of immigration flows, as well as the reinforcement of gender roles due to the interaction between visa policies and labour segregation. To an extent, these effects go back to gender-gaps regarding access to the workforce and higher education, which persist in many sending countries in the Global South. We juxtapose these findings with recent migrant stock data, and explore how the geographic bias of studies on the gendered effects of visa policies on Northern receiving states limits theory development. We argue that future research should be wider in geographical and conceptual scope to include South-South migration, LGBTQ+ migrants, and tourist visas.

L. F. Freier (✉)
Universidad del Pacífico, Lima, Peru
e-mail: lf.freierd@up.edu.pe

N. F. Rodríguez
WZB, Berlin, Germany

C. Mora and N. Piper (eds.), *The Palgrave Handbook of Gender and Migration*, https://doi.org/10.1007/978-3-030-63347-9_31

Introduction

According to the International Organization for Migration (IOM), the share of female international migrants has not changed significantly in the past 60 years. In 2019, this figure declined from 49.1%, in 2000, to 47.9% (UN DESA, 2019). Narrowing down on the gendered geographies of international migration across different world regions, in Latin America, the migration stock is almost equal in gender terms, with men taking a slight lead with 50.1% as of 2019. At the same time, in Asia and Africa, migration remains largely male-dominated with 58.2 and 53% respectively. On the other hand, the migration stock is slightly more female-oriented in Oceania (50.4%), North America (51.8%), and Europe (51.4%).

In spite of the growing recognition of the importance of the gendered nature of international migration, little attention has been paid to legal analyses of the topic (as discussed by DeLaet, 1999; Kofman, Raghuram, & Merefield, 2005; Pedone & Gil, 2008; Piper, 2006), with the notable exception of refugee law (see Calavita, 2006). In this chapter, we critically discuss the existing literature on the gendered effects of work, student, and family reunification visas, juxtaposing the dominant theoretical hypotheses with recent data. More specifically, we focus on two groups of studies: those dealing with the impact of visa policies on the gendered nature of migration flows, and those that analyse the effects visa policies have on the reinforcement of gender roles—within the labour market and the household. We further point out the gendered geographies of international migration and the fact that the effects of visa policies remain understudied in the context of South-South[1] migration. We conclude by pointing out several avenues for further research.

Studying Law from a Gender Perspective

Analysing law—including migration law and visa regulations—from a gender perspective involves the analysis of how legal norms reflect gender roles and power relationships, and, thus, often perpetuate the vulnerability in which women are socially situated (see Crawley, in this volume). This can occur through direct and indirect discrimination. Direct gender discrimination refers to legal norms making explicit distinctions on the basis of gender; whereas indirect discrimination refers to the disparate impact of the application of these norms on women and men, although they appear neutral on the books (García-Castrillón, 2002). Even though democratic constitutions increasingly forbid discrimination on the basis of gender—except in its positive sense when it has a provisory nature—the indirect discrimination of women, through legal norms, still occurs frequently. Often, there is no intention to harm women;

[1]The differentiation between countries in the Global North and South is misleading in that it does not fully correspond to historic and geographic realities, and poses the normative risk of naturalising a developmental divide between the two hemispheres. Nevertheless, this conceptual differentiation is heuristically useful and, thus, applied in this chapter.

rather, discrimination occurs as a result of the disadvantaged social position they find themselves in.

A critical gender perspective dominated the field of legal studies in the 1980s, as a result of radical feminists' critiques. These theorists conceptualised the law, and its interpretation by judges, as a tool to perpetuate patriarchal norms and gender roles. For example, Catharine MacKinnon (1987) states that the pretence of neutrality, characterising law and jurisprudence in liberal systems, tends to hide discrimination towards women. According to this approach, the idea of neutrality is intrinsically male, as power relations between genders position men as knowledge creators (Danielsen, 1988).

Stemming from the aforementioned perspective, legal migration norms also reflect this gender-bias. Immigration regimes determine the opportunities and experiences of both male and female migrants, and, as they originate and operate in unequal societies, contribute to migration's *gendered* nature. Conversely, given that migration is a gendered process, immigration legislation, or policy, is never entirely a gender neutral affair (Dodson & Crush, 2006). Thus, Fitzpatrick and Kelly (1998) argue that migration policy can reproduce and exacerbate the social and cultural inequalities that disempower women migrants, and Iredale (2005) finds that "the selection policies of immigrant-receiving countries are influenced, in part, by gender equity and attitudes to the roles of men and women in society" (p. 158). Furthermore, immigration regulations often limit women's rights and reinforce their financial dependency on men and labour precariousness (Fitzpatrick & Kelly, 1998; van Walsum & Spijkerboer, 2007).

Over the years, the literature has identified two aspects of migration law presenting barriers to women. First, men more readily meet entry and residence requirements under skilled work regimes (see Boucher, in this volume) Second, there tends to be more restrictions on the rights of migrants in categories in which women are overrepresented, such as family migration, which encompasses marriage migration. In the case of asylum seekers and refugees, gendered discrimination has been studied regarding both the experience of forced migration, as well as refugee law (Anker, 2002; Calavita, 2006; Pittaway & Bartolomei, 2001). Overall, this literature focuses on Northern, or Western, receiving states, with very few studies on the gendered effects of immigration policies in South-South contexts.

The Study of Visa Policies from a Gender Perspective

Visa policies are intrinsically selective: they allow the entry of certain individuals and reject others. Thus, "borders are rarely open or closed per se, but only with regard to specific persons and types of mobility" (Mau, 2010: 340). A growing number of scholars point out the persistence of negative selectivity based on nationality, race, and religion, applied to visa policies in the context of the securitisation of migration (Bourbeau, 2011; Freier & Luzes, 2020; Garner, 2007; Lentin, 2011). However, only a few authors have highlighted both the historic and contemporary discriminatory nature of visa policies regarding gender.

Until the mid-twentieth century, visa policies included direct forms of discrimination on the basis of gender, such as the prohibition of entry of economically independent women, stigmatized as potential prostitutes, to countries such as Australia or the United States (Calavita, 2006; Fincher, 1997; Harzig, 2003). Although such cases of explicit discrimination towards women have gradually disappeared, the bulk of the literature on visa policies and gender, written since the 1990s, acknowledges the existence of indirect forms of gender discrimination in immigration norms (Dodson, 2001; Dodson & Crush, 2004; Kofman & Raghuram, 2006, 2011; Pedone & Gil, 2008). Historically, these studies focus on immigrant-receiving countries in the Global North, and highlight that such discrimination occurs through the interaction between legal norms and labour market demands.

Although visa policies, today, are predominantly neutral in gender terms, the social construction of gender roles assigned to women and men, and the division of labour, imply a different impact on both genders. Traditionally, work visas restricted female migration, as women did not have the skills valued in the paid labour market (Harzig, 2003). The transformation of Western societies, towards more egalitarian values, has alleviated some of the discriminatory effects of visa policies. Nonetheless, the literature still finds that visa policies of Northern receiving countries tend to facilitate the presence of immigrant men, over women, as wide gender gaps in women's access to the workforce and higher education persist in many sending countries, often located in the Global South (Iredale, 2005; Kofman et al., 2005).

Visa Policies and Gendered Migration Flows

The existing literature on the impact of visas on gendered migration flows predominantly argues that visa policies determine the *gendered* nature of migration flows in two ways. First, work visa requirements, and the criteria embodied in legal norms, make the access to permanent or temporary visa permits easier for men, especially when they entail points-based or quota systems (Boucher, 2007, 2009; Dauvergne, 2000; Dodson, 2001; Dodson & Crush, 2006; Fincher, 1997; Fincher, Foster, & Wilmot, 1994; Harzig, 2003; Hyndman, 1999; Iredale, 2005; Kofman, 2012; Kofman & Raghuram, 2011; Kofman et al., 2005; Yuval-Davis, Anthias, & Kofman, 2005). Second, there are more restrictions imposed on migrant categories where women are over-represented, such as family migration, both regarding secondary applicants and reunified migrants (Boucher, 2007, 2009; Dauvergne, 2000; Dodson & Crush, 2006; Fincher, 1997; Kofman, 2004; Kofman et al., 2005; Pedone & Gil, 2008).

Gender and Work Visas

Scholars studying work visas, as determinants of the *gendered* nature of migration flows, typically stress the interaction between labour market demands and visas, arguing that states encourage the entry of workers to cover labour demands through the implementation of tailored visa policies. Although this phenomenon has evolved over time, and differs across geographic areas, the literature clearly identifies the dominance of men in the fields of some industries, such as information technology (IT) and engineering and sciences, and that of women in professions related to care-giving, such as health care activities (Kofman, 2000, 2012; Kofman & Raghuram, 2006; Kofman et al., 2005; Raghuram, 2004; Toksöz & Ulutaş, 2012). Thus, if the labour market demands IT workers, visa policies encourage male flows; on the contrary, if the labour market demands health care workers, visa policies facilitate the arrival of female migrants. As Kofman et al. (2005) point out, labour shortages evolve over time. The authors show that, in the UK, for much of the 1990s, the valued skills were those associated with trade and technology, which privileged men, whereas towards the end of the 1990s, labour shortages in welfare sectors, such as health and education, led to increases in work permits for women.

Visa policies never perfectly reflect labour market demands. For example, although there often is labour market demand for low-skilled workers, work visas are more commonly issued to high-skilled workers (Iredale, 2005). The set of criteria for the points-based system, such as education level, previous occupation, income, years of work experience, age, language skills, family ties, or economic resources (Dauvergne, 2000; Fincher, 1997; Fincher et al., 1994) has been referred to as "the principle device" for selecting who will be given a work visa in Northern receiving states (Dauvergne, 2000). The aforementioned set poses another indirect form of discrimination towards women, preventing many from migrating regularly (Boucher, 2007; Dauvergne, 2000; Fincher et al., 1994; Harzig, 2003).

As Fincher (1997) states, the criteria to grant work visas are often determined by trade unions dominated by men, and the occupations labelled as "highly skilled" do not usually include those concerned with raising children. Women, and especially women from countries in the Global South, do not have the same opportunities to access education as men, and are, thus, less likely to classify as "highly skilled workers" in Northern receiving states (Dodson, 2001; Dodson & Crush, 2006; Iredale, 2005). English language skills are another requirement discouraging female migration to Canada and Australia, as these women often carry economic and familiar burdens, which prevent them from attending classes (Boucher, 2007; Hyndman, 1999). The lack of consideration of part-time employment, as previous labour experience, as well as the requirement of earning high salaries, for example under Australian and British legislation, is prejudicial to women (Boucher, 2007; Dauvergne, 2000; Yuval-Davis et al., 2005).

Although women continue to be quasi-exclusively treated as low-skilled labour, some authors find that, since the 1970s, the number of those considered as "highly skilled workers" has markedly increased (Boucher, 2007; Iredale, 2005; Kofman & Raghuram, 2011; Kofman et al., 2005). Kofman (2012) argues that, in these cases, women are more likely to fulfil the criteria of educational background and language skills, rather than occupation, years of work experience, or income levels. Whereas women worldwide have been gradually acquiring higher levels of education, they still often occupy jobs in less well-paid sectors than men, such as health, education, and social work (Kofman, 2012; Kofman & Raghuram, 2011; Kofman et al., 2005).

Moreover, women still have to interrupt their professional lives to raise their children. The intersection between gender income levels, access to education, and other relevant aspects to obtain labour visas differs across country of origin, especially between the Global North and South (Iredale, 2005; Kofman et al., 2005). Nevertheless, as women continue to occupy less skilled jobs than men—partly due to the slower advances in gender equality in some Global South countries—the lack of recognition of the value of less skilled work still harms women's opportunities to migrate under work visas (Yuval-Davis et al., 2005).

Due to its geographic bias on Northern receiving states, existing literature on work visas and gender errs when it conceptualises work visa regimes predominantly as quota- or point-based systems for highly skilled immigration (Bertoli & Fernández-Huertas, 2012). In the Global South, work visas often play out differently. In Asia, the estimated stock of male international migrants grew by over 80%, between 2000 and 2019, mainly due to an increasing demand for low-skilled male workers in oil-producing Western Asian, or Arab countries (UN DESA, 2019). Similar developments have been observed in Africa, which experienced more growth among male migrants during 2000–2017 (41.8%), than female migrants (37.1%) (UN DESA, 2017).

Gender and Student Visas

Regarding international student visas, Myers and Griffin (2019) stress the significant underrepresentation of women among international higher education students, yet with significant variation among different Northern receiving countries (see Raghuram and Sondhi, in this volume). In the case of the United States, men have accounted for a majority of newly enrolled foreign students at US colleges and universities since 2004, with the gender gap widening in recent years (Pew Research Center, 2017). Although, officially, international students are considered temporary migrants, and often start their studies abroad with the intention of returning home upon graduation, many eventually become immigrants (Hazen & Alberts, 2006). This transition is often made through special visas, such as the F1 visa in the US, or the Graduate Job Search Visa in New Zealand (OECD, 2014). The gendered geography of international student visas, thus, feeds into the gender gap of highly skilled migration, likely due to similar structural reasons in important sending countries, such as China and India in the case of the United States (Myers & Griffin, 2019; Pew Research Center, 2017).

Gender and Family Visas

In spite of being the dominant settlement category, research has neglected family-related migration, which encompasses the following main categories: (1) family members sponsored by the primary applicant; and (2) family formation, which comprises migrants' fiancés sponsored by a second-generation migrant or citizen (Kofman, 2004; Kofman et al., 2005). Secondary applicants are granted their visas at the same time as the primary applicant; whereas the process of reunification takes place after the sponsor received his or her visa. At the same time, some studies have pointed out that family formation visas can be used as entry strategies for low-skilled labour migrants, for example in Southeast Asia (Chung, Kim, & Piper, 2016).

Family-related migration tends to be highly feminised. Given the male bias in the issuance of work visas, men disproportionally fall under the category of primary applicants (Boucher, 2007; Dauvergne, 2000; Kofman et al., 2005). Dodson and Crush (2004) argue that the prohibition of work for reunified migrants largely prevents men from migrating as secondary or reunified family migrants, as social norms continue to stigmatise unemployed men. Moreover, it is harder for women to meet the criteria (income level) to bring in their family members (Kofman, 2004). Other authors have noted that, while it remains true that more women immigrate under the spousal or family category than men, women are also starting to sponsor their husbands (Boucher, 2007; Pedone & Gil, 2008). For Calavita (2006), Pedone and Gil (2008), and Toksöz and Ulutaş (2012), this phenomenon is the result of an increasing demand for transnational domestic work, which stems from the process of women's incorporation into the workforce in countries where the welfare state is not consolidated, such as in Spain, Italy, or Turkey. Boucher (2007), in contrast, focuses more on the significant feminisation of skilled migration in the health science professions over the past decades.

Despite the increasing number of high-skilled migrant women worldwide (Boucher, 2007; Iredale, 2005; Kofman & Raghuram, 2011; Kofman et al., 2005), the literature has yet to address the circumstances under which they bring their spouses along. Although such reunification might be discouraged by patriarchal gender roles, the lack of studies does not necessarily imply that high-skilled migrant women do not sponsor their spouses. Kofman (2012: 74) suggests that the absence of such analyses may reflect the lack of an analytical link between women and the "the world of production," and that researchers themselves see women as connected to social welfare and integration.

Regarding the family reunification rights of LGBTQ+ migrants, a number of authors echo reports of human rights organisations denouncing discriminative family reunification policies against same-sex couples in the United States

(Chávez, 2010; Luibhéid, 2008). In the EU, on the other hand, there have been advances in the recognition of family reunification rights to same-sex couples (Tryfonidou, 2019). For instance, the European Court of Justice has recently ruled that the term "spouse"—for the purposes of EU law—includes the same-sex spouse of an EU citizen who has moved between EU Member States.[2]

Visa Policies and Gender Roles

Several authors point out that visa policies can deepen gender hierarchies during the migration experience. In what follows, we will distinguish between two broad phenomena: how visa policies reinforce gender roles and can create situations of vulnerability, and the various ways visa policies reinforce women's financial dependency on men.

Visa Policies and Gender Dependency

Migration can offer women an opportunity to emancipate themselves from men. However, the migration experience can also entail the opposite. Historically, family reunification visa programmes were implemented to facilitate the migration of female spouses of male migrant workers to bring them emotional stability, and, thus, served as a tool for integration. In the case of Europe, Yuval-Davis et al. (2005) argue that family migration is increasingly regarded as negative, as it is associated with traditional practices, such as arranged marriages, authoritarian gender and generational relations, and religious practices.

Along the same lines, Boucher (2007), Chung (2018), Fincher (1997), Kofman et al. (2005), and Pedone and Gil (2008) point out that family migration policies often do not acknowledge changing gender roles. Rather, women are implicitly conceived as in charge of reproduction and raising children; whereas men are seen as the sole breadwinners and, thus, the authority of the core family unit. According to Chung (2018), the marriage migrant visas in South Korea reinforced the idea that foreign female spouses of native Korean men had three principal duties: childbearing, caring for their in-laws, and becoming citizens.

Visa policies can also create gendered vulnerabilities. To countervail marriages of convenience, many countries insist that the spouses live together for a period of time. Separation, divorce, and death therefore affect the spouse's leave to remain, particularly if it happens during a waiting period following effective family reunion. In most countries, divorce means that the spouse's right to stay is withdrawn (OECD, 2020). This can lead to serious rights violations, as it offers an incentive to remain in abusive relationships. In

[2] Coman, Hamilton and Asociatia Accept v Inspectoratul General pentru Imigrări and Ministerul Afacerilor Interne (C-673/16) EU:C:2018:385.

the case of the United States for example, a group of scholars has pointed out the inadequacies of the country's immigration system in ensuring the protection of immigrant spouses who are victims of domestic violence (Anderson, 1993; Olivares, 2011).

Regarding the economic dependency of female migrants on men, although legislation has evolved over the last two centuries, the literature still finds that family reunification visas reinforce women's financial dependency on men, due to restrictions imposed on reunified migrants to work or study, even for high-skilled migrants. Thus, financial dependency on their spouses is often maintained over time. Female spouses' only employment options are, often, in the informal sector with poor pay and working conditions (Birchall, 2016). Due to the difficulties encountered by women to access work visas, described in the previous section, some authors have pointed out the need of some migrant women to enter relationships to obtain regular status, either with natives or with migrant men with regular status (Banerjee, 2013; Bredbenner, 1998; Boucher, 2007; Calavita, 2006; Iredale, 2005).

Even if the right to work is granted to the reunified or secondary applicant, they often encounter difficulties when trying to find a job, in contrast to the first applicant, whose skills have been accurately assessed and accredited. As a result, the reunified or secondary applicants experience "downward mobility" (Boucher, 2007; Dauvergne, 2000; Fincher, 1997; Iredale, 2005). Economic dependency, in turn, together with the absence of emotional support of relatives, can also force migrant women to remain in relationships where they suffer from psychological and physical abuse (Narayan, 1995). In the case of reunified male migrants, the migratory experience can lead to feeling ashamed and humiliated for having *lost their masculinity*, as they become economically dependent on women, which in turn can lead them to become physically and emotionally abusive towards their wives (Banerjee, 2013; Boucher, 2004).

Again, the focus of the literature on Northern receiving countries is problematic, as the development of theory on the impact of visas on gender dependency disregards trends in other regions. For example, in Latin America, several countries, such as Argentina, Bolivia, Ecuador, and Uruguay, have passed exceptionally progressive immigration laws in the past 20 years that extend social and labour rights to all immigrants, irrespective of their migratory status (Acosta & Freier, 2015; Cantor, Freier, & Gauci, 2015).

Further Research

In the 1980s, the literature started to recognise the role of women as agents of international migration, which led to the acknowledgement of the *gendered* nature of migration, i.e. that the social construction of gender interferes in the causes for migration, the make-up of migratory flows, and integration experiences. Despite the attention paid to the *sexist* nature of the law by radical feminists, only a few authors have focused on the role of migration policies in shaping the *gendered* nature of migration. These norms condition the access of migrants to receiving countries, and influence the experience of migration and integration, when limiting migrants' agency by denying them, for instance, entry to the labour market.

Existing studies on the *gendered* nature of migration focus on two core aspects: the gendered nature of migration flows and gender roles, especially women's dependency on men. The literature suggests that the gender make-up of flows largely depends on labour market demands and the sexual division of labour. Although visa policies, in this context, have traditionally favoured men, this phenomenon is gradually changing due to women's incorporation into the labour force and their acquisition of skills. Family reunification has been identified as an institution that tends to balance migration in gender terms, as it allows migrant workers to move with their spouses. However, this institution has also contributed to the reinforcement of gender dependency, as it entails important restrictions on the right to work for spouses, tends to reinforce traditional gender roles, and can enable situations of abuse.

Given the increasing number of high-skilled female migrants worldwide, future studies should focus on this particular social group. For example, the role of women as family sponsors remains understudied. Another related gap in the literature is how family reunification policies apply to same-sex couples. In some contexts, there have been advancements in the recognition of family reunification rights to same-sex couples. Nevertheless, in the Global South, where same-sex marriage has often not yet been recognised, same-sex couples might still encounter discrimination. Thus, further research accounting for the particular situations of LGBTQ+ migrants across world regions is needed.

The literature on the gendered effects of visa policies, although always limited in scope, blossomed between 1990 and 2010. The interest in the impact of visas on the gendered nature of migration, however, has decreased in recent years. This is problematic as the situation of women around the world is changing. We would, thus, like to promote and encourage studies on the gendered effects of visa policies, especially in the context of South-South migration, adding to existing efforts to shift the focus of gendered analysis to intra- and trans-regional migration in South-South contexts (Bastia & Piper, 2019).

Future research should also include the effects of *tourist* visas on the gender make-up and migratory experience, given that past research has shown that many irregular migrants enter their destination countries legally, and then overstay their temporary visas (Fazel-Zarandi, Feinstein, & Kaplan, 2018; Freier & Zubrzycki, 2019; Mau, Gülzau, Laube, & Zaun, 2015). Tourist visa requirements can be discriminatory based on nationality, which correlates with ethnicity, race, religion, and educational levels. However, it is likely that gender discrimination (which might be positive towards women in the context of the securitisation of migration and Islamophobia) occurs, in practice, at the level of policy implementation, at the discretion of migration officials.

To this end, different research methodologies and mixed methods should be employed. Until now, many authors have focused on the analysis of legal norms, largely assuming their impact on the gendered nature of migration. Furthermore, most are case studies with a focus on Anglo-Saxon immigration countries, such as Australia, Canada, and Great Britain. More comparative empirical work should be conducted, with longitudinal quantitative analysis testing the suggested impact of visa policies on migration flows. Qualitative studies on the effect of visa policies on gender relations should be complimented by large N surveys to verify the results drawn mainly from small N individual testimonies. These types of mixed methods could both test and develop new theory on the impact of visa policies on the gendered nature of migration.

Concluding Remarks

Women are increasingly moving independently for work, education, and as heads of households (UN DESA, 2019). This change in the *gendered* nature of international migration is likely due to the removal of restrictions on family reunification, women's access to higher education, and the growing labour demand for *feminized* jobs in both skilled and low-skilled sectors. In acknowledgement of the negative biases against women, legal regulations, oriented to alleviate the negative effects of migration dispositions on women, have been adopted both at the national and international level. Despite such improvements, female migrants continue to face stronger discrimination, and are more vulnerable to mistreatment, both as migrants and as women, in comparison to male migrants. The rise in gender neutral norms will only have an equal impact on men and women if further social change towards gender equality advances and consolidates in both sending and receiving countries.

References

Acosta, D., & Freier, L. (2015). Turning the immigration policy paradox up-side down? Populist liberalism and discursive gaps in South America. *International Migration Review, 49*(3), 659–696.

Anderson, M. (1993). Note, a license to abuse: The impact of conditional status on female immigrants. *102 Yale Law Journal, 102*(6), 1401–1416.

Anker, D. (2002). Refugee law, gender, and the human rights paradigm. *Harvard Human Rights Journal, 15*, 133–154.

Banerjee, P. (2013). *Constructing dependence: Visa regimes and gendered migration in families of Indian professional workers* (Doctoral dissertation).

Bastia, T., & Piper, N. (2019). Women migrants in the global economy: A global overview (and regional perspectives), special issue on migrants in a global economy, guest edited by Ruth Pearson and Caroline Sweetman. *Gender & Development, 27*(1), 15–30.

Bertoli, S., & Fernández-Huertas Moraga, J. (2012). *Visa policies, networks and the cliff at the border* (IZA Discussion Paper No. 7094).

Birchall, J. (2016). *Gender, age and migration: An extended briefing*. Institute of Development Studies.

Boucher, A. (2004). Domestic violence and the application for permanent residency visas to Australia and Germany, unpublished seminar paper for the Universität Potsdam Sociology seminar *Sex and Gender in Germany as a Country of Immigration*.

Boucher, A. (2007). Skill, migration and gender in Australia and Canada: The case of gender-based analysis. *Australian Journal of Political Science, 42*(3), 383–401. https://doi.org/10.1080/10361140701513547.

Boucher, A. (2009). Picking the best and the brightest: Gendering skilled immigration policy in Australia and Canada, presented at First European Conference on Politics and Gender, Queen's University, Belfast, 21–23 January.

Bourbeau, P. (2011). *The securitization of migration: A study of movement and order*. London: Taylor & Francis.

Bredbenner, C. L. (1998). *A nationality of her own: Women, marriage, and the law of citizenship*. Berkeley: University of California Press.

Calavita, K. (2006). Gender, migration, and law: Crossing borders and bridging disciplines. *International Migration Review, 40*(1), 104–132. https://doi.org/10.1111/j.1747-7379.2006.00005.x.

Cantor, D. J., Freier, L. F., & Gauci, J. P. (Eds.). (2015). *A liberal tide? Immigration and asylum law and policy in Latin America*. Institute of Latin American Studies, School of Advanced Study, University of London.

Chávez, K. R. (2010). Border (in) securities: Normative and differential belonging in LGBTQ and immigrant rights discourse. *Communication and Critical/Cultural Studies, 7*(2), 136–155.

Chung, E. A. (2018). Creating hierarchies of noncitizens: Race, gender, and visa categories in South Korea. *Journal of Ethnic and Migration Studies, 46*(12), 1–18.

Chung, C., Kim, K., & Piper, N. (2016). Rethinking marriage migration in Southeast and East Asia: Development, gender and transnationalism. *Critical Asian Studies, 48*(4), 463–472.

Danielsen, D. (1988). Recent publications review of the book *feminism unmodified*, by C. A. MacKinnon. *Harvard Civil Rights-Civil Liberties Law Review, 23*(2), 611–622.

Dauvergne, C. (2000). Citizenship, migration laws and women: Gendering permanent residency statistics. *Melbourne University Law Review, 24*(2), 280–309.

DeLaet, D. L. (1999). Introduction: The invisibility of women in scholarship on international migration. *Gender and Immigration* (pp. 1–17). New York: Springer.

Dodson, B. (2001). Discrimination by default? Gender concerns in South African migration policy. *Africa Today, 48*(3), 73–89.

Dodson, B., & Crush, J. (2004). A report on gender discrimination in South Africa's 2002 immigration act: Masculinizing the migrant. *Feminist Review, 77*(1), 96–119. https://doi.org/10.1057/palgrave.fr.9400158.

Dodson, B., & Crush, J. (2006). South African immigration law: A gender analysis. *SAMP Migration Policy Brief No. 16*. Waterloo, ON: Southern African Migration Programme.

Fazel-Zarandi, M., Feinstein, J., & Kaplan, E. (2018). The number of undocumented immigrants in the United States: Estimates based on demographic modelling with data from 1990 to 2016. *PLoS ONE, 13*(9). https://doi.org/10.1371/journal.pone.0201193.

Fincher, R. (1997). Gender, age, and ethnicity in immigration for an Australian nation. *Environment and Planning A: Economy and Space, 29*(2), 217–236. https://doi.org/10.1068/a290217.

Fincher, R., Foster, L., & Wilmot, R. (1994). *Gender equity and Australian immigration policy*. Canberra: Australian Government Publishing Service.

Fitzpatrick, J., & Kelly, K. R. (1998). Gendered aspects of migration: Law and the female migrant. *Hastings International and Comparative Law Review, 22*(1), 47–112.

Freier, L. F., & Luzes, M. (2020). How humanitarian are humanitarian visas? An analysis of theory and practice in Latin America. In L. Jubilut, G. Mezzanotti, & M. Vera Espinoza (Eds.), *Latin America and refugee protection: Regimes, logic and challenges*. Berghahn: Forthcoming.

Freier, L. F., & Zubrzycki, B. (2019). How do immigrant legalization programs play out in informal labour markets? The case of Senegalese street Hawkers in Argentina. *Migration Studies*. https://doi.org/10.1093/migration/mnz044.

García-Castrillón, C. O. (2002). Igualdad, género y medidas de acción-discriminación positiva en la política social comunitaria. *Revista de Derecho Comunitario Europeo, 6*(12), 489–502.

Garner, S. (2007). The European Union and the racialization of immigration, 1985–2006. *Race/Ethnicity: Multidisciplinary Global Contexts, 1*(1), 61–87.

Harzig, C. (2003). Immigration policies: A gendered historical comparison. In M. Morokvašić, U. Erel, & K. Shinozaki (Eds.), *Crossing borders and shifting boundaries: Vol. I: Gender on the move* (pp. 35–58). VS Verlag für Sozialwissenschaften.

Hazen, H. D., & Alberts, H. C. (2006). Visitors or immigrants? International students in the United States. *Population, Space and Place, 12*(3), 201–216.

Hyndman, J. (1999). Gender and Canadian immigration policy: A current snapshot *Canadian Woman Studies, 19*(3), 6–10.

Iredale, R. (2005). Gender, immigration policies and accreditation: Valuing the skills of professional women migrants. *Geoforum, 36*(2), 155–166.

Kofman, E. (2000). The invisibility of skilled female migrants and gender relations in studies of skilled migration in Europe. *International Journal of Population Geography, 6*(1), 45–59.

Kofman, E. (2004). Family-related migration: A critical review of European studies. *Journal of Ethnic and Migration Studies, 30*(2), 243–262.

Kofman, E. (2012). Gender and skilled migration in Europe. *Cuadernos de Relaciones Laborales, 30*(1), 63–89.

Kofman, E., & Raghuram, P. (2006). Gender and global labour migrations: Incorporating skilled workers. *Antipode: A Radical Journal of Geography, 38*(2), 282–303. https://doi.org/10.1111/j.1467-8330.2006.00580.x.

Kofman, E., & Raghuram, P. (2011). Changing mobility regimes, knowledge and gender equalities in Europe. In S. Ilcan (Ed.), *Mobilities, knowledge and social justice* (pp. 59–75). Montreal: McGill-Queen's University Press.

Kofman, E., Raghuram, P., & Merefield, M. (2005). *Gendered migrations: Towards gender sensitive policies in the UK* (Working Paper, No. 6). Institute for Public Policy Research.

Lentin, A. (2011). *Racism and ethnic discrimination*. New York, NY: Rosen.

Luibhéid, E. (2008). Queer/migration: An unruly body of scholarship. *GLQ, 14*(2–3), 169–190.

MacKinnon, C. A. (1987). *Feminism unmodified: Discourses on life and law*. Cambridge, MA: Harvard University Press.
Mau, S. (2010). Mobility citizenship, inequality, and the liberal state: The case of visa policies. *International Political Sociology, 4*(4), 339–361.
Mau, S., Gülzau, F., Laube, L., & Zaun, N. (2015). The global mobility divide: How visa policies have evolved over time. *Journal of Ethnic and Migration Studies, 41*(8), 1192–1213.
Myers, R. M., & Griffin, A. L. (2019). The geography of gender inequality in international higher education. *Journal of Studies in International Education, 23*(4), 429–450.
Narayan, U. (1995). "Male-order" brides: Immigrant women, domestic violence and immigration law. Hypatia 10(1), *Feminist ethics and social policy*, part 1 (Winter, 1995), 104–119.
OECD. (2014). *Recruiting immigrant workers: New Zealand 2014*. OECD. https://www.oecd-ilibrary.org/social-issues-migration-health/recruiting-immigrant-workers-new-zealand-2014_9789264215658-enI.
OECD. (2020). *Comparative analysis of the legislation and procedures governing the immigration of family members in certain OECD countries*. OECD. http://www.oecd.org/els/mig/41563157.pdf.
Olivares, M. (2011). A final obstacle: Barriers to divorce for immigrant victims of domestic violence in the United States. *34 Hamline Law Review, 149,* 154–156.
Pedone, C., & Gil Araujo, S. (2008). Maternidades transnacionales entre América Latina y el estado español: El impacto de las políticas migratorias en las estrategias de reagrupación familiar. In C. Solé, S. Parella, & L. Cavalcanti (Eds.), *Nuevos retos del transnacionalismo en el estudio de las Migraciones* (pp. 151–176). Ministerio de Trabajo e Inmigración de España.
Pew Research Center. (2017). *Facts on foreign students in the U.S. global attitudes and trends*. https://www.pewresearch.org/global/fact-sheet/foreign-students-in-the-u-s/.
Piper, N. (2006). Gendering the politics of migration. *International Migration Review, 40*(1), 133–164. https://doi.org/10.1111/j.1747-7379.2006.00006.x.
Pittaway, E., & Bartolomei, L. (2001). Refugees, race, and gender: The multiple discrimination against refugee women. *Refuge: Canada's Journal on Refugees, 19*(6), 21–32.
Raghuram, P. (2004). The difference that skills make: Gender, family migration strategies and regulated labour markets. *Journal of Ethnic and Migration Studies, 30*(2), 303–321. https://doi.org/10.1080/1369183042000200713.
Toksöz, G., & Ulutaş, Ç. Ü. (2012). Is migration feminized? A gender- and ethnicity-based review of the literature on irregular migration. In S. Paçaci & T. Straubahaar (Eds.), *Turkey, migration and the EU: Potentials, challenges and opportunities* (pp. 85–112). Hamburg: Hamburg University Press.
Tryfonidou, A. (2019). The EU top court rules that married same-sex couples can move freely between EU member states as "spouses": Case C-673/16, Relu Adrian Coman, Robert Clabourn Hamilton, Asociaţia Accept v Inspectoratul General pentru Imigrări Ministerul Afacerilor Interne. *Feminist Legal Studies, 27*(2), 211–221.
UN DESA. (2017). *International migration report*. https://www.un.org/en/development/desa/population/migration/publications/migrationreport/docs/MigrationReport2017_Highlights.pdf.

UN DESA. (2019). *International migration report*. https://www.un.org/en/development/desa/population/migration/data/estimates2/estimates19.asp.

van Walsum, S., & Spijkerboer, T. (Eds.). (2007). *Women and immigration law: New variations on classical feminist themes*. New York: Routledge-Cavendish.

Yuval-Davis, N., Anthias, F., & Kofman, E. (2005). Secure borders and safe haven: The gendered politics of belonging beyond social cohesion. *Ethnic and Racial Studies, 28*(3), 513–535.

Luisa Feline Freier is Associate Professor of Political Science at the Universidad del Pacífico (Lima, Peru). Her research focuses on migration and refugee policies and laws in Latin America, South-South migration, and the Venezuelan displacement crisis. Prof. Freier has published widely in both academic and media outlets, and has been interviewed on the Venezuelan displacement crisis across international media. Prof. Freier has provided advice to various international organizations, such as Amnesty International, the International Committe of the Red Cross, the Inter-American Development Bank, the International Organization for Migration (IOM), the United Nations High Commissiomer for Refugees, the World Bank, and the European Union. She is part of IOM's Migration Research and Publishing High-Level Advisers, and Associate Editor of the Migration Studies journal.

Nieves Fernández Rodríguez is a PhD candidate at Berlin Graduate School for Global and Transregional Studies (Free University of Berlin), and works as a researcher at the Social Sciences Center of Berlin (WZB). Her PhD dissertation is about the securitisation of migration in Latin America in the face of the Venezuelan exodus. She is also part of the project 'Towards a Typology of Contestations' (TTC) headed by Michael Zürn and Stefan Gosepath, aimed at constructing a typology of actors who contest liberalism.

CHAPTER 32

Women, Borders, and Mobilities in Latin America

Carolina Stefoni, Menara Guizardi, Eleonora López, and Herminia Gonzálvez

Abstract The aim of this chapter is to arrive at a detailed understanding of the manner by which gender systems are intertwined with the reproduction of borders, and how systems of domination are perpetuated while also creating spaces for resistance and contestation. Employing the perspective of feminist economics, we observe how various industries exploit existing inequities at the border to utilize the low cost of women's labour to establish factories that precaritise work in extreme ways. Drawing on contributions from gender studies in Latin America to the theorization of human mobilities in borders, our discussion focuses on the relationship between labour markets and gender. In doing so, we emphasize the raw and brutal structural violence exerted on the bodies and lives of women in Latin American border territories,

We would like to acknowledge the National Agency for Research and Development (ANID) for funding the project "The Boundaries of Gender Violence: Migrant Woman's Experiences in South American Border Territories" (Fondecyt 1190056).

C. Stefoni (✉)
Universidad Mayor; COES/FONDAP, Santiago, Chile

M. Guizardi
National University of San Martín, Buenos Aires, Argentina

E. López
Pontificia Universidad Católica de Chile, Santiago, Chile

H. Gonzálvez
Universidad Central de Chile, Santiago, Chile

C. Mora and N. Piper (eds.), *The Palgrave Handbook of Gender and Migration*, https://doi.org/10.1007/978-3-030-63347-9_32

which find its extreme expression in trafficking, sexual harassment, and femicide. These extreme expressions are fostered by daily manifestations of gender violence that cross the boundaries between the public and the private spheres in borderlands: in homes, in public spaces, and in areas supposedly protected by State officials who, however, often are among the violators of women's human rights.

Introduction

In this chapter, our aim is to highlight two important contributions made by gender studies in Latin America to the theorization of human mobilities in borders. On the one hand, we emphasize the raw and brutal structural violence exerted on the bodies and lives of women in Latin American border territories, which find its extreme expression in trafficking, sexual harassment, or femicide. These extreme expressions are fostered by daily manifestations of gender violence that cross the boundaries between the public and the private spheres in borderlands: inside homes, in the public sphere, and in areas supposedly protected by State officials who, however, often are among the violators of women's human rights. On the other hand, we highlight the relationship between labour markets and gender. Employing the perspective of feminist economics, we observe how various industries exploit existing inequities at the border to utilize the low cost of women's labour to establish factories that precaritise work in extreme ways. Maquilas, or offshore factories, and agro-industrial sectors have become paradigmatic examples of such precarious work, with very high female labour participation. In the informal sector, we also observe how women cross borders daily, carrying merchandise literally on their shoulders to be bought and sold in local cities and markets located on either side of national borders. The smuggling of contraband products is also organized through gender and ethnic inequities (Cardin, 2012; Renoldi, 2013).

Before delving into these debates, it is necessary to establish two key definitions supporting our argument: the first, regarding our definition of borders and, the second, on how we understand the production of gender inequality in border territories.

The border is a polysemic concept that alludes to geopolitical functions and cultural and identity constructs. Therefore, borders have different meanings and are connected to heterogeneous social practices. The way social actors understand them depend both on their particular spatial mobility and on contextual/historical elements. For example, the conceptions and practices of traders who commute across borders on a daily basis differ from those of long distance international/transnational migrants and from the ones developed by internal migrants within a nation-state. It means that different social practices of mobility build up heterogeneous types of borders that can, in often many cases, be superposed: (1) physical (between states); (2) internal (inside a country); (3) identity (related to racial/ethnic, gender or class identifications); and (4) relational (referring to codes of communication and interaction between people that does not recognize themselves as equal). In this chapter, we focus on the practices related to the crossing of international borders. We will specifically address the experiences of social actors whose daily experience entails the crossing of international border; that is, people to which those crossings constitute a way of life.

The theoretical development of the conceptualization of the border occurred alongside the historical development of the nation-state and accelerated changes in the broader context of enhanced globalization (from the 1990's onwards). From its inception, the border was first defined as a space that simultaneously divides and unites, since it establishes the boundary that separates one state from another, and permits or assumes the unification of all who live and reside in the territory from that boundary onwards, under the idea of the nation. At the end of the twentieth century, the social sciences worked to overcome this Euclidean understanding of border territories, in order to attend to the characteristics assumed by an entire area constructed around that geopolitical boundary. This shift in focus enabled more scholarly attention to the many inequities that were concentrated in the border regions, whether in ethnic, national, or economic terms. The latter enabled the observation of how the border articulates and organizes these multiple inequities.

However, all these debates also brought to light the high levels of gender inequality in border spaces. Around the world, the borders are different realities and have different meanings for men and women engaged in crossing their limits on a daily basis around. The practices that these people deploy are also differentiated by gender, impacting the way the inequities between men and women is reproduced.

Despite that, border studies have been dominated by androcentric definitions and perspectives. It has only been in recent decades that we can witness a growing literature that has placed women's experiences at the centre of this debate. The variety of studies on women in Latin American borders allow us to question the naturalization of androcentric perspectives regarding border territories and contribute to theorizing them with dimensions that until recently have been scarcely visible in the dominant literature.

With that in mind, in the following pages, we are interested in understanding how gender systems are intertwined with the reproduction of borders, which in turn perpetuates systems of domination while creating opportunities for resistance and contestation to these systems. To achieve that purpose, we will start by reviewing the different conceptualizations of the border, revealing the absence of a gender perspective—until recently—in the studies on the issue carried out in Latin American contexts. Next, we analyze the inequities observed at the borders and the reasons that, in our view, explain the invisibility of women in this field. We then analyze two significant contributions from gender studies: the visibilization of structural violence that surrounds border areas, and the relationship between labour markets, gender, and borders.

(Androcentric) Definitions of Borders

Some Historical Roots

In its colonial version, the border defined the boundary that separated civilization from the barbaric, natural, and uncivilized world (Mignolo & Tlostanova, 2006). Postcolonial or decolonial criticism (in its Latin American variant) accounts for how the hierarchies established from colonization—civilization/barbarism, modernity/premodern, rationality/irrationality–, are currently maintained and reproduced through what Mignolo (2007), Quijano (2005), Grosfogel (2006), and others have called the "coloniality of being", the "coloniality of power" and the "coloniality of knowledge". Thus, the colonial domination institutionalized a border between "we" (the Europeans) and the "others" (the barbarians) that constitutes the modern notion of otherness. The term "border" today inherits this understanding of a "we" that defines itself as a superior subject to those located on the other side of the dividing line (Grimson, 2000).

This superiority that was expressed in civilizational terms included being in a more advanced stage of the modernization process, thereby validating a position of power over those who were considered to be located in a "lower" position in the social hierarchy (Rivera Cusicanqui, 2008). Women, in this perspective, were subsumed in a subordinate position that confined them to the private space, making their multiple mobilities across borders invisible (Anzaldúa, 1999).

Since its inception, the idea of the border as a boundary between the civilized and the barbaric was linked to the structure of patriarchal domination inherent in the colonial system. However, this was not an explicit topic of investigation or analysis until the late twentieth century, with the emergence of feminist movements and theories (and of the researchers adhering to these developments who were dedicated to studying the border regions). The mutual articulation of the colonial system and the patriarchal order determined that women belonging to the "barbaric world" were considered potential objects of domination and appropriation by the white population (Segato, 2013a).

The development and consolidation of nation-states from the nineteenth century onwards determined a first shift in the understanding of the border. During this period and until the mid-twentieth century, the idea that the border is a fixed and static entity prevails, as can be observed in historical and geopolitical studies of the time. In this view, the border was seen as a dividing line that marked the physical, cultural, and political boundaries of societies, and that which was the subject of dispute and conflict in nascent nation-states (Garduño, 2003).

Such an understanding of the border was a central aspect of methodological nationalism in the social sciences (Wimmer & Glick Schiller, 2003), which prevented many researchers from understanding it as a political construction at the service of the consolidation of the nation-states (Grimson, 2011). In this view, the gender system became a central component—but still invisible for critical analysis—in the configuration of Latin American nations and states. The articulation of the homeland as a female entity that needed to be protected from the enemy, finds a correlation in the understanding of women as male property, who should remain in the private, domestic sphere to ensure their protection and safety (Dore, 2000; Yuval-Davis, 1997).

The Latin American critical feminist perspective seeks to recognize the relationship between this ancestral gender violence and colonial and capitalist gender violence. Speaking in these terms allows us to understand the systemic way in which the original or ancestral patriarchy is readjusted with the incursion of the violence of the modern patriarchy. This implies recognizing that unequal and unfair relationships between men and women are not only a colonial heritage.

In this sense, we rely on Segato (2013b), for whom the colonial invasion in different contexts of Latina America would have caused deep transformations in patriarchate systems on the region. Analyzing the impact of the introduction of colonial public-private hierarchization on the social organization of native groups, Segato (2013b: 86) assumes that this transformation implies for these groups the passage of a *low intensity patriarchy* (in which the domestic space was endowed with political and decision-making power), to a *high intensity patriarchy* that, at the same time, condemns the native man to a masculinity prototypically inferior to that of the colonizer. In other words: "colonization brings with it a radical loss of women's political power, where it existed, while the colonizers negotiated with or invented certain masculine structures in order to achieve allies" (Gautier, 2005: 718 in Segato, 2013b: 86). But, as we have mentioned before, the institutionalization of nation-states and the invention of new border limits on these territories radically aggravates the constitution of patriarchal and masculinized nation-states (Dore, 2000).

Actual Trends

At the end of the twentieth century and the beginning of the twenty-first, a second shift can be observed in the understanding of the border. This turn directly relates to some characteristics provoked by the processes of globalization in recent decades. An extended discussion and definition on globalization is beyond the scope of this chapter; nevertheless, we wish to highlight some elements that are necessary for understanding how this shift affected the new conceptualization of the border. The first element refers to the profound economic changes based on the acceleration, internationalization, and flexibilization of the processes of production and distribution of goods, and its impact in the inequalities faced by migrant women in the international division of labour (Mora, 2008; Sassen, 2003). The second element is the emergence of a new axis of organization of the international system that replaces the East/West division (with the iron curtain as its border) with a division between North and South, and which has its new divisive lines in the Mediterranean and the Mexico-United States border. The third is a technological revolution that definitively changes access to information, forms of communication, the keys to politics and that which establishes simultaneity as an element of reality (Giddens, Bauman, & Beck, 1996). Finally, the fourth element is the growth of the social and cultural diversity of the postcolonial world, establishing multiculturalism and hyper-diversity as a new social reality (Vertovec, 2007).

Consequently, the new theoretical approaches to borders moved away from the static and a-problematic understanding that dominated until the mid-twentieth century. This debate enabled researchers to introduce expanded perspectives on their territorial dimension, arguing for the need to consider regions or areas wider than the simple dividing strip. On the other hand, this debate brought to the centre the tensions, conflicts, paradoxes, and ambivalences that are produced in the interaction of border regions and in the production of daily experiences and identities by the subjects that move across them.

We can identify three perspectives that enrich the current discussion on borders, taking as our starting point that the border is a social, cultural, and political construction. These perspectives can be categorized as: (a) the securitization approach and the global migration management regime, where the border is viewed as a separator of national geopolitical spaces (Balibar, 2010; De Genova, 2017; Mezzadra & Neilson, 2013), experienced in different ways according to gender, class, place of origin or colour of skin (Raghuram & Piper, 2011); (b) the border as a space where a series of power relations that construct national and ethnic identities converge and intersect (Guizardi, Valdebenito, López, & Nazal, 2015); and (c) a series of studies that put gender at the centre of border analysis, revealing experiences and meanings that have until then been invisibilised (Anzaldúa, 1999; Cortés, 2018; Guizardi, Valdebenito, López, & Nazal, 2019; Viteri, Ceja, & Yépez, 2017). It is interesting to notice the lack of dialogue that exists between structural and global perspectives of the border, and the important contributions from gender studies to this field.

A. The Securitization Approach

The global migration management regime approach emphasizes that the increasing mechanisms of control and restriction of human mobility have been established to facilitate the mobility of some and hinder or impede that of others; such differentiation operates as a means to organize and rank different countries in the international arena. A consequence of the global migration regime is that citizens belonging to "privileged" states enjoy global mobility, while those belonging to states defined as "insecure", "risky", or "poor" are hindered from moving across borders (Bauman, 2016).

Although this approach has contributed to understanding the border as a device for global organization, the incorporation of a gender perspective is still lacking. Some questions that could contribute to furthering this proposal and perspective are: How do systems of control affect men and women differently? How are gender relations and women's position reorganized or reconfigured under a global migration regime? Do international systems of protection of the human rights of refugees tend to overrepresent the image of women victims and mothers? Contemporary studies of post-globalization phenomena such as refugeeism and refugee camps give us some important clues associated with mobility. Roldán's work, for example, analyzes how women's mobility continues to be punished by societies of origin, transit and destination, even when they migrate as a result of the situations of violence they face (Roldán, Guerra, & Pérez, 2017). In this way, it is interesting to raise the link between the securitist approach to migration and research on the gendered uses of space.

In the current situation of migration control, human mobility distinguishes between those who are privileged in and by globalization, and those who are excluded from this new global order. Studies on spaces and gender, however, can illuminate the restrictions that women face in terms of moving through and inhabiting public spaces. Thus, it is necessary to conceive of mobility as a right that is doubly affected in the case of women who belong to the world that suffers the costs of globalization (Morokvasic, 2004; Roldán et al., 2017).

B. The Power-Identity Approach

The second border perspective that arises in this new scenario emphasizes the role of national, ethnic, and gender identity constructions, although the latter two have received less attention. This focus incorporated into the analysis the symbolic and subjective meanings of those who lived near or had experiences with a specific border when it was in the processes of becoming more protected and controlled (Garduño, 2003; Velasco & Contreras, 2011). Addressing the border as a process that is continuously constructed enables an understanding of the collective differentiation between social groups, their identifications, and the distribution and organization of cultural features (Barth, 1976). These are transformed into a locus that permits an analysis of the interaction of cultural, political, and social dimensions in the configuration of the nation-state (Kearney, 1991; Vila, 2000). In turn, they provide a key axis for the observation of differences and inequalities articulated around the categories of nationality and ethnicity (Guizardi et al., 2015; Kearney, 2004).

C. The Gender Approach

The third perspective, which emerged in the eighties, pioneered a perspective of the border based on the experience of women. Several women scholars began to write about the borders to reveal several dimensions and realities that had been absent in both historical and more recent theoretical approaches (Anzaldúa, 1999; Fernández-Kelly, 1983; Roldán et al., 2017; Viteri et al., 2017). These elements, however, were not recognized and incorporated initially into relevant academic debates, and the opportunity was lost for the transformation of androcentric theoretical constructions of the border. Nevertheless, in this third perspective, feminist voices from diverse disciplinary fields redefined fundamental aspects of the border. One example from cultural studies are the reflections of Chicano feminists such as Gloria Anzaldúa, with her pioneering work, "Borderlands/La Frontera," published in 1987. This work redefined the border by making visible not only the place occupied by women in construction of national boundaries, but also how gender relates to the identities and mobilities of people and material goods, in addition to how violence is multiplied and naturalized in the everyday construction of border regions.

The extensive work available on this topic has largely been based on the experience of the Mexico-United States border, allowing this region to gain prominence in social thinking and to become a key example in the redefinition of the bordering spaces of nation-states, as well as in the redefinition of gender identities that constitute these spaces. However, as mentioned earlier, this does not imply that many scholars who later became key theorists in the border debate have adopted this perspective. On the contrary, many scholars continue to render invisible the strong entrenchment of gender inequality in borders.

The theoretical and analytical work developed along these three perspectives allows us to affirm that border regions are spaces that concentrate multiple inequalities and the historical accumulation of diverse asymmetries of power. One of them is economic, while another is political-nationalist in nature, and a last asymmetry makes visible the patriarchal violence in border territories. The following section develops this last point in more detail.

Borders and Gender: Tensions, Mobilities, and Power Asymmetries in the Twenty First Century

The development of border studies and critical border studies (CBS) have placed at the centre of the debate various relationships of political, symbolic, and social power that constitute the construction of borders (Rumford, 2006). As Parker and Vaughan states (2012), CBS attempts to decentre the "line in the sand" idea of border, arguing that it is "never simply 'present', nor fully stablished, nor obviously accessible. Rather, it is manifold and in constant state of becoming" (Parker & Vaughan-Williams, 2012: 728). In these analytical fields, we can identify three asymmetries of power that have been more prominent in recent studies on borders: economic inequalities, national hierarchies, and gender inequalities.

Economic inequalities are expressed in territorial terms and are organized from the boundary that separates one State from another. This enables the deployment of transaction and exchange practices that are organized around these inequalities, such as participation in the labour market on the other side of the border; access to specific services; or smuggling and merchandise trade between both territories. This asymmetry has led to studies that analyze how economic exchange and the emergence of commercial activities on and across border areas construct different levels of interdependence between regions. Some scholars employ the idea of border regions (Benedetti, 2014), cross-border regions or urban complexes (Dilla, 2008; Jessop, 2004; Tapia, Liberona, & Contreras, 2017) or binational metropolis (Alegría, 2000) to address different cases.

The second asymmetry refers to the differentiation between the national and the foreign (which is sometimes transformed into a threat) and the establishment of hierarchical relations between states based on the distinction between nations. When both of these asymmetries converge and mutually strengthen each other—such as when strong nationalisms and high economic inequalities generate an increase in migratory flows—states begin to mobilize a series of economic, identity, and military resources aimed at controlling the flow of illicit and irregular goods and managing the movement of people, that is, to impede the movement of some people, and favour that of others (see also Freier & Fernández, in this volume). The Mexico-US border is an emblematic case of the intersection of these two asymmetries and the response of the State and relevant studies also exist on the southern European border (De Genova, 2017; Howell, Altamirano, Totah, & Keles, 2018), and more recently, on the Andean border through the Chilean government's Northern Border Plan, which aimed to increase the presence of military personnel along the country's border with Bolivia and Peru (Ramos & Ovando, 2016)

The third of the asymmetries refers to the entanglement between the political and economic constitution of the borders, and the consequent reproduction of patriarchal violence and gender inequalities in this mutual entrenchment (Monárrez, 2013; Téllez, 2008). In what follows, we highlight some elements that would help to understand the relative invisibility of this third asymmetry in border studies until the end of the twentieth century.

First, gender bias has rendered hegemonic a perspective in the social sciences that defines women as passive, immobile, or with a reduced or transitory mobility, which prevents the documentation, observation, and analysis of the mobilities that they practice and deploy. This bias derives from a series of theoretical constructions and disciplinary approaches that define mobility and ways of inhabiting places based on a public/private dichotomy. Following Bourdieu's analysis (2000) that relates the gender order to the organization of space (women in the private space, men in the public space), human mobility has also been interpreted according to this structure. This view implies according men's mobility social value under the assumption that this mobility would enable the generation of economic resources, while women's mobility is reduced to the private sphere and, therefore, invisible and devoid of social value. Following Massey (1994), a problem arises when the displacement of women is sanctioned because it defies the social norm. Under this approach, the mobilities—and therefore the migrations and crossings of geopolitical borders that women undertake, which defy this hegemonic gender order—are rendered invisible to border scholars in the Latin American contexts.

Second, research inquiring into the place of women in the construction of national spaces and their boundaries was only consolidated in the mid-nineties. As Yuval-Davis (1997) argued, most of the hegemonic historiographies about nations or nationalisms had, until then, ignored gender relations. The marginalization of women in the study of structures considered central in historical terms—the "high" arena of politics, which sets the principles of the State, the nation, and its defining elements, such as borders—constitutes a reproduction of inferiorization of women and of genders considered "non-masculine". The constitution of Latin American republics in the nineteenth century transpired, precisely, in a context of ideological globalization of the association of masculinity to civilization and of femininity to nature. Within the framework of this ideology, which played a fundamental role in the formation of nation-states, territories which were objects of conquest by these states were represented as conducive to male domination and exploitation.

The second element refers to the importance that war or political conflicts acquire in the configuration of borders. The histories of wars, victories, and defeats have been largely written by men. Women have been invisibilized in historiography not because they did not play an important role, but because in a patriarchal and hegemonic order, their role was not considered relevant to those in charge of writing history. Young (2000) points out that institutionalized domination in different models of State had the end of its reproduction anchored in a principle of misogynistic oppression (Phillips, 2002; Song, 2007).

In sum, only in the late eighties did women begin to be visible in border studies and only in relation to two specific situations: (1) their participation in precarious labour circuits; and (2) gender violence, as we will now detail in the next section.

Women's Participation in Circuits of Precarious Labour

Women are visibilised in border areas in their participation in highly precarious, flexible, and informal jobs such as in maquilas, agro-industries, and the "ant trade" (Molina, 1985; Monárrez, 2013; Téllez, 2008). To understand these dynamics, it is important to provide an overview of studies that articulate the relationship between labour markets, gender, and migration, in particular, the work on global care chains, transnational families, and domestic work. Although in the border areas we observe similar dynamics in terms of the precarious conditions of women's work, a distinctive element is the articulation between territorial, national, and gender inequalities when considering their labour market insertion. These multiple inequalities are exploited by certain industries that establish factories in these border areas configuring highly profitable cross-border circuits, which generate profit precisely by subjecting certain social groups to disadvantageous conditions.

This emphasis on the definitive role of the intersectionality between gender, nationality, and ethnic-racial attributions in the exploitation of people within the framework of cross-border circuits, paralleled the reality of increasing levels of violence suffered by women at borders in general (Pickering, 2011: 109–110). However, the specific situation at the Mexico-US border from the nineties onwards made it a paradigmatic place to observe these tensions and for a preliminary analysis of the dynamics and articulations that are produced by the intersection of gender, the border, and mobilities.

The labour vulnerability that women faced also extended to the domestic sphere, in their relations with their partners and male family members (Molina, 1985: 33). One hypothesis that attempts to explain gender violence in border areas is that men perceived and felt displaced from their role as economic providers due to the high percentages of female employability, which then culminated in outbreaks of gender violence (Molina, 1985: 35–36). Likewise, the population explosion caused by industrial development in border areas often surpassed the capacity of social, health, and housing services, as well as infrastructure, resulting in a decline in the living conditions of the border population in general, and women in particular (Molina, 1985: 38).

In agricultural work, it is also possible to identify experiences of violence suffered by women, especially indigenous women. In this case, different elements are juxtaposed. For example, in the case of migration at the Andean triple-border (Bolivia-Peru-Chile), Guizardi et al. (2019) note that since employment opportunities tend to remain within family and ethnic networks, it is very common for systems to reproduce the gender inequality present in communities, and thus exacerbate social exclusion. One of the dimensions analyzed in the labour conditions of the agricultural sector is workers' exposure to the agrochemicals applied to crops, especially in sectors associated with forms of sharecropping and the lease of small farms for vegetable cultivation (and that therefore differ from the agro-industry). These chemicals are used due to poor regulation and checks by health and labour authorities. A second dimension that accounts for labour exploitation is the common use of daily rate workers and the total absence of contracts. When women are incorporated into family networks, they are subjected to extensive working hours that they must continue to make sure are compatible with care work. The combination of both forms of work translates into great physical, emotional, and psychological fatigue (Guizardi et al., 2019; Olivera & Aguiar, 2006).

Another type of work that women participate in border areas is what has been called the "ant trade", one that takes place across the border. While this work can also be carried out by men in more disadvantaged positions, the participation of women shows how women's bodies are transformed into vehicles for transporting merchandise. These women are often the last nexus in a broader organization of the trading of goods across borders. They are known with different names across the region: bagayeras on the border of Argentina and Bolivia (López, 2019), pasadoras on the border of Chile and Peru (Jiménez, 2018), and sacoleiras between Paraguay and Brazil. They visibilise such widespread practices across different borders of Latin America. In these practices associated with the ant trade, commercial integration and interdependence are built on the backs of women.

Women's Experience of Border Violence; Concluding Remarks and Future Research

Diverse studies on women's experience in border territories make visible the violence that women and non-male gendered persons experience in these spaces. They also evidence an intensification of forms of violence due to migration control policies implemented by various countries; due to the consolidation of organized crime; and due to economic and commercial agreements between neighbouring countries that enable the exploitation of wage inequalities for the profit optimization of various economic sectors.

Research conducted in the context of the Mexico-US border that was initiated by the work of Anzaldúa (1999) makes clear that violence against women is a common element in these spaces and that it manifests itself in different ways and by different actors such as, for example, public officials at border crossings, members of criminal gangs, organized crime, families, and employers, among others. Beyond the Mexico-US border, the triple-border of Paraná (Argentina-Brazil-Paraguay) monopolizes many studies due to its high level of contraband and violence. Current research indicate that these phenomena is intricately linked with tourist, commercial, and political circuits in this border area that has undergone extensive globalization processes in recent decades (IOM, 2019: 16). Seaman (2012: 31) describes this triple-border as "a smuggler's dream" because of the laxity of local authorities regarding the control of illicit activity. According to Seaman, smuggling between Paraguay, Brazil, and Argentina involves the traffic of very diverse products (from clothes and toys to cars and drugs). Human trafficking is part of this multifaceted context of institutionalization of smuggling and cross-border trafficking.

Together with the expressions of violence, it is also important to identify women's self-protection strategies that enable them to escape these dynamics and thereby challenge the structural violence in which they find themselves. There are several kinds and systems of self-care strategies that they deploy, often in the absence of security policies to protect them. Some strategies that we have found in different Latin American borders are to avoid walking alone, walking in a group to cross borders, and forms of self-organization to make visible the experiences of violence that they face in everyday life.

From these experiences, it is possible then to understand that the mobility of women constitutes an object of analysis with vast potential for the observation of the paradox between the reproduction of forms of oppression and the possibility of challenging the gender order. The mobility of women, as we have seen in this article, is typically sanctioned by the patriarchal order, which confines them to private spaces. However, the stubbornness that women display and perform in order to mobilize, to cross borders, to migrate and return, illuminate the resistance that is deployed in exercising mobility as a human condition. In this scenario, therefore, mobility arises from women that originates from the duality between oppression and the possibility of change.

Several discussions in this text point us to new agendas for the study of gender and borders. This concluding section reviews some of these elements. Border contexts exceptionally concentrate structural and gender-based violence and are linked to the territorialized local subordination of nation-states to economic, political, and social oppressions and inequalities on a global scale (Morales & Bejarano, 2009: 420). However, despite the magnitude and intensity of these forms of violence, the majority of research conducted in Latin American borders coincide in assuming women's dialectical role in these spaces: on the one hand, their historical and cultural embodiment of the specific submission to their daily contexts and, on the other hand, their simultaneous role as active agents of resistance and empowerment (both personal and communitarian) (Morales & Bejarano, 2009).

First, therefore, we consider it central to illuminate and systematize these practices, because the possibility arises from them to not only empirically but also theoretically challenge androcentric constructions of borders. To do this, it is important to consider the local and microsocial dynamics that women themselves deploy. Women have historically moved and travelled through public and private spaces, and in their displacement, they have crossed not only geographical, but also social (gender, ethnic, and class) borders. The invisibility of these practices has allowed the reproduction of patriarchal structures. Thus, it is central to understand how these mobilities have managed to challenge the prevailing social order.

Second, it is necessary to examine the multiple forms of violence experienced by women living in border areas and to understand how such violence is linked to different kinds of discrimination in terms of, for example, nationality, ethnicity, and class. Consequently, it is important to understand how and why gender-based violence in border areas is exacerbated by the control and securitization policies embedded in global migration management regimes.

Thirdly, we need to further current understanding on how gender and the border relate to the reproduction of economic inequalities experienced by women and how gender inequalities are functional to the economic development of these areas.

References

Alegría, T. (2000). Juntos pero no revueltos: Ciudades en la frontera México-Estados Unidos. *Revista Mexicana de Sociología, 62*(2), 89–107.

Anzaldúa, G. (1999). *Borderlands: La Frontera*. San Francisco, Estados Unidos: Aunt Lute Books.

Balibar, E. (2010). At the borders of citizenship: A democracy in translation? *European Journal of Social Theory, 13*(3), 315–322.

Barth, F. (1976). Introducción. In F. Barth (Ed.), *Los grupos étnicos y sus fronteras* (pp. 9–49). México: Fondo de Cultura Económica.

Bauman, Z. (2016). *Extraños llamando a la puerta*. Barcelona: Paidós.

Benedetti, A. (2014). Espacios fronterizos del sur sudamericano: Propuesta de un modelo conceptual para su estudio. *Estudios Fronterizos, nueva época, 15*(29), 11–47.

Bourdieu, P. (2000). *La dominación masculina*. Barcelona: Anagrama.

Cardin, E. (2012). Trabalho e prácticas de contrabando na fronteira do Brasil com o Paraguai. *Geopolítica, 3*(2), 207–234.

Cortés, A. (2018). Violencia de género y frontera: Migrantes centroamericanas en México hacia los EEUU. *European Review of Latin American and Caribbean Studies, 105*, 39–60.

De Genova, N. (Ed.). (2017). *The borders od «Europe»: Autonomy of migration, tactics of bordering*. Durham: Duke University Press.

Dilla, H. (2008). *Ciudades en la frontera: Aproximaciones críticas a los complejos urbanos transfronterizos*. Santo Domingo: Manatí.

Dore, E. (2000). One step forward, two steps back: Gender and The State in Latin American in the long nineteenth century. In En Dore, E. y Molyneux, M. (Eds.), *Hidden histories of gender and the state in Latin America* (pp. 3–32). Durham: Duke University Press.

Fernández-Kelly, P. (1983). *For we are sold, i and my people: Women and industry in Mexico's Frontier*. Albany: State University of New York Press.

Garduño, E. (2003). Antropología de la frontera, la migración y los procesos transnacionales. *Frontera Norte, 15*(30), 1–30.

Gautier, A. (2005). Mujeres y colonialismo. In M. Ferro (Ed.), *El libro negro del colonialismo: Siglos XVI al XXI: Del exterminio al arrepentimiento*. Madrid: La Espera.

Giddens, A., Bauman, Z., & Beck, U. (1996). *Las consecuencias perversas de la modernidad*. España: Anthropos.

Grimson, A. (2000). *Fronteras, naciones e identidades: La periferia como centro*. Buenos Aires: Ediciones CICCUS. La Crujía.

Grimson, A. (2011). *Los límites de la cultura: Crítica de las teorías de la identidad*. Buenos Aires: Siglo XXI.

Grosfogel, R. (2006). La descolonización de la economía política y los estudios post-coloniales: Transmodernidad, pensamiento fronterizo y colonialidad global. *Tabula Rasa, 4,* 17–48.

Guizardi, M. L., Valdebenito, F., López, E., & Nazal, E. (2015). Condensaciones en el espacio hiperfronterizo: Apropiaciones migrantes en la frontera norte de Chile. En M. L. Guizardi (Ed.), *Las Fronteras del Transnacionalismo: Límites y desbordes de la experiencia migrante en el centro y norte de Chile*. Santiago: Universidad de Tarapacá y Ocho Libros.

Guizardi, M. L., Valdebenito, F., López, E., & Nazal, E. (2019). *Des/venturas de la frontera: Una etnografía sobre las mujeres peruanas entre Chile y Perú*. Santiago: Ediciones Universidad Alberto Hurtado.

Howell, J., Altamirano, D., Totah, F., & Keles, F. (Eds.). (2018). *Porous borders, invisible boundaries? Ethnographic perspectives on the vicissitudes of contemporary migration*. Comittee on Refugees and Immigrants.

International Organization for Migration [IOM]. (2019). *World migration report 2020*. Washington: OIM.

Jessop, B. (2004). La economía política de la escala y la construcción de las regiones transfronterizas. *EURE, XXIX*(89), 25–41.

Jiménez, R. (2018). *Criminalización de la economía informal en las fronteras sudamericanas: Estudio de caso de la triple frontera del Paraná y la triple frontera central surandina* (Maestría). Universidad Federal de Integração Latino-Americana, Foz do Iguaçu.

Kearney, M. (1991). Borders and boundaries of state and self at the end of empire. *Journal of Historical Sociology, 4*(1), 52–74.

Kearney, M. (2004). The classifying and value-filtering missions of borders. *Anthropological Theory, 4*(2), 131–156.

López, A. (2019). Cuerpo, espacio y género: Las mujeres bagayeras en el límite Aguas Blancas, Argentina-Bermejo, Bolivia. *Estudios Fronterizos, 20.*

Massey, D. (1994). *Space, place and gender*. United States of America: University of Minnesota Press.

Mezzadra, S., & Neilson, B. (2013). *Border as method or the multiplication of labour*. Durham, NC: Duke University Press.

Mignolo, W. (2007). *La idea de América Latina: La herida colonial y la opción decolonial*. Barcelona: Gedisa.

Mignolo, W., & Tlostanova, M. (2006). Theorizing from the borders: Shifting to geo- and body-politics of knowledge. *European Journal of Social Theory, 9*(2), 205–221.

Molina, F. (1985). The social impacts of the maquiladora industry on mexican border towns. *Berkeley Planning Journal, 2*(1), 30–40.

Monárrez, J. E. (2013). *Trama de una injusticia. feminicidio sexual sistémico en Ciudad Juárez*. Ciudad Juárez: Colef.

Mora, C. (2008). Globalización, género y migraciones. *Polis, 7*(20), 285–297.

Morales, M. C., & Bejarano, C. (2009). Transnational sexual and gendered violence: An application of border sexual conquest at a Mexico-US border. *Global Network, 9*(3), 420–439.

Morokvasic, M. (2004). Settled in mobility: Engendering post-wall migration in Europe. *Feminist Review, Labour migrations: Women on the Move, 77*, 7–25.

Olivera, M., & Aguilar, M. D. C. G. (2006). Migración y Mujeres en la Frontera Sur: Una agenda de investigación. *El Cotidiano, 21*(139), 31–40.

Parker, N., & Vaughan, W. (2012). Critical border studies: Broadening and deepening the 'lines in the sand' agenda. *Geopolitics 17*(4), 727–733.

Phillips, A. (2002). *Multiculturalism, universalism and the claims of democracy.* Washington: United Nations Research Institute for Social Development.

Pickering, S. (2011). *Women, borders, and violence: Current issues in asylum, forced migration, and trafficking*. New York: Springer.

Quijano, A. (2005). Colonialidad del poder, eurocentrismo y América Latina. En E. Lander (Ed.), *La colonialidad del saber: Eurocentrismo y ciencias sociales: Perspectivas latinoamericanas*. UNESCO, CLACSO.

Raghuram, P., & Piper, N. (2011). Women and global economic migration in Asia -eroding borders, new fixities. In Wastl-Walter (Ed.), *Ashgate research companion to border studies*, 529–48. England: Ashgate.

Ramos, R., & Ovando, C. (2016). La región de Tarapacá: Seguridad fronteriza y múltiples apropiaciones de su espacio. *Polis, Revista Latinoamericana, 15*(44), 1–19.

Renoldi, B. (2013). Fronteras que caminan: Relaciones de movilidad en un límite trinacional. *Revista Transporte y Territorio, 9*, 123–140.

Rivera Cusicanqui, S. (2008). *El potencial epistemológico y teórico de la historia oral: De la lógica instrumental a la descolonización de la historia*. EDIPUCRS: En Teoria crítica dos diretos humanos no século XXI.

Roldán, G., Guerra, M. J., & Pérez, N. (2017). *Las odiseas de Penélope: Feminización de las migraciones y derechos humanos*. UNAM.

Rumford, C. (2006). Theorizing borders. *European Journal of Social Theory, 9*(2), 155–169.

Sassen, S. (2003). *Contrageografías de la globalización: Género y ciudadanía en los circuitos transfronterizos*. Madrid: Traficante de sueños.

Seaman, B. (2012). Cry for me Argentina: The commercial sexual explotation of children in South América. *Law now*. Retreived from https://www.lawnow.org/commercial-sexual-exploitation-of-children-in-south-america/. Accessed 21 June 2019.

Segato, R. L. (2013a). *La escritura en el cuerpo de las mujeres asesinadas en Ciudad Juárez: Territorio, soberanía y crímenes de segundo estado.* Buenos Aires, Argentina: Tinta Limón.

Segato, R. (2013b). *La crítica de la colonialidad en ocho ensayos: Y una antropología por demanda*. Buenos Aires: Prometeo.

Song, S. (2007). *Justice, gender, and the politics of multiculturalism.* Cambridge: Cambridge University Press.

Tapia, M., Liberona, N., & Contreras, Y. (2017). El surgimiento de un territorio circulatorio en la frontera chileno-peruana: Estudio de las prácticas socio-espaciales fronterizas. *Revista de Geografía Norte Grande, 66*, 117–141.

Téllez, M. (2008). Community of struggle: Gender, violence, and resistance on the US/Mexico border. *Gender & Society, 22*(5), 545–567.

Velasco, L., & Contreras, O. (2011). Introducción. En *Mexican voices of the border region*. United States: Temple University Press.

Vertovec, S. (2007). Super-diversity and its implications. *Ethnic and Racial Studies, 30*(6), 1024–1054.

Vila, P. (2000). *Crossing borders, reinforcing borders*. Austin: University of Texas Press.

Viteri, M. A., Ceja, I., & Yépez, C. (2017). *Corpografías: Género y fronteras en América Latina*. Ecuador: Flacso Ecuador, IDRC.

Wimmer, A., & Glick Schiller, N. (2003). Methodological nationalism, the social sciences, and the study of migration: An essay in historical epistemology. *International Migration Review, 37*(3), 576–610.
Young, I. M. (2000). *La injusticia y la política de la diferencia*. Madrid: Cátedra.
Yuval-Davis, N. (1997). *Gender and nation*. London: Sage.

Carolina Stefoni is a Sociologist by Pontificia Universidad Católica de Chile, Master in Cultural Studies at the University of Birmingham and Ph.D. in Sociology at Alberto Hurtado University. Full Professor at Universidad Mayor, Chile, associate researcher at the Centre for Social Conflict and Cohesion Studies (COES). Consultant of international organizations in the area of migrations for ECLAC / CELADE; ILO, IOM, UNESCO and UNHCR. She has published in national and international academic journals, books and book chapters on migration studies. Her main research topics are migration policy, intercultural education, labour and gender migration. She currently coordinates the project "Routes and trajectories of Venezuelan migrants throughout South America. When borders start closing".

Menara Lube Guizardi holds a MA degree in Latin American Studies (2007) and a Ph.D. in Social Anthropology (2011), both from the Autonomous University of Madrid (UAM, Spain). She is currently an associate researcher at CONICET (Argentina) and an associate researcher at the University of Tarapacá (Arica, Chile). Her main research interests include borders, gender, migrations, inter-ethnic relations, racism and social exclusion. Her most recent books include "*Des/venturas de las Fronteras: una etnografía sobre las mujeres peruanas entre Chile y Perú*", published in 2019 and "*Capoeira: Etnografía de una historia transnacional entre Brasil y Madrid*", published in 2017, both by the Editions of the Alberto Hurtado University (Santiago, Chile). She is also the editor of the volume "*Las fronteras del transnacionalismo.Límites y desbordes de la experiencia migrante en el centro y norte de Chile*" published in 2015 by Ocho Libros (Santiago, Chile).

Eleonora López Contreras obtained a Bachelor's degree in Sociology at Universidad Nacional Autónoma de México (UNAM), as well as a Master's degree in Social Sciences at Universidad de Chile and a Master's degree in Sociology at Pontificia Universidad Católica de Chile. She is currently a Ph.D. student in Sociology and professor of Immigration Law at Pontificia Universidad Católica de Chile. Her lines of research are social suffering, racism, migrations, decoloniality and interculturality. Email: eleonoralopezcontreras@gmail.com

Herminia Gonzálvez Torralbo holds a BA Degree in Social y Cultural Anthropology (2002) at Universidad Miguel Hernández (UMH), as well as a Master's degree in Migration, Refuge and Intercommunity Relations (2007) at the Autonomous University of Madrid (UAM, Spain). She also holds a Ph.D. in Social Anthropology (2010), at the Granada University (UGR, Spain). She is currently a researcher at the Central University of Chile (Santiago, Chile). Her main research topics are care, gender, migrations, kinship relations, and transnational family. Her most recent books are "*Epistemologías feministas desde el Sur. Aportes, tensiones y perspectivas*", published in

2018 by Ril Editores and "*Diversidades familiares, cuidados y migración. Nuevos enfoques y viejos dilemas*", published in 2015 by the Editions of the Alberto Hurtado University (Santiago, Chile).

Index

C. Mora and N. Piper (eds.), *The Palgrave Handbook of Gender and Migration*, https://doi.org/10.1007/978-3-030-63347-9

CPSIA information can be obtained
at www.ICGtesting.com
Printed in the USA
LVHW080729270622
722171LV00003B/108

9 783030 633462